ANALYSIS IN VECTOR SPACES

ANALYSIS IN VECTOR SPACES

A Course in Advanced Calculus

Mustafa A. Akcoglu
University of Toronto

Paul F. A. Bartha
University of British Columbia

Dzung M. Ha
Ryerson University

WILEY-INTERSCIENCE

A JOHN WILEY & SONS, INC., PUBLICATION

Published by John Wiley & Sons, Inc., Hoboken, New Jersey.
Published simultaneously in Canada.

Library of Congress Cataloging-in-Publication Data:

Analysis in Vector Spaces / Mustafa Akcoglu, Paul Bartha, Dzung Minh Ha.

10 9 8 7 6 5 4 3 2 1

CONTENTS

PREFACE

Every mathematician needs to know advanced calculus. Some courses on the subject emphasize practical applications and problem solving. Others stress the rigorous exposition of a standard set of topics: basic topology and continuity, differentiation and integration of functions of several variables, and calculus on manifolds. This book combines a strong theoretical grounding in each of these topics with applications to a wide variety of problems and examples.

The book is aimed at the second-year undergraduate level. Indeed, the material presented here evolved over many years of teaching a course at this level at the University of Toronto. Students in the course tend to be specialists in mathematics, computer science, physics, and related areas. Thus, the book presupposes a good understanding of first-year (one-variable) calculus and linear algebra, as well as a certain level of comfort with a rigorous style of proof.

The most distinctive characteristic of the book is its geometric approach to central concepts, theorems, and applications. Geometric intuition is essential for both the theoretical and practical aspects of advanced calculus and for the subsequent study of analysis. Our aim throughout the book is to cultivate this intuition as we present the theorems and their applications.

As the title suggests, we believe that the geometric character of advanced calculus is most effectively conveyed in the vector space setting. Following two introductory chapters that supply background information on set theory and basic properties of the real numbers, Chapter 3 provides a thorough review of linear algebra. A notable feature of the chapter is a geometric version of the spectral theorem. This version gives a direct proof of an important property of determinants: they are the "volume multipliers" associated with linear transformations. The spectral theorem also gives a simple geometric picture of orthogonal projections between the subspaces of a Euclidean space. This has applications in the later study of manifolds.

Chapter 4 discusses normed vector spaces. A normed vector space is an ordinary vector space equipped with a norm, a well-behaved function that assigns a nonnegative "length" to each vector. Basic topological concepts— convergence, continuity, and compactness—are presented in this setting. There is often more than one natural choice for a norm, but it is shown that all norms on a finite-dimensional vector space are equivalent: they define the same topological notions. This fact is useful in later applications.

This preparatory material allows us to define derivatives, in Chapter 5, in the general setting of normed spaces rather than just on $\mathbb{R}^n$. Differential calculus is the study of functions between normed vector spaces that behave locally like linear transformations. In fact, the derivative at any point is just a linear transformation. The chapter highlights the idea of approximation through mean value theorems, which allow us to estimate the increments of a function in terms of increments of the approximating linear transformation.

In Chapter 6 we prove the inverse function theorem, a fundamental result of differential calculus. The theorem also provides an excellent illustration of the proper role and limitations of geometric intuition in analysis. The theorem states a fact that may seem obvious: if the derivative of a function at a point is an invertible linear transformation, then the function itself is invertible in a neighborhood of that point. This fact should seem obvious because the one-dimensional special case is easily proven. Yet the general proof of this "obvious" fact, in n dimensions, requires careful and involved analysis. Thus, while our geometric intuition may point us in the direction of a correct result and may even give us hints as to the proof, the actual proof often requires hard work and ideas that are not at all obvious. Chapter 6 also introduces manifolds, defined here as generalizations of graphs.

Approximation by a linear transformation corresponds to approximation by a first-degree polynomial. Better approximations require higher-degree polynomials. These considerations lead to higher-order derivatives, Taylor polynomials, and Taylor series. These concepts are introduced in Chapter 5 for vector-valued functions of a single variable and in Chapter 7 for functions of a vector variable. The definition of higher-order derivatives is considerably more complex for vector variables than for single variables. Chapter 7 may be considered optional, as later material does not depend upon this chapter in any essential way.

Our discussion of the theory of integration begins in Chapter 8. We take volume as the fundamental concept here, and our strategy is based on Archimedes' approach to defining volume more than two thousand years ago. The same approach leads to a rigorous definition of the integral of a function as the volume under its graph. A central result of this chapter is the change of variable theorem in integration. Like the inverse function theorem, this is another highly plausible result that requires hard work to prove.

The final two chapters deal with integration on manifolds and Stokes' theorem. We restrict our attention to manifolds in Euclidean spaces. The content (or volume) of subsets of a manifold is usually defined using change of variables. Chapter 9 develops this idea in detail and explains how it is naturally extended to the integration of vector and tensor fields on manifolds. The final part of the chapter offers an alternative, geometrically motivated definition of content on a manifold. Content of subsets of a manifold can be defined directly from the volume function on the larger Euclidean space in which the manifold is embedded. This geometric approach to content on manifolds agrees with the usual definition in important cases, and it also extends to cases not easily dealt with by the 'change of variables' approach.

Chapter 10 presents two distinct approaches to Stokes' theorem. The first approach shows how a special case of the theorem is a direct generalization of the fundamental theorem of calculus. The second approach, directed towards the same special case, is offered in the spirit of the original classical analysis in terms of the flows generated by vector fields. The remainder of the chapter shows how the special case can be transformed into more general tensor and vector formulations of Stokes' theorem. It should be noted that both of our approaches are different from the formulation of Stokes' theorem in terms of differential forms. That approach is elegant and very general, but we leave it for a later course.

Many of our important proofs and examples favor a more lengthy explanatory style than is common in other mathematics texts. Working through these arguments and solving the many problems in the book is the key to mastering a subject which is both a source of interesting and enjoyable problems and central to more advanced work in mathematics.

This book has taken shape over many years, during the course of which many individuals have made significant contributions. We received many valuable suggestions from our colleagues, including Edward Bierstone and Andrés del Junco (who has our special thanks for using early versions of the book in his courses). We acknowledge with gratitude the supportive staff of the mathematics department at the University of Toronto, and in particular we thank Marie Bachtis, Pat Broughton, Ida Bulat, Anu Mohindra, Betsy Moshopoulos, and Karin Smith for their help. We appreciate the encouragement of the editors at Wiley and the many helpful suggestions of several anonymous reviewers. Two former students, Karhan Akcoglu and Dennis Hui, have our deepest gratitude for revising and contributing to early versions. Finally, this book would not have been possible without input from our long-suffering students. It is to them, and to future mathematics students, that we dedicate this book.

PART I

BACKGROUND MATERIAL

CHAPTER 1

SETS AND FUNCTIONS

This first chapter introduces the notation, terminology and basic concepts needed for what lies ahead. We review some basic facts about sets in general, about sets of numbers, and about functions. We also take the opportunity to introduce some elementary functions of several (mainly two or three) variables that will be used in several examples in later chapters.

1.1 SETS IN GENERAL

Why should we begin our discussion with sets and not with numbers? After all, most of the sets we deal with are sets of numbers. Furthermore, the mathematical concept of number is older than that of set and is probably more intuitive.

Even though both concepts seem to be primitive (we shall not define either), sets are, in fact, more fundamental than numbers and can be used to generate number systems. Today, most mathematics is based on a solid set theoretic foundation, too lengthy to

Analysis in Vector Spaces.
By M. A. Akcoglu, P. F. A. Bartha and D. M. Ha

present here. Instead, we will confine ourselves to the elements of "naive set theory," doing little more than reviewing some standard notation and presenting a few basic facts about sets.

Definition 1.1.1 Terminology for sets. The following are basic notations and terms involving sets and set relations.

1. A *set* is a collection of objects. If A is a set, then the objects in A are the *elements* or *members* or *points* of A. The notation $x \in A$ means that x is a member of A and $x \notin A$ means that x is not a member of A.

2. The set having no elements is the *empty set*, denoted by $\emptyset$. A nonempty set is a set that contains at least one element.

3. Let A and B be sets. Then A is a *subset of* B if $x \in A$ implies $x \in B$. The notation $A \subset B$ means that A is a subset of B and $A \not\subset B$ means that A is not a subset of B. Thus, $A \not\subset B$ if and only if there is an $x \in A$ such that $x \notin B$. Hence, it follows that $\emptyset \subset B$ for every set B, since $\emptyset$ has no elements, and hence no element x for which $x \notin B$. We say that A is a *proper subset* of B if $A \subset B$ and there is an $x \in B$ such that $x \notin A$.

4. Let X be a set. Then the *power set of* X, denoted by $\mathcal{P}(X)$, is the set of all subsets of X.

5. If $A \subset B$ and $B \subset A$, then we write $A = B$.

6. Let S be a set. For each $x \in S$, let $P(x)$ be a statement about x that is either true or false. We define

$$\{ x \in S \mid P(x) \}$$

   to be the subset of S consisting of those x in S for which $P(x)$ is true. When S is clear from the context, this set is also expressed as $\{ x \mid P(x) \}$.

Example 1.1.2 Let $A = \{1, 5, \{1, 5\}\}, B = \{1, 5, \emptyset\}$.

1. The members of A are $1, 5$ and $\{1, 5\}$. The members of B are $1, 5$ and $\emptyset$.

2. $A \not\subset B$ because $\{1, 5\} \in A$ but $\{1, 5\} \notin B$. Also, $B \not\subset A$ because $\emptyset \in B$ but $\emptyset \notin A$.

3. $\{1, 5\}$ is a proper subset of both A and B.

4. We have

$$\mathcal{P}(A) = \{\emptyset, \{1\}, \{5\}, \{\{1, 5\}\}, \{1, 5\}, \{1, \{1, 5\}\}, \{5, \{1, 5\}\}, \{1, 5, \{1, 5\}\}\}.$$

Note that $\{1,5\}$ is an element in A so that $\{\{1,5\}\}$ is a subset of A. Thus, $\{\{1,5\}\}$ is an element in $\mathcal{P}(A)$ and $\{\{1,5\}\} \neq \{1,5\}$. Note that A has 3 elements and $\mathcal{P}(A)$ has $8 = 2^3$ elements. In general, whenever X is a finite set with n elements, $\mathcal{P}(X)$ has 2^n elements. We can see this by observing that every subset of X is formed by considering each element of X and either including it or omitting it. $\triangle$

Definition 1.1.3 Operations on sets. Let A, B be sets.

1. We define

$$\begin{aligned} A \cup B &= \{ x \mid x \in A \text{ or } x \in B \} \\ A \cap B &= \{ x \mid x \in A \text{ and } x \in B \}. \end{aligned}$$

The sets $A \cup B$ and $A \cap B$ are called, respectively, the *union of A and B* and the *intersection of A and B*.

2. If $A \cap B = \emptyset$, then A and B are *disjoint* sets. If $A \cap B \neq \emptyset$, then we say that A and B *intersect*, or A *intersects* B, or B *intersects* A. The sets in a collection of sets are called *pairwise disjoint* if any two (different) sets in this collection are disjoint.

3. The *complement of B in A*, denoted by $A \setminus B$, is defined by

$$A \setminus B = \{ x \mid x \in A \text{ and } x \notin B \}.$$

When all sets $A, B, \ldots$ under discussion are subsets of a fixed set S, we write A^c for $S \setminus A$ and B^c for $S \setminus B$. If S is left implicit, then $A^c = S \setminus A$ is called the *complement of A* rather than the complement of A in S.

4. The *symmetric difference* of A and B is defined by

$$A \triangle B = (A \setminus B) \cup (B \setminus A).$$

5. The *Cartesian product of A and B*, denoted by $A \times B$, is defined by

$$A \times B = \{ (a, b) \mid a \in A, b \in B \}.$$

An element (a, b) of $A \times B$ is an *ordered pair*. Note that $A \times B \neq B \times A$ unless $A = B$. Similarly, $A \times B \times C = \{(a, b, c) \mid a \in A, b \in B, c \in C\}$ consists of ordered 3-tuples. The Cartesian product of any finite number of sets is defined in a similar way.

Basic properties of set operations are summarized in the following lemma.

Lemma 1.1.4 *The following are true for all subsets A, B and C of a set X.*

(1) $A \cup B = B \cup A$, $A \cap B = B \cap A$.

(2) $(A \cup B) \cup C = A \cup (B \cup C)$, $(A \cap B) \cap C = A \cap (B \cap C)$.

(3) $A \cap (B \cup C) = (A \cap B) \cup (A \cap C)$.

(4) $(A \cup B)^c = A^c \cap B^c$.

(5) $(A \cap B)^c = A^c \cup B^c$.

In parts (4) and (5), the complements are with respect to X.

Proof. We will prove part (3) to illustrate the general method; the other assertions are of comparable difficulty or even easier. They are left as exercises.

Let $x \in A \cap (B \cup C)$. Then $x \in A$ and $x \in B \cup C$. Thus, either $x \in B$ or $x \in C$. If $x \in B$, then $x \in A \cap B$. If $x \in C$, then $x \in A \cap C$. Hence, $x \in A \cap B$ or $x \in A \cap C$. That is, $x \in (A \cap B) \cup (A \cap C)$. Thus,

$$A \cap (B \cup C) \quad \subset \quad (A \cap B) \cup (A \cap C) \,.$$

Conversely, let $y \in (A \cap B) \cup (A \cap C)$. Then either $y \in A \cap B$ or $y \in A \cap C$. If $y \in A \cap B$, then of course, $y \in A \cap (B \cup C)$. Similarly, if $y \in A \cap C$, then $y \in A \cap (B \cup C)$. Thus, in both cases, $y \in A \cap (B \cup C)$. Hence,

$$(A \cap B) \cup (A \cap C) \quad \subset \quad A \cap (B \cup C).$$

These two inclusions imply that $(A \cap B) \cup (A \cap C) = A \cap (B \cup C)$. □

Definition 1.1.5 General unions and intersection. Let $\mathcal{G}$ be a collection of sets. Hence $\mathcal{G}$ is a set whose elements G are also sets. The union and intersection of the sets in this collection are defined in an obvious way as

$$\bigcup_{G \in \mathcal{G}} G \quad = \quad \{\, x \mid \text{there is a } G \in \mathcal{G} \text{ such that } x \in G \,\},$$
$$\bigcap_{G \in \mathcal{G}} G \quad = \quad \{\, x \mid x \in G \text{ for all } G \in \mathcal{G} \,\}.$$

Additional concepts and notation for dealing with collections of sets will be introduced in the examples below.

Relations and Equivalences

Definition 1.1.6 Relations. Let A and B be sets. A *relation* between the elements of A and the elements of B is a subset R of $A \times B$. If $(a, b) \in R$, then one says that R is *satisfied by a and b* or that *a is related to b by R*. One may omit explicit mention of R if this relation is understood from the context. There is an all-important class of relations: *functions*. They are introduced in Section 1.3. Another important class of relations is formed by *equivalences*, introduced below in Definition 1.1.9.

Example 1.1.7 Let A be a set and let $B = \mathcal{P}(A)$ be the power set of A. Then

$$R = \{ (a, S) \in A \times \mathcal{P}(A) \mid a \in S \}$$

defines a relation between the elements of A and the subsets of A. This relation is satisfied by a and S if and only if $a \in S$. △

Example 1.1.8 Let A be a set. Then

$$R = \{ (U, V) \in \mathcal{P}(A) \times \mathcal{P}(A) \mid U \subset V \}$$

defines a relation R between elements of $\mathcal{P}(A)$. An ordered pair (U, V) of subsets of A satisfies this relation if and only if $U \subset V$. △

Definition 1.1.9 Equivalences. Let A be a set. Let $E \subset A \times A$. Then E is called an *equivalence relation on A* (or an equivalence on A, or simply an equivalence when A is clear from the context) if it has the following properties.

Reflexivity If $a \in A$, then $(a, a) \in E$.

Symmetry If $(a, b) \in E$, then $(b, a) \in E$.

Transitivity If $(a, b) \in E$ and $(b, c) \in E$, then $(a, c) \in E$.

If $E \subset A \times A$ is an equivalence, then one usually writes $a \sim b$ to indicate that $(a, b) \in E$. With this notation the properties above are expressed as follows. Let $a, b, c \in A$. Then (1) $a \sim a$, (2) if $a \sim b$ then $b \sim a$, and (3) if $a \sim b$ and $b \sim c$ then $a \sim c$. If $a \sim b$, then one also says that a and b are *equivalent* (with respect to the underlying equivalence).

Example 1.1.10 Let A be a set and let $C \subset A$. Define a relation $E \subset A \times A$ such that, for any two points a and b in A, $(a, b) \in E$ if and only if both a and b are in C or both a and b are not in C. Hence let

$$E = \{ (a, b) \in A \times A \mid \text{Either } [\, a \in C \text{ and } b \in C \,] \text{ or } [\, a \notin C \text{ and } b \notin C \,] \} .$$

It is easy to check that E is an equivalence . $\triangle$

Definition 1.1.11 Equivalence classes. Let $\sim$ be an equivalence on A. Let $p \in A$. Then

$$[p] = \{ a \in A \mid a \sim p \}$$

is called *the equivalence class of* p. A subset P of A is called *an equivalence class* if there is a $p \in A$ such that $P = [p]$.

Theorem 1.1.12 *Let $\sim$ be an equivalence relation on A.*

(1) Let x, y in A. Then $[x] = [y]$ if and only if $x \sim y$.

(2) Two different equivalence classes are disjoint.

(3) The union of all equivalence classes is A.

Proof. Assume that $x \sim y$. If $a \in [x]$, then $a \sim x$. By transitivity, $a \sim y$. Thus, $[x] \subset [y]$. Similarly, $[y] \subset [x]$. Hence, $[x] = [y]$. Conversely, if $[x] = [y]$, then of course, $x \in [y]$ and $x \sim y$. This proves (1).

Now assume that $[x] \cap [y] \neq \emptyset$. We will show that $[x] = [y]$. Let $a \in [x] \cap [y]$. Then $a \sim x$ and $a \sim y$. Hence, by symmetry and transitivity of $\sim$, we have $x \sim y$. Thus, by part (1), $[x] = [y]$. Hence two different equivalence classes can not intersect. This proves (2).

Finally, $a \in [a]$ for all $a \in A$. Hence, every element of A belongs to an equivalence class. This proves (3). $\square$

Definition 1.1.13 Complete set of representatives. Let A be a set with an equivalence. Let P be an equivalence class. Any point $p \in P$ is called a *representative* for P. A subset R of A is called a *complete set of representatives* if each equivalence class has exactly one point in R as its representative.

Example 1.1.14 Let A be a set and let C be a subset of A. Let $\sim$ be the equivalence defined in Example 1.1.10. Then C and $A \setminus C$ are the only two equivalence classes. If both C and $A \setminus C$ are nonempty, then any two-point set consisting one point from C and one point from $A \setminus C$ is a complete set of representatives. $\triangle$

Example 1.1.15 Let X and Y be sets and let $Z = X \times Y$. Define a relation $\sim$ on Z as follows: $(x, y) \sim (x, y')$ for all $x \in X$ and all y, y' in Y. We verify easily that this relation is an equivalence on Z. Let $b \in Y$ be a fixed point in Y. It is easy to see that $X \times \{b\}$ is a complete set of representatives for this equivalence. $\triangle$

Problems

1.1 Give an example of a family of sets such that any two sets in the family intersect (that is, they have nonempty intersection) but the intersection of all the sets in this family is empty.

1.2 Let $\mathcal{A}$ be a collection of subsets of a set X. Show that

$$(\bigcup_{A\in\mathcal{A}} A)^c = \bigcap_{A\in\mathcal{A}} A^c \text{ and } (\bigcap_{A\in\mathcal{A}} A)^c = \bigcup_{A\in\mathcal{A}} A^c.$$

Hence, $(A \cup B)^c = A^c \cap B^c$ and $(A \cap B)^c = A^c \cup B^c$.

1.3 If $\mathcal{A}$ is a collection of subsets of a set X and if $B \subset X$, then show that

$$\begin{aligned}\left(\bigcup_{A\in\mathcal{A}} A\right) \cap B &= \bigcup_{A\in\mathcal{A}} (A \cap B) \text{ and}\\ \left(\bigcap_{A\in\mathcal{A}} A\right) \cup B &= \bigcap_{A\in\mathcal{A}} (A \cup B).\end{aligned}$$

1.4 Show that $A \triangle B = (A \cup B) \setminus (A \cap B)$. Deduce that $A \triangle B = \emptyset$ if and only if $A = B$ and $A \triangle B = X$ if and only if $B = A^c$.

1.5 Show that $A \triangle B \subset (A \triangle C) \cup (C \triangle B)$ for any three sets. Give an example to show that in general the inclusion is a proper inclusion.

1.6 A collection of subsets of a set X is called an *algebra of sets* if it satisfies the following three conditions:

1. $X \in \mathcal{A}$;
2. if $A \in \mathcal{A}$, then also $A^c \in \mathcal{A}$;
3. if $A, B \in \mathcal{A}$, then also $A \cup B \in \mathcal{A}$.

Show that if $\mathcal{A}$ is an algebra of sets and if $A,\ B \in \mathcal{A}$, then $A \cap B$, $A \setminus B$, and $A \triangle B$ are also in $\mathcal{A}$.

1.7 A collection of nonempty subsets of a set X is called a *partition* of X if the sets in this collection are pairwise disjoint and if their union is X. Show that any partition is the family of equivalence classes with respect to an equivalence on X.

1.8 Let $\mathcal{A}$ be a partition of X and $\mathcal{B}$ a partition of Y. Show that the family

$$\mathcal{C} = \{\, A \times B \mid A \in \mathcal{A},\ B \in \mathcal{B} \,\}$$

is a partition of $X \times Y$.

1.2 SETS OF NUMBERS

In this course we will deal mainly with sets of numbers. We shall assume that the set of *natural numbers*, $\mathbb{N}$, the set of *integers*, $\mathbb{Z}$, the set of *nonnegative integers*, $\mathbb{Z}^+$, and the set of *rational numbers*, $\mathbb{Q}$, are familiar. These sets are

$$\begin{aligned} \mathbb{N} &= \{1,2,3,\ldots\}, \\ \mathbb{Z} &= \{\ldots,-2,-1,0,1,2,\ldots\}, \\ \mathbb{Z}^+ &= \{0,1,2,\ldots\}, \\ \mathbb{Q} &= \{\, a/b \mid a,b \in \mathbb{Z}, b \neq 0 \,\}\ . \end{aligned}$$

We shall assume that the reader is acquainted with the addition and multiplication operations and the order relations on these sets. Another familiar set of numbers is $\mathbb{R}$, the set of *real numbers*. Note that $\mathbb{N} \subset \mathbb{Z}^+ \subset \mathbb{Z} \subset \mathbb{Q} \subset \mathbb{R}$. We will discuss $\mathbb{R}$ in some detail in Chapter 2. However, we shall assume that the basic properties of the real numbers are familiar. These properties are used below to give other examples of sets of numbers.

Intervals in $\mathbb{R}$

Example 1.2.1 Intervals. Four basic types of unbounded interval are defined in terms of a number $p \in \mathbb{R}$. These intervals are, in standard notation,

$$\begin{aligned} [\,p,\,\infty) &= \{\, r \in \mathbb{R} \mid p \leq r \,\}, & (\,p,\,\infty) &= \{\, r \in \mathbb{R} \mid p < r \,\}, \\ (-\infty,\,p\,] &= \{\, r \in \mathbb{R} \mid r \leq p \,\}, & (-\infty,\,p\,) &= \{\, r \in \mathbb{R} \mid r < p \,\}. \end{aligned}$$

Intersections of these intervals give other types of intervals. For example, again in standard notation,

$$[\,a,\,b\,) = (-\infty,\,b\,) \cap [\,a,\,\infty) = \{\, t \in \mathbb{R} \mid a \leq t < b \,\}.$$

Here a and b are two fixed numbers in $\mathbb{R}$. Note that $[\,a,\,b\,) = \emptyset$ if $b \leq a$. $\triangle$

Example 1.2.2 Collections of intervals. Let $r > 0$ be a fixed number. For each $a \in \mathbb{R}$, let $I_a = [a - r,\, a + r)$. Then $\mathcal{I} = \{\, I_a \mid a \in \mathbb{R} \,\}$ is a collection of intervals. Denote this collection as $\{I_a\}_{a\in\mathbb{R}}$ and the union and intersection of the intervals in this collection as $\cup_{a\in\mathbb{R}} I_a$ and $\cap_{a\in\mathbb{R}} I_a$. Obviously,

$$\bigcup\nolimits_{a\in\mathbb{R}} I_a = \mathbb{R} \quad \text{and} \quad \bigcap\nolimits_{a\in\mathbb{R}} I_a = \emptyset.$$

We can also consider subcollections of this collection. For example, $\{I_a\}_{0\leq a\leq 1}$ is such a subcollection. We see easily that

$$\bigcup\nolimits_{0\leq a\leq 1} I_a = [-r,\, 1 + r) \quad \text{and} \quad \bigcap\nolimits_{0\leq a\leq 1} I_a = [1 - r,\, r).$$

Hence $\bigcap_{0 \leq a \leq 1} I_a = \emptyset$ if $1 - r \geq r$, that is, if $r \leq 1/2$. $\triangle$

Examples 1.2.3 Lines and half-planes in $\mathbb{R}^2$. We consider $\mathbb{R}^2 = \mathbb{R} \times \mathbb{R}$ as the usual xy-plane. It consists of all ordered pairs (x, y), with $x, y \in \mathbb{R}$. A line in $\mathbb{R}^2$ is a set of the form $L = \{ (x, y) \in \mathbb{R}^2 \mid Ax + By + C = 0 \}$, where A, B, and C are three fixed numbers in $\mathbb{R}$ and at least one of A or B is nonzero. A line divides $\mathbb{R}^2$ into three pairwise disjoint sets

$$\begin{aligned} L &= \{ (x, y) \in \mathbb{R}^2 \mid Ax + By + C = 0 \}, \\ H_1 &= \{ (x, y) \in \mathbb{R}^2 \mid Ax + By + C > 0 \}, \\ H_2 &= \{ (x, y) \in \mathbb{R}^2 \mid Ax + By + C < 0 \}. \end{aligned}$$

Here H_1 and H_2 are the two half-planes bounded by the line L. We may refer to them as the *lower* and *upper* half-planes or as the *left-hand* and *right-hand* half-planes, depending on the position of L. Finally, one may also refer to the equation $Ax + By + C = 0$ as a *line*. $\triangle$

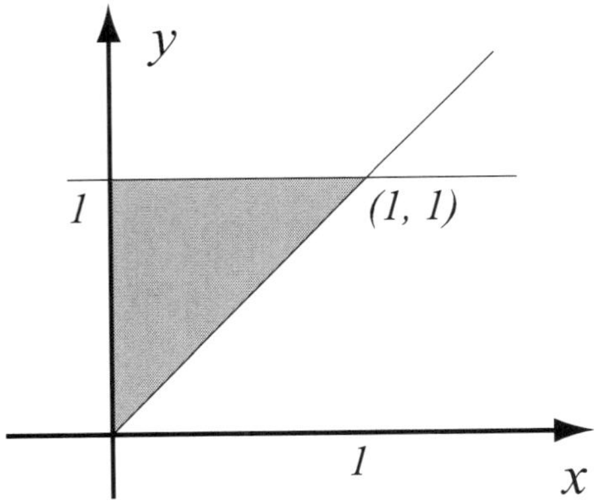

Figure 1.1. Triangle in Example 1.2.4.

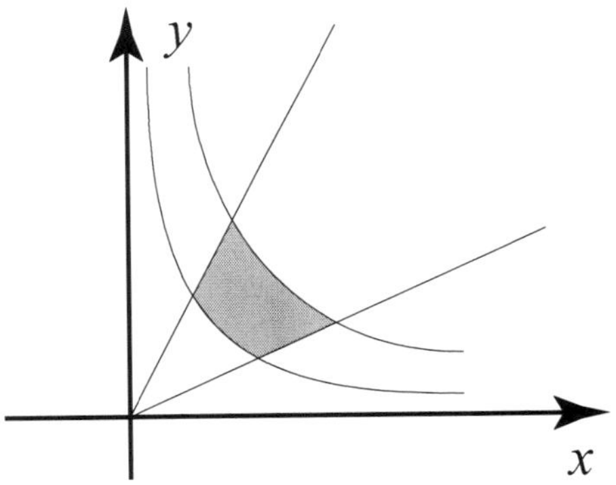

Figure 1.2. Region in Example 1.2.5.

Example 1.2.4 Let $R \subset \mathbb{R}^2$ be the triangle in the xy-plane bounded by the lines $x = 0$, $y = 1$, and $x = y$ (see Figure 1.1). Specify R by a set of inequalities.

Solution. We see that R is the intersection of the following three half-planes. (1) The right-hand side of $x = 0$, (2) the lower part of $y = 1$, and (3) the upper part of $x = y$. Hence $(x, y) \in R$ if and only if $x \geq 0$ and $y \leq 1$ and $y \geq x$. We can express this more concisely as

$$(x, y) \in R \text{ if and only if } 0 \leq x \leq y \leq 1.$$

Here we have assumed that R contains the inside, the edges, and the vertices of this triangle. The inside of R without the vertices and without the edges corresponds to the relation $0 < x < y < 1$. $\triangle$

Example 1.2.5 Let R be the region in the xy-plane that lies in the first quadrant (that is, $x \geq 0$ and $y \geq 0$) and is between the hyperbolas $xy = 1$, $xy = 2$, and bounded by the lines $2y = x$, $y = 2x$ (see Figure 1.2). Specify R by a set of inequalities.

Solution. The region in the first quadrant that is between the hyperbolas $xy = 1$ and $xy = 2$ is specified by the conditions that $x > 0$ and $1 \leq xy \leq 2$. Similarly, the region in the first quadrant that is between the lines $2y = x$ and $y = 2x$ is specified by the conditions that $x > 0$ and $1/2 \leq y/x \leq 1$. Hence we see that $(x, y) \in R$ if and only if

$$1 \leq xy \leq 2 \text{ and } 1/2 \leq y/x \leq 2 \text{ and } x > 0. \quad \triangle$$

Remark 1.2.6 Note that in the last two examples we have used undefined (yet intuitive) terms like *bounded by* and *upper part*. In such cases, the final formal statements in terms of inequalities may be considered as the definition of these intuitive terms.

Discs in $\mathbb{R}^2$

Example 1.2.7 Collections of discs. Let $r > 0$. If (a, b) is a point in the xy-plane (i.e., a point in $\mathbb{R}^2$), then let

$$D_r(a, b) = \{ (x, y) \in \mathbb{R}^2 \mid (x - a)^2 + (y - b)^2 < r^2 \}$$

be the (open) disc of radius r about the point (a, b). This notation is convenient because we shall often need to refer to open discs, just as we often refer to intervals in $\mathbb{R}$. We see easily that

$$\bigcup_{(a, b) \in \mathbb{R}^2} D_r(a, b) = \mathbb{R}^2 \text{ and } \bigcap_{(a, b) \in \mathbb{R}^2} D_r(a, b) = \emptyset.$$

It may require some work to identify

$$U(G, r) = \bigcup_{(a, b) \in G} D_r(a, b) \text{ and } I(G, r) = \bigcap_{(a, b) \in G} D_r(a, b)$$

for various regions G in the xy-plane. If

$$G = \{ (a, b) \in \mathbb{R}^2 \mid a^2 + b^2 = 1 \}$$

is the unit circle about the origin, for example, then

$$U(G, r) = \begin{cases} \{ (x, y) \mid (1 - r)^2 < (x^2 + y^2) < (1 + r)^2 \} & \text{if } r \leq 1 \\ \{ (x, y) \mid (x^2 + y^2) < (1 + r)^2 \} & \text{if } r > 1, \end{cases}$$

$$I(G, r) = \begin{cases} \emptyset & \text{if } r \leq 1 \\ \{ (x, y) \mid (x^2 + y^2) < (r - 1)^2 \} & \text{if } r > 1. \end{cases}$$

In obtaining these results, it is helpful to keep in mind the geometric interpretation of the above sets. For example, $U(G, r)$ is the set of all points in the plane that have a distance less than r to a point in G. $\triangle$

The Induction Principle

There is an "obvious" property of $\mathbb{N}$ called the *well-ordering property*. Since this property does not follow formally from the rules of the order relation, it is stated as an axiom for $\mathbb{N}$.

Axiom 1.2.8 The well-ordering axiom. Any nonempty subset of $\mathbb{N}$ contains a smallest element. More explicitly, if $T \subset \mathbb{N}$ and if T is not empty, then there is an $n \in T$ such that $n \leq m$ for all $m \in T$.

The well-ordering of $\mathbb{N}$ implies the induction principle for $\mathbb{N}$.

Theorem 1.2.9 Induction principle. *Let $S \subset \mathbb{N}$. Suppose that $1 \in S$ and that $k+1 \in S$ whenever $k \in S$. Then $S = \mathbb{N}$.*

Proof. Assume, on the contrary, that $S \neq \mathbb{N}$. Let $T = \mathbb{N} \setminus S$. Then T is a nonempty subset of $\mathbb{N}$. Hence, by the well-ordering axiom, there is a smallest element a in T. Since $a \notin S$ and $1 \in S$, we must have $a > 1$. Thus, $a - 1 \in \mathbb{N}$. Since a is the smallest element in T, we must have $a - 1 \notin T$. Hence, $a - 1 \in S$. By property 2, we have $a = (a-1) + 1 \in S$, a contradiction. $\square$

Example 1.2.10 Let $S \subset \mathbb{Z}$. Assume that $a \in S$ and that $k+1 \in S$ whenever $k \in S$. Show that $\{\, a + k \mid k \in \mathbb{N} \,\} \subset S$.

Solution. Set $T = \{\, k \in \mathbb{N} \mid a + k \in S \,\}$. We will show that $T = \mathbb{N}$. First, since $a \in S$, we have $a + 1 \in S$. Hence, $1 \in T$. Assume that $k \in T$. Then $a + k \in S$. Hence, by assumption, $a + (k+1) = (a+k) + 1 \in S$. Thus, $k + 1 \in T$. By the induction principle, $T = \mathbb{N}$. $\triangle$

The following example shows how the induction principle gives us the familiar method of proof by induction. The idea is to prove that some result holds for $n = 1$ (the *base case*) and to show that if it holds for n, then it holds for $n+1$ (the *inductive step*).

Example 1.2.11 Show that $1 + 2 + \cdots + n = n(n+1)/2$ for all $n \in \mathbb{N}$.

Solution. Let $S = \{ n \in \mathbb{N} \mid 1 + 2 + \cdots + n = n(n+1)/2 \}$. Then $1 \in S$ (the base case). For the inductive step, suppose that $n \in S$. Then

$$\begin{aligned} 1 + 2 + \cdots + n + (n+1) &= (1 + 2 + \cdots + n) + (n+1) \\ &= (n(n+1)/2) + (n+1) \\ &= (n+1)(n+2)/2 \end{aligned}$$

shows that $(n+1) \in S$. Hence $S = \mathbb{N}$ by the induction principle 1.2.9. $\triangle$

Remarks 1.2.12 Limitations of the induction principle. The induction principle is useful in stating certain arguments in a clear and concise way. But this principle may not be very helpful in obtaining new results. For example, the principle will not help you to guess the result $1 + \cdots + n = n(n+1)/2$. To obtain this result, you need other methods.

Definition 1.2.13 Binomial coefficients. Let $r \in \mathbb{R}$ and $k \in \mathbb{N}$. Define

$$\begin{pmatrix} r \\ 0 \end{pmatrix} = 1, \quad \begin{pmatrix} r \\ 1 \end{pmatrix} = r, \quad \text{and} \quad \begin{pmatrix} r \\ k+1 \end{pmatrix} = \frac{r(r-1)\cdots(r-k)}{(k+1)!}.$$

These are the general binomial coefficients. These expressions will be used in examples and later in the discussion of multilinear functions. If $r \in \mathbb{N}$, then $\begin{pmatrix} r \\ k \end{pmatrix}$ is the number of ways to select k objects from a collection of r objects.

Problems

1.9 Express the set

$$C = \left\{ x \in \mathbb{R} \;\middle|\; 0 < x^2 - 5x + 4 \leq 10 \right\} \subset \mathbb{R}$$

in terms of intervals.

1.10 There are four bounded regions in the xy-plane bounded by the lines $y = x$ and $y = 2x + 1$, and by the ellipse $x^2 + 4y^2 = 16$. One of these regions is

$$\left\{ (x,\, y) \in \mathbb{R}^2 \;\middle|\; x^2 + 4y^2 \leq 16 \text{ and } y \geq x \text{ and } y \geq 2x + 1 \right\}.$$

Express the other three regions similarly.

1.11 Let $G = \left\{ (x,\, y) \in \mathbb{R}^2 \mid -1 \leq x \leq 1,\ -1 \leq y \leq 1 \right\}$. Consider the sets $U(G,\, 1)$ and $I(G,\, 1)$ defined in Example 1.2.7. Express these sets in simpler terms.

1.12 Let $(a_1, b_1), (a_2, b_2)$ be in $U(C, r)$. Consider $\mathbb{R}^2$ as the xy-plane and use the customary vectorial notations. A subset C of the xy-plane is called *convex* if whenever C contains two points $(a_1,\, b_1)$ and $(a_2,\, b_2)$, then C also contains all the points on the line segment joining these points. Let C be the set of all points $(x,\, y)$ that can be expressed as

$$(x,\, y) = p\,(1,\, 0) + q\,(2,\, 0) + r\,(0,\, 2) + s\,(-1,\, -1)\,,$$

where p, q, r and s are all nonnegative and $p + q + r + s = 1$. Show that C is a convex set and describe it in simpler terms.

1.13 Define a relation $D \subset \mathbb{Z} \times \mathbb{Z}$ as

$$D = \{\, (a,\, b) \in \mathbb{Z} \times \mathbb{Z} \mid \text{There is a } k \in \mathbb{Z} \text{ such that } a = kb \,\}\,.$$

This relation is called *divisibility*: $(a,\, b) \in D$ just in case a is *divisible by* b. Show that divisibility is reflexive and transitive but not symmetric.

1.14 Let $p \in \mathbb{N}$. Define a relation $C_p \subset \mathbb{Z} \times \mathbb{Z}$ by

$$C_p = \{\, (a,\, b) \in \mathbb{Z} \times \mathbb{Z} \mid (a - b,\, p) \in D \,\}\ ,$$

where D is the divisibility relation defined in Problem 1.13. This relation among the integers is called *congruence modulo p.*

1. Show that congruence modulo p is an equivalence on $\mathbb{Z}$.
2. Show that there are exactly p equivalence classes for this equivalence.
3. Show that the set of integers

   $$R = \{\, 1,\, 2,\, \ldots,\, p\,\}$$

   is a complete set of representatives for congruence modulo p.
4. What is the equivalence class represented by $k \in \mathbb{Z}$?

1.15 Define a relation $C_1 \subset \mathbb{R} \times \mathbb{R}$ by the condition that $(r,\, s) \in C_1$ if and only if $(r - s) \in \mathbb{Z}$. This relation among the real numbers is called *congruence modulo* 1.

1. Show that congruence modulo 1 is an equivalence on $\mathbb{R}$.
2. Show that the interval

   $$R = [\,0,\, 1) = \{\, t \in \mathbb{R} \mid 0 \le t < 1\,\}$$

is a complete set of representatives for congruence modulo 1.

3. What is the equivalence class represented by $t \in R$?

Note. Let $p \in \mathbb{N}$, $p \geq 2$. Congruence modulo p can be defined on $\mathbb{R}$, but this is not customary.

1.16 Define a relation among the points $(x,\, y) \in \mathbb{R}^2$ as follows. A point $(x,\, y)$ is related to a point $(x',\, y')$ if and only if $x^2 + y^2 = x'^2 + y'^2$. Show that this is an equivalence. What are the equivalence classes? What is a complete set of representatives? Is the x-axis

$$\{\, (x, 0) \in \mathbb{R}^2 \mid x = 0 \,\}$$

a complete set of representatives? Why or why not?

1.17 Define a relation among the points $(x,\, y) \in \mathbb{R}^2$ as follows. A point $(x,\, y)$ is related to a point $(x',\, y')$ if and only if $xy = x'y'$. Show that this is an equivalence. What are the equivalence classes? What is the equivalence class containing the origin $(0,\, 0)$? What is a complete set of representatives? Is the line

$$\{\, (x,\, y) \in \mathbb{R}^2 \mid x = y \,\}$$

a complete set of representatives? Why or why not?

1.18 Let C be a convex set in the xy-plane (see Problem 1.12) and let $r > 0$. Show that the sets $U(C,\, r)$ and $I(C,\, r)$ are also convex, with the notation of Example 1.2.7. (Hint for the convexity of $I(C,\, r)$: show that the intersection of any family of convex sets is convex.)

1.19 Let $n \in \mathbb{N}$. Show by induction that $4^n - 3n - 1$ is divisible by 9. (Divisibility is defined in Problem 1.13.)

1.20 Let $r \in \mathbb{R}$ and $n \in \mathbb{Z}^+$.

1. Show that

$$\binom{r}{n} + \binom{r}{n+1} = \binom{r+1}{n+1}.$$

2. Use the induction principle to show that for all integers $n \geq 0$,

$$\sum_{k=0}^{n} \binom{r+k}{k} = \binom{r+n+1}{n}.$$

1.21 (*Binomial Theorem*) Let $a,\ b \in \mathbb{R}$ and $n \in \mathbb{Z}^+$. Use the induction principle to show that

$$(a+b)^n = \sum_{k=0}^{n} \binom{n}{k} a^{n-k} b^k .$$

1.22 Let $r,\ s \in \mathbb{R}$ and $n \in \mathbb{Z}^+$. Show that

$$\sum_{k=0}^{n} \binom{r}{k}\binom{s}{n-k} = \binom{r+s}{n} .$$

1.23 Let $n \in \mathbb{N}$. Show that $\sum_{k=1}^{n} k^2 = (1/6)n(n+1)(2n+1)$.

1.3 FUNCTIONS

The concept of a function is one of the most important ideas in mathematics. Functions are certainly of paramount importance in analysis. A function from a set X to a set Y is a special type of relation between the elements of X and Y. A subset of $X \times Y$ that defines a function is called a *graph.*

Definition 1.3.1 Graphs. Let X and Y be nonempty sets. A subset Γ of $X \times Y$ is called a *graph* if, whenever $(x,\ y)$ and $(x,\ y')$ are both in Γ, then $y = y'$. Hence Γ is a graph if for each $x \in X$ there is at most one point $(x,\ y) \in \Gamma$. *The domain of* Γ, denoted by $\operatorname{Dom}\Gamma$, is defined by

$$\operatorname{Dom}\Gamma = \{\, x \in X \mid \text{there is a } y \in Y \text{ such that } (x,\ y) \in \Gamma \,\}.$$

Also, *the domain space of* Γ is X and *the range space of* Γ is Y.

Definition 1.3.2 Functions. Let $\Gamma \subset X \times Y$ be a graph and let $D = \operatorname{Dom}\Gamma$. The relation defined by this graph Γ is called a *function* $f : D \to Y$ from D to Y. It is customary to denote graphs and functions by different symbols. What distinguishes a function f from a general relation is the condition that each $x \in D$ is related to a *unique* point $y \in Y$ by f. One calls y *the value of* f *at* x, or *the image of* x *under* f, and one writes this as $y = f(x)$. If $x \in D$, then one also says that f *is defined at* x. The domain D of Γ is also called *the domain* of f and denoted as $\operatorname{Dom} f$. *The domain space of* f is X and *the range space of* f is Y.

Remark 1.3.3 Roles of X and Y. Note that the sets X and Y are not uniquely determined by the function f. In fact, X can be any set containing the domain of f, and Y can be any set containing all the values of f. They do determine, however, the nature of the function under consideration. If $X = Y = \mathbb{R}$, for example, then we are dealing with a real-valued function defined on a set of real numbers. Also, the sets X and Y are important for defining bijections in Definition 1.3.20 below.

Remarks 1.3.4 Role of the graph. The formal definition of a function given above in Definition 1.3.2 is satisfactory but rarely used in the actual statement of a function. One usually defines $f(x)$ for a general point x by an explicit rule or computational formula. Nevertheless, the graphical definition of a function is an important idea that will have applications later.

Remarks 1.3.5 Notation for functions. When a function is defined by an explicit rule, then one denotes this function by $y = f(x)$ or by $f(x)$. This notation is not strictly correct: $y = f(x)$ is a point in Y rather than the function $f : D \to Y$ itself. Nevertheless, $y = f(x)$ is convenient notation which causes no confusion in practice. When a function is given as $y = f(x)$, $D = \text{Dom } f$ is understood to be the set of x for which $f(x)$ is defined.

We shall sometimes express a function as $y = y(x)$. This notation indicates that we are dealing with a function the points of whose domain space are denoted by x and points in the range space by y. It eliminates the unnecessary symbol f. Still, the most common notation in practice is $y = f(x)$.

Definition 1.3.6 Restrictions of a function. Let $f : D \to Y$ be a function and $A \subset D$. The *restriction of f to A* is a new function that has the value $f(x)$ for $x \in A$ but is undefined if $x \notin A$. In general, it is not necessary to use a different notation for the restricted function. This is understood from the context.

Definition 1.3.7 Identity functions. Let X be a set. Define a function $I_X : X \to X$ as $I_X(x) = x$ for all $x \in X$. It is called the *identity function* on X, or simply the identity on X. We also write I instead of I_X if X is understood from the context.

Definition 1.3.8 Sequences. Let $\mathbb{K}$ be a subset of $\mathbb{Z}$ such that $\mathbb{K}$ is bounded below but not above. A function defined on $\mathbb{K}$ is called a *sequence*. One usually takes $\mathbb{K}$ as $\mathbb{N}$ or an unbounded subset of $\mathbb{N}$. The range space of a sequence can be any nonempty set Y. By a *sequence in Y*, we mean a sequence with the range space Y. $\mathbb{K}$ is called an *index set*. The value of a sequence $a : \mathbb{K} \to Y$ at $k \in \mathbb{K}$ is called the kth term of the sequence and is denoted by a_k. The sequence itself may be denoted as $a : \mathbb{K} \to Y$, as a_k, $k \in \mathbb{K}$, as $\{a_k\}$, or simply as a_k if the domain $\mathbb{K}$ is understood. This last notation is not strictly correct, but it is convenient to use when the meaning is clear from the context. A sequence is usually given by a formula involving n. Such a sequence is defined for all $n \in \mathbb{N}$ for which this formula is meaningful. For example, $a_n = (n - 5)$ defines a sequence $a : \mathbb{N} \to \mathbb{Z}$ and $b_n = 1/(n - 5)$ defines a sequence $b : \mathbb{K} \to \mathbb{Q}$, where $\mathbb{K} = \mathbb{N} \setminus \{5\}$.

Examples 1.3.9 Graphs in $\mathbb{R} \times \mathbb{R} = \mathbb{R}^2$. The graphs introduced in Definition 1.3.1 above become ordinary graphs in the plane when $X = Y = \mathbb{R}$.

1. Let $G = \{\, (x, x^2) \in \mathbb{R}^2 \mid x \in \mathbb{R} \,\}$. Then G is a graph and $\operatorname{Dom} G = \mathbb{R}$. The function defined by G is $y = x^2$. As noted above in Remarks 1.3.4, one defines this function directly by the formula $y = x^2$, without mentioning the graph G. In this case, G is a parabola.

2. Let $G' = \{\, (x,\, y) \in \mathbb{R}^2 \mid x = y^2 \,\}$. Then G' is once again a parabola, but G' is not a graph. If $x > 0$, then both $(x,\, x^{1/2})$ and $(x,\, -x^{1/2})$ are in G' and $x^{1/2} \neq -x^{1/2}$. The uniqueness requirement of Definition 1.3.1 is not satisfied.

   Note that G' is obtained from G by interchanging x and y. As x and y do not appear symmetrically in the definition of a graph, it turns out that G is a graph but G' is not.

3. Let $G_1 = \{\, (x,\, y) \in \mathbb{R}^2 \mid x = y^2,\, y \geq 0 \,\}$. G_1 is the upper branch of the parabola G' above. G_1 is a graph and $\operatorname{Dom} G = [0,\, \infty)$, the positive part of the x-axis. The function g_1 defined by G_1 is $y = x^{1/2}$. Note that this formula is meaningful only if $x \geq 0$. Hence $D = [0,\, \infty)$ is the domain of g_1. The lower branch G_2 of G' is another graph. The function g_2 defined by G_2 is $y = -x^{1/2}$. Also, $\operatorname{Dom} g_2 = [0,\, \infty)$.

4. Let $H = \{\, (x,\, y) \in \mathbb{R}^2 \mid xy = 1 \,\}$ be a hyperbola. We see that H is a graph. In fact, if $(x,\, y)$ and $(x,\, y')$ are both in H, then $xy = xy' = 1$ implies that $x \neq 0$ and $y = y'$. We see easily that the domain of H is

   $$D = \operatorname{Dom} H = (\mathbb{R} \setminus \{0\}) = (-\infty,\, 0) \cup (0,\, \infty).$$

   The function defined by H is $y = 1/x$, $x \neq 0$. △

Examples 1.3.10 More general functions. In this course we deal mainly with functions for which the domain space is $\mathbb{R}^m$ and the range space is $\mathbb{R}^n$, where $m,\, n \in \mathbb{N}$. An efficient way to work with these functions in specific examples is to define them directly by a set of formulas.

1. Let $m = 2$, $n = 1$. Identify the domain space $\mathbb{R}^2$ with the xy-plane and the range space $\mathbb{R}$ with the z-axis. A formula $z = f(x,\, y)$ gives us a real-valued function defined on a subset D of the xy-plane. For example, let $f(x,\, y) = xy/(x^2 + y^2)$. Then the domain of f is the set D of all $(x,\, y) \in \mathbb{R}^2$ for which the expression $(xy)/(x^2 + y^2)$ is meaningful. We see that D contains all points in $\mathbb{R}^2$ except the origin $(0,\, 0)$ of $\mathbb{R}^2$. The graph Γ of f is a subset of $\mathbb{R}^2 \times \mathbb{R} = \mathbb{R}^3$. More explicitly,

   $$\Gamma = \{\, (x,\, y,\, z) \in \mathbb{R}^3 \mid z = xy/(x^2 + y^2) \text{ and } x^2 + y^2 \neq 0 \,\}.$$

Geometrically, Γ is a surface in $\mathbb{R}^3$.

2. Let $m = n = 2$. Identify the domain space $\mathbb{R}^2$ with the xy-plane and the range space $\mathbb{R}^2$ with the uv-plane. A set of formulas

$$u = u(x,\, y), \quad v = v(x,\, y)$$

gives us a function defined on a subset of the xy-plane and taking values in the uv-plane. For example

$$u = x^2 + y^2, \quad v = y/x$$

is such a set of formulas. The domain of this function is the set

$$D = \left\{ (x,\, y) \;\middle|\; (x,\, y) \in \mathbb{R}^2,\; x \neq 0 \right\}.$$

Hence one obtains D by removing the y-axis from the xy-plane. $\triangle$

Remark 1.3.11 Coordinate changes. Certain functions for which both the domain and range spaces are $\mathbb{R}^n$ are called *coordinate changes* in $\mathbb{R}^n$. Coordinate changes are considered in later chapters in some detail. Here we provide a few examples of functions $\mathbb{R}^n \to \mathbb{R}^n$ that are used as coordinate changes. $\triangle$

Example 1.3.12 Polar coordinates. Polar coordinates are given as

$$x = x(r,\, \theta) = r\cos\theta, \quad y = y(r,\, \theta) = r\sin\theta.$$

This is a coordinate change in $\mathbb{R}^2$. Both the domain and range spaces for this function are $\mathbb{R}^2$. The domain space $\mathbb{R}^2$ is identified with the $r\theta$-plane and the range space $\mathbb{R}^2$ with the xy-plane. This function takes the point $(r,\, \theta) \in \mathbb{R}^2$ in the $r\theta$-plane to the point $(r\cos\theta,\, r\sin\theta) \in \mathbb{R}^2$ in the xy-plane. In the present context, $r \in \mathbb{R}$ and $\theta \in \mathbb{R}$ are two real numbers and $(r,\, \theta) \in \mathbb{R}^2$ is an ordered pair of real numbers. We see that $(r\cos\theta,\, r\sin\theta) \in \mathbb{R}^2$ is defined for all $(r,\, \theta) \in \mathbb{R}^2$. Hence the domain of this function is also $\mathbb{R}^2$. $\triangle$

Example 1.3.13 Cylindrical coordinates. These coordinates are given as

$$x = r\cos\theta, \quad y = r\sin\theta, \quad z = \zeta.$$

This is a coordinate change in $\mathbb{R}^3$. The domain space $\mathbb{R}^3$ is identified with the $r\theta\zeta$-space and the range space $\mathbb{R}^3$ with the xyz-space. Actually, the standard notation for ζ is also z. Hence the domain space is the $r\theta z$-space and the range space is the xyz-space. The domain of this function is also $\mathbb{R}^3$, since

$$(r\cos\theta,\, r\sin\theta,\, \zeta) \in \mathbb{R}^3$$

is defined for all $(r,\, \theta,\, \zeta) \in \mathbb{R}^3$. $\triangle$

Example 1.3.14 Spherical coordinates. These coordinates are given as

$$x = \rho \sin\varphi \cos\theta, \quad y = \rho \sin\varphi \sin\theta, \quad z = \rho \cos\varphi.$$

This is a coordinate change in $\mathbb{R}^3$. The domain space $\mathbb{R}^3$ is identified with the $\rho\varphi\theta$-space and the range space $\mathbb{R}^3$ with the xyz-space. The domain of this function is also $\mathbb{R}^3$, since

$$(\rho \sin\varphi \cos\theta,\ \rho \sin\varphi \sin\theta,\ \rho \cos\varphi) \in \mathbb{R}^3$$

is defined for all $(\rho,\, \varphi,\, \theta) \in \mathbb{R}^3$. $\triangle$

Images Under Functions

Definition 1.3.15 Images of sets. Let $f : D \to Y$ be a function with domain space X and range space Y.

1. **Direct images under a function.** Let $U \subset X$. Then the *direct image of* U under f is defined as the set

   $$f(U) = \{\, y \in Y \mid y = f(x),\ \ x \in U \cap D \,\}.$$

   Hence $f(U)$ is the set of all values $f(x)$, where $x \in U \cap D$. Note that $f(U) = \emptyset$ if and only if $U \cap D = \emptyset$. Also note that $f(U) = f(U \cap D)$ for any set $U \subset X$.

2. **The range of a function.** The direct image of the domain space is called the *range of* f and denoted as Range f. Hence

   $$\text{Range}\, f = f(X) = f(D)\,.$$

   The range of f is the set of all points $y \in Y$ in the range space that are the images of points $x \in D$.

3. **Inverse images under a function.** Let $V \subset Y$. Then the inverse image of V under f is the set

   $$f^{-1}(V) = \{\, x \in X \mid f(x) \in V \,\} = \{\, x \in D \mid f(x) \in V \,\}\,.$$

   Here the first equality is the definition of $f^{-1}(V)$ as the set of all points $x \in X$ which have images $f(x)$ in V. Since $f(x)$ exists only for $x \in D$, we need to consider only points $x \in D$. This is expressed by the second equality.

Example 1.3.16 Images under polar coordinates. Let $f : \mathbb{R}^2 \to \mathbb{R}^2$ be the polar coordinates defined in Example 1.3.12. The domain space is represented by the

$r\theta$-plane and the range space by the xy-plane. The value of f at the point (r, θ) in the domain space is the point

$$(x, y) = f(r, \theta) = (r\cos\theta, r\sin\theta)$$

in the range space. Note that

$$\begin{aligned} f(r, \theta) &= f(r, \theta + 2k\pi) \\ &= f(-r, \theta + (2k+1)\pi) \text{ for all } k \in \mathbb{Z}. \end{aligned}$$

Hence, if an inverse image $f^{-1}(V)$ contains a point (r, θ), then it also contains all the points of the form

$$(r, \theta + 2k\pi) \text{ and } (-r, \theta + (2k+1)\pi), \text{ where } k \in \mathbb{Z}.$$

Vertical lines in the $r\theta$-plane correspond to the constant values of r. We see that the direct images of these lines in the $r\theta$-plane are concentric circles about the origin in the xy-plane. The images of the horizontal lines in the $r\theta$-plane are lines passing through the origin in the xy-plane. Let C be a circle of radius a about the origin in the xy-plane. Then the inverse image $f^{-1}(C)$ of this circle consists of two vertical lines $r = \pm a$ in the $r\theta$-plane. Let L be a line in the xy-plane passing through the origin and making an angle of φ with the positive x-axis. Then the inverse image $f^{-1}(L)$ of this lines consists of infinitely many horizontal lines in the $r\theta$-plane given as $\theta = \varphi + k\pi$, $k \in \mathbb{Z}$.

Figure 1.3 shows a part of the inverse image of the shaded region in the xy-plane. This region is bounded by two circles about the origin and two lines passing through the origin. Its inverse image under f is the union of infinitely many rectangles in the $r\theta$-plane. Figure 1.3 shows four of these rectangles. The direct image of each of these rectangles is the same shaded region in the xy-plane. △

This last example shows that a function may have the same value at many different points. Functions for which this does not happen are important. They are called *one-to-one* functions.

Definition 1.3.17 One-to-one functions. Let f be a function with $D = \operatorname{Dom} f$. Let $A \subset D$. Then f is said to be *one-to-one* (or *injective*) on A if $f(x_1) \neq f(x_2)$ whenever $x_1, x_2 \in A$ and $x_1 \neq x_2$. Equivalently, f is one-to-one on A if $x_1 = x_2$ whenever $x_1, x_2 \in A$ and $f(x_1) = f(x_2)$.

Theorem 1.3.18 *Let f be a function with $D = \operatorname{Dom} f$. Let $A \subset D$ and $B = f(A)$. Then f is one-to-one on A if and only if there is a function*

$$g : B \to A \quad \textit{such that } g(f(x)) = x \textit{ for all } x \in A.$$

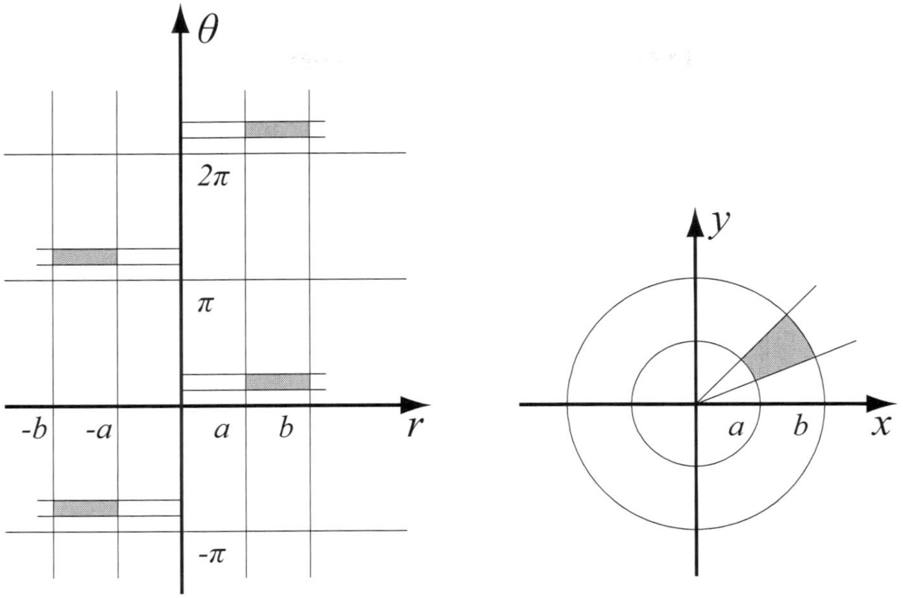

Figure 1.3. Images under polar coordinates.

Also, if such a function g exists, then it is unique and $f(g(y)) = y$ for each $y \in B$.

Proof. Assume that f is one-to-one on A. Let $B = f(A)$. Then for each $y \in B$, there is a unique $x \in A$ such that $f(x) = y$. Hence, we can define a function $g : B \to A$ by letting $x = g(y)$ whenever $f(x) = y$. Hence, $x = g(y) = g(f(x))$ for all $x \in A$.

Conversely, assume the existence of g. If $x_1, x_2 \in A$ and $f(x_1) = f(x_2)$, then

$$x_1 = g(f(x_1)) = g(f(x_2)) = x_2.$$

Hence f is one-to-one on A.

Finally, if g exists, then it is unique. In fact, the previous argument shows that if g exists, then for each $b \in B$ there is a unique $a \in A$ such that $b = f(a)$ and $g(b) = g(f(a)) = a$. If h is another function on B such that $h(f(x)) = x$ for all $x \in A$, then $h(b) = h(f(a)) = a = g(b)$. Hence $h = g$. Finally, $g(y) = x$ and $f(x) = y$ show that $f(g(y)) = y$. □

Definition 1.3.19 Inverse functions. Let f be a function with $D = \text{Dom}\, f$. Let $A \subset D$ and $B = f(A)$. A function defined on B is called an *inverse function of*

f *on* A if $g(f(x)) = x$ for all $x \in A$. Theorem 1.3.18 above shows that f has an inverse function on A if and only if f is one-to-one on A. Also, if an inverse function g exists, then it is unique. Its value at $y \in B$ is the unique solution of the equation $y = f(x)$. Finally, the same theorem also shows that if g is the inverse of f on A, then f is the inverse of g on $B = f(A)$.

Definition 1.3.20 Invertible functions. A function $f : A \to B$ is called an *invertible function*, or a *bijection*, or a *one-to-one correspondence* between A and B if it has an inverse function $g : B \to A$. Theorem 1.3.18 shows that $f : A \to B$ is an invertible function between A and B if and only if f is one-to-one on A and $B = f(A)$.

Definition 1.3.21 One-to-one and onto functions. When a function $f : A \to B$ is said to be invertible, it is understood that it is invertible between A and B. Hence, in the case of an invertible function f, the sets A and B in the notation $f : A \to B$ become important. The invertibility of $f : A \to B$ means that f is one-to-one on A and that $B = f(A)$. An invertible function $f : A \to B$ is also called a *one-to-one and onto function.* Here the sets A and B are again important. It is understood that f is one-to-one on A and that it maps A *onto* $B = f(A)$.

Example 1.3.22 Let $y = f(x) = (2x - 1)/(3x + 1)$. This function is defined for all $x \neq -1/3$. Hence $D = \operatorname{Dom} f = (\mathbb{R} \setminus \{-1/3\})$. The equation

$$y = f(x) = \frac{2x - 1}{3x + 1}$$

is uniquely solved as

$$x = g(y) = \frac{1 + y}{2 - 3y}$$

for each $y \neq 2/3$. Hence f is one-to-one on D, and its inverse on D is g. Also, $\operatorname{Dom} g = (\mathbb{R} \setminus \{2/3\})$. $\triangle$

Example 1.3.23 Let $y = f(x) = x^2 + x + 1$. Then $f(x)$ is defined for all $x \in \mathbb{R}$. Hence $D = \operatorname{Dom} f = \mathbb{R}$. The solution of the equation $x^2 + x + 1 = y$ is given by the formula

$$x = -(1/2)(1 \pm (4y - 3)^{1/2}).$$

This formula defines x if and only if $4y - 3 \geq 0$, that is, if and only if $y \geq 3/4$. Hence $f(D) = f(\mathbb{R}) = [3/4, \infty)$. Finally, the equation $x^2 + x + 1 = y$ has two solutions for each $y > 3/4$. These two solutions are symmetrical with respect to the point $x = -1/2$. There is only one solution in $A = [-1/2, \infty)$ and also only one solution in $A' = (-\infty, -1/2]$. Hence f is one-to-one on A and also one-to-one on A'. The inverse of f on A is

$$g(y) = (1/2)(-1 + (4y - 3)^{1/2})$$

and the inverse of f on A' is

$$g'(y) = (1/2)(-1 - (4y - 3)^{1/2}).$$

Note that $f(A) = f(A') = f(\mathbb{R}) = [3/4, \infty)$. There are (infinitely many) other sets $C \subset \mathbb{R}$ such that f is one-to-one on C and such that $f(C) = f(\mathbb{R})$. For example, $C = (-\infty, -1) \cup [-1/2, 0]$ is such a set. $\triangle$

Example 1.3.24 Polar coordinates were discussed in Examples 1.3.12 and 1.3.16. They are defined as a function $f : \mathbb{R}^2 \to \mathbb{R}^2$ such that

$$f(r, \theta) = (x, y), \text{ where } x = r\cos\theta \text{ and } y = r\sin\theta.$$

As observed in 1.3.16, f is not one-to-one on $\mathbb{R}^2 = \operatorname{Dom} f$. Hence f does not have an inverse function on $\mathbb{R}^2$. But there are many sets $A \subset \operatorname{Dom} f$ such that f is one-to-one on A. For example any function is one-to-one on a singleton set. The important point is to find a set $A \subset \mathbb{R}^2 = \operatorname{Dom} f$ such that f is one-to-one on A and such that $f(A) = f(\mathbb{R}^2)$. There are many such sets. Let, for example, $\alpha \in \mathbb{R}$ be a fixed number and define

$$\begin{aligned} A_\alpha &= \{\, (r, \theta) \in \mathbb{R}^2 \mid 0 < r,\, \alpha \le \theta < \alpha + 2\pi \,\} \cup \{(0, 0)\}, \\ A'_\alpha &= \{\, (r, \theta) \in \mathbb{R}^2 \mid r < 0,\, \alpha \le \theta < \alpha + 2\pi \,\} \cup \{(0, 0)\}. \end{aligned}$$

It is easy to check that f is one-to-one on each of these sets and also that

$$f(A_\alpha) = f(A'_\alpha) = f(\mathbb{R}^2) = \mathbb{R}^2 \text{ for each } \alpha \in \mathbb{R}.$$

Note that f is not one-to-one, for example, on

$$C_0 = \{\, (r, \theta) \in \mathbb{R}^2 \mid 0 \le r,\, 0 \le \theta < 2\pi \,\}.$$

In fact, in this case f maps all the points on the vertical segment

$$S_0 = \{\, (r, \theta) \in \mathbb{R}^2 \mid r = 0,\, 0 \le \theta < 2\pi \,\} \subset C_0$$

to $(0, 0)$. This is a triviality, and it is ignored in many cases. One defines the polar coordinates of (x, y) as (r, θ) such that $0 \le r$, $\alpha \le \theta < \alpha + 2\pi$, and such that $x = r\cos\theta$ and $y = r\sin\theta$. Here α is a fixed specific number like $\alpha = 0$ or $\alpha = -\pi$, depending on the problem. Everyone knows that this does not determine the θ value for $(x, y) = (0, 0)$. But everyone also knows that this is not an important point in most cases. $\triangle$

Composition of Functions

Definition 1.3.25 Composition of functions. If f and g are two functions, then the *composition* $g \circ f$ is defined by $(g \circ f)(x) = g(f(x))$. The domain of $g \circ f$ is specified by this definition in an obvious way. It is the set of all x for which $g(f(x))$ is defined. Let $F = \text{Dom}\, f$ and $G = \text{Dom}\, g$. We see that

$$\text{Dom}\,(g \circ f) = F \cap f^{-1}(G) = f^{-1}(G).$$

In fact, F is the set of all x for which $f(x)$ is defined and

$$f^{-1}(G) = \{\, x \in F \mid f(x) \in G \,\} \subset F$$

is the set of all $x \in F$ for which $f(x)$ is in the domain of g. Hence $g(f(x))$ is defined if and only if $x \in f^{-1}(G)$.

The composition of more than two functions is defined similarly. If f, g, and h are three functions, for example, then $h \circ g \circ f$ is defined as

$$(h \circ g \circ f)(x) = h(g(f(x))).$$

The domain of $h \circ g \circ f$ is the set of all x such that $h(g(f(x)))$ is defined. If

$$F = \text{Dom}\, f\,, \ G = \text{Dom}\, g\,, \text{ and } H = \text{Dom}\, h,$$

then we see that

$$\text{Dom}\,(h \circ g \circ f) = f^{-1}(g^{-1}(H)).$$

In fact, $x \in \text{Dom}\,(h \circ g \circ f)$ if and only if $g(f(x)) \in H$. This happens if and only if $f(x) \in g^{-1}(H)$, which is equivalent to $x \in f^{-1}(g^{-1}(H))$.

Lemma 1.3.26 Images under compositions. *Let f and g be two functions. Then*

$$(g \circ f)^{-1}(E) = f^{-1}(g^{-1}(E))$$

for any set E. Also, if $A \subset$ Dom $(g \circ f)$, then

$$(g \circ f)(A) = g(f(A)).$$

Proof. For the first part, note that

$$\begin{aligned} x \in (g \circ f)^{-1}(E) &\iff (g \circ f)(x) \in E \iff g(f(x)) \in E \\ &\iff f(x) \in g^{-1}(E) \iff x \in f^{-1}(g^{-1}(E)). \end{aligned}$$

Also,

$$\begin{aligned} (g \circ f)(A) &= \{\, (g \circ f)(x) \mid x \in A \,\} = \{\, g(f(x)) \mid x \in A \,\} \\ &= \{\, g(y) \mid y \in f(A) \,\} = g(f(A)). \quad \square \end{aligned}$$

Remarks 1.3.27 Compositions and inverses. Let X be a set. The identity function $I_X : X \to X$ on X was defined in Definition 1.3.7 as $I_X(x) = x$ for all $x \in X$. Let f be a function with $A = \operatorname{Dom} f$ and $B = f(A)$. Let g be a function with $B = \operatorname{Dom} g$. Then g is the inverse of f on A if and only if $g \circ f = I_A$. In this case also, $f \circ g = I_B$. These remarks follow directly from Theorem 1.3.18.

Problems

1.24 Let $f : D \to Y$ be a function with the domain $D \subset X$ and the domain space X. Let $\mathcal{E}$ be a collection of subsets of X. Show that

$$f\left(\bigcup_{E\in\mathcal{E}} E)\right) = \bigcup_{E\in\mathcal{E}} f(E) \quad \text{and} \quad f(\bigcap_{E\in\mathcal{E}} E) \subset \bigcap_{E\in\mathcal{E}} f(E).$$

Give an example to show that $f(\bigcap_{E\in\mathcal{E}} E) \neq \bigcap_{E\in\mathcal{E}} f(E)$ is possible.

1.25 Let $f : D \to Y$ be a function with the range space Y. Let $\mathcal{F}$ be a collection of subsets of Y. Show that

$$f^{-1}\left(\bigcup_{F\in\mathcal{F}} F\right) = \bigcup_{F\in\mathcal{F}} f^{-1}(F) \quad \text{and} \quad f^{-1}\left(\bigcap_{F\in\mathcal{F}} F\right) = \bigcap_{F\in\mathcal{F}} f^{-1}(F).$$

1.26 Let $f : D \to Y$ be a function with the domain $D \subset X$, the domain space X, the range space Y, and the range $R = f(X) = f(D) \subset Y$.

1. Show that $f(f^{-1}(B)) = B \cap R$ for all $B \subset Y$.
2. Show that $A \cap D \subset f^{-1}(f(A))$ for all $A \subset X$.
3. Give examples to show that $f^{-1}(f(A)) \neq A \cap D$ is possible.
4. Show that if f is one-to-one on D, then $f^{-1}(f(A)) = A \cap D$.

1.27 Let $f(x) = x^2 - 6x - 7$. What is the range $f(D)$ of f? Is f one-to-one on its domain? If not, find two different sets P, $Q \subset \mathbb{R}$ such that f is one-to-one on P and one-to-one on Q and such that $f(P) = f(Q) = f(\mathbb{R})$. Find the inverse function of f on P and the inverse function of f on Q.

1.28 Let $f : D \to Y$ be a function with the domain $D \subset X$, the domain space X, the range space Y, and the range $R = f(X) = f(D) \subset Y$. Define a relation on D by the condition that $a \in D$ is related to $b \in D$ if and only if $f(a) = f(b)$.

1. Show that this is an equivalence in D.

2. Let $P \subset D$ be an equivalence class and let $V \subset Y$ be any subset of Y. Show that either $P \subset f^{-1}(V)$ or $P \cap f^{-1}(V) = \emptyset$.

3. Let $A \subset D$ be a complete set of representatives for this equivalence. Show that f is one-to-one on A and $R = f(A)$.

1.29 Let $f : \mathbb{R}^2 \to \mathbb{R}^2$ be the polar coordinates defined by

$$(x,\, y) = f(r,\, \theta) = (r \cos\theta,\, r \sin\theta)\,.$$

Find $f^{-1}(A)$ for

$$A = \left\{\, (x,\, y) \in \mathbb{R}^2 \;\middle|\; 1 \le (x^2 + y^2) \le 4 \text{ and } 0 \le (y/x) \le 1 \,\right\}.$$

1.30 Define the function f from the xy-plane to the uv-plane by

$$(u,\, v) = f(x,\, y) = (3x + 2y,\, 6x + 4y).$$

1. What is the domain $D \subset \mathbb{R}^2$ of f?

2. What is the range $R = f(\mathbb{R}^2) = f(D) \subset \mathbb{R}^2$ of f?

3. Let $a,\, b \in \mathbb{R}$, and let L_a be the line $u = a$ and M_b the line $v = b$ in the uv-plane. What are the inverse images $f^{-1}(L_a)$ and $f^{-1}(M_b)$?

4. Let $(a,\, b) \in R$. What is $f^{-1}(\{\, (a,\, b) \,\})$?

5. Find some examples of $A \subset D$ such that f is one-to-one on A and such that $f(A) = f(D)$.

1.31 Define the function f from the xy-plane to the uv-plane by

$$(u,\, v) = f(x,\, y) = (3x + 2y,\, 6x - 4y).$$

Repeat the parts of Problem 1.30 for this example.

1.32 Define the function f from the xy-plane to the uv-plane by

$$(u,\, v) = f(x,\, y) = (xy,\, y/x).$$

Repeat the parts of Problem 1.30 for this example.

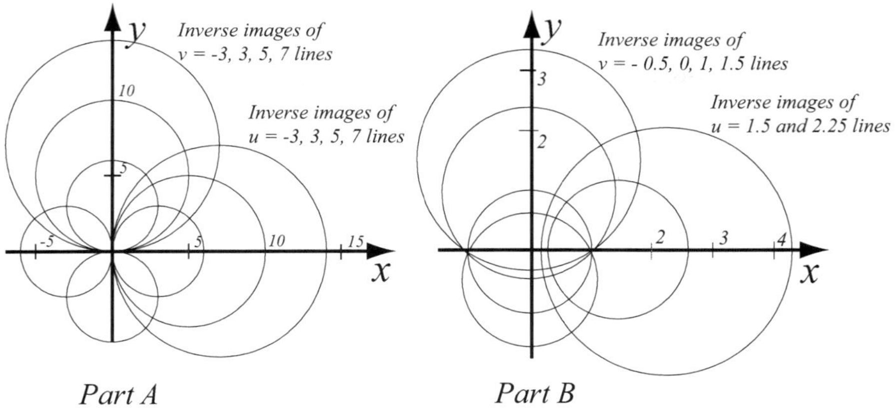

Figure 1.4. Hints for Problems 1.33 and 1.34.

1.33 Define the function f from the xy-plane to the uv-plane by

$$(u,\, v) = f(x,\, y) = ((x^2 + y^2)/(2x),\, (x^2 + y^2)/(2y)).$$

Repeat the parts of Problem 1.30 for this example. (*Hint.* In Part A of Figure 1.4 we see the inverse images of the lines $u = -3,\, 3,\, 5,\, 7$ and the lines $v = -3,\, 3,\, 5,\, 7$.)

1.34 Define the function f from the xy-plane to the uv-plane by

$$(u,\, v) = f(x,\, y) = ((x^2 + y^2 + 1)/(2x),\, (x^2 + y^2 - 1)/(2y)).$$

Repeat the parts of Problem 1.30 for this example. (*Hint.* In Part B of Figure 1.4 we see the inverse images of the lines $u = 1.5,\, 2.25$ and the lines $v = -0.5,\, 0,\, 1,\, 1.5$.)

1.35 Define the function f from the xy-plane to the uv-plane by

$$(u,\, v) = f(x,\, y) = (p(x,\, y) + q(x,\, y),\, p(x,\, y) - q(x,\, y)),$$

where $p(x,\, y) = ((x+1)^2 + y^2)^{1/2}$ and $q(x,\, y) = ((x-1)^2 + y^2)^{1/2}$. Repeat the parts of Problem 1.30 for this example.

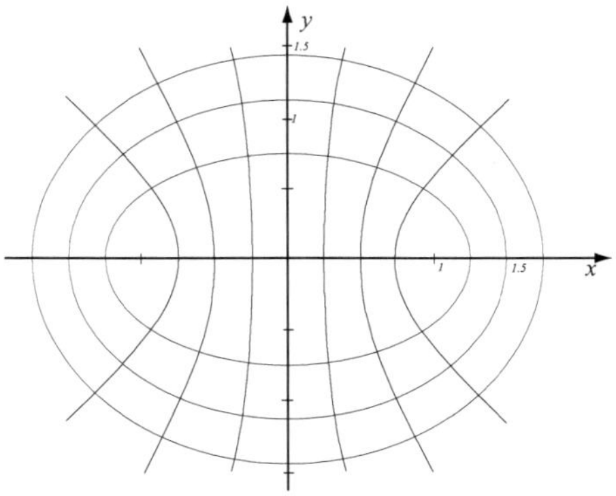

Figure 1.5. Hint for Problem 1.35.

CHAPTER 2

REAL NUMBERS

This chapter reviews key facts about the set of real numbers, denoted by $\mathbb{R}$. Some important subsets of $\mathbb{R}$ are the set of natural numbers $\mathbb{N} = \{1, 2, \dots\}$, the set of integers $\mathbb{Z} = \{0, \pm 1, \pm 2, \dots\}$, and the set of rational numbers $\mathbb{Q} = \{\, p/q \mid p, q \in \mathbb{Z}, q \neq 0 \,\}$. The set of integers and the set of rational numbers are constructed easily, starting with the set of natural numbers. By contrast, there is no easy construction for the set of real numbers. Appendix A provides a construction of $\mathbb{R}$.

We shall be selective in our discussion of $\mathbb{R}$, for we shall assume a working knowledge of the real numbers. Specifically, we shall assume that all the rules of working with the arithmetic operations and with the order relations are known. By the arithmetic operations we mean addition, multiplication, subtraction, and division. By the rules of order relations we mean the rules of working with equalities, inequalities, and absolute values. Section 2.1 summarizes the basic facts about order relations, which may not be quite so familiar as the arithmetic operations.

Analysis in Vector Spaces.
By M. A. Akcoglu, P. F. A. Bartha and D. M. Ha

An essential property of the real numbers is *completeness*. Most of the basic results in this course rely upon this property. Section 2.2 explains the completeness property, and subsequent sections of the chapter explore some of its implications for analysis.

The final sections of the chapter review basic facts about series of real numbers, as well as essential results about the *topology* of $\mathbb{R}$, i.e., open, closed, and connected sets.

2.1 REVIEW OF THE ORDER RELATIONS

We begin with the set of positive real numbers,

$$P = \{ x \in \mathbb{R} \mid 0 < x \} \subset \mathbb{R}.$$

Other order relations may be defined in terms of P. The basic inequality $x < y$ means that $(y - x) \in P$. Other inequalities are defined in terms of this first one: $y > x$, $x \leq y$, and $y \geq x$. We assume that the rules of working with inequalities are known, noting only that all of them can be derived easily from the following two properties of positive numbers. (Some examples are given below in Example 2.1.5 and also in the problems.)

Remark 2.1.1 Two properties of P. The set P of positive numbers has the following properties.

1. For each $x \in \mathbb{R}$ exactly one of the following three cases is true: $x = 0$, $x \in P$, or $-x \in P$.

2. If $x,\ y \in P$, then $x + y \in P$ and $xy \in P$.

Remark 2.1.2 A property of $\mathbb{Z}$. The set of integers has the following special property, stated in terms of the order relations on $\mathbb{R}$.

For every real number $s \in \mathbb{R}$, there is an integer $k \in \mathbb{Z}$ such that

$$k \leq s < k + 1.$$

More generally, let $n \in \mathbb{N}$. Then for every real number $t \in \mathbb{R}$, there is an integer $k \in \mathbb{Z}$ such that

$$(k/n) \leq t < (k + 1)/n.$$

To see this, apply the first result to $s = tn$: there is a $k \in \mathbb{Z}$ such that $k \leq tn < k+1$. Since $0 < n$, we can divide these inequalities by n to obtain $(k/n) \leq t < (k+1)/n$.

Remark 2.1.3 A property of $\mathbb{N}$. Let $r \in \mathbb{R}$ and $0 < r$. Then there is an $n \in \mathbb{N}$ such that $(1/n) < r$. For if $s = 1/r$, then there is a $k \in \mathbb{Z}$ such that $k \leq s < k + 1$, and we may take $n = k + 1$. Since $0 < s < n$, we have $(1/n) < r$. It is clear that if $r \in \mathbb{R}$ is such that $0 \leq r \leq (1/n)$ for all $n \in \mathbb{N}$, then $r = 0$.

Density of $\mathbb{Q}$. The following lemma shows that the rational numbers are *dense* in $\mathbb{R}$: between any two distinct real numbers, there is a rational number.

Lemma 2.1.4 *Let a, $b \in \mathbb{R}$ and $a < b$. Then there is a rational number $p \in \mathbb{Q}$ such that $a < p \leq b$.*

Proof. Let $t = b - a$. Then $0 < t$. Use property 2.1.3 of $\mathbb{N}$ to find an $n \in \mathbb{N}$ such that $(1/n) < t$. Then use property 2.1.2 of $\mathbb{Z}$ to find a $k \in \mathbb{Z}$ such that $(k/n) \leq b < (k+1)/n$. Let $p = (k/n)$. Then $p \in \mathbb{Q}$ and $p \leq b$. Also

$$b - p < [((k+1)/n) - (k/n)] = (1/n) < t = b - a,$$

so that $a < p$. Hence $a < p \leq b$. □

Example 2.1.5 Inequalities. Prove the following inequalities by applying the two properties of P stated in 2.1.1.

1. If $x \neq 0$, then $0 < x^2$.

   We are given that either $x \in P$ or $(-x) \in P$. In the first case $x^2 = xx \in P$. In the second case $x^2 = (-x)(-x) \in P$.

2. If $x < y$ and $0 < z$, then $zx < zy$.

   We are given that $(y - x) \in P$ and $z \in P$. Hence $(y - x)z = (yz - xz) \in P$. Therefore $xz < yz$.

3. If $0 < x$ and $0 < xy$, then $0 < y$.

   We are given that $x \in P$ and $(xy) \in P$. If $y = 0$, then $xy = 0$, which contradicts $xy \in P$. Hence $y \neq 0$. If $(-y) \in P$, then $x(-y) = -(xy) \in P$. This contradicts $xy \in P$. Hence $(-y) \notin P$. Then $y \in P$ is the only remaining possibility. △

Example 2.1.6 We want to show that $n < 2^n$ for all $n \in \mathbb{N}$. We use an induction argument. Let $G = \{ n \in \mathbb{N} \mid n < 2^n \}$. Then we see that $1 \in G$. Assume that $n \in G$. Then also $(n+1) \in G$, since

$$(n+1) = (1 + (1/n))n \leq 2n < 2\,2^n = 2^{n+1}.$$

Here the first inequality follows from the fact that $(1/n) \leq 1$ for all $n \in \mathbb{N}$, and the second inequality follows from the induction hypothesis. Hence $G = \mathbb{N}$. Also note that for any integer $K \geq 2$, $n < 2^n \leq K^n$ for all $n \in \mathbb{N}$. △

Example 2.1.7 Let $r \in \mathbb{N}$. If $0 \leq r \leq 2^{-n}$ for all $n \in \mathbb{N}$, then $r = 0$. In fact, in this case we also have $0 \leq r \leq 2^{-n} < (1/n)$ for all $n \in \mathbb{N}$. The last inequality follows from Example 2.1.6 above. Hence $r = 0$, as in 2.1.3. $\triangle$

Definition 2.1.8 Absolute values. Let $x \in \mathbb{R}$. Then the *absolute value of* x, denoted by $|x|$, is defined by

$$|x| = \begin{cases} -x & \text{if } x < 0\,, \\ 0 & \text{if } x = 0\,, \\ x & \text{if } 0 < x\,. \end{cases}$$

Hence $|x| = -x$ if $x \leq 0$ and $|x| = x$ if $0 \leq x$. Note that $0 \leq |x|$ in every case. In arguments involving $|x|$, we usually separate the cases $x \leq 0$ and $0 \leq x$.

Lemma 2.1.9 *The following are true for all* $x,\ y \in \mathbb{R}$.

(1) $|-x| = |x|$.

(2) $-|x| \leq x \leq |x|$.

(3) $|x| \leq y$ *if and only if* $-y \leq x \leq y$.

(4) $|xy| = |x|\,|y|$.

(5) $|x + y| \leq |x| + |y|$. *(The triangle inequality)*

Proof. (1) If $x \leq 0$, then $0 \leq -x$. Hence $|x| = -x$ and $|-x| = -x$. Therefore $|x| = |-x|$ in this case. The other case is similar.

(2) If $x \leq 0$ then $0 \leq (-x)$ and $|x| = -x$. Hence

$$-|x| = -(-x) = x \leq 0 \leq (-x) = |x|.$$

The other case is similar.

(3) Assume that $|x| \leq y$. Then $-y \leq -|x|$. Hence, by the previous case,

$$-y \leq -|x| \leq x \leq |x| \leq y.$$

Conversely, assume that $-y \leq x \leq y$. Multiply these inequalities by -1 to obtain $-y \leq -x \leq y$. Since $|x|$ is either x or $-x$, we see that $-y \leq |x| \leq y$.

(4) Let $0 \leq x$ and $0 \leq y$. Then $0 \leq (xy)$. Hence $|xy| = xy = |x|\,|y|$ in this case. The general case follows from this case by observing that $|xy| = |\,(|x|\,|y|)\,|$.

(5) By part 2, $-|x| \leq x \leq |x|$ and $-|y| \leq y \leq |y|$. Hence,

$$-(|x| + |y|) \leq x + y \leq |x| + |y|.$$

Hence, by part 3, $|x+y| \leq |x| + |y|$. $\square$

Example 2.1.10 Find all real numbers x such that $|2x+1| + |1-3x| < 5$.

Solution. Call the given inequality (A). The expression for $|2x+1|$ changes at $x = -1/2$. Similarly, the expression for $|1-3x|$ changes at $x = 1/3$. Hence (A) have different expressions in the intervals

$$I = (-\infty,\ -1/2],\ \ J = [-1/2,\ 1/3],\ \text{and}\ \ K = [1/3,\ \infty).$$

If $x \in I = (-\infty,\ -1/2]$, then (A) becomes

$$|2x+1| + |1-3x| = (-2x-1) + (1-3x) = -5x < 5.$$

This gives $-1 < x$. Hence (A) is satisfied in $I \cap (-1,\ \infty) = (-1,\ -1/2]$.

If $x \in J = [-1/2,\ 1/3]$, then (A) becomes

$$|2x+1| + |1-3x| = (2x+1) + (1-3x) = 2 - x < 5.$$

This gives $-3 < x$. Hence (A) is satisfied in $J \cap (-3,\ \infty) = [-1/2,\ 1/3] = J$.

If $x \in K = [1/3,\ \infty)$, then (A) becomes

$$|2x+1| + |1-3x| = (2x+1) + (-1+3x) = 5x < 5.$$

This gives $x < 1$. Hence (A) is satisfied in $K \cap (-\infty,\ 1) = [1/3,\ 1)$.

Combining these results, we see that (A) is satisfied if and only if

$$x \in (-1,\ -1/2] \cup [-1/2,\ 1/3] \cup [1/3,\ 1) = (-1,\ 1). \quad \triangle$$

Problems

2.1 Show that if $0 < a < b$, then $0 < (1/b) < (1/a)$. Use only the two properties of positive numbers listed in Remark 2.1.1.

2.2 Let S be the set of all real numbers $x \in \mathbb{R}$ such that

$$-2 \leq x^2 - x + |x^2 - 1| \leq 2.$$

Express S in terms of intervals.

2.3 Let S be the set of all points $(x,\ y)$ in the xy-plane such that

$$|x-2| + |y-3| \leq 1.$$

Show that S is a region bounded by four lines. Find these lines and describe S geometrically.

2.4 Show that for all real x, y, we have $||x| - |y|| \leq |x - y|$.

2.5 Let a, b, and c be real numbers with $a > 0$. Show that there is an $M \in \mathbb{R}$ such that if $|x| > M$, then $ax^2 + bx + c > 0$.

2.6 Show that for all positive a, we have

$$a + \frac{1}{a} \geq 2.$$

2.7 Let a, b, and c be real numbers. Show that

$$a^2 + b^2 + c^2 \geq ab + bc + ca.$$

2.8 Find all $x \in \mathbb{R}$ such that

$$(x - 1)(x + 3)(x + 5)(x + 2) < 0.$$

2.2 COMPLETENESS OF REAL NUMBERS

Completeness is a basic property of the real numbers. The rational numbers do not have this property. In fact, the real numbers can be obtained by adding new elements to the set of rational numbers to make it complete. Details of this construction are found in Appendix A. Here we simply treat the completeness of the real numbers as an axiom, since our objective is to explore its implications.

Upper and Least Upper Bounds

Definition 2.2.1 Upper bounds. Let $A \subset \mathbb{R}$. If there is a number $M \in \mathbb{R}$ such that $a \leq M$ for all $a \in A$, then A is is said to be *bounded above*. Any number M such that $a \leq M$ for all $a \in A$ is called an *upper bound* of A. Hence a subset of $\mathbb{R}$ is bounded above if and only if it has an upper bound. Also, if M is an upper bound of A and if $M \leq M'$, then M' is also an upper bound of A.

Definition 2.2.2 Least upper bounds. Let A be bounded above. An upper bound L of A is called a *least upper bound* of A if all upper bounds for A are greater than or equal to L.

For example, the interval (1, 4) = $\{ x \in \mathbb{R} \mid 0 < x < 4 \}$ has least upper bound 4.

Axiom 2.2.3 Completeness of $\mathbb{R}$. Let A be a *nonempty* subset of $\mathbb{R}$. If A has an upper bound, then A has a least upper bound.

Remark 2.2.4 Uniqueness of the least upper bound. A set can have at most one least upper bound. For if L and L' are both least upper bounds of A, then they are also upper bounds of A. Therefore $L \leq L'$ and $L' \leq L$. Hence $L = L'$.

Remark 2.2.5 Completeness axiom and the empty set. The completeness axiom excludes the empty set $\emptyset$. In fact, any number $M \in \mathbb{R}$ is an upper bound for $\emptyset$, since the condition that $a \leq M$ for all $a \in \emptyset$ is vacuously satisfied for any $M \in \mathbb{R}$. Hence there are upper bounds but no least upper bound for the empty set.

Remark 2.2.6 Incompleteness of $\mathbb{Q}$. The set of rational numbers is not complete: $S = \{ x \in \mathbb{Q} \mid x^2 < 2 \}$ has an upper bound but no least upper bound. There is no rational number r such that $r^2 = 2$, as proved in Example 2.2.7 below. But there is a real number ρ such that $\rho^2 = 2$. We show this in Example 2.2.8, by using the completeness of the real numbers. Hence the set of rational numbers cannot be complete.

Example 2.2.7 There is no rational number r such that $r^2 = 2$.

To see this, assume that there are integers a and b such that $2 = (a/b)^2$. By reducing to least terms, we may assume that a and b have no common factors except ± 1. Now $2b^2 = a^2$ so that 2 is a factor of a^2. It follows that 2 is also a factor of a, i.e, $a = 2m$ for some integer m. Hence, $2b^2 = 4m^2$ so that $b^2 = 2m^2$. This implies that 2 is also a factor of b. This contradicts our assumption that a and b have no common (nontrivial) factors. Hence, no such a and b exist. △

Example 2.2.8 There is a real number ρ such that $\rho^2 = 2$.

To obtain such a ρ, let $A = \{ x \in \mathbb{R} \mid 0 \leq x \text{ and } x^2 \leq 2 \}$. Then A is nonempty since, for example, $1 \in A$. Also, A is bounded above (by 2). By completeness, A has a least upper bound, ρ. Note that $1 \leq \rho \leq 2$. We will show that $\rho^2 = 2$.

Suppose that $\rho^2 < 2$. Then $\rho^2 = 2 - \varepsilon$ for some $\varepsilon > 0$. So $(\rho + \delta)^2 = \rho^2 + 2\rho\delta + \delta^2$ will be less than 2, provided we take $\delta > 0$ small enough that $2\rho\delta + \delta^2 < \varepsilon$. But then $\rho + \delta \in A$, so that ρ is not an upper bound for A, a contradiction. Similarly, if $\rho^2 > 2$, we can find $\delta > 0$ such that $(\rho - \delta)^2 > 2$, proving that ρ is not the least upper bound for A. Once again, we have a contradiction. Hence, $\rho^2 = 2$. △

Definition 2.2.9 Irrational numbers. A real number is called an *irrational number* if it is not a rational number. The previous examples show that there are irrational numbers. In fact, the following example shows that the set of irrational numbers is also dense in $\mathbb{R}$.

Example 2.2.10 Density of irrational numbers. There is an irrational number between any two distinct real numbers. We see easily that if a is a rational number, then $a\sqrt{2}$ is an irrational number. Now given any $r,\ s \in \mathbb{R}$, if $r < s$, then we have $(r/\sqrt{2}) < (s/\sqrt{2})$. Then, by Lemma 2.1.4, there is a rational number a such that $(r/\sqrt{2}) < a < (s/\sqrt{2})$. Hence $r < a\sqrt{2} < s$ and $a\sqrt{2}$ is irrational. $\triangle$

Lower and Greatest Lower Bounds

Upper and least upper bounds have symmetrical counterparts: lower and greatest lower bounds. There is an equivalent statement of the completeness axiom in terms of lower bounds.

Definition 2.2.11 Lower bounds. Let $A \subset \mathbb{R}$. If there is a number $m \in \mathbb{R}$ such that $m \leq a$ for all $a \in A$, then A is is said to be *bounded below*. Any number m such that $m \leq a$ for all $a \in A$ is called a *lower bound* of A. Hence a subset of $\mathbb{R}$ is bounded below if and only if it has a lower bound. Also, if m is a lower bound of A and if $m' \leq m$, then m' is also a lower bound of A.

Definition 2.2.12 Greatest lower bounds. Let A be bounded below. A lower bound ℓ for A is called a *greatest lower bound* of A if all lower bounds for A are less than or equal to ℓ.

Lemma 2.2.13 *Let $A \subset \mathbb{R}$ and $M,\ L \in \mathbb{R}$. Let $-A = \{\,-a \in \mathbb{R} \mid a \in A\,\}$. Then the following are true.*

(1) M is an upper bound of A if and only if $-M$ is a lower bound of $-A$.

(2) L is the least upper bound of A if and only if $-L$ is the greatest lower bound of $-A$.

(3) A is bounded above if and only if $-A$ is bounded below.

Proof. Trivial. $\square$

Remark 2.2.14 The completeness axiom 2.2.3 is equivalent to the following axiom: if a nonempty set of real numbers has a lower bound, then it has a greatest lower bound.

Definition 2.2.15 Supremum and infimum. Let $A \subset \mathbb{R}$. The least upper bound of A, if it exists, is called the *supremum* of A and is denoted as $\sup A$. Similarly, the greatest lower bound of A, if it exists, is called the *infimum* of A and is denoted as $\inf A$.

Definition 2.2.16 Bounded sets. A set of real numbers is called a *bounded set* if it is both bounded above and bounded below. By the completeness axiom, a nonempty bounded set has both a greatest lower bound (an infimum) and a least upper bound (a supremum). Hence, if A is a nonempty bounded set of real numbers, then $\sup A$ and $\inf A$ both exist.

Theorem 2.2.17 *Let* $c = \sup P$. *Then*

$$(c, \infty) \cap P = \emptyset \text{ and } (s, c] \cap P \neq \emptyset \text{ for all } s < c.$$

Proof. Since c is an upper bound for P, we have $p \leq c$ for all $p \in P$. Hence $(c, \infty) \cap P = \emptyset$. If $(s, c] \cap P$ were empty for some $s < c$, then we would have $(s, \infty) \cap P = \emptyset$ and $p \leq s$ for all $p \in P$. Hence s would be an upper bound for P that is smaller than the least upper bound c. □

Problems

2.9 If $a \in \mathbb{R}$ and $S \subset \mathbb{R}$, then let aS denote the set of all real numbers of the form $x = as$, where $s \in S$. Show that if $\sup S$ exists and if $a > 0$, then $\sup(aS) = a \sup S$. If $a < 0$, then show that $\inf(aS) = a \sup S$.

2.10 Let S, T be two nonempty bounded sets of real numbers. Let $S + T = \{ s + t \mid s \in S, t \in T \}$ and $ST = \{ st \mid s \in S, t \in T \}$. Show that

$$\begin{aligned} \sup(S + T) &= \sup S + \sup T \\ \inf(S + T) &= \inf S + \inf T. \end{aligned}$$

Is it true that $\sup(ST) = (\sup S)(\sup T)$?

2.11 Let $E \subset \mathbb{R}$ be nonempty and bounded. Show that $\mathbb{R} \setminus E$ is bounded neither from above nor below.

2.12 Let

$$S = \left\{ \frac{1}{k} - \frac{1}{k+1} \mid k \in \mathbb{Z}, k \neq 0, -1 \right\}.$$

Find $\inf S$ and $\sup S$.

2.13 Let a, b, c, and d be in $\mathbb{R}$ with $a < b$ and $c < d$. Set

$$T = \{ |2x - 3y| \mid x \in [a, b], y \in [c, d] \}.$$

Show that T is bounded. Find $\sup T$ and $\inf T$.

2.14 Let S, T be nonempty sets of real numbers that are bounded from above. Is it true that $\{ s - t \mid s \in S, t \in T \}$ is also bounded from above?

2.3 SEQUENCES OF REAL NUMBERS

Sequences were defined in Definition 1.3.8. Here we consider sequences of real numbers. Such a sequence is typically a function $x : \mathbb{N} \to \mathbb{R}$, where x_n is the value of the sequence at $n \in \mathbb{N}$. We also commonly denote the sequence itself as x_n. This slight abuse of notation is convenient when the meaning is clear from the context. Sequences may also be defined only on subsets of $\mathbb{N}$.

Definition 2.3.1 Bounded sequences. A sequence x_n of real numbers is called a *bounded sequence* if there is a number A such that $|x_n| \le A$ for all $n \in \mathbb{N}$. Otherwise, x_n is an *unbounded sequence.*

Definition 2.3.2 Convergent sequences. A sequence x_n is called a *convergent sequence* if there is a number $a \in \mathbb{R}$ with the following property: for each $\varepsilon > 0$ there is an $N \in \mathbb{N}$ such that $|x_n - a| < \varepsilon$ for all $n \ge N$. In this case the number a is called the *limit of the sequence* x_n and one says that x_n *converges to* a. The convergence of x_n to a is indicated as $\lim_n x_n = a$, or as $\lim_{n \in \mathbb{N}} x_n = a$, or simply as $x_n \longrightarrow a$.

Lemma 2.3.3 *A sequence cannot converge to two different points.*

Proof. Assume that $x_n \to a$ and $x_n \to b$ and $|a - b| = p > 0$. Find an $M \in \mathbb{N}$ such that $|x_n - a| < p/3$ for all $n \ge M$. Also find an $N \in \mathbb{N}$ such that $|x_n - b| < p/3$ for all $n \ge N$. If $n > \max(M, N)$, then

$$|a - b| \le |a - x_n| + |b - x_n| < (2/3)p < p.$$

This contradicts $|a - b| = p > 0$. Hence, if $x_n \to a$ and $x_n \to b$, then $a = b$. □

Lemma 2.3.4 *Any convergent sequence is bounded.*

Proof. Let $x_n \to a$. Find an $N \in \mathbb{N}$ such that $|x_n - a| < 1$ for all $n > N$. Hence

$$|x_n| \le |a| + |x_n - a| = |a| + 1$$

for all $n > N$. Let $A = \max(|a_1|, \ldots, |a_N|)$ and let $M = A + |a| + 1$. Then we see that $|x_n| \leq M$ for all $n \in \mathbb{N}$. □

Remarks 2.3.5 A bounded sequence need not be convergent. $x_n = (-1)^n$ is a bounded but not a convergent sequence.

Definition 2.3.6 Zero sequences. A sequence x_n in $\mathbb{R}$ is called a *zero sequence* if it converges to zero.

Lemma 2.3.7 *A sequence x_n in $\mathbb{R}$ converges to $a \in \mathbb{R}$ if and only if $x_n = a + z_n$, where z_n is a zero sequence.*

Proof. This a reformulation of the definition of convergence. □

Theorem 2.3.8 *The sum of two zero sequences is a zero sequence. The product of a zero sequence and a bounded sequence is a zero sequence. In particular, the product of a zero sequence and a convergent sequence is a zero sequence. Also, the product of two zero sequences is a zero sequence.*

Proof. Let r_n and s_n be two zero sequences. Given $\varepsilon > 0$ find $P \in \mathbb{N}$ and $Q \in \mathbb{N}$ such that $|r_n| < \varepsilon/2$ if $n \geq P$ and $|s_n| < \varepsilon/2$ if $n \geq Q$. Let $N = \max(P, Q)$. Then $|r_n + s_n| \leq |r_n| + |s_n| < (\varepsilon/2) + (\varepsilon/2) = \varepsilon$ whenever $n \geq N$. Hence $(r_n + s_n)$ is a zero sequence.

Now let r_n be a zero sequence and u_n a bounded sequence in $\mathbb{R}$. Hence there is a $K > 0$ such that $|u_n| \leq K$ for all $n \in \mathbb{N}$. Given $\varepsilon > 0$, find an $N \in \mathbb{N}$ such that $|r_n| < \varepsilon/K$ for all $n \geq N$. Then

$$|u_n r_n| = |u_n|\,|r_n| \leq K\,|r_n| < K(\varepsilon/K) = \varepsilon$$

for all $n \geq N$. Hence $u_n r_n$ is a zero sequence. □

Theorem 2.3.9 *Let u_n and v_n be two sequences in $\mathbb{R}$. Assume that $u_n \to a$ and $v_n \to b$. Then $(u_n + v_n) \to (a + b)$ and $(u_n\, v_n) \to (a\, b)$. If $a \neq 0$, then also $(v_n/u_n) \to (b/a)$.*

Proof. Let $r_n = (u_n - a)$ and $s_n = (v_n - b)$. Both are zero sequences. Then Theorem 2.3.8 shows that

$$\begin{aligned}
(u_n + v_n) - (a + b) &= r_n + s_n, \\
u_n v_n - ab &= (a + r_n)(b + s_n) - ab \\
&= a s_n + b r_n + + r_n s_n
\end{aligned}$$

are zero sequences. Hence $(u_n + v_n) \to (a + b)$ and $(u_n\, v_n) \to (a\, b)$.

Now $1/u_n$ is defined only if $u_n = (a + r_n) \neq 0$. Since $r_n \to 0$, there is an $N \in \mathbb{N}$ such that $|r_n| < |a|/2$ for all $n \geq N$. Let

$$p_n = \frac{1}{a + r_n} - \frac{1}{a} = \frac{-r_n}{a(a + r_n)}.$$

But $|a + r_n|^{-1} < 2/|a|$ and $|p_n| < (2/|a|^2)\,|r_n|$ for all $n \geq N$. Therefore, p_n is a zero sequence. Hence $(1/u_n) \to (1/a)$. Then

$$(v_n/u_n) = v_n(1/u_n) \to b(1/a)$$

by the first part. □

Theorem 2.3.10 *Let u_n and v_n be two sequences in $\mathbb{R}$. If $u_n \to a$ and if $(v_n - u_n)$ is a zero sequence then also $v_n \to a$.*

Proof. We see that $(v_n - a) = (u_n - a) + (v_n - u_n)$ is a zero sequence, as the sum of two zero sequences. □

Monotone Convergence Theorem

The Monotone Convergence Theorem gives a sufficient condition for the convergence of a sequence. This condition does not refer to the limit of the sequence, but only to the terms of the sequence.

Notations 2.3.11 If s_n is a bounded sequence in $\mathbb{R}$, the set

$$S = \{\, s_n \mid n \in \mathbb{N} \,\}$$

is a bounded set in $\mathbb{R}$. Hence $\inf S$ and $\sup S$ exist. For this particular set S we write $\inf S = \inf_n s_n$ and $\sup S = \sup_n s_n$.

Definition 2.3.12 Monotone sequences in $\mathbb{R}$. Let s_n be a sequence in $\mathbb{R}$. Then s_n is called a *monotone sequence* if either $s_n \leq s_{n+1}$ for all $n \in \mathbb{N}$ or $s_{n+1} \leq s_n$ for all $n \in \mathbb{N}$. In the first case, s_n is also called an *increasing sequence*; in the second case, it is called a *decreasing sequence*.

Theorem 2.3.13 Monotone Convergence Theorem. *Every monotone and bounded sequence is convergent. More specifically, let s_n be a bounded and monotone sequence. Then $s_n \to \inf_n s_n$ if s_n is decreasing and $s_n \to \sup_n s_n$ if s_n is increasing.*

Proof. Let s_n be a decreasing and bounded sequence. Let $a = \inf_n s_n$. Then $a \le s_n$ for all $n \in \mathbb{N}$, but if $a' > a$, there is an $N \in \mathbb{N}$ such that $s_N < a'$. We claim that $s_n \to a$. Let $\varepsilon > 0$. Then $a' = (a+\varepsilon) > a$, and therefore there is an $N \in \mathbb{N}$ such that $s_N < (a+\varepsilon)$. But $s_n \le s_N$ for all $n \ge N$ since s_n is a decreasing sequence. Therefore, if $n \ge N$, then

$$|s_n - a| = s_n - a \le s_N - a \le (a+\varepsilon) - a = \varepsilon.$$

Hence $s_n \to a = \inf S$. The proof for increasing and bounded sequences is similar. In this case we see that $s_n \to b = \sup_n s_n$. □

Example 2.3.14 Geometric series. Let p be a number, $0 \le p < 1$. Define $x_n = 1 + p + p^2 + \cdots + p^n$. We claim that x_n is a convergent sequence. Clearly, $x_n \le x_{n+1}$ for all $n \in \mathbb{N}$, so x_n is an increasing sequence. Also, by familiar algebraic manipulations we see that

$$x_n = 1 + p + p^2 + \cdots + p^n = (1-p^{n+1})/(1-p) \le 1/(1-p).$$

Hence x_n is also a bounded sequence. Therefore, x_n converges by the Monotone Convergence Theorem. In this case we can also show that $x_n \to (1-p)^{-1}$, but this point is not important for the following application. △

Examples 2.3.15 An application of geometric series. Let s_n be a sequence in $\mathbb{R}$. Assume that there is a constant M, an integer N, and a number p such that $0 \le p < 1$ and such that $0 \le s_n \le Mp^n$ for all $n \ge N$. Let $x_n = s_1 + s_2 + \cdots + s_n$. Then x_n is monotone increasing. It is also bounded. In fact, if $m \in \mathbb{N}$, then

$$\begin{aligned} x_{N+m} &= x_{N-1} + x_N + \cdots + x_{N+m} \\ &\le x_{N-1} + Mp^N(1 + p + \cdots + p^m) \\ &\le x_{N-1} + Mp^N(1-p)^{-1}. \end{aligned}$$

Therefore x_n converges in $\mathbb{R}$.

As a specific case, let $s_n = r^n/(n!)$, where $r \in \mathbb{R}$ is a fixed number, $0 < r$, and $n! = 1\cdot 2\cdot 3 \cdots n$, as usual. Find $N \in \mathbb{N}$ such that $p = (r/N) < 1$. Then we see easily that $s_{N+m} \le s_N p^m = (s_N p^{-N})p^{N+m}$ for all $m \in \mathbb{N}$. Hence our requirements are satisfied with $M = s_N p^{-N} = s_N (N/r)^N$. Therefore we know that

$$x_n = 1 + \frac{r}{1!} + \frac{r^2}{2!} + \cdots + \frac{r^n}{n!}$$

converges for all $r > 0$. The limit is well defined, but it cannot be expressed (at least not in an obvious way) in terms of r and the previously defined operations. Of course, this limit is e^r. △

Problems

2.15 Let $|a| < 1$. Show that $\lim_n a^n = 0$.

2.16 Let $0 < a$. Show that $\lim_n a^{1/n} = 1$.

2.17 Show that each irrational number is the limit of a sequence of rational numbers. Also show that any rational number is the limit of a sequence of irrational numbers.

2.18 Let $(x_n), (y_n)$ be sequences of real numbers with $\lim_{n\to\infty} x_n = -2$ and $\lim_{n\to\infty} y_n = 3$. If a, b are real numbers such that

$$\begin{aligned} \lim_{n\to\infty} (ax_n - by_n) &= \lim_{n\to\infty} (bx_n + ay_n) \\ \lim_{n\to\infty} (bx_n - ay_n) &= \lim_{n\to\infty} (ax_n + by_n), \end{aligned}$$

what are the values of a and b?

2.19 Let $|a| < 1$ and $r \in \mathbb{R}$. Show that the sequence

$$x_n = \sum_{k=1}^{n} \binom{r}{k} a^k$$

is convergent. (The limit of this sequence is $(1 + a)^r$, but the proof of this fact requires more work.)

2.20 Let $a > 0$. Define a sequence x_n recursively by $x_1 = a$ and

$$x_{n+1} = 1 + (1 + x_n)^{-1}, \ n \in \mathbb{N}.$$

Show that x_n is a convergent sequence. Show that $\lim_n x_n = \sqrt{2}$.

2.21 Let $a > 0$ and $c > 0$. Define a sequence x_n recursively by $x_1 = a$ and

$$x_{n+1} = 1 + c(1 + x_n)^{-1}, \ n \in \mathbb{N}.$$

Show that x_n is a convergent sequence. What is $\lim_n x_n$?

2.22 Let $a > 0$ and $c > 0$. Define a sequence x_n recursively by $x_1 = a$ and

$$x_{n+1} = 1 + (c/x_n), \ n \in \mathbb{N}.$$

Show that x_n is a convergent sequence. What is $\lim_n x_n$?

2.4 SUBSEQUENCES

Lemma 2.3.4 shows that every convergent sequence is a bounded sequence. It is clear that not every bounded sequence is a convergent sequence. Nevertheless, there is a basic relation between bounded sequences and convergent sequences: every bounded sequence has a convergent subsequence. This is called the Bolzano-Weierstrass theorem, and it is a very important consequence of the completeness of the real numbers.

Definition 2.4.1 Subsequences. Let $x : \mathbb{N} \to \mathbb{R}$ be a sequence of real numbers. A *subsequence* of x is the restriction of the domain $\mathbb{N}$ to an unbounded subset $\mathbb{K} \subset \mathbb{N}$. Hence x_k, $k \in \mathbb{K}$, is a subsequence of x_n, $n \in \mathbb{N}$. We write $x_k \to a$ or $\lim_{k \in \mathbb{K}} x_k = a$ to signify the convergence of the subsequence x_k to $a \in \mathbb{R}$. A more familiar way to denote a subsequence is the following. Let x_k, $k \in \mathbb{K}$, be a subsequence. Order the integers in $\mathbb{K}$ as a sequence $k_1 < k_2 < \cdots$ and denote this subsequence as x_{k_n}, where $n \in \mathbb{N}$, or simply as x_{k_n}. We then write $x_{k_n} \to a$ or $\lim_{n \in \mathbb{N}} x_{k_n} = a$ for the convergence of x_{k_n} to a.

Remarks 2.4.2 Note that if a sequence is convergent, then every subsequence of it is also convergent and converges to the same point as the full sequence. This follows easily from the definitions.

Lemma 2.4.3 *Let x_n be a sequence in $\mathbb{R}$. Let $a \in \mathbb{R}$. Assume that the set*

$$\mathbb{K}_r = \{\, n \in \mathbb{N} \mid a - r < x_n < a + r \,\}$$

is an infinite (unbounded) set of integers for each $r > 0$. Then x_n has a subsequence that converges to a.

Proof. Define a sequence of integers $k_1 < k_2 < \cdots$ as follows. Let k_1 be the smallest integer in $\mathbb{K}_1$. If $k_1 < \cdots < k_n$ are defined, then let k_{n+1} be the smallest integer in $\mathbb{K}_{1/(n+1)}$ such that $k_n < k_{n+1}$. Such an integer exists since $\mathbb{K}_{1/(n+1)}$ is an infinite subset of $\mathbb{N}$. Then an induction argument shows that k_n is defined for each $n \in \mathbb{N}$. Now given any $r > 0$, there is an $N \in \mathbb{N}$ such that $(1/N) < r$. Hence we see that

$$|\, x_{k_n} - a \,| < 1/n \leq 1/N < r \text{ whenever } n \geq N.$$

This means that the subsequence x_{k_n} converges to a. □

Lemma 2.4.4 *Let x_n be a sequence in $\mathbb{R}$. For each $t \in \mathbb{R}$, let*

$$Q(t) = \{\, n \in \mathbb{N} \mid x_n \leq t \,\}.$$

Let $a \in \mathbb{R}$ be a number such that $Q(u)$ is a finite set of integers whenever $u < a$ and $Q(v)$ is an infinite set of integers whenever $a < v$. Then there is a subsequence of x_n that converges to a.

Proof. Given $r > 0$, find $u,\ v \in \mathbb{R}$ such that

$$a - r \le u < a < v < a + r.$$

Then we have

$$\begin{aligned} Q(v) \setminus Q(u) &= \{ n \in \mathbb{N} \mid u < x_n \le v \} \\ &\subset \{ n \in \mathbb{N} \mid a - r < x_n < a + r \} = \mathbb{K}_r . \end{aligned}$$

We see that $(Q(v) \setminus Q(u)) \subset \mathbb{N}$ is an infinite set, since $Q(v)$ is an infinite set and $Q(u)$ is a finite set. Therefore $\mathbb{K}_r$ is an infinite set for each $r > 0$. Then Lemma 2.4.3 shows that there is a subsequence of x_n that converges to a. □

Theorem 2.4.5 Bolzano-Weierstrass Theorem. *Every bounded sequence in $\mathbb{R}$ has a convergent subsequence.*

Proof. Let x_n be a bounded sequence in $\mathbb{R}$. Assume that $p \le x_n \le q$ for all $n \in \mathbb{N}$. For each $t \in \mathbb{R}$, let

$$Q(t) = \{ n \in \mathbb{N} \mid x_n \le t \} .$$

Let $T \subset \mathbb{R}$ be the set of all $t \in \mathbb{R}$ for which $Q(t)$ is a finite set of integers. That is, a real number t belongs to T just in case all but finitely many members of the sequence x_n are larger than t.

If $t < p$, then $Q(t) = \emptyset$ is a finite (bounded) set. Hence $t \in T$ for all $t < p$. Therefore T is not an empty set. If $t \ge q$, then $Q(t) = \mathbb{N}$ is an infinite (unbounded) set. Hence $t \notin T$ for all $t \ge q$. Therefore T is bounded above. Hence $a = \sup T$ exists by the completeness axiom.

If $v > a$, then $v \notin T$, since a is an upper bound for T. Then $Q(v)$ is infinite. If $u < a$, then u is not an upper bound for T. Therefore there is a $w \in T$ such that $u < w$. Hence $Q(w)$ is finite. Then $Q(u)$ is also finite since $Q(u) \subset Q(w)$. Hence $Q(u)$ is finite for all $u < a$ and $Q(v)$ is infinite for all $v > a$. Lemma 2.4.4 shows that there is a subsequence of x_n converging to a. □

Cauchy Sequences

Definition 2.4.6 Cauchy sequences. A sequence of real numbers x_n is called a *Cauchy sequence* if for each $\varepsilon > 0$ there is an integer $N \in \mathbb{N}$ such that $|x_n - x_m| < \varepsilon$ for all $m,\ n \ge N$.

Theorem 2.4.7 *A sequence of real numbers is a Cauchy sequence if and only if it is a convergent sequence.*

Proof. Assume that x_n is a convergent sequence and let $x_n \to a$. Given $\varepsilon > 0$, find $N \in \mathbb{N}$ such that $|x_n - a| < \varepsilon/2$ for all $n \geq N$. Then

$$|x_n - x_m| \leq |x_n - a| + |a - x_m| < (\varepsilon/2) + (\varepsilon/2) = \varepsilon$$

for all $m,\, n \geq N$. Hence x_n is a Cauchy sequence.

Conversely, assume that x_n is a Cauchy sequence. Then x_n is clearly bounded. Use the Bolzano-Weierstrass theorem to find a convergent subsequence x_k, $k \in \mathbb{K}$. Let $a = \lim_{k\in\mathbb{K}} x_k$. Then for each $\varepsilon > 0$ there is a $K \in \mathbb{K}$ such that $|x_k - a| < \varepsilon/2$ for all $k \geq K$, $k \in \mathbb{K}$. Also, find an $N \in \mathbb{N}$ such that $|x_n - x_m| < \varepsilon/2$ for all $m,\, n \geq N$. Let $k \geq \max(K,\, N)$ and $k \in \mathbb{K}$. If $n \geq N$, then

$$|x_n - a| \leq |x_n - x_k| + |x_k - a| < \varepsilon/2 + \varepsilon/2 = \varepsilon,$$

since $n, k \geq N$ and $k \geq K$. Hence $x_n \to a$. □

Theorem 2.4.8 *Let x_n and y_n be Cauchy sequences. Then $x_n + y_n$ and $x_n y_n$ are also Cauchy sequences. If $\lim_n x_n \neq 0$, then $1/x_n$ is also a Cauchy sequence.*

Proof. Theorem 2.3.9 shows that the sum and the product of two convergent sequences is again a convergent sequence. Theorem 2.4.7 shows that a sequence is a Cauchy sequence if and only if it is convergent. These two results show that the sum and the product of two Cauchy sequences is again a Cauchy sequence. The last part of the theorem follows in the same way from the corresponding part of Theorem 2.3.9. Here $1/x_n$ is defined, of course, only when $x_n \neq 0$, as mentioned in the proof of Theorem 2.3.9. □

Remarks 2.4.9 Convergent sequences and Cauchy sequences. Theorem 2.4.7 shows that, in $\mathbb{R}$, the notions of Cauchy sequence and convergent sequence are equivalent. This is an important fact about $\mathbb{R}$ – it does not hold for the rational numbers. The concept of a Cauchy sequence is the simpler of the two, for the definition is stated in terms of the sequence only and does not refer to the limit point, which may or may not belong to the sequence. The definition of Cauchy sequences can be simplified even further, as the following theorem shows.

Theorem 2.4.10 *Let x_n be a sequence in $\mathbb{R}$. Then the following are equivalent.*

(1) There is a number $a \in \mathbb{R}$ such that $x_n \to a$; i.e., for each $\varepsilon > 0$ there is an $N \in \mathbb{N}$ such that $|x_n - a| < \varepsilon$ for all $n \geq N$.

(2) *For each* $\varepsilon > 0$ *there is an* $N \in \mathbb{N}$ *such that* $|x_n - x_N| < \varepsilon$ *for all* $n \geq N$.

(3) *For each* $\varepsilon > 0$ *there is an* $N \in \mathbb{N}$ *such that* $|x_n - x_m| < \varepsilon$ *for all* $m,\ n \geq N$.

Proof. The equivalence of (1) and (3) is Theorem 2.4.7 above. Also, it is clear that (3) implies (2). Now assume (2). Given $\varepsilon > 0$, find $N \in \mathbb{N}$ such that $|x_n - x_N| < \varepsilon/2$ for all $n \geq N$. If $m,\ n \geq N$, then

$$|x_n - x_m| \leq |x_n - x_N| + |x_m - x_N| < (\varepsilon/2) + (\varepsilon/2) = \varepsilon.$$

Hence (3) follows. □

Problems

2.23 Let x_n be a sequence in $\mathbb{R}$. Let $a \in \mathbb{R}$. Assume that

$$\mathbb{L}_r = \{\, n \in \mathbb{N} \mid a - r < x_n \leq a \,\}$$

is an infinite (unbounded) set of integers for each $r > 0$. Show that x_n has a monotone increasing subsequence converging to a.

2.24 Let x_n be a sequence in $\mathbb{R}$. Let $a \in \mathbb{R}$. Assume that

$$\mathbb{S}_r = \{\, n \in \mathbb{N} \mid a \leq x_n < a + r \,\}$$

is an infinite (unbounded) set of integers for each $r > 0$. Show that x_n has a monotone decreasing subsequence converging to a.

2.25 Let x_n be a bounded sequence in $\mathbb{R}$. For each $n \in \mathbb{N}$, let

$$S_n = \{\, x \in \mathbb{R} \mid x = x_k,\ n \leq k \,\}.$$

Then S_n is a nonempty and bounded set and $s_n = \sup S_n$ exists. Show that s_n is a convergent sequence. Show that if $s_n \to a$, then x_n has a subsequence converging to a. (This gives another proof for the Bolzano-Weierstrass theorem.)

2.26 Let x_n be a bounded sequence in $\mathbb{R}$. Let s_n be the sequence obtained in Problem 2.25. Assume that there is an $N \in \mathbb{N}$ such that $s_N = s_{N+k}$ for all $k \in \mathbb{N}$. Show that x_n has a monotone increasing convergent subsequence.

2.27 Let x_n be a bounded sequence in $\mathbb{R}$. Let s_n be the sequence obtained in Problem 2.25. Assume that there is no $N \in \mathbb{N}$ such that $s_N = s_{N+k}$ for all $k \in \mathbb{N}$. Show that x_n has a monotone decreasing convergent subsequence.

2.28 Show that every sequence (bounded or not) has a monotone subsequence.

2.29 Show that a sequence x_n is a Cauchy sequence if and only if there is a zero sequence z_n such that $|x_n - x_m| \leq z_n$ for all $m, n \in \mathbb{N}, m \geq n$.

2.30 Let x_n and y_n be Cauchy sequences. Show directly, without using Theorem 2.4.7, that $x_n + y_n$ and $x_n y_n$ are also Cauchy sequences. Also show directly that if there is an $a > 0$ such that $|x_n| \geq a$, then $1/x_n$ is a Cauchy sequence. Give an example of a Cauchy sequence x_n such that $x_n \neq 0$ for all $n \in \mathbb{N}$ but $1/x_n$ is not a Cauchy sequence.

2.31 Give an example of a nonconvergent sequence x_n such that

$$|x_{n+2} - x_{n+1}| < |x_{n+1} - x_n| \text{ for all } n \in \mathbb{N}.$$

2.32 Let x_n be a sequence in $\mathbb{R}$. Define a new sequence

$$s_n = |x_2 - x_1| + |x_3 - x_2| + \cdots + |x_n - x_{n-1}|, \; n \geq 2.$$

Show that if s_n is a bounded sequence then x_n is a Cauchy sequence.

2.33 Give an example of a Cauchy sequence x_n for which the sequence

$$s_n = |x_2 - x_1| + |x_3 - x_2| + \cdots + |x_{n+1} - x_n|, \; n \in \mathbb{N},$$

is an unbounded sequence.

2.34 Let p_n be any sequence of numbers such that $p_n > 0$ for all $n \in \mathbb{N}$. Show that any Cauchy sequence x_n has a subsequence x_{k_n} such that

$$|x_{k_{n+1}} - x_{k_n}| \leq p_n \text{ for all } n \in \mathbb{N}.$$

2.35 Show that every Cauchy sequence x_n has a subsequence x_{k_n} for which

$$s_n = |x_{k_2} - x_{k_1}| + |x_{k_3} - x_{k_2}| + \cdots + |x_{k_{n+1}} - x_{k_n}|, \; n \in \mathbb{N},$$

is a bounded sequence.

2.36 Let c_n be a sequence. Assume that the sequence

$$C_n = |c_1| + \cdots + |c_n|$$

is a bounded sequence. Show that there is a zero sequence z_n such that

$$|\, c_{n+1} + \cdots + c_{n+k}| \leq z_n$$

for all $n,\ k \in \mathbb{N}$.

2.37 Let $p,\ q \in \mathbb{N}$ and $p < q$.

1. With the general binomial coefficients as defined in Definition 1.2.13, show that
$$q^k \begin{pmatrix} p \\ k \end{pmatrix} \leq p^k \begin{pmatrix} q \\ k \end{pmatrix}$$
for all $k \in \mathbb{N}$.

2. Show that $(1 + (q/p)x)^p \leq (1 + x)^q$ for all $x \geq 0$.

3. Let $x = 1/n$ and $(q/p) - 1 = r > 0$. Transform the last inequality algebraically to obtain
$$\frac{1}{(n+1)^{1+r}} \leq \frac{1}{r}\left(\frac{1}{n^r} - \frac{1}{(n+1)^r}\right)$$
for all $n \in N$.

2.5 SERIES OF REAL NUMBERS

Definition 2.5.1 Series. Let x_n, $n \in \mathbb{Z}^+$, be a sequence. Then the sequence

$$s_n = x_0 + x_1 + \cdots + x_n = \sum\nolimits_{k=0}^{n} x_k,\ n \in \mathbb{Z}^+,$$

is called a *series* (or an *infinite series*). The terms s_n of this series are called its *partial sums*. Note that there is no great difference between series and sequences: just as a series is a sequence of partial sums, any sequence s_n may be considered as the series $s_n = \sum_{k=0}^{n} x_k$ with $x_0 = s_0$ and $x_n = s_n - s_{n-1}$, $n \in \mathbb{N}$.

Definition 2.5.2 Convergence of series. If $\lim_n s_n = \lim_n \sum_{k=0}^{n} x_k$ exists, then s_n is called a *convergent series*. If the limit does not exist, then the series *diverges*. We denote the limit of a convergent series as

$$s = \lim\nolimits_n s_n = \lim\nolimits_n \sum\nolimits_{k=0}^{n} x_k = \sum\nolimits_{k=0}^{\infty} x_k.$$

If $p_n \geq 0$ and if $s_n = p_0 + p_1 + \cdots + p_k$ is a bounded sequence then we express this as

$$\sum\nolimits_{k=0}^{\infty} p_k = \sum\nolimits_{n} p_n < \infty.$$

Note that although expressions like $s = \sum_{k=0}^{\infty} x_k$ and $\sum_{k=0}^{\infty} p_k$ involve a summation sign, they are not proper summations. That is because, in general, they lack important properties of ordinary finite summation (e.g., insensitivity to the order of summation).

Lemma 2.5.3 *If $\sum_n |x_n| < \infty$, then $S = \sum_{k=0}^{\infty} |x_k|$ and $s = \sum_{k=0}^{\infty} x_k$ exist.*

Proof. The sequence $S_n = \sum_{k=0}^{n} |x_k|$ is monotone increasing. If $\sum_n |x_n| < \infty$, then S_n is also a bounded sequence. Hence $S = \lim_n S_n = \sum_{k=0}^{\infty} |x_k|$ exists. This follows from the Monotone Convergence Theorem, 2.3.13. Then we know that S_n is a Cauchy sequence. Therefore, given $\varepsilon > 0$, there is an $N \in \mathbb{N}$ such that $S_{n+k} - S_n = |S_{n+k} - S_n| < \varepsilon$ for all $n \geq N$ and for all $k \in \mathbb{N}$. Then

$$|s_{n+k} - s_n| = |x_{n+1} + \cdots + x_{n+k}| \leq |x_{n+1}| + \cdots + |x_{n+k}| = S_{n+k} - S_n$$

shows that s_n is a Cauchy sequence. Hence $s = \lim_n s_n = \sum_{k=0}^{\infty} x_k$ exists. $\square$

Definition 2.5.4 Absolute convergence. A series $s_n = \sum_{k=0}^{n} x_k$ is called *absolutely convergent* if $\sum_n |x_n| < \infty$.

Lemma 2.5.3 shows that if a series is absolutely convergent, then it is also convergent. It also shows that a series $s_n = \sum_{k=0}^{n} x_k$ is absolutely convergent if and only if the series $S_n = \sum_{k=0}^{n} |x_k|$ is convergent.

A convergent series is not necessarily absolutely convergent. A counterexample is $s_n = \sum_{k=0}^{n} (-1)^n/(n+1)$. The limit of an absolutely convergent series is insensitive to rearrangements of the terms of the series, but the same is not true for a series that is convergent but not absolutely convergent. Using Problem 2.42, one can show that the terms of the series $s_n = \sum_{k=0}^{n} (-1)^n/(n+1)$ can be rearranged to converge to any desired limit or to diverge.

Tests for the Convergence of Series

We review some familiar and very useful tests to investigate the convergence of a series.

Theorem 2.5.5 The root test. *Let $s_n = \sum_{k=0}^{n} x_k$ be a series. Assume that there are M, $r \in \mathbb{R}$ and $N \in \mathbb{N}$ such that $0 < r < 1$ and such that*

$$|x_n| \leq M\, r^n$$

for all $n \geq N$. Then $S = \sum_{k=0}^{\infty} |x_k|$ and $s = \sum_{k=0}^{\infty} x_k$ both exist.

Proof. We have

$$\begin{aligned} \sum_{k=0}^{N+n} |x_k| &= \sum_{k=0}^{N} |x_k| + \sum_{k=N+1}^{N+n} |x_k| \\ &\leq \sum_{k=0}^{N} |x_k| + \sum_{k=N+1}^{N+n} M\, r^k \\ &\leq \sum_{k=0}^{N} |x_k| + M/(1-r) \end{aligned}$$

for all $n \in \mathbb{N}$. Hence we see that $\sum_n |x_n| < \infty$. Then Lemma 2.5.3 shows that $S = \sum_{k=0}^{\infty} |x_k|$ and $s = \sum_{k=0}^{\infty} x_k$ exist. □

Corollary 2.5.6 The ratio test. *Let $s_n = \sum_{k=0}^{n} x_k$ be a series. Assume that there is an $r \in \mathbb{R}$ and $N \in \mathbb{N}$ such that $0 < r < 1$ and such that*

$$(|x_{n+1}|/|x_n|) \leq r$$

for all $n \geq N$. Then $S = \sum_{k=0}^{\infty} |x_k|$ and $s = \sum_{k=0}^{\infty} x_k$ both exist.

Proof. Let $M = |x_N|$. Then, by an induction, $|x_{N+k}| \leq M\, r^k$ for all $k \in \mathbb{N}$. Hence $|x_n| \leq (M/r^N)r^n$ for all $n \geq N$. Then an application of Theorem 2.5.5 completes the proof. □

Theorem 2.5.7 *Let $a_n \geq 0$ be a sequence. Assume that there is an $r > 0$ such that $\sum_n |a_n|\, r^n < \infty$. Then $\sum_n a_n n^k x^n < \infty$ for all $k \in \mathbb{N}$ and $|x| < r$.*

Proof. Let $y_1 = (x/r)$ and $y_n = (n/(n-1))^k (x/r)$, $n \geq 2$. Note that

$$n^k x^n = y_1 \cdot y_2 \cdots y_n\, r^n.$$

We see that $\lim_n (n/(n-1))^k = 1$. Also, $|x/r| < 1$. Hence there is a $N \in \mathbb{N}$ such that $|y_n| < 1$ for all $n \geq N$. Let $M = |y_1 \cdot y_2 \cdots y_N|$. Then

$$|a_n n^k x^n| \leq M\, |a_n| r^n$$

for all $n \geq N$. This shows that $\sum_n |a_n n^k x^n| < \infty$. Hence $s_n(x) = \sum_{m=1}^{n} a_m m^k x^m$ is absolutely convergent. □

Problems

2.38 If $\sum a_k$ converges, show that $\lim_{n\to\infty} a_n = 0$.

2.39 For each $n \geq 1$, let

$$a_n = \frac{n!(3n-1)^2}{1 \cdot 3 \cdot \cdots \cdot (2n+1)}.$$

Determine if $\sum a_n$ converges.

2.40 Suppose that $\sum a_n$ and $\sum b_n$ are infinite series with positive terms. Assume that

$$\lim_{n\to\infty} \frac{a_n}{b_n} = \ell > 0.$$

Show that both $\sum a_n$ and $\sum b_n$ converge or diverge together.

2.41 Let a_n be a decreasing sequence of positive numbers such that $\sum a_n$ converges. Show that $\lim na_n = 0$.

2.42 Show that $\sum_{n=1}^{\infty} \frac{1}{n}$ diverges. Hence, show that if $p(n)$ and $q(n)$ are polynomials and $q(n) \neq 0$ for any $n \in N$, then

$$\sum \frac{p(n)}{q(n)}$$

diverges if the degree of p is 1 less than the degree of q.

2.43 Suppose that $\sum |a_k|$ converges. Show that $\sum |a_k|^p$ converges for all $p > 1$.

2.44 Let x_n be a sequence in $\mathbb{R}$ such that $\lim_{n\to\infty} x_n$ exists. Decide whether or not

$$\sum_{n=1}^{\infty} (x_{n+1} - x_n)$$

converges.

2.45 Suppose that $\sum a_k$ converges. What can we say about the convergence of

$$\sum \frac{1}{1 + a_k^2}?$$

2.46 Suppose that a_n converges and $\sum x_k$ converges. If $x_k \geq 0$ for all k, show that $\sum a_k x_k$ converges absolutely.

2.47 Does the series $\sum \sin(1/k)$ converge?

2.48 Show that there is a unique function $\tau : \mathbb{N} \to \mathbb{Z}^+$ such that

$$2^{\tau(n)} \leq n < 2^{\tau(n)+1}$$

for each $n \in \mathbb{N}$. Define $x_n = (2^{-\tau(n)}n - 1)$ for $n \in \mathbb{N}$. Write the first fifteen terms of this sequence. Show that for each $a \in \mathbb{R}$, $0 \leq a \leq 1$, there is a convergent subsequence of x_n converging to r.

2.49 Let r_n be a sequence in $\mathbb{R}$. Assume that there is an $R \in \mathbb{R}$, $0 < R$, such that $|r_n| \leq R^n$ for all $n \in \mathbb{N}$. Show that

$$s_n = \sum_{k=0}^{n} \frac{r_k}{k!}$$

is a convergent sequence.

2.50 Let x_n be a sequence of positive real numbers with

$$x_n = \frac{1}{x_1 \ldots x_{n-1}}$$

for all $n \geq 2$. Determine whether or not $\sum_{k=1}^{\infty} x_k$ converges.

2.51 **Binary expansions.** Let $r \in \mathbb{R}$ and $0 \leq r \leq 1$. A sequence b_n is called a *binary expansion* of r if $b_n = 0$ or $b_n = 1$ for each $n \in \mathbb{N}$ and if the sequence

$$s_n = \sum\nolimits_{k=1}^{n} b_n 2^{-n}$$

converges to r. Show that each $r \in [0, 1]$ has a binary expansion. Show that the binary expansion is unique, except for the numbers of the form $r = k2^{-n}$, for some $n, k \in \mathbb{N}$ with $0 \leq k \leq 2^n$. Show that for numbers of this type there are exactly two binary expansions.

2.52 **Ternary expansions.** Let $r \in \mathbb{R}$ and $0 \leq r \leq 1$. A sequence t_n is called a *ternary expansion* of r if $t_n = 0$ or $t_n = 1$ or $t_n = 2$ for each $n \in \mathbb{N}$ and if the sequence

$$s_n = \sum\nolimits_{k=1}^{n} t_n 3^{-n}$$

converges to r. Show that each $r \in [0, 1]$ has a ternary expansion. Show that the ternary expansion is unique, except for the numbers of the form $r = k3^{-n}$, for some $n, k \in \mathbb{N}$ with $0 \leq k \leq 3^n$. Show that for numbers of this type there are exactly two ternary expansions.

2.6 INTERVALS AND CONNECTED SETS

Intervals were introduced in Example 1.2.1. Here we define intervals rigorously and prove that they are the only *connected* subsets of $\mathbb{R}$. The concept of a connected set, explained below, will be important (and much less simple) in multidimensional vector spaces.

Definition 2.6.1 Intervals. A set of real numbers I is called an *interval* if it satisfies the following condition: if $a, b \in I$ and $a < t < b$, then $t \in I$. Hence I is an interval if whenever I contains two numbers, then it also contains all the numbers in between.

Definition 2.6.2 End points of bounded intervals. Let I be a bounded nonempty interval. Hence $a = \inf I$ and $b = \sup I$ both exist. They are called the *end points* of I. In particular, $a = \inf I$ is the *initial point* and $b = \sup I$ is the *final point* of I.

Definition 2.6.3 Open intervals, closed intervals. We see that

$$(a,\, b) = \{\, t \in \mathbb{R} \mid a < t < b \,\} \quad \text{and} \quad [\, a,\, b\,] = \{\, t \in \mathbb{R} \mid a \leq t \leq b \,\}$$

are both intervals with end points a and b. We call $(a,\, b)$ an *open interval* and $[\, a,\, b\,]$ a *closed interval.*

Lemma 2.6.4 *Let I be a bounded interval with the initial point a and the final point b. Then $(a,\, b) \subset I \subset [\, a,\, b\,]$.*

Proof. Let $t \in (a,\, b)$. Then $a < t$ implies that t is not a lower bound for I. Therefore, there is an $a' \in I$ such that $a' < t$. Similarly, b is not an upper bound for I, and therefore there is a $b' \in I$ such that $t < b'$. Hence $a' < t < b'$ with $a',\, b' \in I$. So $t \in I$ by the definition 2.6.1 of an interval. This shows that $(a,\, b) \subset I$. The other inclusion is obvious. If $t \in I$, then $\inf I = a \leq t \leq b = \sup I$, since a is a lower bound and b is an upper bound for the points in I. □

Remarks 2.6.5 Bounded intervals. Let I be a bounded interval with the initial point a and the final point b. Lemma 2.6.4 shows that there are only four possible forms for I. These are $(a,\, b)$, $[\, a,\, b\,]$, and the half-open intervals

$$[\, a,\, b) = \{\, t \in \mathbb{R} \mid a \leq t < b \,\} \quad \text{and} \quad (a,\, b\,] = \{\, t \in \mathbb{R} \mid a < t \leq b \,\}.$$

Incidentally, we note that the infimum and the supremum of a bounded set may or may not be elements of the set.

Closures of Sets in $\mathbb{R}$

Definition 2.6.6 Closure of a set. Let $A \subset \mathbb{R}$. The *closure of A* is defined as the set

$$\overline{A} = \{\, x \in \mathbb{R} \mid \text{there is a sequence } x_n \in A,\, n \in \mathbb{N}, \text{ such that } x = \lim_n x_n \,\}.$$

Hence $\overline{A}$ consists of all those points $x \in \mathbb{R}$ that are the limits of convergent sequences in A. Intuitively, one may say that $\overline{A}$ is the set of all points that are approachable by sequences from A. Note that $A \subset \overline{A}$. In fact, every $x \in A$ is also the limit of a constant sequence $x_n = x$ in A.

Lemma 2.6.7 Points in the closure. *Let $A \subset \mathbb{R}$ and $u \in \mathbb{R}$. Then $u \in \overline{A}$ if and only if for each $r > 0$ there is an $x \in A$ such that $|x - u| < r$.*

Proof. Assume that $u \in \overline{A}$. Hence there is a sequence x_n in A such that $x_n \to u$. Hence $|x_n - u|$ is a zero sequence. Therefore, for each $r > 0$, there is an $N \in \mathbb{N}$ such that $|x_n - u| < r$ for all $n \geq N$.

Conversely, assume that for each $r > 0$, there is an $x \in A$ such that $|x - u| < r$. For each $n \in \mathbb{N}$, choose $x_n \in A$ such that $|x_n - u| < 1/n$. Then we see that x_n is a sequence in A and $x_n \to u$. Hence $u \in \overline{A}$. □

Remarks 2.6.8 Reformulations of the above lemma. We may also state Lemma 2.6.7 as follows. A point $u \in \mathbb{R}$ is in the closure of $A \subset \mathbb{R}$ if and only if for each $r > 0$ the open interval $(u - r,\, u + r)$ intersects A. Equivalently: a point $u \in \mathbb{R}$ is in the closure of $A \subset \mathbb{R}$ if and only if every open interval containing u intersects A.

Lemma 2.6.9 *Let $A \subset \mathbb{R}$ and $t \in \mathbb{R}$. Assume that $a \leq t$ for all $a \in A$. Then also $x \leq t$ for all $x \in \overline{A}$.*

Proof. Assume that $u > t$. Then $r = (u - t) > 0$. For any $a \in A$,

$$(u - a) \geq (u - t) = r.$$

Hence there is no $a \in A$ such that $|a - u| < r$. Therefore, by Lemma 2.6.7, $u \notin \overline{A}$. □

Corollary 2.6.10 *Let $c \in \mathbb{R}$ and $A \subset (-\infty, c]$, $B \subset [\,c,\, \infty)$. Then $\overline{A} \cap \overline{B}$ is either empty or the single point set $\{\,c\}$.*

Proof. Lemma 2.6.9 shows that if $A \subset (-\infty, c\,]$, then also $\overline{A} \subset (-\infty, c\,]$. Similarly, we see that $\overline{B} \subset [\,c,\, \infty)$. Hence $\overline{A} \cap \overline{B} \subset \{\,c\}$. □

The proof of the following lemma contains the main argument of this section.

Lemma 2.6.11 *Let $a < b$. Let A and B be two sets such that $A \cup B = [\,a,\, b\,]$, $a \in A$, $b \in B$. Then $[\,a,\, b\,] \cap \overline{A} \cap \overline{B}$ is not empty.*

Proof. The set A is nonempty, since $a \in A$, and bounded, since $A \subset [\,a,\, b\,]$. Hence $c = \sup A$ exists and $a \leq c \leq b$. We show that every open interval containing c intersects both A and B.

Now $(s,\, c\,]$ intersects A for all $s < c$ and $(c,\, t)$ is disjoint from A for all $t > c$. This follows easily from the the definition of c as $\sup A$ or directly from Theorem 2.2.17. Hence we see that each open interval $(s,\, t)$ containing c intersects both A and $A^c = \mathbb{R} \setminus A$. We want to show that $(s,\, t)$ also intersects B. If $c = b$, then this is trivial, because $b \in B$. If $c < b$, then $(c,\, b) \subset [\,a,\, b\,]$ but $(c,\, b) \cap A = \emptyset$. Since $A \cup B = [\,a,\, b\,]$, this means that $(c,\, b) \subset B$. Hence $(s,\, t)$ intersects both A and B whenever $s < c < t$. Therefore $c \in \overline{A} \cap \overline{B}$ by the remarks above in 2.6.8. Hence $[\,a,\, b\,] \cap \overline{A} \cap \overline{B}$ contains $c \in [\,a,\, b\,]$ and therefore is nonempty. □

Connected Sets in $\mathbb{R}$

Definition 2.6.12 Connected sets. A set C in $\mathbb{R}$ is called a *connected set* if the following condition is satisfied: for any two nonempty sets A and B such that $C = A \cup B$, $C \cap \overline{A} \cap \overline{B}$ is nonempty. An intuitive formulation of this condition is that whenever C is the union of two nonempty sets A and B, then C contains points that are approachable both by sequences from A and by sequences from B.

Theorem 2.6.13 Connectedness in $\mathbb{R}$. *A set in $\mathbb{R}$ is connected if and only if it is an interval.*

Proof. The empty set is (trivially) an interval and (again trivially) a connected set. Now assume that C is a nonempty connected set. We will show that C is an interval. Let $a,\, b \in C$ and $a < c < b$. Let $A = C \cap (-\infty, c]$ and $B = C \cap [c,\, \infty)$. Then $a \in A$ and $b \in B$. Hence A and B are both nonempty and $C = A \cup B$. Therefore, $C \cap \overline{A} \cap \overline{B}$ is nonempty. But by Corollary 2.6.10, $\overline{A} \cap \overline{B}$ may contain only c. Therefore $c \in C$. This shows that C is an interval.

Conversely, assume that C is a nonempty interval. Let P and Q be two nonempty disjoint sets and $C = P \cup Q$. Assume $a \in P, b \in Q$, and $a < b$. Note that $[a,\, b] \subset C$, since C is an interval and $a,\, b \in C$. Let $A = P \cap [a,\, b]$ and $B = Q \cap [a,\, b]$. Lemma 2.6.11 shows that $[a,\, b] \cap \overline{A} \cap \overline{B}$ is nonempty. But we see easily that $\overline{A} \subset \overline{P}$ and $\overline{B} \subset \overline{Q}$. Hence $C \cap \overline{P} \cap \overline{Q}$ is also nonempty. □

Closed Sets and Open Sets in $\mathbb{R}$

Definition 2.6.14 Closed sets. A set C in $\mathbb{R}$ is called *a closed set* if C contains the limits of all convergent sequences in C. More explicitly, C is closed if $x \in C$ whenever there is a sequence x_n in C such that $\lim_n x_n = x$.

Recall (Definition 2.6.17) that the closure $\overline{A}$ of A consists of all limit points of convergent sequences in A.

Theorem 2.6.15 Closures and closed sets. *Let $A \subset \mathbb{R}$. Then the closure $\overline{A}$ of A is the smallest closed set that contains A. That is:*

(1) The closure $\overline{A}$ of A is a closed set and $A \subset \overline{A}$.

(2) If C is a closed set and if $A \subset C$, then $\overline{A} \subset C$.

Proof. (1) First, we prove that $\overline{A}$ is closed. Let $y_n \in \overline{A}$, $n \in \mathbb{N}$, be a sequence in $\overline{A}$. Assume that $\lim_n y_n = y$. We show that $y \in \overline{A}$. For each $n \in \mathbb{N}$ there is an $x_n \in A$

such that $|y_n - x_n| < 1/n$. This follows from the fact that $y_n \in \overline{A}$ is the limit of a sequence in A. Then $(y_n - x_n)$ is a zero sequence. So by Theorem 2.3.10, x_n is also convergent and $\lim y_n = \lim x_n = y$. Hence $y \in \overline{A}$ and therefore $\overline{A}$ is closed. Also, clearly, $A \subset \overline{A}$. In fact, every $x \in A$ is the limit of the constant sequence $x_n = x$ in A.

(2) Let C be a closed set and $A \subset C$. If $x \in \overline{A}$, then there is a sequence x_n in A such that $x_n \to x$. But x_n is also a sequence in C. Therefore $x \in C$ since C is closed. This shows that $\overline{A} \subset C$. □

Corollary 2.6.16 *A set A in $\mathbb{R}$ is closed if and only if $A = \overline{A}$.*

Proof. We always have that $\overline{A}$ is closed and $A \subset \overline{A}$, by Part (1) of Theorem 2.6.15 above. Hence, if $A = \overline{A}$, then A is closed. Conversely, if A is closed, then, by Part (2) of the same theorem, $\overline{A} \subset A$. Since the other inclusion is always true, we obtain $A = \overline{A}$. □

Recall that $A^c = \mathbb{R} \setminus A$ is the complement of A.

Definition 2.6.17 Boundary of a set. Let A be a set in $\mathbb{R}$. Then $\partial A = \overline{A} \cap \overline{A^c}$ is defined as the *boundary of* A. Points in ∂A are the *boundary points of* A. Since $(A^c)^c = A$, we see that $\partial A = \partial A^c$.

Definition 2.6.18 Open sets. A set G in $\mathbb{R}$ is said to be an *open set* if its complement $G^c = \mathbb{R} \setminus G$ is a closed set.

Remarks 2.6.19 Topology of $\mathbb{R}$. The family of open sets in $\mathbb{R}$ is called the *topology* of $\mathbb{R}$. Informally, the topology of $\mathbb{R}$ refers to the collection of results about open sets, closed sets, and boundaries. In Chapter 4, on normed vector spaces, we will discuss the topology of a normed vector space, which includes the topology of $\mathbb{R}$ as a special case. For now, we limit our exposition to the basic definitions and results just presented. Some further results are stated as problems.

Problems

2.53 Let E be a finite subset of $\mathbb{R}$. Show that $\overline{E} = E$.

2.54 Let $E = \{ 1 - 1/n \mid n \in \mathbb{N} \}$. Find $\overline{E}$.

2.55 What is $\partial\mathbb{Q}$?

2.56 Let A, B be nonempty connected subsets of $\mathbb{R}$ such that $A \cap B = \emptyset$. Is $A \cup B$ connected?

2.57 Show that an intersection of finitely many closed sets in $\mathbb{R}$ is closed.

2.58 Prove or disprove: If A, B are subsets of $\mathbb{R}$, then $\partial(A \cup B) = \partial A \cup \partial B$.

2.59 Show that the intersection of any family of intervals is again an interval. Also show that the intersection of any family of closed intervals is again a closed interval. Give examples to show that the intersection of a family of open intervals can be any type of interval (open, closed, or half-open).

2.60 **Nested intervals.** A sequence of intervals I_n is called *nested* if $I_{n+1} \subset I_n$ for all $n \in \mathbb{N}$. Let I_n be a nested sequence of intervals. If I_1 is bounded and if each I_n is closed and nonempty, then show that the intersection of this family of intervals is also nonempty. Give an example to show that a nested sequence of bounded and nonempty intervals may have an empty intersection (so that the result only applies to sequences of closed intervals).

2.61 Let I_n be a sequence of bounded and closed intervals. Assume that the intersection $\cap_{k=1}^{n} I_k$ is nonempty for each $k \in \mathbb{N}$. Show that $\cap_{n\in\mathbb{N}} I_n$ is also nonempty.

2.62 **Bisection sequences.** Let $I_n = [\,a_n,\, b_n]$ be a sequence of closed and bounded intervals with the middle points $c_n = (a_n + b_n)/2$. Then I_n is called a *bisection sequence* of intervals if $I_{n+1} = [\,a_n,\, c_n]$ or $I_{n+1} = [\,c_n,\, b_n]$ for each $n \in \mathbb{N}$.

(1) Show that the intersection $\cap_{n\in\mathbb{N}} I_n$ of a bisection sequence of intervals contains exactly one point $r \in \mathbb{R}$.

(2) Show that a bisection sequence is determined by the first interval I_1 and by a sequence b_n such that $b_n = 0$ or $b_n = 1$ for each $n \in \mathbb{N}$ and such that $I_{n+1} = [\,a_n,\, c_n]$ if and only if $b_n = 0$ and $I_{n+1} = [\,c_n,\, b_n]$ if and only if $b_n = 1$.

(3) If $I_1 = [\,0,\, 1\,]$, then show that the sequence b_n obtained in Part (2) is a binary expansion for the number $r \in \mathbb{R}$ obtained in Part (1).

2.63 Let I be an interval. Let $A \subset \mathbb{R}$. If I contains points both from A and from its complement $A^c = \mathbb{R} \setminus A$, then show that I also contains points from the boundary of A.

CHAPTER 3

VECTOR FUNCTIONS

We assume that the reader has a working knowledge of vector spaces. The first section of the chapter reviews basic results and notation. In particular, this initial section summarizes the essential facts about linear functions between vector spaces. These functions are essential tools for analysis in vector spaces. Much of this material is treated in an elementary course in linear algebra, although an important exception is the dimension theorem 3.1.14 (proved in Appendix B).

After this review, we introduce first bilinear and then multilinear functions, which operate on Cartesian products of vector spaces. As far as possible, the section on multilinear functions is structured to parallel the preceding section on the special (and simpler) case of bilinear functions. An understanding of bilinear and multilinear functions is also crucial for analysis. In particular, polynomials in vector variables are defined in terms of multilinear functions. As we shall see in later chapters, the subject of differential calculus is functions that can be approximated by such polynomials.

The last two sections of the chapter review the most important facts about Euclidean spaces, orthogonal projections, and linear transformations between Euclidean spaces.

Analysis in Vector Spaces.
By M. A. Akcoglu, P. F. A. Bartha and D. M. Ha

3.1 VECTOR SPACES: THE BASICS

A vector space X is a nonempty set with two operations, *scalar multiplication* and *vector addition*. Scalar multiplication is a function $\mathbb{R} \times X \to X$ written $(a, \mathbf{x}) \to a\mathbf{x}$. Vector addition is a function $X \times X \to X$ written $(\mathbf{x}, \mathbf{x}') \to \mathbf{x} + \mathbf{x}'$. The two operations satisfy a set of conditions known as the *vector space axioms* (omitted here). Elements of a vector space are called *vectors*. The zero vector of X is denoted as $\mathbf{0}_X$ or simply as $\mathbf{0}$ if X is understood. In general, we denote vectors by boldface letters like $\mathbf{x}$. If X and Y are two vector spaces and if $f : X \to Y$ is a function, then the value of f at $\mathbf{x} \in X$ is denoted by $f(\mathbf{x}) \in Y$. Boldface letters are not used for the vector $f(\mathbf{x})$.

Example 3.1.1 Let D be any nonempty set and let X be any vector space. Let $\mathcal{F}(D, X)$ be the set of all functions from D into X. If f, g are in $\mathcal{F}(D, X)$ and t is any scalar, define $f + g$ and tf as functions from D to X by the following formulas: for all $d \in D$,

$$\begin{aligned} (f+g)(d) &= f(d) + g(d) \\ (tf)(d) &= tf(d). \end{aligned}$$

In the first equation above, the " + " on the right-hand side is the addition operation in the vector space X. Thus, $f(d) + g(d)$ is the sum of the vectors $f(d)$ and $g(d)$ in X. Similarly, $tf(d)$ is the scalar multiple of $f(d)$ by t.

Thus, $f + g$ and tf are in $\mathcal{F}(D, X)$. The zero vector of $\mathcal{F}(D, X)$ is the function that maps each d in D to $\mathbf{0}$, the zero vector in X. It's easy to check that $\mathcal{F}(D, X)$, together with these operations, is a vector space. In particular, $\mathcal{F}(D, \mathbb{R}^n)$ is a vector space. △

Spans, Subspaces, Linear Independence, Bases

Definition 3.1.2 Linear combinations. Let $A = \{\mathbf{a}_1, \ldots, \mathbf{a}_n\} \subset X$ be a finite nonempty set of vectors. A *linear combination* of $\{\mathbf{a}_1, \ldots, \mathbf{a}_n\}$ is any vector of the form $r_1\mathbf{a}_1 + \cdots + r_n\mathbf{a}_n$, where $r_1, \ldots, r_n \in \mathbb{R}$.

Definition 3.1.3 Span of a finite set. Let $A = \{\mathbf{a}_1, \ldots, \mathbf{a}_n\} \subset X$ be a finite nonempty set of vectors. The *span of A* is defined as the set of all linear combinations of vectors in A:

$$\text{Span } A = \{ r_1\mathbf{a}_1 + \cdots + r_n\mathbf{a}_n \mid r_1, \ldots, r_n \in \mathbb{R} \}.$$

We define the span of the empty set as Span $\emptyset = \{\mathbf{0}\}$, the set consisting of the zero vector alone.

Definition 3.1.4 Subspaces. A set U in a vector space X is called a *subspace* if it is *nonempty* and if it is *closed under linear combinations*. This last condition means that if $\mathbf{u}$, $\mathbf{u}' \in U$ and r, $r' \in \mathbb{R}$, then $r\mathbf{u} + r'\mathbf{u}' \in U$. Note that $\mathbf{0} \in U$ for any subspace U. Also, the set $\{\mathbf{0}\}$ is a subspace. It is the smallest subspace in the sense that $\{\mathbf{0}\} \subset U$ for any subspace U.

Examples 3.1.5

1. Let $U = \{ (x_1, x_2, x_3) \in \mathbb{R}^3 \mid 2x_1 - |x_2| + x_3 = 0 \}$. Clearly, $(0,0,0) \in U$. If $\mathbf{x} \in U$ and $t \in \mathbb{R}$, then $t\mathbf{x} \in U$. However, U it is not a subspace of $\mathbb{R}^3$ because U is not closed under vector addition. For example, $(0,1,1)$ and $(0,-1,1)$ are in U, but their sum, $(0,0,2)$, is not in U.

2. Let $U = \{ (x_1, x_2, x_3) \mid |x_1| \geq |x_2 x_3| \}$. Then $(0,0,0) \in U$. If $\mathbf{x} = (x_1, x_2, x_3) \in U$ and $t \in \mathbb{R}$, then $t\mathbf{x} \in U$ since $|tx_1| = |t|\,|x_1| \geq |t|\,|x_2x_3| = |tx_2x_3|$. However, U is not a subspace of $\mathbb{R}^3$ since $(1,0,1)$ and $(-1,1,0)$ are in U , but their sum, $(0,1,1)$, is not in U.

3. Let $n \in \mathbb{N}$ and let $a_1, \ldots, a_n$ be any real numbers. Set
$$U = \{ (x_1, \ldots, x_n) \mid a_1x_1 + \cdots + a_nx_n = 0 \}.$$
Then U is a subspace of $\mathbb{R}^n$.

4. Let X be the set of all functions from $[-1,1]$ into $\mathbf{R}$. As explained in Example 3.1.1, this is a vector space. Let
$$E = \{ f \in X \mid f(-t) = f(t) \text{ for all } t \in [-1,1] \} \tag{3.1}$$
and let
$$O = \{ f \in X \mid f(-t) = -f(t) \text{ for all } t \in [-1,1] \}. \tag{3.2}$$
Then E and O are subspaces of X. For instance, to verify that E is a subspace of X, note that the zero vector of X (which is the function that takes each $t \in [-1,1]$ to 0) is in E. If f, g are in E and a, b are scalars, then
$$(af + bg)(-t) = a(f(-t)) + b(g(-t)) = af(t) + bg(t) = (af + bg)(t).$$

5. Let D be a nonempty subset of $\mathbb{R}$. A *polynomial* on D is a function $f : D \to \mathbb{R}$ for which there is some nonnegative integer n and real numbers $a_0, \ldots, a_n$ such that
$$f(t) = a_0 + a_1t + \cdots + a_nt^n \quad \text{for all } t \in D.$$
Let $\mathcal{P}(D)$ denote the collection of all polynomials on D. Clearly, $\mathcal{P}(D) \subset \mathcal{F}(D, \mathbb{R})$. It is easy to verify that $\mathcal{P}(D)$ is a subspace of $\mathcal{F}(D, \mathbb{R})$. △

Example 3.1.6 Intersection of subspaces. Let X be a vector space. Then the intersection of a family of subspaces of X is also a subspace.

To see this, let $\mathcal{V}$ be a collection of subspaces and set $U = \bigcap_{V \in \mathcal{V}} V$. Hence, U consists of all $\mathbf{x} \in X$ such that $\mathbf{x} \in V$ for each $V \in \mathcal{V}$. Now $\mathbf{0} \in V$ for each $V \in \mathcal{V}$. Therefore $\mathbf{0} \in U$. Let $\mathbf{x}$, $\mathbf{y} \in U$ and a, $b \in \mathbb{R}$. Then $\mathbf{x}$, $\mathbf{y} \in V$ for each $V \in \mathcal{V}$. Therefore $a\mathbf{x} + b\mathbf{y} \in V$ for each $V \in \mathcal{V}$. Hence $a\mathbf{x} + b\mathbf{y} \in U$. $\triangle$

Remarks 3.1.7 The span of any set is a subspace. In fact, Span A is the smallest subspace that contains A. This means that if U is a subspace and if $A \subset U$, then Span $A \subset U$.

Definition 3.1.8 Sums of subspaces. Let U and V be two subspaces in X. Then the *sum of U and V* is the set of all vectors of the form $\mathbf{u} + \mathbf{v}$ with $\mathbf{u} \in U$ and $\mathbf{v} \in V$. We denote this set by $U + V$. It is another subspace of X.

Definition 3.1.9 Linear independence. A finite set $A = \{\mathbf{a}_1, \ldots, \mathbf{a}_n\}$ of distinct vectors is *linearly independent* if whenever $r_1, \ldots, r_n$ are real numbers and

$$r_1\mathbf{a}_1 + \cdots + r_n\mathbf{a}_n = \mathbf{0},$$

then $r_1 = \ldots = r_n = 0$. This is equivalent to saying that each vector in the span of A is a *unique* linear combination of vectors in A.

Definition 3.1.10 Bases. A linearly independent set A is called a (finite) *basis for* Span A. In particular, if $B = \{\mathbf{e}_1, \ldots, \mathbf{e}_n\}$ is linearly independent and if Span $B = X$, then B is called a *basis for* X. Equivalently, B is a basis for X if each $\mathbf{x} \in X$ is a unique linear combination of vectors in B.

Example 3.1.11 Standard basis for $\mathbb{R}^n$. The set $\mathbb{R}^n$ consisting of all n-tuples of real numbers $(x_1, \ldots, x_n)$ is a vector space with the usual definitions of linear operations. The set $E = \{\mathbf{e}_1, \ldots, \mathbf{e}_n\}$ consisting of n vectors

$\mathbf{e}_1 = (1, 0, \ldots, 0), \mathbf{e}_2 = (0, 1, \ldots, 0), \ldots, \mathbf{e}_n = (0, 0, \ldots, 1)$ is a basis for $\mathbb{R}^n$. It is called the *standard basis of* $\mathbb{R}^n$.

Examples 3.1.12 Bases for some vector spaces.

1. Each zero vector space $\{\mathbf{0}\}$ has $\emptyset$ as a basis.

2. Let $W = \{ (x, y, z) \in \mathbb{R}^3 \mid x + y - z = 0 \}$. Then W is a subspace of $\mathbb{R}^3$. Let us find a basis for W. Let $\mathbf{v} = (x, y, z) \in \mathbb{R}^3$. Then $\mathbf{v} \in W$ if and only if $x = -y + z$. Thus,

$$\begin{aligned} W &= \{ (-y + z, y, z) \mid y, z \text{ are arbitrary real numbers} \} \\ &= \{ -y(1, 1, 0) + z(1, 0, 1) \mid y, z \text{ are arbitrary real numbers} \} \\ &= \text{Span } \{(-1, 1, 0), (1, 0, 1)\}. \end{aligned}$$

   It's easy to verify that $E = \{(-1, 1, 0), (1, 0, 1)\}$ is a linearly independent set; hence, it is a basis for W.

3. Let $E = \{(1, 1, 2, 1), (2, -1, 0, 4), (4, 1, 4, 6)\}$ and let $W = \text{Span } E$. Obviously, E satisfies the first condition for being a basis for W. But E is not linearly independent since

$$2(1, 1, 2, 1) + (2, -1, 0, 4) - (4, 1, 4, 6) = \mathbf{0}.$$

   Hence, we cannot conclude that E is a basis for W. However, it can be shown that the smaller set $\{(1, 1, 2, 1), (2, -1, 0, 4)\}$ is a basis for W. △

Remarks 3.1.13 Finite dimensional vector spaces. A vector space may or may not have a basis in the sense of Definition 3.1.10 above. Unless otherwise stated, we consider only vector spaces with a (finite) basis. These spaces are also called *finite dimensional vector spaces*. Hence, in this course, by a vector space we mean a finite dimensional vector spaceunless otherwise stated. The following is a major theorem about vector spaces.

Theorem 3.1.14 The dimension theorem. *Any two bases for a vector space contain the same number of vectors.*

See Appendix B for a proof of this theorem.

Definition 3.1.15 The dimension of a vector space. The number of vectors in a basis for X is called the *dimension* of X. This number is denoted by $\dim X$. The dimension of X is independent of the choice of basis because of the dimension theorem 3.1.14 above.

Examples 3.1.16

1. Since $\emptyset$ is a basis for $\{\mathbf{0}\}$, we have $\dim\{\mathbf{0}\} = 0$.

2. Example 3.1.11 shows that $\mathbb{R}^n$ has a basis consisting of n vectors. Hence, $\dim \mathbb{R}^n = n$.

3. Let $n \in N$ and let $\mathcal{P}_n$ be the vector space of all polynomials on $\mathbb{R}$ of degree no more than n. Let $p_k(x) = x^k$ for all $x \in \mathbb{R}$ and all $k = 0, 1, \ldots, n$. Then $\{p_0, \ldots, p_n\}$ is a basis for $\mathcal{P}_n$. Hence,

$$\dim \mathcal{P}_n = n + 1.$$

Direct Sums and Complementary Subspaces

Lemma 3.1.17 *Let $U_1, \ldots, U_k$ be subspaces of X. Then the following are equivalent.*

(1) For each $\mathbf{x} \in X$ there are unique $\mathbf{u}_i \in U_i$ such that $\mathbf{x} = \mathbf{u}_1 + \cdots + \mathbf{u}_k$.

(2) If $A_i = \{ \mathbf{a}_1^i, \ldots, \mathbf{a}_{n_i}^i \}$ is a basis for U_i, then $A = \cup_{i=1}^k A_i$ is a basis for X.

Proof. This is left as an exercise. □

Definition 3.1.18 Direct sums. Let U_i, $i = 1, \ldots, k$, be subspaces of X. Then X is said to be the *direct sum* of $U_1, \ldots, U_k$ if the conditions of Lemma 3.1.17 above are satisfied. This is expressed as $X = U_1 \oplus \cdots \oplus U_k$. Note that in this case $\dim X = (\dim U_1) + \cdots + (\dim U_k)$.

Definition 3.1.19 Complementary subspaces. A set of subspaces U_i is called a *complementary set of subspaces in X* if they satisfy the conditions in Lemma 3.1.17 above or, equivalently, if $X = U_1 \oplus \cdots \oplus U_k$.

Lemma 3.1.20 *The subspaces U and V in X are complementary if and only if $X = U + V$ and $U \cap V = \{ \mathbf{0} \}$.*

Proof. This is left as an exercise. □

Examples 3.1.21

1. Let X be any vector space. Then $\{\mathbf{0}\}$ and X are always complementary subspaces in X.

2. Let $U = \{ (a + b, a + 2b, a + b) \mid a, b \in \mathbb{R} \}$ and let $V = \{ (0, 0, c) \mid c \in \mathbb{R} \}$. Then $U \cap V = \{\mathbf{0}\}$ and for any $\mathbf{x} = (x_1, x_2, x_3) \in \mathbb{R}^3$, we have

$$\mathbf{x} = (a + b, a + 2b, a + b) + (0, 0, c),$$

where $a = 2x_1 - x_2, b = x_2 - x_1, c = x_3 - x_1$. Hence, $\mathbb{R}^3 = U + V$. Thus, $\mathbb{R}^3 = U \oplus V$. So, U and V are complementary subspaces in $\mathbb{R}^3$.

3. Let $U = \{ (a, b, 0, 0) \mid a, b \in \mathbb{R} \}$ and let $V = \{ (0, 0, 0, c) \mid c \in \mathbb{R} \}$. Then U and V are complementary subspaces in $\{ (x, y, 0, z) \mid x, y, z \in \mathbb{R} \}$. However, U and V are not complementary subspaces in $\mathbb{R}^4$ because $U + V \neq \mathbb{R}^4$.

4. Consider X, E and O as in part 4 of Examples 3.1.5. Let us see that

$$X = E \oplus O. \tag{3.3}$$

First, let $f \in X$. Define $f_1(x) = f(x) + f(-x)$ and $f_2(x) = f(x) - f(-x)$ for all $x \in [-1, 1]$. Then $f_1 \in E$ and $f_2 \in O$. Thus,

$$f = \frac{1}{2} f_1 + \frac{1}{2} f_2 \in E + O.$$

Hence, we have shown that $X = E + O$. Also, if $h \in E \cap O$, then for all $x \in [-1, 1]$, we have $-h(x) = h(-x) = h(x)$ so that $h(x) = 0$ for all $x \in [-1, 1]$. Thus, $E \cap O$ contains only the zero vector of X. So, (3.3) holds. △

Cartesian Products and Direct Sums

Definition 3.1.22 Cartesian products. The Cartesian product of k vector spaces $U_1, \ldots, U_k$ is defined as a new vector space $X = U_1 \times \cdots \times U_k$. The elements of X are the k-tuples $\mathbf{x} = (\mathbf{u}_1, \ldots, \mathbf{u}_k)$ of vectors $\mathbf{u}_i \in U_i$. Linear operations on these k-tuples are defined as

$$r\mathbf{x} + s\mathbf{y} = r(\mathbf{u}_1, \ldots, \mathbf{u}_k) + s(\mathbf{v}_1, \ldots, \mathbf{v}_k) = (r\mathbf{u}_1 + s\mathbf{v}_1, \ldots, r\mathbf{u}_k + s\mathbf{v}_k)$$

for all $r, s \in \mathbb{R}$ and for all $\mathbf{x} = (\mathbf{u}_1, \ldots, \mathbf{u}_k)$, $\mathbf{y} = (\mathbf{v}_1, \ldots, \mathbf{v}_k) \in X$. It is easy to check that X is a vector space with these operations.

Remarks 3.1.23 Cartesian products and direct sums. For simplicity we consider only the products of two factors. Generalizations to more than two factors will be obvious. Let $X = U \times V$ and put

$$\begin{aligned} U' &= \{ \mathbf{u}' = (\mathbf{u}, \mathbf{0}) \mid \mathbf{u} \in U \} \subset X, \\ V' &= \{ \mathbf{v}' = (\mathbf{0}, \mathbf{v}) \mid \mathbf{v} \in V \} \subset X. \end{aligned}$$

We see that each $\mathbf{x} \in X$ has a unique representation as

$$\mathbf{x} = (\mathbf{u}, \mathbf{v}) = \mathbf{u}' + \mathbf{v}'$$

with $\mathbf{u}' \in U'$, $\mathbf{v}' \in V'$. Hence $X = U' \oplus V'$ by Definition 3.1.18. This is the standard representation of a Cartesian product as a direct sum.

Notation 3.1.24 Identification of direct sums and Cartesian products. Note that U and V above are separate vector spaces, but U' and V' are the subspaces of $X = U \times V$. We will usually ignore the difference between U and U' and between V and V'. The meaning will be clear from the context. This amounts to the following. In a direct sum $X = U \oplus V$, one can consider U and V as separate vector spaces. If this is done, then X becomes the Cartesian product space $U \times V$. The difference between $U \oplus V$ and $U \times V$ is only notational. We will ignore this difference and treat these spaces as the same space.

Linear Transformations

Definition 3.1.25 Linear functions. Let X and Y be two vector spaces. A function $T : X \to Y$ is said to be a *linear function* if for all $\mathbf{u}, \mathbf{v} \in X$ and for all $r, s \in \mathbb{R}$,

$$T(r\mathbf{u} + s\mathbf{v}) = rT(\mathbf{u}) + sT(\mathbf{v}).$$

A linear function is also called a *linear transformation*, a *linear operator*, or a *linear map* – these terms are interchangeable. The values of a linear map $T : X \to Y$, denoted as $T(\mathbf{x})$ or $T\mathbf{x}$, are vectors in Y.

Definition 3.1.26 The vector space $L(X, Y)$ of linear maps. Let $L(X, Y)$ be the set of all linear functions $T : X \to Y$. Then $L(X, Y)$ is itself a vector space. The linear operations on $L(X, Y)$ are defined in a natural way. If $S, T \in L(X, Y)$ and if $a, b \in \mathbb{R}$, then $R = aS + bT$ is the function

$$R\mathbf{x} = aS\mathbf{x} + bT\mathbf{x} \text{ for } \mathbf{x} \in X.$$

An easy verification shows that $R : X \to Y$ is also linear. Hence

$$R = (aS + bT) \in L(X, Y).$$

Another easy verification shows that these linear operations on $L(X, Y)$ satisfy the axioms for a vector space.

Examples 3.1.27

1. The identity function $I_A : A \to A$ on any set A is defined by $I_A(a) = a$ for all $a \in A$. In particular, there is an identity function $I_X : X \to X$ on any vector space X. It is easy to see that $I_X : X \to X$ is actually a linear transformation. If X is understood from the context, then we simply write I instead of I_X.

2. Let $T \in L(X, Y)$ and assume that T is invertible: T one-to-one on X and $T(X) = Y$. Let $S : Y \to X$ be the inverse of T. Thus, $S(\mathbf{y}) = \mathbf{x}$ if and only

if $T(\mathbf{x}) = \mathbf{y}$. Then S is also linear. To see this, let $\mathbf{y}_1, \mathbf{y}_2$ be in Y and let a, b be scalars. Suppose that $S(\mathbf{y}_1) = \mathbf{x}_1$ and $S(\mathbf{y}_2) = \mathbf{x}_2$. Then

$$T(a\mathbf{x}_1 + b\mathbf{x}_2) = aT\mathbf{x}_1 + \mathbf{t}\mathbf{x}_2 = a\mathbf{y}_1 + b\mathbf{y}_2.$$

Hence,

$$S(a\mathbf{y}_1 + b\mathbf{y}_2) = a\mathbf{x}_1 + b\mathbf{x}_2 = aS(\mathbf{y}_1) + bS(\mathbf{y}_2).$$

This shows that S is linear.

Definition 3.1.28 Range and the kernel of a linear transformation. Let $T : X \to Y$ be a linear transformation. Then the *range* and the *kernel* of T are defined as

$$\begin{aligned} \text{Range}\, T &= \{ T\mathbf{x} \mid \mathbf{x} \in X \} \subset Y \text{ and} \\ \text{Ker}\, T &= \{ \mathbf{x} \mid T\mathbf{x} = \mathbf{0} \} \subset X. \end{aligned}$$

Lemma 3.1.29 Images of subspaces. *The images and inverse images of subspaces under linear transformations are also subspaces. In particular, the range and the kernel of T,*

$$\text{Range}\, T = T(X) \text{ and } \text{Ker}\, T = T^{-1}\{\mathbf{0}\},$$

are subspaces of Y and X, respectively.

Proof. This is left as an exercise. □

Lemma 3.1.30 *Let $T : X \to Y$ be a linear map. If T is one-to-one on a subspace $U \subset X$, then* $\dim U = \dim T(U)$.

Proof. Assume that T is one-to-one on U. Then it is easy to show that T maps linearly independent sets in U to linearly independent sets in $T(U)$. Hence T maps bases of U to bases of $T(U)$. Hence we see that $\dim U = \dim T(U)$. □

Lemma 3.1.31 *Let $T : X \to Y$ be a linear transformation. If a subspace U of X is complementary to* Ker T, *then T is one-to-one on U and*

$$T(U) = T(X) = \text{Range}\, T.$$

Proof. Let $\mathbf{u} \in U$ and $T\mathbf{u} = \mathbf{0}$; that is, $\mathbf{u} \in U \cap (\text{Ker}\, T)$. Then $\mathbf{u} = \mathbf{0}$, since U and Ker T are complementary. From this we see that

$$\text{if } \mathbf{u}, \mathbf{u}' \in U \text{ and if } T\mathbf{u} = T\mathbf{u}', \text{ then } \mathbf{u} = \mathbf{u}'$$

by applying the preceding observation to $\mathbf{u} - \mathbf{u}'$. Hence T is one-to-one on U. Also, any $\mathbf{x}$ can be expressed as $\mathbf{x} = \mathbf{u} + \mathbf{v}$ with $\mathbf{u} \in U$ and $\mathbf{v} \in \text{Ker } T$, since $X = U \oplus (\text{Ker } T)$. Hence $T\mathbf{x} = T\mathbf{u}$. This shows that $T(X) \subset T(U)$. But $T(U) \subset T(X)$, since $U \subset X$. Therefore $T(U) = T(X)$. □

Theorem 3.1.32 *Let $T : X \to Y$ be a linear transformation. Then*

$$\dim(\text{Range } T) + \dim(\text{Ker } T) = \dim X.$$

Proof. Let U be a complementary subspace to Ker T. Hence

$$\dim U + \dim(\text{Ker } T) = \dim X.$$

Now T is one-to-one on U and $T(U) = T(X) = \text{Range } T$ by Lemma 3.1.31. Lemma 3.1.30 shows that $\dim U = \dim T(U)$. Then the result follows. □

Examples 3.1.33

1. Suppose that $T : \mathbb{R}^3 \to \mathbb{R}^5$ is linear and $T(1,1,1) = \mathbf{0} = T(2,1,0)$. Then at least one of $\mathbf{u}_1 = (1,1,1,1,1)$ or $\mathbf{u}_2 = (3,1,1,0,2)$ is not in $T(\mathbb{R}^3)$. This is because $(1,1,1)$ and $(2,1,0)$ are in Ker T, so $\dim \text{Ker } T \geq 2$. Hence, $\dim T(\mathbb{R}^3) \leq 1$, so $\mathbf{u}_1, \mathbf{u}_2$ cannot both belong to $T(\mathbb{R}^3)$.

2. Let $T : X \to R$ be linear, where $\dim X = n$ for some $n \in \mathbb{N}$. Then either $T = \mathbf{0}$ or $\dim \text{Ker } T = n - 1$. To see this, suppose that $T \neq \mathbf{0}$. Then $T\mathbf{x} \neq \mathbf{0}$ for some $\mathbf{x} \in X$. Hence,

$$1 \leq \dim(T(X)) \leq \dim \mathbb{R} = 1.$$

   Thus, $\dim(T(X)) = 1$. So, $\dim(\text{Ker } T) = n - 1$.

3. Let X, Y be finite-dimensional vector spaces and let $T : X \to Y$ be linear. Then

$$\dim \text{Ker } T = \dim X - \dim T(X) \geq \dim X - \dim Y.$$

   Hence, if $\dim X > \dim Y$, then $\text{Ker } T \neq \{\mathbf{0}\}$. It follows easily that T cannot be one-to-one on X.

Remarks 3.1.34 Linear transformations and bases. Let X and Y be two vector spaces with the bases

$$A = \{\mathbf{a}_1, \ldots, \mathbf{a}_n\} \quad \text{and} \quad B = \{\mathbf{b}_1, \ldots, \mathbf{b}_m\},$$

respectively. Any linear map $T : X \to Y$ is uniquely determined by its values $\mathbf{c}_i = T\mathbf{a}_i \in Y$ on the basis vectors. In fact, if the values $\mathbf{c}_i$ are known, then

$$\begin{aligned} T\mathbf{x} &= T(x_1\mathbf{a}_1 + \cdots + x_n\mathbf{a}_n) \\ &= x_1 T\mathbf{a}_1 + \cdots + x_n T\mathbf{a}_n \\ &= x_1\mathbf{c}_1 + \cdots + x_n\mathbf{c}_n \end{aligned}$$

is determined for all $\mathbf{x} \in X$. Here, the x_i s are the coordinates of $\mathbf{x}$ with respect to the basis A. There is no restriction on the $\mathbf{c}_i$ s: they can be n arbitrary vectors in Y. In particular, referring to the basis vectors in A and B, for each pair $(\mathbf{a}_i, \mathbf{b}_j)$, define $T_{ij} : X \to Y$ by

$$T_{ij}\, \mathbf{a}_k = \begin{cases} \mathbf{b}_j & \text{if } i = k \\ \mathbf{0}_Y & \text{if } i \neq k \end{cases}$$

for each $k = 1, \ldots, n$. Hence T_{ij} maps $\mathbf{a}_i$ to $\mathbf{b}_j$ and maps all other $\mathbf{a}_k$ to $\mathbf{0}_Y$. We see easily that the set of these nm maps is a basis for $L(X, Y)$. Therefore we conclude that $\dim L(X, Y) = (\dim X)(\dim Y)$.

Remarks 3.1.35 Composition of linear transformations. Let $R : X \to Y$ and $S : Y \to Z$ be two linear transformations. Let us verify that the composition

$$S \cdot R = SR : X \to Z \tag{3.4}$$

is a linear transformation. In fact, setting $T = S \circ R$, we have

$$\begin{aligned} T(\alpha\mathbf{x} + \alpha'\mathbf{x}') &= S(R(\alpha\mathbf{x} + \alpha'\mathbf{x}')) = S(\alpha R\mathbf{x} + \alpha' R\mathbf{x}') \\ &= \alpha S(R\mathbf{x}) + \alpha' S(R\mathbf{x}') = \alpha T\mathbf{x} + \alpha' T\mathbf{x}' \end{aligned}$$

for all $\mathbf{x}, \mathbf{x}' \in X$ and for all $\alpha, \alpha' \in \mathbb{R}$.

Forming the compositions of linear transformations is a kind of multiplication. This multiplication is associative, in the sense that $R(ST) = (RS)T$, but not necessarily commutative. In general, $RS \neq SR$. If $T : X \to X$, then one can form powers T^n of T. They are defined inductively by $T^1 = T$, $T^2 = T \cdot T$, $T^{n+1} = T \cdot T^n$.

Projections

Projections are an especially simple, but important, category of linear transformations. A close connection between projections and direct sums (or Cartesian products) is brought out in Theorem 3.1.37.

Definition 3.1.36 Projections. A linear map $P : X \to X$ is called a *projection* if $P^2 = PP = P$.

Projections are closely connected with complementary subspaces.

Theorem 3.1.37 *Let U_i be a set of complementary subspaces of X. Then there is a set of projections $P_i : X \to X$ such that $P_i X = U_i$ and such that $\sum_i P_i = I$, the identity on X*

Proof. Let U_i s be k complementary subspaces of X. Then for each $\mathbf{x} \in X$ there are k uniquely defined vectors $\mathbf{u}_i \in U_i$ such that $\mathbf{x} = \sum_i \mathbf{u}_i$. This follows from Definition 3.1.19 of complementary subspaces. Hence there are k functions $P_i : X \to X$ such that $P_i\mathbf{x} = \mathbf{u}_i$. An easy check shows that each P_i is a linear transformation and $P_i^2 = P_i$, that is, $P_i\mathbf{u}_i = \mathbf{u}_i$. Hence each P_i is a projection. Also, $\sum_i P_i = I$ since $\sum_i P_i\mathbf{x} = \sum_i \mathbf{u}_i = \mathbf{x}$ for each $\mathbf{x} \in X$. □

The converse is also true. Let P_i s be a finite set of projections $X \to X$ such that $\sum_i P_i = I$. Then their ranges $U_i = P_i X$ constitute a set of complementary subspaces of X. This is left as an exercise. We will consider only the following special case.

Theorem 3.1.38 *If $P : X \to X$ is a projection, then $Q = (I - P)$ is also a projection. The ranges $U = PX$ and $V = QX$ are complementary subspaces of X. Also, $U = \text{Ker } Q$ and $V = \text{Ker } P$.*

Proof. If P is a projection, then $P^2 = P$ and therefore

$$(I - P)^2 = (I - P)(I - P) = I - 2P + P^2 = (I - P).$$

Hence $Q = (I - P)$ is also a projection. Here we have used some composition rules, such as $(R + S)T = RT + ST$. These identities have easy verifications. Also,

$$PQ = P(I - P) = P - P^2 = P - P = \mathbf{0}.$$

Now

$$\mathbf{x} = (P + Q)\mathbf{x} = P\mathbf{x} + Q\mathbf{x} = \mathbf{u} + \mathbf{v} \text{ with } \mathbf{u} \in U \text{ and } \mathbf{v} \in V$$

for any $\mathbf{x} \in X$. Also, if $\mathbf{x} = \mathbf{u}' + \mathbf{v}'$ with $\mathbf{u}' \in U$ and $\mathbf{v}' \in V$, then $\mathbf{u}' = P\mathbf{a}$ and $\mathbf{v}' = Q\mathbf{b}$ for some $\mathbf{a}$, $\mathbf{b} \in X$. Therefore

$$\mathbf{u} = P(\mathbf{x}) = P(P\mathbf{a} + Q\mathbf{b}) = P^2\mathbf{a} + PQ\mathbf{a} = P\mathbf{a} = \mathbf{u}'.$$

Similarly, $\mathbf{v}' = \mathbf{v}$. This shows that the representation

$$\mathbf{x} = \mathbf{u} + \mathbf{v} \text{ with } \mathbf{u} \in U \text{ and } \mathbf{v} \in V$$

is unique. Therefore U and V are complementary subspaces. Finally, $P\mathbf{x} = \mathbf{0}$ if and only if $\mathbf{x} = Q\mathbf{x} = \mathbf{v} \in V$. Hence $\text{Ker } P = V = QX$. Also, $\text{Ker } Q = U = PX$. □

Definition 3.1.39 Complementary projections. If $P : X \to X$ is a projection, then $Q = I - P$ is also a projection by Theorem 3.1.38. The projections P and Q and are called *complementary projections*. Note that if P and Q are complementary projections, then their ranges $U = PX$ and $V = QX$ are complementary subspaces in X.

Remarks 3.1.40 Projections on a subspace. Let $P : X \to X$ be a projection. If $U = PX$ is the range of P, then we denote P also as $P : X \to U$ and call it a projection on U. Note that a subspace U does not define a projection. If $U \neq X$, then there are many projections of X on U. A projection on U is specified if one also chooses a subspace V which is complementary to U. In this case there is a unique projection $P : X \to U$ such that $V = \text{Ker } P$. Also, $Q = (I - P) : X \to V$ is the projection on V corresponding to the choice of U as a complementary subspace to V.

Example 3.1.41 Let $X = \mathbb{R}^3$ be the xyz-space. Let U be the xy-plane. Let V be the line spanned by the vector $(1, 1, 1) \in X$. Show that U and V are complementary subspaces. What are the associated projections P and Q?

Solution. A general vector in U is of the form $\mathbf{u} = (u_1, u_2, 0)$ with $u_1, u_2 \in \mathbb{R}$. A general vector in V is of the form $\mathbf{v} = (v, v, v)$ with $v \in \mathbb{R}$. If $\mathbf{w} \in U \cap V$, then $\mathbf{w} = (u_1, u_2, 0) = (v, v, v)$. Hence $v = 0$ and $\mathbf{w} = \mathbf{0}$. To obtain the associated projections, we find the unique decomposition of $\mathbf{w} = (x, y, z)$ as $\mathbf{w} = \mathbf{u} + \mathbf{v}$ with $\mathbf{u} \in U$ and $\mathbf{v} \in V$. Setting

$$(x, y, z) = (u_1, u_2, 0) + (v, v, v)$$

gives $v = z$, $u_1 = x - v = x - z$, $u_2 = y - v = y - z$. Therefore

$$P(x, y, z) = (x - z, y - z, 0) \text{ and } Q(x, y, z) = (z, z, z)$$

define the associated projections P and Q. We check that we have indeed $P^2 = P$, $Q^2 = Q$, $P + Q = I$, and $\text{Range } P = U$, $\text{Range } Q = V$. △

Coordinate Systems

Definition 3.1.42 Coordinate systems. Let U and V be a pair of complementary subspaces of X. Then (U, V) is called a *coordinate system* in X. The projections

$P : X \to U$ and $Q : X \to V$ are the *coordinate projections* of this coordinate system. In this case $X = U \times V$. Any $\mathbf{x} \in X$ is represented as

$$\mathbf{x} = (P\mathbf{x},\, Q\mathbf{x}) = (\mathbf{u},\, \mathbf{v})$$

in terms of its *coordinates* $P\mathbf{x} = \mathbf{u} \in U$ and $Q\mathbf{x} = \mathbf{v} \in V$ in this system. The generalization to more than two components is obvious. If $X = U_1 \times \cdots \times U_k$, then $(U_1, \ldots, U_k)$ is a coordinate system in X with the corresponding set of coordinate projections $P_i : X \to U_i$.

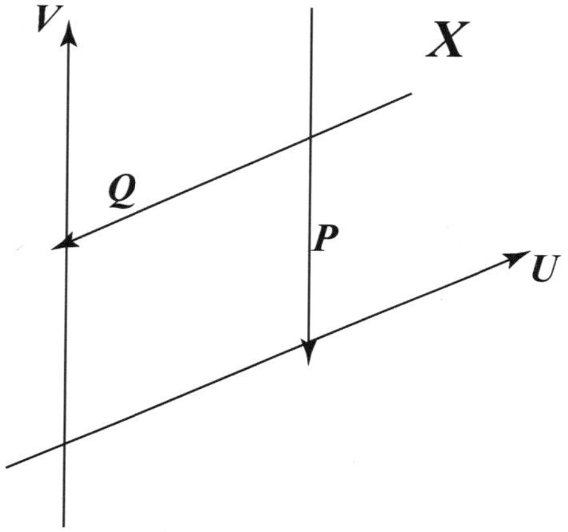

Figure 3.1. Complementary projections P and Q.

Definition 3.1.43 Coordinate functions of a basis. Let X be a vector space with the basis $A = \{\mathbf{a}_1, \ldots, \mathbf{a}_n\}$. Let U_i be the one-dimensional space spanned by $\mathbf{a}_i$. Then the U_i s form a set of complementary subspaces. The coordinate system $(U_1, \ldots, U_n)$ is called the *coordinate system defined by the basis A*. In this case, the coordinate projections $P_i : X \to U_i$ are projections on one-dimensional spaces. They can be represented by functions $x_i : X \to \mathbb{R}$ such that $P_i\mathbf{x} = x_i(\mathbf{x})\mathbf{a}_i$. This gives, as before, the unique expression for $\mathbf{x} \in X$,

$$\mathbf{x} = x_1(\mathbf{x})\mathbf{a}_1 + \cdots + x_n(\mathbf{x})\mathbf{a}_n, \tag{3.5}$$

as a linear combination of basis vectors. The functions $x_i : X \to \mathbb{R}$ are called the *coordinate functions (of the basis A)*. Note that $x_i(\mathbf{a}_j) = 0$ if $i \neq j$ and $x_i(\mathbf{a}_i) = 1$. We see that the earlier form

$$\mathbf{x} = x_1\mathbf{a}_1 + \cdots + x_n\mathbf{a}_n \tag{3.6}$$

was not quite precise, as the coefficients are actually functions $x_i : X \to \mathbb{R}$ and should not have been confused with their values $x_i(\mathbf{x})$. Nevertheless, we will continue to use the form (3.6) for convenience. Also, an easy verification shows that $x_i \in L(X, \mathbb{R})$.

Remarks 3.1.44 Coordinate functions as a basis for $L(X, \mathbb{R})$. Coordinate functions themselves are vectors in the vector space $L(X, \mathbb{R})$ (even though they are not written in boldface letters). We show that they are linearly independent. Suppose

$$f = r_1 x_1 + \cdots + r_n x_n = 0. \tag{3.7}$$

Here the meaning is that f is the 0-function, i.e., $f(\mathbf{x}) = 0$ for all $\mathbf{x} \in X$. In particular, $f(\mathbf{a}_i) = r_i = 0$, since $x_j(\mathbf{a}_i) = 0$ if $j \neq i$. Hence x_is are linearly independent in the vector space $L(X, \mathbb{R})$. They also form a basis for $L(X, \mathbb{R})$. In fact, let $f \in L(X, \mathbb{R})$ be arbitrary. Let $f(\mathbf{a}_i) = r_i$. Then

$$\begin{aligned} f(\mathbf{x}) &= f(x_1(\mathbf{x})\mathbf{a}_1 + \cdots + x_n(\mathbf{x})\mathbf{a}_n) \\ &= x_1(\mathbf{x}) f(\mathbf{a}_1) + \cdots + x_n(\mathbf{x}) f(\mathbf{a}_n) \\ &= r_1 x_1(\mathbf{x}) + \cdots + r_n x_n(\mathbf{x}) \end{aligned}$$

for all $\mathbf{x} \in X$. This means that $f = r_1 x_1 + \cdots r_n x_n$, as a linear combination in the vector space $L(X, \mathbb{R})$. This shows that the set

$$A^* = \{ x_1, \ldots, x_n \} \subset L(X, \mathbb{R})$$

is linearly independent and spans $L(X, \mathbb{R})$. Hence it is a basis for $L(X, \mathbb{R})$.

Isomorphic Spaces

Definition 3.1.45 Isomorphisms. A linear map $T : X \to Y$ is called an *isomorphism from X to Y* if it is one-to-one and onto, i.e., $T(X) = Y$. Equivalently, an isomorphism is an invertible linear map from X to Y. An easy check shows that the inverse map $T^{-1} : Y \to X$ is also an invertible linear map.

Example 3.1.46 Assume that $\dim X = n = \dim Y$ for some $n \in \mathbb{N}$. Let $T : X \to Y$ be linear. Then T is one-to-one on X if and only if T is an isomorphism. To see this, assume that T is one-to-one on X. Then $\text{Ker } T = \{\mathbf{0}\}$. Hence, by the dimension theorem, $\dim X = \dim T(X)$. Thus, $\dim Y = \dim(T(X))$. Since $T(X)$ is a subspace of Y, we have $T(X) = Y$. Conversely, if T is an isomorphism, then of course, T is one-to-one.

Definition 3.1.47 Isomorphic spaces. Two vector spaces X and Y are called *isomorphic* if there is an isomorphism $T : X \to Y$ from X to Y. The fact that X and Y

are isomorphic spaces is expressed as $X \sim Y$. *Being isomorphic* is an equivalence relation between vector spaces.

The next two results establish that for two vector spaces to be isomorphic amounts to having the same dimension.

Lemma 3.1.48 *If* $\dim X = n$*, then* X *is isomorphic to* $\mathbb{R}^n$*.*

Proof. Let $A = \{\mathbf{a}_1, \ldots, \mathbf{a}_n\}$ be a basis for X. Define $T : X \to \mathbb{R}^n$ by

$$T(\mathbf{x}) = (x_1(\mathbf{x}), \ldots, x_n(\mathbf{x})) \in \mathbb{R}^n$$

for $\mathbf{x} \in X$. Here $x_i : X \to \mathbb{R}$ are the coordinate functions of the basis A. We see easily that T is an isomorphism. Let $E = \{\mathbf{e}_1, \ldots, \mathbf{e}_n\}$ be the standard basis of $\mathbb{R}^n$, as in Definition 3.1.11. Then T takes $\mathbf{a}_i \in X$ to $\mathbf{e}_i \in \mathbb{R}^n$. □

Theorem 3.1.49 *Two (finite dimensional) vector spaces* X *and* Y *are isomorphic if and only if* $\dim X = \dim Y$*.*

Proof. Assume that X and Y are isomorphic. Let $T : X \to Y$ be an isomorphism from X to Y. Then T is one-to-one on X and $T(X) = Y$. Therefore, by Lemma 3.1.30, $\dim X = \dim T(X) = \dim Y$. Conversely, assume that $\dim X = \dim Y = n$. Then X and Y are both isomorphic to $\mathbb{R}^n$. □

Matrices

Definition 3.1.50 Matrices. Let $m, n \in \mathbb{N}$. An *$m \times n$ matrix* $\mathbf{A}$ is a function

$$\mathbf{A} : \{1, \ldots, m\} \times \{1, \ldots, n\} \to \mathbb{R}$$

that takes the pair $(i, j) \in \{1, \ldots, m\} \times \{1, \ldots, n\}$ to the number $A_{ij} \in \mathbb{R}$. The numbers A_{ij} are called the *entries* or *components* or *coordinates* of the matrix $\mathbf{A}$. Such a matrix is defined by arranging its entries as a table

$$\mathbf{A} = \begin{bmatrix} A_{11} & \ldots & A_{1n} \\ \vdots & \ddots & \vdots \\ A_{m1} & \ldots & A_{mn} \end{bmatrix}$$

consisting of m rows and n columns. We also denote a matrix $\mathbf{A}$ in terms of its entries as $\{A_{ij}\}$, or simply as A_{ij}, if the meaning is clear from the context.

Definition 3.1.51 The vector space $\mathbb{M}_{mn}$. Let $\mathbb{M}_{mn}$ be the set of all $m \times n$ matrices. Then $\mathbb{M}_{mn}$ is a vector space under the usual (componentwise) definition of linear operations. The set of matrices with all components equal to 0 except one component equal to 1 forms a basis for $\mathbb{M}_{mn}$. Since this set contains mn matrices, we obtain

$$\dim(\mathbb{M}_{mn}) = mn.$$

Another way of seeing this is as follows. There is an obvious isomorphism between $\mathbb{M}_{mn}$ and $\mathbb{R}^{mn}$. In fact, both spaces are the set of all functions $F \to \mathbb{R}$, where F is a finite set of mn elements. In each case, the set of all functions $F \to \mathbb{R}$ that take the value 1 at one point and vanish at all other points forms a basis. Hence we see that the difference between $\mathbb{M}_{mn}$ and $\mathbb{R}^{mn}$ is only notational.

Remarks 3.1.52 Matrices and linear transformations. Let $\dim X = n$ and $\dim Y = m$. Then we see that $\mathbb{M}_{mn}$ and $L(X, Y)$ are of the same dimension. Therefore they are isomorphic spaces. There is a standard isomorphism between $L(\mathbb{R}^n, \mathbb{R}^m)$ and $\mathbb{M}_{mn}$. Given a matrix $\mathbf{T} = \{T_{ij}\} \in \mathbb{M}_{mn}$, define a transformation $T : \mathbb{R}^n \to \mathbb{R}^m$ as follows. If

$$T(x_1, \ldots, x_n) = (y_1, \ldots, y_m),$$

then $y_i = \sum_{j=1}^{n} T_{ij}\, x_j$ for all $i = 1, \ldots, m$.

There is one further basic connection between matrices and linear transformations: composition of linear transformations corresponds to matrix multiplication. We shall establish this in the next section.

Problems

3.1 Let $T : X \to Y$ be a linear map. Let U be a subspace of X. Show that if $\dim U = \dim T(U)$, then T is one-to-one on U.

3.2 Let X be a vector space and let $\mathbf{x} \in X$. Show that for all scalars a, b, we have $(a - b)\mathbf{x} = a\mathbf{x} - b\mathbf{x}$.

3.3 Let X be a vector space. Show that if $\mathbf{x} \in X$ is nonzero and s, t are distinct scalars, then $s\mathbf{x} \neq t\mathbf{x}$.

3.4 Let $\mathbf{x}_1, \ldots, \mathbf{x}_n$ be vectors in a vector space X. Assume that $\mathbf{u}, \mathbf{v}$ are linear combinations of $\mathbf{x}_1, \ldots, \mathbf{x}_n$. Show that any linear combination of $\mathbf{u}$ and $\mathbf{v}$ is also a linear combination of $\mathbf{x}_1, \ldots, \mathbf{x}_n$.

3.5 Let $\mathbf{u} = (1, 2, 3), \mathbf{v} = (2, 5, -4)$. Find all real numbers a such that $(-2, a, 7)$ is a linear combination of $\mathbf{u}$ and $\mathbf{v}$.

3.6 Let

$$A = \begin{bmatrix} -2 & 1 & 3 \\ 3 & 0 & 7 \\ 11 & -5 & 9 \\ 2 & 2 & 0 \end{bmatrix}, \quad B = \begin{bmatrix} 1 & 4 & 3 \\ -2 & 0 & 5 \\ 7 & -1 & 3 \\ 0 & 3 & -1 \end{bmatrix}.$$

Find a matrix X such that $A - 3X = 2B$.

3.7 Give an example of a nonempty subset A of $\mathbb{R}^2$ such that $A + A \neq 2A$. Find all subsets B of $\mathbb{R}^2$ with $tB + sB \subset B$ for all scalars s and t.

3.8 Given any $\mathbf{u}$ and $\mathbf{v}$ in $\mathbb{R}^3$, show that

$$\mathbb{R}^3 \neq \{ a\mathbf{u} + b\mathbf{v} \mid a \in \mathbb{R},\, b \in \mathbb{R} \}.$$

(The solution is easy if one uses the results on dimensions: a three-dimensional space cannot be spanned by two vectors. Try to give a solution that uses only the basic definitions.)

3.9 For each $a \in \mathbb{R}$, let $U_a = \{ (x,\, y,\, z) \mid a|x| = x + y + z \}$. Find all a for which U_a is a subspace of $\mathbb{R}^3$.

3.10 Let U, V be subspaces of a vector space X. When is it true that $U \cup V$ is also a subspace of X?

3.11 Let $\mathbf{u}_1, \ldots, \mathbf{u}_n$ be vectors in a vector space X. Let

$$U = \{ \mathbf{a} = (a_1, \ldots, a_n) \in \mathbb{R}^n \mid a_1\mathbf{u}_1 + \cdots + a_n\mathbf{u}_n = \mathbf{0} \},$$

where $\mathbf{0}$ is the zero vector of X. Show that U is a subspace of $\mathbb{R}^n$.

3.12 Let $W = \{ (x,\, y,\, z) \in \mathbb{R}^3 \mid 2x - z = y \}$. Show that W is a subspace of $\mathbb{R}^3$ and find a basis for W.

3.13 Let A be a nonempty subset of a vector space X. Let t be a nonzero scalar. Is it true that Span $(tA) =$ Span A?

3.14 Find a finite subset S of $\mathbb{R}^4$ such that Span $S = A + B$, where

$$\begin{aligned} A &= \{ (x,\, y,\, z,\, t) \in \mathbb{R}^4 \mid x - y + t = 0 \} \text{ and} \\ B &= \{ (x,\, y,\, z,\, t) \in \mathbb{R}^4 \mid x + 4z - 2t = 0 \}. \end{aligned}$$

3.15 Let A and B be two subsets in a vector space X. Does Span $A \subset$ Span B imply that $A \subset B$?

3.16 Find mutually disjoint subsets A, B, and C of $\mathbb{R}^2$ such that

$$\mathbb{R}^2 = \text{Span } A = \text{Span } B = \text{Span } C.$$

3.17 Let A and B be two subsets in a vector space X. Show that

$$\text{Span } (A + B) \subset \text{Span } A + \text{Span } B.$$

3.18 Let U, V be two subspaces of X such that $\dim U = \dim V$ and $U \neq V$.

1. Show that $U + V \neq U$ and $U + V \neq V$.
2. Show that if $(\dim U) = (\dim V) = (\dim X) - 1$, then $X = U + V$.

3.19 Let $T : \mathbb{R}^2 \to \mathbb{R}^3$ be defined by

$$T(x,\ y) = (x - y, y - 3x, x + |y|) \text{ for all } (x,\ y) \in \mathbb{R}^2.$$

Is T linear?

3.20 Is there a linear $T : \mathbb{R}^3 \to \mathbb{R}^2$ such that

$$T(1,2,1) = (1,1),\ \ T(-2,0,3) = (2,5),\ \text{ and } T(-4,-4,1) = (-3,2)?$$

3.21 Let $T \in L(X, Y)$ and let $\mathbf{y}_0 \in Y$ be such that

$$\{\,\mathbf{x} \in X \mid T\mathbf{x} = \mathbf{y}_0\,\}$$

is a subspace of X. Show that $\mathbf{y}_0 = \mathbf{0}$.

3.22 Suppose that $E \subset X$ and Span $E = X$. Let T and S be in $L(X, Y)$. Show that $T = S$ if and only if $T\mathbf{u} = S\mathbf{u}$ for all $\mathbf{u} \in E$.

3.23 Let T, S be in $L(X, Y)$. Let $U = \{\,\mathbf{x} \in X \mid T\mathbf{x} = S\mathbf{x}\,\}$. Show that U is a subspace of X.

3.24 Show that if $X \sim Y$ and $Y \sim Z$, then $X \sim Z$, recalling that $X \sim Y$ means that X is isomorphic to Y.

3.25 Let $a \in \mathbb{R}$. Consider the linear map $T : \mathbb{R}^3 \to \mathbb{R}^3$ given by

$$T(x,\, y,\, z) = (x,\, ax + y,\, z) \text{ for all } (x,\, y,\, z) \in \mathbb{R}^3.$$

Show that T is invertible and find its inverse map $T^{-1} : \mathbb{R}^3 \to \mathbb{R}^3$.

3.26 Let X and Y be vector spaces and $\mathbf{a} \in X$. Define $f : L(X, Y) \to Y$ by

$$f(T) = T\mathbf{a} \text{ for all } T \in L(X, Y).$$

Show that f is linear.

3.27 Let $R : X \to Y$ and $S : Y \to Z$ be isomorphisms. Show that $SR : X \to Z$ is also an isomorphism and $(SR)^{-1} = R^{-1}S^{-1} : Z \to X$.

3.28 Let $(\dim X) \geq 2$. Give an example of a nonzero $T \in L(X,\, X)$ such that T^2 is the zero transformation.

3.29 Let X be a vector space. Let $T \in L(X,\, X)$ be non-invertible. Show that $CT = \mathbf{0}$ for a nonzero $C \in L(X,\, X)$.

3.30 Let $T : \mathbb{R}^5 \to \mathbb{R}^5$ be linear. Assume that whenever $\mathbf{x} \in \mathbb{R}^5$ and $T\mathbf{x} = \mathbf{x}$, then $\mathbf{x} = \mathbf{0}$. Show that for any $\mathbf{y} \in \mathbb{R}^5$, there exists $\mathbf{x} \in \mathbb{R}^5$ with

$$\mathbf{x} = T\mathbf{x} + \mathbf{y}.$$

3.31 Is there a linear map $T : X \to X$ such that T is not one-to-one on X but $T^k : X \to X$ is one-to-one for some $k \geq 2$?

3.32 Let $A,\, B \in L(X,\, X)$. If AB is invertible, show that both A and B are invertible.

3.33 Let $T \in L(X,\, Y)$. Let U be a subspace of X such that $U \cap \operatorname{Ker} T = \{\mathbf{0}\}$. Show that $\dim U = \dim T(U)$.

3.34 Let $T \in L(X,\, Y)$. Let U be a subspace of X such that $X = U \oplus \operatorname{Ker} T$. Show that $\dim X = \dim T(U) + \dim \operatorname{Ker} T$.

3.35 Let X and Y be two vector spaces. Let W be a subspace of X. If

$$\dim Y \geq (\dim X) - (\dim W),$$

then show that there is a $T \in L(X,\, Y)$ such that $W = \operatorname{Ker} T$.

3.36 Let $f,\, g \in L(X, \mathbb{R})$. Show that Ker $f \subset$ Ker g if and only if $g = cf$ for some scalar c.

3.37 Let U, V be subspaces of a finite-dimensional vector space. Let $W = \{ (-\mathbf{x}, \mathbf{x}) \mid \mathbf{x} \in U \cap V \}$.

1. Show that W is a subspace of $U \times V$ and W is isomorphic to $U \cap V$.
2. Define $f : U \times V \to U + V$ by $f(\mathbf{u}, \mathbf{v}) = \mathbf{u} + \mathbf{v}$ for all $(\mathbf{u}, \mathbf{v}) \in U \times V$. Show that f is linear. Hence, deduce that

$$\dim(U + V) = \dim U + \dim V - \dim(U \cap V).$$

3.38 Let W and Z be two vector spaces with $\dim W = \dim Z$. Let X be a subspace of W and let $T : X \to Z$ be a one-to-one linear map. Show that there is an isomorphism $R : W \to Z$ such that the restriction of R to X is T.

3.39 Let X and Z be two vector spaces with $(\dim X) \leq (\dim Z)$. Show that there is a vector space Y with the following property. Given any one-to-one linear map $T : X \to Z$, there is an isomorphism $R : (X \times Y) \to Z$ such that the restriction of R to X is T. Here X is identified with the subspace of $X \times Y$ consisting of vectors of the form $(\mathbf{x},\, \mathbf{0}) \in X \times Y$, with $\mathbf{x} \in X$.

3.40 Let W and Z be two vector spaces with $\dim W = \dim Z$. Let $S : Z \to W$ be a linear map and let $X = S(Z)$ be the range of S, which is a subspace of W. Show that there is an isomorphism $L : Z \to W$ and a projection $P : W \to W$ such that $S = PL$.

3.41 Let X and Z be two vector spaces with $(\dim X) \leq (\dim Z)$. Show that there is a vector space Y with the following property. Given any linear map $S : Z \to X$ that maps Z onto X (that is, $T(Z) = X$), there is an isomorphism $L : Z \to (X \times Y)$ such that $S = PL$, where $P : (X \times Y) \to X$ is the coordinate projection onto X. Recall that $P(\mathbf{x},\, \mathbf{y}) = (\mathbf{x},\, \mathbf{0})$ for all $(\mathbf{x},\, \mathbf{y}) \in X \times Y$. Here X is again identified with a subspace of $X \times Y$.

3.42 Let $(U,\, V)$ be a coordinate system in X, in the sense of Definition 3.1.42. A set $\Gamma \subset X$ is a *graph* in this system if there is a set $A \subset U$ and a function $f : A \to V$ such that Γ is the set of points $(\mathbf{a},\, f(\mathbf{a}))$ with $\mathbf{a} \in A$. Let $(U_i,\, V)$, $i = 1,\, 2$, be two coordinate systems for X. Show that Γ is a graph in one system if and only if it is a graph in the other system.

3.43 Let $(U,\, V_i)$, $i = 1,\, 2$, be two coordinate systems in X, in the sense of Definition 3.1.42. Consider the following proposition: a set $\Gamma \subset X$ is a graph in one system if and only if it is a graph in the other system. Show either that this proposition is true or that it is false.

3.2 BILINEAR FUNCTIONS

Ordinary multiplication is an operation that takes a pair of real numbers (a, b) to their product $M(a, b) = ab$. Multiplication is *bilinear*: for all $a, b, c, r, s \in \mathbb{R}$,

$$\begin{aligned} M(ra + sb, c) &= rM(a, c) + sM(b, c) \quad \text{and} \\ M(a, rb + sc) &= rM(a, b) + sM(a, c). \end{aligned}$$

Bilinearity deserves special attention, for it will play an important part in our discussion of polynomials.

Definition 3.2.1 Bilinear operations. Let X, Y, and Z be three vector spaces. A function $B : X \times Y \to Z$ is called a *bilinear function* (or a *bilinear operation*, or a *bilinear map*) if

$$\begin{aligned} B(\alpha\mathbf{x} + \alpha'\mathbf{x}', \mathbf{y}) &= \alpha B(\mathbf{x}, \mathbf{y}) + \alpha' B(\mathbf{x}', \mathbf{y}) \quad \text{and} \\ B(\mathbf{x}, \beta\mathbf{y} + \beta'\mathbf{y}') &= \beta B(\mathbf{x}, \mathbf{y}) + \beta' B(\mathbf{x}, \mathbf{y}') \end{aligned}$$

for all $\mathbf{x}, \mathbf{x}' \in X$, $\mathbf{y}, \mathbf{y}' \in Y$, and $\alpha, \alpha', \beta, \beta' \in \mathbb{R}$.

Thus, a bilinear function can be considered as a function of two variables that is linear in each variable separately when the other variable is kept constant.

Remarks 3.2.2 Linear and bilinear functions. The term *bilinear* may perhaps suggest that bilinear functions are some type of special linear function. This is not the case. Bilinear functions are completely different from linear functions. Consider, for example, the linear and bilinear functions from $\mathbb{R} \times \mathbb{R} = \mathbb{R}^2$ to $\mathbb{R}$. Here $\mathbb{R}$ is considered as a (one-dimensional) vector space. We see that a linear function $L : \mathbb{R}^2 \to \mathbb{R}$ is of the form

$$L(x, y) = Ax + By, \text{ where } A = L(1, 0) \text{ and } B = L(0, 1) \text{ are two constants.}$$

On the other hand a bilinear function $B : \mathbb{R}^2 \to \mathbb{R}$ is of the form

$$B(x, y) = Pxy, \text{ where } P = B(1, 1) \text{ is a constant.}$$

This follows from the observation that

$$B(x, y) = B(x1, y) = xB(1, y) = xB(1, y1) = xyB(1, 1).$$

Remarks 3.2.3 General form of bilinear functions. Let X and Y be vector spaces with bases $A = \{\mathbf{a}_1, \ldots, \mathbf{a}_n\}$ and $B = \{\mathbf{b}_1, \ldots, \mathbf{b}_m\}$, respectively. Then a function $R : X \times Y \to Z$ is a bilinear function if and only if

$$R(\mathbf{x}, \mathbf{y}) = \sum_i \sum_j x_i(\mathbf{x})\, y_j(\mathbf{y})\, \mathbf{c}_{ij}. \tag{3.8}$$

Here $x_i : X \to \mathbb{R}$ and $y_j : Y \to \mathbb{R}$ are the coordinate functions with respect to the bases A and B, and $\mathbf{c}_{ij} = R(\mathbf{a}_i, \mathbf{b}_j)$ are arbitrary vectors in Z. Equation (3.8) follows by expressing $\mathbf{x}$ and $\mathbf{y}$ as linear combinations of $\mathbf{a}_i$ and $\mathbf{b}_j$, respectively, and then expanding $R(\mathbf{x}, \mathbf{y})$ using bilinearity.

Example 3.2.4 Standard dot product. Define $B : \mathbb{R}^n \times \mathbb{R}^n \to \mathbb{R}$ by

$$B(\mathbf{x}, \mathbf{y}) = x_1 y_1 + \cdots + x_n y_n \quad \text{for all } (\mathbf{x}, \mathbf{y}) \in \mathbb{R}^n \times \mathbb{R}^n.$$

This function is the *standard dot product* on $\mathbb{R}^n$. It is easy to verify that $B : \mathbb{R}^n \times \mathbb{R}^n \to \mathbb{R}$ is a bilinear function. △

Example 3.2.5 Bilinear functions $(\mathbb{R}^3 \times \mathbb{R}) \to \mathbb{R}$. Denote the points in $\mathbb{R}^3$ as (x, y, z) and the points in $\mathbb{R}$ as t. Hence the points in $\mathbb{R}^3 \times \mathbb{R}$ are $((x, y, z), t)$. Let $A, B, C \in \mathbb{R}$ be arbitrary. An easy check shows that

$$R((x, y, z), t) = Axt + Byt + Czt, \quad ((x, y, z), t) \in \mathbb{R}^3 \times \mathbb{R},$$

defines a bilinear function $\mathbb{R}^3 \times \mathbb{R} \to \mathbb{R}$. Also, this is the general form of a bilinear function $\mathbb{R}^3 \times \mathbb{R} \to \mathbb{R}$. We see that

$$R((1, 0, 0), 1) = A, \; R((0, 1, 0), 1) = B, \; R((0, 0, 1), 1) = C. \qquad \triangle$$

Example 3.2.6 Bilinear functions $(\mathbb{R}^2 \times \mathbb{R}^2) \to \mathbb{R}$. Denote the points in $\mathbb{R}^2 \times \mathbb{R}^2$ as $((x, y), (u, v))$. Let A, B, C, D be arbitrary in $\mathbb{R}$. An easy check shows that

$$R((x, y), (u, v)) = Axu + Byu + Cxv + Dyv, \quad ((x, y), (u, v)) \in \mathbb{R}^2 \times \mathbb{R}^2,$$

defines a bilinear function $\mathbb{R}^2 \times \mathbb{R}^2 \to \mathbb{R}$. Also, this is the general form of a bilinear function $\mathbb{R}^2 \times \mathbb{R}^2 \to \mathbb{R}$. We see that

$$\begin{array}{rclrcl} R((1, 0), (1, 0)) & = & A, & R((0, 1), (1, 0)) & = & B, \\ R((1, 0), (0, 1)) & = & C, & R((0, 1), (0, 1)) & = & D. \end{array} \qquad \triangle$$

Remarks 3.2.7 Dependence on the factorization. In Examples 3.2.5 and 3.2.6 above, the domain space of the bilinear functions is the same space

$$\mathbb{R}^4 = \mathbb{R}^3 \times \mathbb{R} = \mathbb{R}^2 \times \mathbb{R}^2.$$

However, the two classes of bilinear functions are different. Hence, bilinearity depends on the way of expressing the domain space as a Cartesian product.

Definition 3.2.8 The space of bilinear functions. The set of all bilinear maps $X \times Y \to Z$ will be denoted by $BL(X \times Y, Z)$. An easy verification shows that $BL(X \times Y, Z)$ is actually a vector space itself, by the natural definitions of linear operations. The following lemma states this explicitly.

Lemma 3.2.9 *Let P and Q be two bilinear functions, and let $r, s \in \mathbb{R}$. Define $(rP + sQ) : X \times Y \to Z$ as*

$$(rP + sQ)(\mathbf{x}, \mathbf{y}) = rP(\mathbf{x}, \mathbf{y}) + sQ(\mathbf{x}, \mathbf{y})$$

for all $(\mathbf{x}, \mathbf{y}) \in X \times Y$. Then $(rP + sQ) : X \times Y \to Z$ is also a bilinear function. The set of all bilinear functions $X \times Y \to Z$ becomes a vector space $BL(X, Y; Z)$ with these linear operations.

Proof. This is left as an exercise. □

Example 3.2.10 Composition of linear transformations. If $R : X \to Y$ and $S : Y \to Z$ are two linear transformations, then their composition

$$C(R, S) = S \cdot R = SR : X \to Z$$

is another linear transformation. We see that the operation of composition itself,

$$C : L(X, Y) \times L(Y, Z) \to L(X, Z),$$

is a bilinear operation. In fact, we have

$$\begin{aligned} C(\alpha R + \alpha' R', S)(\mathbf{x}) &= S((\alpha R + \alpha' R')(\mathbf{x})) = S(\alpha R\mathbf{x} + \alpha' R'\mathbf{x}) \\ &= \alpha S(R\mathbf{x}) + \alpha' S(R'\mathbf{x}) = (\alpha SR + \alpha' SR')(\mathbf{x}) \\ &= (\alpha C(R, S) + \alpha' C(R', S))(\mathbf{x}) \end{aligned}$$

for all $\mathbf{x} \in X$. Similarly, we can verify the linearity of C in its second factor. Thus, C is a bilinear operation $L(X, Y) \times L(Y, Z) \to L(X, Z)$. △

Matrix Multiplication

Matrix multiplication is defined as follows. If $\mathbf{A} = \{A_{ik}\} \in \mathbb{M}_{m\ell}$ is an $m \times \ell$ matrix, and $\mathbf{B} = \{B_{kj}\} \in \mathbb{M}_{\ell n}$ is an $\ell \times n$ matrix, then their product

$$M(\mathbf{A}, \mathbf{B}) = \mathbf{AB} = \mathbf{C} = \{C_{ij}\} \in \mathbb{M}_{mn}$$

is the $m \times n$ matrix whose entry in row i and column j is given by $C_{ij} = \sum_{k=1}^{\ell} A_{ik} B_{kj}$. We see that this operation defines a bilinear function

$$M : \mathbb{M}_{m\ell} \times \mathbb{M}_{\ell n} \to \mathbb{M}_{mn}$$

and therefore a product between the matrices.

An easy check shows that matrix multiplication is associative. If $\mathbf{A} \in \mathbb{M}_{m\ell}$, $\mathbf{B} \in \mathbb{M}_{\ell p}$, and $\mathbf{C} \in \mathbb{M}_{pn}$, then $\mathbf{A}(\mathbf{BC})$ and $(\mathbf{AB})\mathbf{C}$ define the same matrix in $\mathbb{M}_{mn}$, which we may consequently denote as $\mathbf{ABC}$. Significantly, matrix multiplication is noncommutative. In fact, $\mathbf{BA}$ may not be defined, even if $\mathbf{AB}$ is defined. This happens, for instance, if $\mathbf{A}$ is a 2×3 matrix and $\mathbf{B}$ is a 3×5 matrix.

Matrices can be used to represent linear maps, and a particularly simple kind of matrix multiplication can be used to represent the effect of applying a linear map to a vector.

Example 3.2.11 Linear maps as matrix multiplication. There is a standard isomorphism between $\mathbb{M}_{mn}$ and $L(\mathbb{R}^n, \mathbb{R}^m)$, as pointed out in Remarks 3.1.52. It takes the matrix $\mathbf{T} = \{T_{ij}\} \in \mathbb{M}_{mn}$ to the transformation $T : \mathbb{R}^n \to \mathbb{R}^m$ defined as follows: if $T(x_1, \ldots, x_n) = (y_1, \ldots, y_m)$, then $y_i = \sum_{j=1}^{n} T_{ij}\, x_j$. T can be expressed as matrix multiplication in the following way. Let $C_n : \mathbb{R}^n \to \mathbb{M}_{n1}$ take vectors in R^n to $n \times 1$ matrices as follows:

$$C_n(x_1, \ldots, x_n) = \begin{bmatrix} x_1 \\ \vdots \\ x_n \end{bmatrix} \in \mathbb{M}_{n1}$$

for $(x_1, \ldots, x_n) \in \mathbb{R}^n$. We usually call the matrices in $\mathbb{M}_{n1}$ *column vectors* and ignore any differences between $\mathbb{R}^n$ and $\mathbb{M}_{n1}$. We denote the elements of both spaces by symbols like $\mathbf{x}$ and $\mathbf{y}$. If $\mathbf{T} \in \mathbb{M}_{mn}$, with corresponding operator $T : \mathbb{R}^n \to \mathbb{R}^m$, then $\mathbf{y} = T\mathbf{x}$ if and only if $\mathbf{y} = \mathbf{T}\mathbf{x}$, or, more explicitly,

$$\mathbf{y} = \begin{bmatrix} y_1 \\ \vdots \\ y_m \end{bmatrix} = \begin{bmatrix} T_{11} & \ldots & T_{1n} \\ \vdots & \ddots & \vdots \\ T_{m1} & \ldots & T_{mn} \end{bmatrix} \begin{bmatrix} x_1 \\ \vdots \\ x_n \end{bmatrix} = \mathbf{T}\,\mathbf{x}.$$

Thus, any linear transformation can be represented as matrix multiplication.

Example 3.2.12 Compositions as matrix multiplication. Let

$$R \in L(\mathbb{R}^n, \mathbb{R}^\ell) \text{ and } S \in L(\mathbb{R}^\ell, \mathbb{R}^m).$$

Then their composition $T = SR$ belongs to $L(\mathbb{R}^n, \mathbb{R}^m)$, as mentioned in Example 3.2.10. Let the maps R, S, T correspond to the matrices

$$\mathbf{R} \in \mathbb{M}_{\ell n}, \mathbf{S} \in \mathbb{M}_{m\ell}, \text{ and } \mathbf{T} \in \mathbb{M}_{mn},$$

as above in Example 3.2.11. Then $\mathbf{T} = \mathbf{S}\,\mathbf{R}$. In fact, by the associativity of matrix multiplication,

$$\mathbf{T}\mathbf{x} = \mathbf{S}(\mathbf{R}\mathbf{x}) = (\mathbf{S}\mathbf{R})\mathbf{x}\,.$$

Second-degree Homogeneous Polynomials

Second-degree real homogeneous polynomials in two variables have the form

$$f(x,\, y) = Ax^2 + 2Bxy + Cy^2 \;\text{ with } A,\, B,\, C \in \mathbb{R}.$$

As noted at the start of the chapter, this class of functions will have an important role later. This function is still defined if the coefficients are vectors from a vector space Y. Such polynomials are defined in terms of bilinear functions.

Definition 3.2.13 Second-degree homogeneous polynomials. Let X and Y be two vector spaces. A *second-degree homogeneous polynomial* $f : X \to Y$ is a function of the form $f(\mathbf{x}) = B(\mathbf{x},\, \mathbf{x})$, $\mathbf{x} \in X$, where $B : X \times X \to Y$ is a bilinear function.

Definition 3.2.14 Symmetric bilinear functions. If the bilinear function

$$B : X \times X \to Y \text{ satisfies } B(\mathbf{u},\, \mathbf{v}) = B(\mathbf{v},\, \mathbf{u}) \text{ for all } \mathbf{u},\, \mathbf{v} \in X,$$

then it is called a *symmetric bilinear function.* If $B : X \times X \to Y$ is any bilinear function, then

$$\widetilde{B}(\mathbf{u},\, \mathbf{v}) = (1/2)(B(\mathbf{u},\, \mathbf{v}) + B(\mathbf{v},\, \mathbf{u}))$$

is a symmetric bilinear function. It is called the *symmetric part of* B. Note that

$$B(\mathbf{x},\, \mathbf{x}) = \widetilde{B}(\mathbf{x},\, \mathbf{x}) \;\text{ for all } \mathbf{x} \in X.$$

Hence a bilinear function and its symmetric part define the same second-degree homogeneous polynomial.

Example 3.2.15 Second-degree homogeneous polynomials $f : \mathbb{R}^2 \to \mathbb{R}$. A general bilinear function $B : \mathbb{R}^2 \times \mathbb{R}^2 \to \mathbb{R}$ is of the form

$$B((x,\, y),\, (u,\, v)) = Axu + Byu + Cxv + Dyv, \;\; ((x,\, y),\, (u,\, v)) \in \mathbb{R}^2 \times R^2,$$

as shown in Example 3.2.6. Hence a general second-degree homogeneous polynomial $f : \mathbb{R}^2 \to \mathbb{R}$ is of the from

$$f(x,\, y) = B((x,\, y),\, (x,\, y)) = Ax^2 + (B + C)xy + Dy^2, \;\; (x,\, y) \in \mathbb{R}^2.$$

Note that the symmetric part of B is

$$\widetilde{B}((x,\,y),\,(u,\,v)) = Axu + (1/2)(B+C)(xv+yu) + Dyv.$$

Hence we see that the relation

$$B((x,\,y),\,(x,\,y)) = \widetilde{B}((x,\,y),\,(x,\,y)) = Ax^2 + (B+C)xy + Dy^2$$

is verified. △

Problems

3.44 Let $X,\,Y,\,Z$ be three vector spaces. Show that the vector spaces

$$BL(X,\,Y;\,Z),\ \ L(X,\,L(Y,\,Z)),\ \ \text{and}\ \ L(Y,\,L(X,\,Z))$$

are isomorphic to each other.

3.45 The *cross product* of two vectors is defined as

$$(x_1,\,y_1,\,z_1)\times(x_2,\,y_2,\,z_2) = (y_1z_2 - y_2z_1,\,z_1x_2 - x_1z_2,\,x_1y_2 - y_1x_2).$$

Show that the cross product, as a function $\mathbb{R}^3\times\mathbb{R}^3\to\mathbb{R}^3$, is a bilinear operation. Hence, the cross product is also a product in the sense defined here.

3.46 Show that $B((u_1,\,\ldots,\,u_n),\,(v_1,\,\ldots,\,v_m)) = \sum_{i=1}^{n}\sum_{j=1}^{m}u_iv_j\mathbf{a}_{ij}$, with arbitrary $\mathbf{a}_{ij}\in Y$, is a general bilinear function $B:\mathbb{R}^n\times\mathbb{R}^m\to Y$.

3.47 Show that $f(x_1,\,\ldots,\,x_n) = \sum_{i=1}^{n}\sum_{j=1}^{i}x_ix_j\mathbf{a}_{ij}$ is a general second-degree homogeneous polynomial $f:\mathbb{R}^n\to Y$. Here $\mathbf{a}_{ij}\in Y$ are arbitrary.

3.48 Suppose that $T:\,X\times Y\to Z$ is bilinear. Must its kernel

$$W = \{\,(\mathbf{x},\mathbf{y})\in X\times Y \mid T(\mathbf{x},\mathbf{y}) = \mathbf{0}\,\}$$

be a subspace? What about the range of T?

3.49 Let X be a vector space with basis $\{\mathbf{x}_1,\ldots,\mathbf{x}_k\}$. Then every bilinear map $S:\,X\times\mathbb{R}^n\to\mathbb{R}^m$ is of the form

$$S(\mathbf{x},\mathbf{y}) = c_{x1}A_1\mathbf{y} + \cdots + c_{xk}A_k\mathbf{y}\quad\text{for all } \mathbf{y}\in\mathbb{R}^n,$$

where $A_1,\ldots,A_k$ are $m\times n$ matrices and $c_{x1},\ldots,c_{xk}$ are scalars such that

$$\mathbf{x} = c_{x1}\mathbf{x}_1 + \cdots + c_{xk}\mathbf{x}_k.$$

3.50 Let X, Y, and Z be vector spaces. Then both $BL(X \times Y, Z)$ and $L(X \times Y, Z)$ are subspaces of $\mathcal{F}(X \times Y, Z)$, the vector space of all functions from $X \times Y$ into Z. Here, $L(X \times Y, Z)$ is the subspace of all linear maps from $X \times Y$ to Z, where $X \times Y$ is equipped with the usual vector space structure. Show that

$$BL(X \times Y, Z) \cap L(X \times Y, Z) = \{\mathbf{0}\}.$$

3.51 Let $T : X \times Y \to Z$ be bilinear. Suppose that

$$(\dim Y)(\dim Z) < \dim X.$$

Show that there is a nonzero $\mathbf{x}_0 \in X$ such that

$$T(\mathbf{x}_0, \mathbf{y}) = \mathbf{0} \quad \text{for all } \mathbf{y} \in Y.$$

3.52 Suppose that f is a homogeneous polynomial of degree 2 on $\mathbb{R}^2$ such that $f(1,0) = 1, f(0,1) = 1$. What is the value of $f(1,1)$ so that $f(x,y) \geq 0$ for all $(x,y) \in \mathbb{R}^2$?

3.3 MULTILINEAR FUNCTIONS

Products between two vector spaces are defined in terms of bilinear functions. Products between finitely many vector spaces are expressed in terms of multilinear functions. Before we define multilinear functions, we review the coordinate systems defined by Cartesian products, as described in Definition 3.1.42.

Notation 3.3.1 Review of coordinate systems. Consider $(U_1, \ldots, U_k)$ as a coordinate system in $X = U_1 \times \cdots \times U_k$. Let $P_i : X \to U_i$ be the associated coordinate projections and let $Q_i = I - P_i$. Here $I : X \to X$ is the identity. Hence P_i and Q_i are complementary projections as in Definition 3.1.39.

Remarks 3.3.2 Multilinearity as componentwise linearity. Let $M : X \to Y$ be a function, where $X = U_1 \times \cdots \times U_k$. Multilinearity of this function is defined as its linearity in each component (or coordinate) taken separately, with all other components kept fixed. That is,

$$M(\alpha_1 \mathbf{x}_1 + \alpha_1' \mathbf{x}_1', \mathbf{x}_2, \ldots, \mathbf{x}_k) = \alpha_1 M(\mathbf{x}_1, \mathbf{x}_2, \ldots, \mathbf{x}_k) + \alpha_1' M(\mathbf{x}_1', \mathbf{x}_2, \ldots, \mathbf{x}_k)$$

for all $\mathbf{x}_1, \mathbf{x}_1' \in U_1$, $\mathbf{x}_i \in U_i$, and $\alpha_1, \alpha_1' \in \mathbb{R}$; and similarly for each of the other components.

To formulate this concisely, note the following. If $\mathbf{a} = (\mathbf{a}_1, \ldots, \mathbf{a}_k)$ and $\mathbf{x} = (\mathbf{x}_1, \ldots, \mathbf{x}_k)$, then

$$Q_i\mathbf{a} + P_i\mathbf{x} = (\mathbf{a}_1, \ldots, \mathbf{x}_i, \ldots, \mathbf{a}_k)$$

is obtained from $\mathbf{a}$ by replacing only its ith component by the ith component of $\mathbf{x}$. Therefore, if we let $T_i(\mathbf{a}) : X \to Y$ be the function $T_i(\mathbf{a})(\mathbf{x}) = M(Q_i\mathbf{a} + P_i\mathbf{x})$, then the linearity of all of the functions $T_i(\mathbf{a}) : X \to Y$ is equivalent to the linearity of $M : X \to Y$ in each component separately, when all other components are kept constant. Hence we introduce the definition of multilinearity as follows.

Definition 3.3.3 Multilinear functions. Let $M : X \to Y$ be a function, where $X = U_1 \times \cdots \times U_k$. Let $\mathbf{a} \in X$ and let $i = 1, \ldots, k$ be fixed. Let

$$T_i(\mathbf{a})\mathbf{x} = M(Q_i\mathbf{a} + P_i\mathbf{x}) \text{ for } \mathbf{x} \in X.$$

Then $M : X \to Y$ is called a *multilinear* (or *k-linear*) function if $T_i(\mathbf{a}) : X \to Y$ is a linear function for each $\mathbf{a} \in X$ and for each $i = 1, \ldots, k$.

Example 3.3.4 Suppose that $T : \mathbb{R}^2 \times \mathbb{R} \times \mathbb{R} \to \mathbb{R}^2$ is multilinear. Given that $T((1,0),1,1) = (2,3), T((0,1),1,1) = (5,-1)$, let us find a formula for $T((x,y),u,v)$ where $((x,y),u,v) \in \mathbb{R}^2 \times \mathbb{R} \times \mathbb{R}$.

Let $((x,y),u,v) \in \mathbb{R}^2 \times \mathbb{R} \times \mathbb{R}$ be arbitrary. Then

$$\begin{aligned} T((x,y),u,v) &= T(x(1,0) + y(0,1),u,v) \\ &= xT((1,0),u,v) + yT((0,1),u,v) \\ &= xuT((1,0),1,v) + yuT((0,1),1,v) \\ &= xuvT((1,0),1,1) + yuvT((0,1),1,1) \\ &= xuv(2,3) + yuv(5,-1) \\ &= ((2x+5y)uv, (3x-y)uv). \end{aligned}$$

Lemma 3.3.5 *If $M : X \to Y$ is multilinear, then*

$$T_i(\mathbf{a})(\mathbf{x}) = M(Q_i\mathbf{a} + P_i\mathbf{x}) = M(\mathbf{a} + P_i\mathbf{x}) - M(\mathbf{a}), \quad \mathbf{x} \in X. \tag{3.9}$$

Proof. We have

$$\begin{aligned} M(\mathbf{a} + P_i\mathbf{x}) &= M(Q_i\mathbf{a} + P_i\mathbf{a} + P_i\mathbf{x}) = M(Q_i\mathbf{a} + P_i(\mathbf{x} + \mathbf{a})) && (3.10) \\ &= M(Q_i\mathbf{a} + P_i\mathbf{x}) + M(Q_i\mathbf{a} + P_i\mathbf{a}) && (3.11) \\ &= M(Q_i\mathbf{a} + P_i\mathbf{x}) + M(\mathbf{a}). && (3.12) \end{aligned}$$

Hence (3.9) follows. Here (3.11) uses the linearity of $T_i(\mathbf{a}) : X \to Y$. □

Definition 3.3.6 Spaces of multilinear functions. Let U_i and Y be vector spaces. Let $X_k = U_k \times \cdots \times U_1$, $k \in N$. (The U_is are listed in descending order for notational convenience in the arguments below.) Let

$$ML_k(U_k \times \cdots \times U_1,\, Y) = ML_k(X_k,\, Y)$$

be the set of all k-linear functions. We see easily that $ML_k(X_k,\, Y)$ is a vector space under the natural definitions of linear operations. That is, if $A,\, B : X_k \to Y$ are multilinear functions and if $r,\, s \in \mathbb{R}$, then

$$(rA + sB)(\mathbf{x}) = rA(\mathbf{x}) + sB(\mathbf{x}) \text{ where } \mathbf{x} = (\mathbf{u}_k,\, \ldots,\, \mathbf{u}_1) \in X_k$$

defines another multilinear function $rA + sB : X_k \to Y$.

Definition 3.3.7 An identification of multiple products. There is a natural isomorphism between $ML_{k+1}(X_{k+1},\, Y)$ and $L(U_{k+1},\, ML_k(X_k,\, Y))$. For each $F \in ML_{k+1}(X_{k+1},\, Y)$, define $\vartheta F \in L(U_{k+1},\, ML_k(X_k,\, Y))$ by

$$\vartheta F(\mathbf{u}_{k+1})(\mathbf{u}_k,\, \ldots,\, \mathbf{u}_1) = F(\mathbf{u}_{k+1},\, \mathbf{u}_k,\, \ldots,\, \mathbf{u}_1). \tag{3.13}$$

Here ϑF is a linear function $U_{k+1} \to ML_k(X_k,\, Y)$. It maps $\mathbf{u}_{k+1} \in U_{k+1}$ to $\vartheta F(\mathbf{u}_{k+1}) \in ML_k(X_k,\, Y)$. Equation (3.13) defines $\vartheta F(\mathbf{u}_{k+1}) : X_k \to Y$ at each $(\mathbf{u}_k,\, \ldots,\, \mathbf{u}_1) \in X_k$. An easy verification shows that

$$\vartheta : ML_{k+1}(X_{k+1},\, Y) \to L(U_{k+1},\, ML_k(X_k,\, Y))$$

is an isomorphism. That is, ϑ is linear and invertible. In practice, we ignore the difference between F and ϑF. Hence the values of $F \in ML_{k+1}(X_{k+1},\, Y)$ may be denoted as

$$F(\mathbf{u}_{k+1},\, \mathbf{u}_k,\, \ldots,\, \mathbf{u}_1)$$

or as

$$F(\mathbf{u}_{k+1})(\mathbf{u}_k,\, \ldots,\, \mathbf{u}_1),$$

depending on the context. In the basic case of the usual product of $(k+1)$ numbers, this corresponds to identifying

$$r_{k+1} \cdot r_k \cdots r_1 \text{ and } r_{k+1} \cdot (r_k \cdots r_1).$$

The main significance of this identification is as follows. General results about $ML_k(X_k,\, Y)$ are usually proved by mathematical induction on k. Our identification simplifies these inductive proofs, as the isomorphism ϑ is used in the induction step to pass from k to $(k+1)$.

Note also that

$$ML_1(X_1,\, Y) = L(U_1,\, Y)$$

is the space of linear functions $U_1 \to Y$ and

$$ML_2(X_2,\, Y) = BL(U_2 \times U_1,\, Y) \sim L(U_2,\, L(U_1,\, Y))$$

is the space of bilinear functions $(U_2 \times U_1) \to Y$.

General Polynomials in Vector Variables

Here we generalize our earlier discussion of second-degree homogeneous polynomials. We assume some familiarity with permutations. Appendix C on determinants also contains a review of permutations. For each $k \in \mathbb{N}$ we let $\mathbb{N}_k = \{1, \ldots, k\}$ and denote the set of all permutations of $\mathbb{N}_k$ by $\mathcal{S}_k$. Note that $\mathcal{S}_k$ contains $k!$ elements.

Notation 3.3.8 If $U_1 = \cdots = U_n = X$, then we let

$$X_k = U_k \times \cdots \times U_1 = X^k.$$

Also, $ML_k(X^k, Y)$ denotes the space of all k-linear functions (or *k-products*) with all factors from X.

Definition 3.3.9 Homogeneous polynomials of degree k. Each

$$M \in ML_k(X^k, Y)$$

defines a function $f : X \to Y$ by

$$f(\mathbf{x}) = M(\mathbf{x}, \ldots, \mathbf{x}), \quad \mathbf{x} \in X.$$

Such a function is called a *homogeneous polynomial* (of a vector variable and of degree k). A sum of homogeneous polynomials is called a *polynomial.*

Definition 3.3.10 Symmetric multilinear functions. A multiple product $M \in ML_k(X^k, Y)$ is called *symmetric* if it is independent of the ordering of its k arguments. More explicitly, M is symmetric if

$$M(\mathbf{x}_1, \ldots, \mathbf{x}_k) = M(\mathbf{x}_{\sigma(1)}, \ldots, \mathbf{x}_{\sigma(k)}) \tag{3.14}$$

for all permutations $\sigma \in \mathcal{S}_k$.

Definition 3.3.11 Space of symmetric multilinear functions. Denote the set of all symmetric k-products as $SML_k(X^k, Y)$. Hence

$$SML_k(X^k, Y) \subset ML_k(X^k, Y).$$

We see easily that $SML_k(X^k, Y)$ is a subspace of $ML_k(X^k, Y)$.

Definition 3.3.12 The symmetric part of a multilinear function. Given $M \in ML_k(X^k, Y)$, define

$$\widetilde{M}(\mathbf{x}_1, \ldots, \mathbf{x}_k) = \frac{1}{k!} \sum_{\sigma \in \mathcal{S}_k} M(\mathbf{x}_{\sigma(1)}, \ldots, \mathbf{x}_{\sigma(k)}) \tag{3.15}$$

for all $(\mathbf{x}_1, \ldots, \mathbf{x}_k) \in X^k$. We see that $\widetilde{M} \in SML_k(X^k, Y)$. In fact, a permutation of the arguments of $\widetilde{M}$ results only in a change of the order of summation in (3.15). Hence the value of $\widetilde{M}$ does not change under a permutation of its arguments. (Note that this definition is a generalization of Definition 3.2.14.)

Lemma 3.3.13 *Let $M \in ML_k(X^k, Y)$ and let $\widetilde{M} \in SML_k(X^k, Y)$ be the symmetric part of M. Then M and $\widetilde{M}$ define the same polynomial. That is,*

$$f(\mathbf{x}) = M(\mathbf{x}, \ldots, \mathbf{x}) = \widetilde{M}(\mathbf{x}, \ldots, \mathbf{x})$$

for all $\mathbf{x} \in X$.

Proof. If $\mathbf{x}_i = \mathbf{x}$ for all $i = 1, \ldots, k$, then

$$(\mathbf{x}_1, \ldots, \mathbf{x}_k) = (\mathbf{x}_{\sigma(1)}, \ldots, \mathbf{x}_{\sigma(k)})$$

for any permutation $\sigma \in \mathbb{S}_k$. Then the conclusion follows from the definition of $\widetilde{M}$ given above in Definition 3.3.12. □

Example 3.3.14 Compositions of operators. Let $L = L(X, X)$ be the vector space of all linear operators $R : X \to X$. The composition of two operators is a 2-product, as shown in Example 3.2.10. If $R_1, \ldots, R_k \in L$ are k linear operators $X \to X$, then their composition

$$M(R_k, \ldots, R_1) = (R_k \cdots R_1) \in L$$

defines a k-product $M \in ML_k(L^k, L)$. In general, this is not a symmetric (commutative) product. Note that the corresponding polynomial is

$$f(T) = M(T, \ldots, T) = T^k,$$

with the definition of powers in Remarks 3.1.35. Note that T^k may be induced by several different k-products. For example, T^3 is induced by $M_1(P, Q, R) = PQR$ or by $M_2((P, Q, R) = (1/2)(PQR + PRQ)$. These are both non-symmetric products. Their symmetric parts are

$$\widetilde{M_1} = \widetilde{M_2} = (1/6)(PRS + RSP + SPR + PSR + SRP + RPS).$$

The equality of the symmetric parts is not accidental. If $M_1, M_2 \in ML_k(X^k, Y)$ induce the same polynomial

$$f(\mathbf{x}) = M_1(\mathbf{x}, \ldots, \mathbf{x}) = M_2(\mathbf{x}, \ldots, \mathbf{x}),$$

then $\widetilde{M_1} = \widetilde{M_2}$.

Problems

3.53 Define $\varphi : \mathbb{R}^3 \times \mathbb{R}^3 \times \mathbb{R}^3 \to \mathbb{R}$ as $\varphi(\mathbf{x}, \mathbf{y}, \mathbf{z}) = \langle \mathbf{x} \times \mathbf{y}, \mathbf{z} \rangle$, where $\mathbf{x} \times \mathbf{y}$ is the usual cross-product of $\mathbf{x}$ and $\mathbf{y}$, and $\langle , \rangle$ is the usual inner product operation. Show that φ is a multilinear function. Find a linear function $T : \mathbb{R}^3 \to BL(\mathbb{R}^3 \times \mathbb{R}^3, \mathbb{R})$ such that $\varphi(\mathbf{x}, \mathbf{y}, \mathbf{z}) = (T\mathbf{x})(\mathbf{y}, \mathbf{z})$ for all $(\mathbf{x}, \mathbf{y}, \mathbf{z}) \in \mathbb{R}^3 \times \mathbb{R}^3 \times \mathbb{R}^3$.

3.54 Define $\varphi : \mathbb{R}^3 \times \mathbb{R}^3 \times \mathbb{R}^3 \times \mathbb{R}^3 \to \mathbb{R}$ as $\varphi(\mathbf{x}, \mathbf{y}, \mathbf{z}, \mathbf{u}) = \langle \mathbf{x} \times \mathbf{y}, \mathbf{z} \times \mathbf{u} \rangle$. Show that φ is a multilinear function. Find a linear function

$$T : \mathbb{R}^3 \to ML_3(\mathbb{R}^3 \times \mathbb{R}^3 \times \mathbb{R}^3, \mathbb{R})$$

such that $\varphi(\mathbf{x}, \mathbf{y}, \mathbf{z}, \mathbf{u}) = (T\mathbf{x})(\mathbf{y}, \mathbf{z}, \mathbf{u})$ for all $(\mathbf{x}, \mathbf{y}, \mathbf{z}, \mathbf{u}) \in \mathbb{R}^3 \times \mathbb{R}^3 \times \mathbb{R}^3 \times \mathbb{R}^3$.

3.55 Define $\varphi : \mathbb{R}^3 \times \mathbb{R}^3 \times \mathbb{R}^3 \to \mathbb{R}^3$ as $\varphi(\mathbf{x}, \mathbf{y}, \mathbf{z}) = (\mathbf{x} \times \mathbf{y}) \times \mathbf{z}$. Show that φ is a multilinear function. Find a linear function $T : \mathbb{R}^3 \to BL(\mathbb{R}^3 \times \mathbb{R}^3, \mathbb{R}^3)$ such that $\varphi(\mathbf{x}, \mathbf{y}, \mathbf{z}) = (T\mathbf{x})(\mathbf{y}, \mathbf{z})$ for all $(\mathbf{x}, \mathbf{y}, \mathbf{z}) \in \mathbb{R}^3 \times \mathbb{R}^3 \times \mathbb{R}^3$.

3.56 Let $T : X_1 \times \cdots \times X_m \to Z$ be multilinear, where $X_1, \ldots, X_m, Z$ are vector spaces. Suppose that $\mathbf{x} = (\mathbf{x}_1, \ldots, \mathbf{x}_m) \in X_1 \times \cdots \times X_m$ and $\mathbf{x}_k = \mathbf{0}$ for some k. Show that $T\mathbf{x} = \mathbf{0}$.

3.57 Let $X_1, \ldots, X_k$ be vector spaces and let $T : X_1 \times \cdots \times X_k \to X_1 \times \cdots \times X_k$ be multilinear. Suppose that there are linear maps $T_j : X_j \to X_j$ such that $T(\mathbf{x}_1, \ldots, \mathbf{x}_k) = (T_1\mathbf{x}_1, \ldots, T_k\mathbf{x}_k)$. If $k \geq 2$, must $T = \mathbf{0}$?

3.58 Let $m > 1$ be an integer and let $X_1, \ldots, X_m, Z$ be vector spaces. Let $T_k : X_k \to Z$ be linear and define $S : X_1 \times \cdots \times X_m \to Z$ by

$$S(\mathbf{x}_1, \ldots, \mathbf{x}_m) = T_1\mathbf{x}_1 + \cdots + T_m\mathbf{x}_m \quad \text{for all } (\mathbf{x}_1, \ldots, \mathbf{x}_n) \in X^n.$$

Show that S is multilinear if and only if $T_k = \mathbf{0}$ for all $k = 1, \ldots, m$.

3.59 Let $T : \mathbb{R}^3 \times \mathbb{R}^2 \times \mathbb{R}^5 \to \mathbb{R}^4$ be multilinear. Show that there is a multilinear map $\varphi : \mathbb{R}^3 \times \mathbb{R}^2 \to M_{4\times 5}$ such that

$$T(\mathbf{x}, \mathbf{y}, \mathbf{z}) = \varphi(\mathbf{x}, \mathbf{y})\mathbf{z} \quad \text{for all } (\mathbf{x}, \mathbf{y}, \mathbf{z}) \in \mathbb{R}^3 \times \mathbb{R}^2 \times \mathbb{R}^5.$$

Here, we identify each vector in $\mathbb{R}^m$ with an $m \times 1$ column matrix. Thus, the right-hand side of the equation above is interpreted as the product of a 4×5 matrix and a 5×1 column matrix. The product obtained is a 4×1 matrix, which is identified with the corresponding vector in $\mathbb{R}^4$.

3.60 Let $M : X_1 \times \cdots \times X_n \to Z$ be multilinear, and let $T_k : X_k \to X_k$ be linear for each $k = 1, \ldots, n$. Define $U : X_1 \times \cdots \times X_n \to Z$ by

$$U(\mathbf{x}_1, \ldots, \mathbf{x}_n) = M(T_1\mathbf{x}_1, \ldots, T_n\mathbf{x}_n) \quad \text{for all } (\mathbf{x}_1, \ldots, \mathbf{x}_n) \in X_1 \times \cdots \times X_n.$$

Show that U is multilinear.

3.61 Let $T \in ML(X_1 \times \cdots \times X_k, Z)$ and let $B_1, \ldots, B_k$ be nonempty bases for $X_1, \ldots, X_k$, respectively. Show that if $T(\mathbf{b}_1, \ldots, \mathbf{b}_k) = \mathbf{0}$ for all $(\mathbf{b}_1, \ldots, \mathbf{b}_k) \in B_1 \times \cdots \times B_k$, then $T = \mathbf{0}$.

3.62 Let $X_1, \ldots, X_k, Z$ be vector spaces, and let $L : X_1 \times \cdots \times X_k \to Z$ be multilinear. Let Y be any any nonzero vector space, and let $\mathbf{y}_0$ be any nonzero vector in Y. Let $D \subset Y$ with $\mathbf{y}_0 \notin D$ be such that $\{\mathbf{y}_0\} \cup D$ is a basis for Y. Show that there is a multilinear map $S : X_1 \times \cdots \times X_k \times Y \to Z$ such that for all $(\mathbf{x}_1, \ldots, \mathbf{x}_k) \in X_1 \times \cdots \times X_k$,

$$S(\mathbf{x}_1, \ldots, \mathbf{x}_k, \mathbf{u}) = \begin{cases} L(\mathbf{x}_1, \ldots, \mathbf{x}_k) & \text{if } \mathbf{u} = \mathbf{y}_0 \\ \mathbf{0} & \text{if } \mathbf{u} \in D. \end{cases}$$

3.63 Let $X_1, \ldots, X_k, Z$ be vector spaces and suppose that $B_1, \ldots, B_k$ are bases for $X_1, \ldots, X_k$, respectively. Show that

1. Given $\mathbf{b} \in B_1 \times \cdots \times B_k$ and $\mathbf{z} \in Z$, there is a unique multilinear map $T_{\mathbf{b},\mathbf{z}} : X_1 \times \cdots \times X_k \to Z$ such that

$$T_{\mathbf{b},\mathbf{z}}(\mathbf{x}) = \begin{cases} \mathbf{0} & \text{if } \mathbf{x} \in B_1 \times \cdots \times B_k, \mathbf{x} \neq \mathbf{b} \\ \mathbf{z} & \text{if } \mathbf{x} = \mathbf{b}. \end{cases}$$

2. If B is a basis for Z, then a basis for $ML(X_1 \times \cdots \times X_k, Z)$ is the set

$$\{\, T_{\mathbf{b},\mathbf{z}} \mid \mathbf{b} \in B_1 \times \cdots \times B_k, \mathbf{z} \in B \,\},$$

where $T_{\mathbf{b},\mathbf{z}}$ is the unique multilinear map in part 1. It follows that

$$\dim ML(X_1 \times \cdots \times X_k, Z) = (\dim X_1) \cdots (\dim X_k)(\dim Z).$$

3. If $k \geq 2$, then

$$ML(X_1 \times \cdots \times X_k, Z) \sim L(X_1, ML(X_2 \times \cdots \times X_k, Z)).$$

3.64 Give an example of a bilinear map $T : \mathbb{R}^2 \times \mathbb{R}^2 \to \mathbb{R}$ such that the map $S : \mathbb{R} \times \mathbb{R}^3 \to \mathbb{R}$ defined by $S(x, (u, v, w)) = T((x, u), (v, w))$ is not bilinear.

3.65 Find all homogeneous polynomials of degree 3 from $\mathbb{R}^2$ to $\mathbb{R}$.

3.4 INNER PRODUCTS

Definition 3.4.1 Inner products. Let W be a vector space. An *inner product* on W is a function $B : W \times W = W^2 \to \mathbb{R}$ that satisfies the following three conditions for all $\mathbf{x}, \mathbf{y}, \mathbf{z} \in W$ and for all $\alpha, \beta \in \mathbb{R}$.

(1) $B(\mathbf{x}, \mathbf{x}) \geq 0$ for all $\mathbf{x} \in W$ and $B(\mathbf{x}, \mathbf{x}) = \mathbf{0}$ if and only if $\mathbf{x} = 0$.

(2) $B(\mathbf{x}, \mathbf{y}) = B(\mathbf{y}, \mathbf{x})$.

(3) $B(\alpha\mathbf{x} + \beta\mathbf{y}, \mathbf{z}) = \alpha B(\mathbf{x}, \mathbf{z}) + \beta B(\mathbf{y}, \mathbf{z})$.

In working with a particular inner product, it is customary to denote this inner product by $\langle\ ,\ \rangle$ and to write $\langle \mathbf{x}, \mathbf{y} \rangle$ instead of $B(\mathbf{x}, \mathbf{y})$.

Remarks 3.4.2 Inner products and bilinear functions. Properties (2) and (3) of inner products imply that an inner product B is a symmetric bilinear function. Property (1) is referred to as *positive definiteness*. Hence, an inner product on X is a symmetric and positive definite bilinear function $B : X^2 \to \mathbb{R}$.

Definition 3.4.3 Inner product spaces and Euclidean spaces. If B is an inner product on W, then (W, B) is called an *inner product space*. The inner product B is usually understood from the context, and W itself is also called an inner product space. A finite dimensional inner product space is called a *Euclidean space*. As we consider only finite dimensional spaces, all inner product spaces we consider are Euclidean spaces. We assume that Euclidean spaces are nontrivial, i.e., their dimension is at least 1.

Definition 3.4.4 The standard Euclidean space. The standard dot product on $\mathbb{R}^n$ was defined in Example 3.2.4 as

$$\langle \mathbf{x}, \mathbf{y} \rangle = x_1 y_1 + \cdots + x_n y_n \in \mathbb{R},$$

where $\mathbf{x} = (x_1, \ldots, x_n)$ and $\mathbf{y} = (y_1, \ldots, y_n)$. We see that this is an inner product on $\mathbb{R}^n$, so that $\mathbb{R}^n$ becomes a Euclidean space with this inner product. This is the *standard inner product* on $\mathbb{R}^n$.

Definition 3.4.5 Norms. The norm of a vector $\mathbf{x}$ in a Euclidean space W is defined as

$$\|\mathbf{x}\| = \sqrt{\langle \mathbf{x}, \mathbf{x} \rangle}. \tag{3.16}$$

Note that for all $\mathbf{x} \in W$,

$$\|\mathbf{x}\| \geq 0 \text{ and } \|\mathbf{x}\| = 0 \text{ if and only if } \mathbf{x} = \mathbf{0}. \tag{3.17}$$

This property is referred to as *positive definiteness* of the norm. Also, for any $\mathbf{x} \in W$ and $t \in \mathbb{R}$,

$$\|t\mathbf{x}\|^2 = \langle t\mathbf{x},\, t\mathbf{x}\rangle = t^2\langle \mathbf{x},\, \mathbf{x}\rangle = t^2\|\mathbf{x}\|,$$

so that

$$\|t\mathbf{x}\| = |t|\|\mathbf{x}\|.$$

This property of the norm is referred to as *homogeneity*. A third important property of norms, the triangle inequality, is established below as Theorem 3.4.7.

Theorem 3.4.6 The Cauchy-Schwartz inequality. *If* $\mathbf{x}$ *and* $\mathbf{y}$ *are two vectors in a Euclidean space* W*, then* $|\langle \mathbf{x},\, \mathbf{y}\rangle| \le \|\mathbf{x}\|\, \|\mathbf{y}\|$.

Proof. Let $\mathbf{x}, \mathbf{y} \in W$. Then, for all $t \in \mathbb{R}$,

$$0 \le \langle t\mathbf{x} + \mathbf{y},\, t\mathbf{x} + \mathbf{y}\rangle = \|\mathbf{x}\|^2\, t^2 + 2\langle \mathbf{x},\, \mathbf{y}\rangle\, t + \|\mathbf{y}\|^2.$$

The conclusion follows from the following elementary fact: if $A,\ B,\ C \in \mathbb{R}$ and if $At^2 + 2Bt + C \ge 0$ for all $t \in \mathbb{R}$, then $B^2 \le AC$. In not, the quadratic equation would have two distinct real roots, implying the existence of negative values. □

Theorem 3.4.7 The triangle inequality. *Let* W *be a Euclidean space and* $\mathbf{x},\ \mathbf{y} \in W$*. Then* $\|\mathbf{x} + \mathbf{y}\| \le \|\mathbf{x}\| + \|\mathbf{y}\|$.

Proof. We have

$$\begin{aligned}
\|\mathbf{x} + \mathbf{y}\|^2 &= \langle \mathbf{x} + \mathbf{y},\, \mathbf{x} + \mathbf{y}\rangle \\
&= \langle \mathbf{x},\, \mathbf{x}\rangle + 2\langle \mathbf{x},\, \mathbf{y}\rangle + \langle \mathbf{y},\, \mathbf{y}\rangle \\
&= \|\mathbf{x}\|^2 + 2\langle \mathbf{x},\, \mathbf{y}\rangle + \|\mathbf{y}\|^2 \\
&\le (\|\mathbf{x}\| + \|\mathbf{y}\|)^2,
\end{aligned}$$

where the last step follows from the Cauchy-Schwartz inequality. □

Corollary 3.4.8 *If* $\mathbf{x},\ \mathbf{y} \in W$*, then* $|\ \|\mathbf{x}\| - \|\mathbf{y}\|\ | \le \|\mathbf{x} - \mathbf{y}\|$.

Proof. This is left as an exercise. □

Example 3.4.9 Inner products and homogeneous polynomials. For $k = 1, \ldots, n$, let $T_k : X \to X$ be a linear map and let $\langle\ ,\rangle_k$ be an inner product on X. Define $S : X \to \mathbb{R}$ by

$$S\mathbf{x} = \|T_1\mathbf{x}\|_1^2 + \cdots + \|T_n\mathbf{x}\|_n^2 \quad \text{for all } \mathbf{x} \in X.$$

Then S is a homogeneous polynomial of degree 2 on X. To see this, define $U_k : X \times X \rightarrow \mathbb{R}$ by

$$U_k(\mathbf{x}, \mathbf{y}) = \langle T_k\mathbf{x}, T_k\mathbf{y}\rangle_k \quad \text{for all } (\mathbf{x}, \mathbf{y}) \in X \times X.$$

Then U_k is bilinear. Hence, the sum $U = U_1 + \cdots + U_n$ is also a bilinear map from $X \times X$ into $\mathbb{R}$. Hence, the map $\mathbf{x} \mapsto U(\mathbf{x}, \mathbf{x})$ is a homogeneous polynomial of degree 2 on X. Clearly,

$$U(\mathbf{x}, \mathbf{x}) = \sum_{k=1}^{n} U_k(\mathbf{x}, \mathbf{x}) = \sum_{k=1}^{n} \|T_k\mathbf{x}\|_k^2 = S\mathbf{x} \quad \text{for all } \mathbf{x} \in X.$$

Orthogonality

Definition 3.4.10 Orthonormal sets and orthonormal bases. Let W be an inner product space.

1. Two vectors $\mathbf{x}$ and $\mathbf{y}$ in W are said to be *orthogonal* (or perpendicular) if $\langle \mathbf{x}, \mathbf{y}\rangle = 0$. To indicate that $\mathbf{x}$ and $\mathbf{y}$ are orthogonal, we write $\mathbf{x} \perp \mathbf{y}$.
2. Let $A \subset W$. Then A is called an *orthogonal set* if $\mathbf{u} \perp \mathbf{v}$ whenever $\mathbf{u}$ and $\mathbf{v}$ are distinct vectors in A.
3. Let $A \subset W$. Then A is called an *orthonormal set* if $\|\mathbf{u}\| = 1$ for each $\mathbf{u} \in A$ and $\mathbf{u} \perp \mathbf{v}$ whenever $\mathbf{u}$ and $\mathbf{v}$ are distinct members in A.
4. If an orthonormal set is also a basis for W, then it is called an *orthonormal basis* for W.

Example 3.4.11 Let X be an inner product space and $\mathbf{x}, \mathbf{y} \in X$. Then $\langle \mathbf{x}, \mathbf{y}\rangle = 0$ if and only if $\|\mathbf{x} + \alpha\mathbf{y}\|^2 = \|\mathbf{x} - \alpha\mathbf{y}\|^2$ for all scalars α. To see this, let $\mathbf{x}, \mathbf{y}$ be in X and let $t \in \mathbb{R}$. Then

$$\|\mathbf{x} + t\mathbf{y}\|^2 = \langle \mathbf{x} + t\mathbf{y}, \mathbf{x} + t\mathbf{y}\rangle = \|\mathbf{x}\|^2 + 2t\langle \mathbf{x}, \mathbf{y}\rangle + t^2 \|\mathbf{y}\|^2.$$

Thus, $\|\mathbf{x} + \alpha\mathbf{y}\|^2 = \|\mathbf{x} - \alpha\mathbf{y}\|^2$ if and only if $4\alpha\langle \mathbf{x}, \mathbf{y}\rangle = 0$. Hence, if $\alpha \neq 0$, then $\|\mathbf{x} + \alpha\mathbf{y}\|^2 = \|\mathbf{x} - \alpha\mathbf{y}\|^2$ implies that $\langle \mathbf{x}, \mathbf{y}\rangle = 0$. The converse is obvious.

Theorem 3.4.12 Pythagorean Theorem. *Let* $\mathbf{a}$ *and* $\mathbf{b}$ *be two vectors in a Euclidean space. Then* $\mathbf{a} \perp \mathbf{b}$ *if and only if*

$$\|\mathbf{a} + \mathbf{b}\|^2 = \|\mathbf{a}\|^2 + \|\mathbf{b}\|^2. \tag{3.18}$$

Proof. $\|\mathbf{a} + \mathbf{b}\|^2 = \langle \mathbf{a} + \mathbf{b}, \mathbf{a} + \mathbf{b}\rangle = \|\mathbf{a}\|^2 + 2\langle \mathbf{a}, \mathbf{b}\rangle + \|\mathbf{b}\|^2$. □

Orthonormal bases for a Euclidean space have particular importance. We restate their definition separately.

Definition 3.4.13 Orthonormal bases. Let W be a Euclidean space. Let

$$E = \{ \mathbf{e}_1, \ldots, \mathbf{e}_n \}$$

be a basis for W. Then E is called an *orthonormal basis* if

$$\langle \mathbf{e}_i, \mathbf{e}_j \rangle = \begin{cases} 0 & \text{if } i \neq j, \\ 1 & \text{if } i = j. \end{cases}$$

We also write this last condition as $\langle \mathbf{e}_i, \mathbf{e}_j \rangle = \delta_{ij}$.

Example 3.4.14 Let X be a Euclidean space. Let $\{ \mathbf{u}_1, \ldots, \mathbf{u}_n \}$ and $\{ \mathbf{e}_1, \ldots, \mathbf{e}_n \}$ be two orthonormal bases for X. Define $T \in L(X, X)$ by $T\mathbf{e}_i = \mathbf{u}_i$ for $i = 1, \ldots, n$. Let us verify that T is invertible and then compute $T^{-1}\mathbf{e}_i$ in terms of the $\mathbf{e}_j$ s and the $\mathbf{u}_k$ s.

Since $\{\mathbf{u}_1, \ldots, \mathbf{u}_n\}$ is a basis for X and $T\mathbf{e}_j = \mathbf{u}_j$, we have

$$T(X) = \text{Span}\,\{T\mathbf{e}_1, \ldots, T\mathbf{e}_n\} = \text{Span}\,\{\mathbf{u}_1, \ldots, \mathbf{u}_n\} = X.$$

Hence T is invertible. We have

$$\mathbf{e}_j = \langle \mathbf{e}_j, \mathbf{u}_1 \rangle \mathbf{u}_1 + \cdots + \langle \mathbf{e}_j, \mathbf{u}_n \rangle \mathbf{u}_n \quad \text{for all } j = 1, \ldots, n,$$

since $\{\mathbf{u}_1, \ldots, \mathbf{u}_n\}$ is orthonormal. Hence

$$\begin{aligned} T^{-1}\mathbf{e}_j &= \langle \mathbf{e}_j, \mathbf{u}_1 \rangle T^{-1}\mathbf{u}_1 + \cdots + \langle \mathbf{e}_j, \mathbf{u}_n \rangle T^{-1}\mathbf{u}_n \\ &= \langle \mathbf{e}_j, \mathbf{u}_1 \rangle \mathbf{e}_1 + \cdots + \langle \mathbf{e}_j, \mathbf{u}_n \rangle \mathbf{e}_n. \end{aligned}$$

Remark 3.4.15 Inner products in an orthonormal basis. Let

$$E = \{ \mathbf{e}_1, \ldots, \mathbf{e}_n \}$$

be an orthonormal basis for a Euclidean space W. Let

$$\mathbf{u} = u_1\mathbf{e}_1 + \cdots + u_n\mathbf{e}_n \quad \text{and} \quad \mathbf{v} = v_1\mathbf{e}_1 + \cdots + v_n\mathbf{e}_n$$

be two vectors in W. Then we see easily that

$$\langle \mathbf{u}, \mathbf{v} \rangle = u_1v_1 + \cdots + u_nv_n\,.$$

Theorem 3.4.16 *Given any basis A for a vector space W, there is a unique inner product on W with respect to which A is an orthonormal basis.*

Proof. Let $A = \{\mathbf{a}_1, \ldots, \mathbf{a}_n\}$ be a basis for W. If

$$\mathbf{x} = a_1\mathbf{a}_1 + \cdots + a_n\mathbf{a}_n \in W \quad \text{and} \quad \mathbf{y} = b_1\mathbf{a}_1 + \cdots + b_n\mathbf{a}_n \in W,$$

then define

$$\langle \mathbf{x}, \mathbf{y} \rangle = a_1 b_1 + \cdots + a_n b_n .$$

We see that this is an inner product on W. The basis A becomes an orthonormal basis in this inner product. Also, this is the only inner product on W that makes A an orthonormal basis. This follows from the expression of an inner product in terms of an orthonormal basis, as given above in Remark 3.4.15. □

Theorem 3.4.20 below shows that there are orthonormal bases for any Euclidean space. That is, given an inner product on W, we can find an orthonormal basis. Hence any inner product on a finite dimensional vector space is of the form described above, with an appropriate choice of the basis.

Lemma 3.4.17 *An orthonormal set A is a linearly independent set.*

Proof. If $c_1\mathbf{u}_1 + \cdots + c_n\mathbf{u}_n = \mathbf{0}$ where $\mathbf{u}_i$ s are distinct members of A, then $0 = \langle (c_1\mathbf{u}_1 + \cdots + c_n\mathbf{u}_n), \mathbf{u}_i \rangle = c_i$ for all $i = 1, \ldots, n$. □

Lemma 3.4.18 *In an n-dimensional Euclidean space, any orthonormal set with n vectors is an orthonormal basis.*

Proof. Orthonormal sets are linearly independent, by Lemma 3.4.17. Also, in an n-dimensional vector space, any linearly independent set of n vectors is a basis, by the dimension theorem 3.1.14. □

Lemma 3.4.19 Gram-Schmidt process. *Let $A = \{\mathbf{a}_1, \ldots, \mathbf{a}_k\}$ be an orthonormal set in an n-dimensional Euclidean space W. If $k < n$, then there is an $\mathbf{a} \in W$ such that $A' = \{\mathbf{a}_1, \ldots, \mathbf{a}_k, \mathbf{a}\}$ is also an orthonormal set.*

Proof. We have Span $A \neq W$. In fact, A contains fewer than n vectors and therefore cannot be a basis for the n-dimensional space W. Hence there is a $\mathbf{b} \in W$ such that $\mathbf{b} \notin$ Span A. Let

$$\mathbf{b}' = \langle \mathbf{b}, \mathbf{a}_1 \rangle \mathbf{a}_1 + \cdots + \langle \mathbf{b}, \mathbf{a}_k \rangle \mathbf{a}_k$$

and set $\mathbf{c} = \mathbf{b} - \mathbf{b}'$. We see that $\mathbf{c} \perp \mathbf{a}_i$ for all $i = 1, \ldots, n$, since

$$\langle \mathbf{c}, \mathbf{a}_i \rangle = \langle \mathbf{b} - \mathbf{b}', \mathbf{a}_i \rangle = \langle \mathbf{b}, \mathbf{a}_i \rangle - \langle \mathbf{b}', \mathbf{a}_i \rangle = 0.$$

Hence $\{\mathbf{a}_1, \ldots, \mathbf{a}_k, \mathbf{c}\}$ is an orthogonal set. Also, $\mathbf{c} = (\mathbf{b} - \mathbf{b}') \neq \mathbf{0}$, since $\mathbf{b} \notin$ Span A and $\mathbf{b}' \in$ Span A. Now let $\mathbf{a} = (\|\mathbf{c}\|)^{-1}\,\mathbf{c}$. An easy check shows that $A' = \{\mathbf{a}_1, \ldots, \mathbf{a}_k, \mathbf{a}\}$ is an orthonormal set. □

Theorem 3.4.20 Existence of orthonormal bases. *Any Euclidean space X has an orthonormal basis.*

Proof. Let $n = \dim X$, $n \geq 1$ (since we assume that X is nontrivial, as in Definition 3.4.3). Lemma 3.4.18 shows that any orthonormal set with n elements is an orthonormal basis for X. To see the existence of such an orthonormal set, take any nonzero vector $\mathbf{a} \in X$ and set $\mathbf{e}_1 = (\|\mathbf{a}\|)^{-1}\mathbf{a}$. Then $E = \{\mathbf{e}\}$ is an orthonormal set in X. By Lemma 3.4.19 (the Gram-Schmidt process), if $1 < n$, we can find $\mathbf{e}_2$ such that $\{\mathbf{e}_1, \mathbf{e}_2\}$ is also an orthonormal set. Continuing to apply the Gram-Schmidt process, after a finite number of steps, we obtain an orthonormal set with n elements, which must be a basis. □

Theorem 3.4.20 on the existence of an orthonormal basis is important. One of its consequences is another basic result about the representation of linear functions.

Theorem 3.4.21 Representation of linear functions. *Let $f : X \to \mathbb{R}$ be a linear function on a Euclidean space X. Then there is a unique vector $\mathbf{a} \in X$ such that $f(\mathbf{x}) = \langle \mathbf{a}, \mathbf{x} \rangle$ for each $\mathbf{x} \in X$.*

Proof. Let $E = \{\mathbf{e}_1, \ldots, \mathbf{e}_n\}$ be an orthonormal basis for X. Let

$$\mathbf{a} = a_1\mathbf{e}_1 + \cdots + a_n\mathbf{e}_n$$

with $a_j = f(\mathbf{e}_j)$. Let $\mathbf{x} = x_1\mathbf{e}_1 + \cdots + x_n\mathbf{e}_n$ be any vector in X. Then

$$\begin{aligned} f(\mathbf{x}) &= f(x_1\mathbf{e}_1 + \cdots + x_n\mathbf{e}_n) \\ &= x_1 f(\mathbf{e}_1) + \cdots + x_n f(\mathbf{e}_n) \\ &= x_1 a_1 + \cdots + x_n a_n \\ &= \langle \mathbf{a}, \mathbf{x} \rangle . \end{aligned}$$

This shows the existence of an $\mathbf{a} \in X$ such that $f(\mathbf{x}) = \langle \mathbf{a}, \mathbf{x} \rangle$ for all $\mathbf{x} \in X$. To see the uniqueness of $\mathbf{a}$, let $\mathbf{a}' \in X$ be another vector such that $f(\mathbf{x}) = \langle \mathbf{a}', \mathbf{x} \rangle$ for all $\mathbf{x} \in X$. Then

$$\langle \mathbf{a}, \mathbf{x} \rangle - \langle \mathbf{a}', \mathbf{x} \rangle = \langle \mathbf{a} - \mathbf{a}', \mathbf{x} \rangle = 0$$

for all $\mathbf{x} \in X$. Hence $\langle \mathbf{a} - \mathbf{a}', \mathbf{a} - \mathbf{a}' \rangle = 0$, and therefore $\mathbf{a} - \mathbf{a}' = \mathbf{0}$. □

Example 3.4.22 Let B_1, B_2 be inner products on a vector space W. Then for any $\mathbf{a} \in W$, there is an $\mathbf{a}' \in W$ such that

$$B_1(\mathbf{a}, \mathbf{x}) = B_2(\mathbf{a}', \mathbf{x}) \quad \text{for all } \mathbf{x} \in W.$$

To see this, let $\mathbf{a} \in W$. Define $f(\mathbf{x}) = B_1(\mathbf{a}, \mathbf{x})$ for all $\mathbf{x} \in W$. Then $f : W \to \mathbb{R}$ is a linear map on the Euclidean space (W, B_2). Hence, by Theorem 3.4.21, there is an $\mathbf{a}' \in W$ such that

$$B_1(\mathbf{a}, \mathbf{x}) = f(\mathbf{x}) = B_2(\mathbf{a}', \mathbf{x}) \quad \text{for all } \mathbf{x} \in W. \qquad \triangle$$

Example 3.4.23 Let X be a finite dimensional inner product space with inner product $\langle\ ,\ \rangle$. Let us determine all bilinear maps from $X \times X$ into $\mathbb{R}$ in terms of the inner product $\langle\ ,\ \rangle$.

Suppose that $T : X \times X \to \mathbb{R}$ is multilinear. For each $\mathbf{x} \in X$, define $T_{\mathbf{x}}(\mathbf{y}) = T(\mathbf{x}, \mathbf{y})$ for all $\mathbf{y} \in X$. Then each $T_{\mathbf{x}}$ is a linear functional on X, and hence, there exists a unique $\mathbf{z}_{\mathbf{x}}$ in X such that

$$T_{\mathbf{x}}(\mathbf{y}) = \langle \mathbf{y}, \mathbf{z}_{\mathbf{x}} \rangle \quad \text{for all } \mathbf{y} \in X.$$

Since T is also linear in the first variable, it follows that

$$\mathbf{z}_{a\mathbf{u}+b\mathbf{v}} = a\mathbf{z}_{\mathbf{u}} + b\mathbf{z}_{\mathbf{v}}.$$

Define $f : X \to X$ by $f(\mathbf{x}) = \mathbf{z}_{\mathbf{x}}$. Then f is a linear map. Let A be the standard matrix for f with respect to some basis B of X. Then $f(\mathbf{x}) = A[\mathbf{x}]$ for all $\mathbf{x} \in X$, where $[\mathbf{x}]$ is the coordinate vector of $\mathbf{x}$ with respect to B. Thus,

$$T(\mathbf{x}, \mathbf{y}) = \langle \mathbf{y}, A[\mathbf{x}] \rangle \quad \text{for all } (\mathbf{x}, \mathbf{y}) \in X \times X. \qquad \triangle$$

The Cauchy-Schwartz inequality 3.4.6 and the representation theorem 3.4.21 above have the following important consequence.

Theorem 3.4.24 Boundedness of linear functions. *Let W be a Euclidean space. Let $f : W \to \mathbb{R}$ be a linear function. Then there is a constant K such that $|f(\mathbf{w})| \le K\|\mathbf{w}\|$ for all $\mathbf{w} \in W$.*

Proof. The representation theorem 3.4.21 shows that for any linear $f : W \to \mathbb{R}$, there exists $\mathbf{a} \in W$ such that $f(\mathbf{w}) = \langle \mathbf{a}, \mathbf{w} \rangle$ for all $\mathbf{w} \in W$. Hence, by the Cauchy-Schwartz inequality (Theorem 3.4.6),

$$|f(\mathbf{w})| = |\langle \mathbf{a}, \mathbf{w} \rangle| \le \|\mathbf{a}\|\, \|\mathbf{w}\|$$

for all $\mathbf{w} \in W$. Put $K = \|\mathbf{a}\|$. $\square$

Theorem 3.4.25 Boundedness of linear transformations. *Let X and Y be two Euclidean spaces. Let $T : X \to Y$ be a linear transformation. Then there is an M such that $\|T\mathbf{x}\|_Y \le M\|\mathbf{x}\|_X$ for all $\mathbf{x} \in X$.*

Proof. Let $(\mathbf{u}_1, \ldots, \mathbf{u}_n)$ be an orthonormal basis for Y. For each $i = 1, \ldots, n$, let $f_i(\mathbf{x}) = \langle T\mathbf{x}, \mathbf{u}_i \rangle$. Then $f_i : X \to \mathbb{R}$ is a linear function. Theorem 3.4.24 shows that there is a K_i such that $|f_i(\mathbf{x})| \le K_i\|\mathbf{x}\|$ for all $\mathbf{x} \in X$. Hence,

$$\begin{aligned}
\|T\mathbf{x}\| &= \|\langle T\mathbf{x}, \mathbf{u}_1 \rangle \mathbf{u}_1 + \cdots + \langle T\mathbf{x}, \mathbf{u}_n \rangle \mathbf{u}_n\| \\
&= \|f_1(\mathbf{x})\mathbf{u}_1 + \cdots + f_n(\mathbf{x})\mathbf{u}_n\| \\
&\le |f_1(\mathbf{x})| + \cdots + |f_n(\mathbf{x})| \\
&\le (K_1 + \cdots + K_n)\|\mathbf{x}\|.
\end{aligned}$$

The conclusion follows by setting $M = K_1 + \cdots + K_n$. $\square$

Example 3.4.26 Let X be a Euclidean space and $f \in L(X, \mathbb{R})$. Theorem 3.4.24 shows that there exists a constant K for which $|f(\mathbf{x})| \leq K\|\mathbf{x}\|$ for all $\mathbf{x} \in X$. The smallest such K is $\|\mathbf{a}\|$, where $\mathbf{a} \in X$ is such that $f(\mathbf{x}) = \langle \mathbf{a}, \mathbf{x} \rangle$ for all $\mathbf{x} \in X$.

To see this, let K_0 be the smallest K as defined above. The Cauchy-Schwartz inequality shows that $|f(\mathbf{x})| = |\langle \mathbf{a}, \mathbf{x} \rangle| \leq \|\mathbf{a}\| \, \|\mathbf{x}\|$ for all $\mathbf{x} \in X$. Hence $K_0 \leq \|\mathbf{a}\|$. But $|f(\mathbf{a})| = \|\mathbf{a}\| \, \|\mathbf{a}\|$ shows that we cannot have $K_0 < \|\mathbf{a}\|$. Hence $K_0 = \|\mathbf{a}\|$.

Problems

3.66 Let $\langle \ , \rangle$ be an inner product on a vector space X. Let T, S be elements of $L(X, X)$. Define $U : X \times X \to \mathbb{R}$ by

$$U(\mathbf{x}, \mathbf{y}) = \langle S\mathbf{x}, T\mathbf{y} \rangle \quad \text{for all } (\mathbf{x}, \mathbf{y}) \in X \times X.$$

Show that U is bilinear.

3.67 Suppose that X is any finite dimensional vector space. Show that given a homogeneous polynomial $f : X \to \mathbb{R}$ of degree 2 and any inner product $\langle \ , \ \rangle$ on X, there is a linear map $L : X \to X$ such that

$$f(\mathbf{x}) = \langle \mathbf{x}, L\mathbf{x} \rangle \quad \text{for all } \mathbf{x} \in X.$$

3.68 Let $B_1, \ldots, B_k$ be inner products on a vector space X. For any positive scalars $c_1, \ldots, c_k$, show that $c_1 B_1 + \cdots + c_k B_k$ is an inner product on X.

3.69 Let X be an inner product space. Let $\mathbf{x}, \mathbf{y} \in X$ and $\mathbf{x} \neq \mathbf{0}$. Show that $\|\mathbf{x} + \mathbf{y}\| = \|\mathbf{x}\| + \|\mathbf{y}\|$ if and only if there is an $\alpha \geq 0$ such that $\mathbf{y} = \alpha\mathbf{x}$.

3.70 If $\mathbf{x}, \mathbf{y}$ are vectors in an inner product space, show that

$$\|\mathbf{x} + \mathbf{y}\|^2 + \|\mathbf{x} - \mathbf{y}\|^2 = 2(\|\mathbf{x}\|^2 + \|\mathbf{y}\|^2).$$

3.71 Suppose that $\mathbf{u}, \mathbf{v}$ are vectors in an inner product space X and $\|\mathbf{u}\| = \|\mathbf{v}\|$. Show that $\mathbf{u} - \mathbf{v} \perp \mathbf{u} + \mathbf{v}$.

3.72 Suppose that $\{\mathbf{u}_1, \ldots, \mathbf{u}_n\}$ is an orthogonal set of distinct vectors in an inner product space. Show that for all scalars $c_1, \ldots, c_n$, we have

$$\|c_1\mathbf{u}_1 + \cdots + c_n\mathbf{u}_n\|^2 = c_1^2\|\mathbf{u}_1\|^2 + \cdots + c_n^2\|\mathbf{u}_n\|^2.$$

3.73 Let X be a Euclidean space. Let $f \in L(X, \mathbb{R})$. Then, by Theorem 3.4.21, there is a unique $\mathbf{a}_f \in W$ such that $f(\mathbf{x}) = \langle \mathbf{a}_f, \mathbf{x} \rangle$ for all $\mathbf{x} \in X$. Define $T : L(X, \mathbb{R}) \to X$ by $Tf = \mathbf{a}_f$. Show that T is an isomorphism.

3.74 Let B be an orthonormal basis for a Euclidean space X, and let S be a nonempty subset of B with $S \neq B$. Define $T : X \to X$ by

$$T\mathbf{x} = \sum\nolimits_{\mathbf{v} \in B \setminus S} \langle \mathbf{x}, \mathbf{v} \rangle \mathbf{v} \quad \text{for all } \mathbf{x} \in X.$$

Show that $\text{Ker } T = \text{Span } S$.

3.75 Let X be a Euclidean space. Let $f_1, \ldots, f_m$ be in $L(X, \mathbb{R})$, If $m > \dim X$, then show that some f_k is a linear combination of the remaining f_js.

3.76 Let $S = \{\mathbf{u}_1, \ldots, \mathbf{u}_m\}$ be an orthonormal subset in a Euclidean space X. Define $T : X \to \mathbb{R}^m$ by

$$T(\mathbf{x}) = (\langle \mathbf{x}, \mathbf{u}_1 \rangle, \ldots, \langle \mathbf{x}, \mathbf{u}_m \rangle) \quad \text{for all } \mathbf{x} \in \mathbb{R}^m.$$

Show that $\dim \text{Ker } T = \dim X - m$.

3.77 Let X be an inner product space, and let $\mathbf{a} \in X$ be nonzero. Show that for any scalar c, there is an $f \in L(X, \mathbb{R})$ and a $\mathbf{u} \in X$ such that

$$\{\, \mathbf{x} \in X \mid \langle \mathbf{x}, \mathbf{a} \rangle = c \,\} = \text{Ker } f + \mathbf{u}.$$

3.5 ORTHOGONAL PROJECTIONS

Recall the definition of orthogonality given in Definition 3.4.10: $\mathbf{u}$ and $\mathbf{v}$ are orthogonal, written $\mathbf{u} \perp \mathbf{v}$, if $\langle \mathbf{u}, \mathbf{v} \rangle = 0$.

Definition 3.5.1 Orthogonal subspaces. Let U and V be two subspaces of a Euclidean space X. Then U and V are said to be *orthogonal* to each other if $\mathbf{u} \perp \mathbf{v}$ for all $\mathbf{u} \in U$ and $\mathbf{v} \in V$. We write $U \perp V$ to indicate that U and V are orthogonal to each other.

Lemma 3.5.2 *If $U \perp V$, then $U \cap V = \{\mathbf{0}\}$.*

Proof. Let $\mathbf{w} \in U \cap V$. Then $\langle \mathbf{w}, \mathbf{w} \rangle = 0$. Hence $\mathbf{w} = \mathbf{0}$. □

Corollary 3.5.3 *If $U \perp V$, then U and V are complementary in $W = U + V$.*

Proof. By Lemma 3.1.20, U and V are complementary in W if $U + V = W$ and if $U \cap V = \{\mathbf{0}\}$. Hence the conclusion follows from Lemma 3.5.2. □

Definition 3.5.4 Orthogonal complements. If $U \perp V$ and $U + V = X$, then U and V are called the *orthogonal complements* of each other. In this case the pair (U, V) is called an *orthogonal decomposition of* X. Corollary 3.5.3 shows that orthogonal complements are indeed complementary subspaces in X. Hence U and V are orthogonal complements if and only if $U \perp V$ and $X = U \oplus V$.

Remark 3.5.5 Uniqueness of the orthogonal complements. In general, a subspace U of X has many complementary subspaces. If X is a Euclidean space, then the orthogonal complement of U is a special complementary subspace. It is uniquely determined by U, as the following theorem shows.

Theorem 3.5.6 *Let U be a subspace of a Euclidean space X and let*

$$\begin{aligned} V &= \{\mathbf{v} \in X \mid \mathbf{v} \perp \mathbf{u} \textit{ for all } \mathbf{u} \in U\} \\ &= \{\mathbf{v} \in X \mid \langle \mathbf{v}, \mathbf{u} \rangle = 0 \textit{ for all } \mathbf{u} \in U\}. \end{aligned}$$

Then U and V are orthogonal complements. Also, the orthogonal complement of U is uniquely defined by U.

Proof. We have to show that V is a subspace, $U \perp V$, and $X = U + V$. Let $\mathbf{v}, \mathbf{v}' \in V$ and $r, r' \in \mathbb{R}$. If $\mathbf{u} \in U$, then $\langle \mathbf{v}, \mathbf{u} \rangle = 0$ and $\langle \mathbf{v}', \mathbf{u} \rangle = 0$. Hence, if $\mathbf{u} \in U$, then

$$\langle r\mathbf{v} + r'\mathbf{v}', \mathbf{u} \rangle = r\langle \mathbf{v}, \mathbf{u} \rangle = r'\langle \mathbf{v}', \mathbf{u} \rangle = 0.$$

This shows that $r\mathbf{v} + r'\mathbf{v}' \in V$. Hence V is a subspace of X. Also, if $\mathbf{v} \in V$ and $\mathbf{u} \in U$, then $\langle \mathbf{v}, \mathbf{u} \rangle = 0$, and therefore $\mathbf{u} \perp \mathbf{v}$. Hence $U \perp V$.

Now we show that $X = U + V$. Let $\{\mathbf{e}_1, \ldots, \mathbf{e}_k\}$ be an orthonormal basis for U. For each $\mathbf{x} \in X$, define

$$\mathbf{a} = \langle \mathbf{x}, \mathbf{e}_1 \rangle \mathbf{e}_1 + \cdots + \langle \mathbf{x}, \mathbf{e}_k \rangle \mathbf{e}_k,$$

and let $\mathbf{b} = \mathbf{x} - \mathbf{a}$. We see that $\langle \mathbf{b}, \mathbf{e}_i \rangle = \langle \mathbf{x} - \mathbf{a}, \mathbf{e}_i \rangle = \langle \mathbf{x}, \mathbf{e}_i \rangle - \langle \mathbf{x}, \mathbf{e}_i \rangle = 0$ for all $i = 1, \ldots, k$. But any $\mathbf{u} \in U$ is a linear combination $\mathbf{u} = u_1\mathbf{e}_1 + \cdots + u_k\mathbf{e}_k$ of $\mathbf{e}_i$ s. Hence $\langle \mathbf{b}, \mathbf{u} \rangle = 0$ for all $\mathbf{u} \in U$. Therefore $\mathbf{b} \in V$. Also, clearly, $\mathbf{a} \in U$ and $\mathbf{x} = \mathbf{a} + \mathbf{b}$. Hence $X = U + V$.

To see the uniqueness of V, let V' be another orthogonal complement of U. If $\mathbf{v}' \in V'$, then $\mathbf{v}' \perp \mathbf{u}$ for all $\mathbf{u} \in U$. Hence $\mathbf{v}' \in V$. Therefore $V' \subset V$. Similarly, we obtain $V \subset V'$. Hence $V' = V$. □

Notation 3.5.7 Orthogonal complements. One denotes the orthogonal complement of U as $U^{\perp}$. Note that $(U^{\perp})^{\perp} = U$ by the preceding theorem.

Example 3.5.8 Application to linear functionals. Let E and F be subspaces of a Euclidean space W. Let $U = \text{Span}\,(E \cup F)$. Then $E \subset U$ and $F \subset U$. Hence, $U^{\perp} \subset E^{\perp}$ and $U^{\perp} \subset F^{\perp}$. Thus, $U^{\perp} \subset E^{\perp} \cap F^{\perp}$. So,

$$(E^{\perp} \cap F^{\perp})^{\perp} \subset (U^{\perp})^{\perp} = U. \tag{3.19}$$

We will next show that if $f_1, \ldots, f_m$ and g are linear functionals on X, then

$$g \in \text{Span}\,\{f_1, \ldots, f_m\} \tag{3.20}$$

if and only if

$$\bigcap_{k=1}^{m} \text{Ker}\ f_k \subset \text{Ker}\ g. \tag{3.21}$$

It is clear that (3.20) implies (3.21). Conversely, assume (3.21). By Theorem 3.4.21, there are $\mathbf{a}_1, \ldots, \mathbf{a}_m$ and $\mathbf{b}$ in X such that $\text{Ker}\ f_k = (\text{Span}\,\{\mathbf{a}_k\})^{\perp}$ and $\text{Ker}\ g = (\text{Span}\,\{\mathbf{b}\})^{\perp}$. Put $E_k = \text{Span}\,\{\mathbf{a}_k\}$ and $U = \text{Span}\,\{\mathbf{b}\}$. Then by our assumption (3.21),

$$E_1^{\perp} \cap \cdots \cap E_m^{\perp} \subset U^{\perp}.$$

Hence, by (3.19),

$$U = (U^{\perp})^{\perp} \subset (E_1^{\perp} \cap \cdots \cap E_m^{\perp})^{\perp} \subset \text{Span}\,(E_1 \cup \cdots \cup E_m).$$

This implies that $\mathbf{b} \in \text{Span}\,\{\mathbf{a}_1, \ldots, \mathbf{a}_m\}$, and thus, $g \in \text{Span}\,\{f_1, \ldots, f_m\}$.

Definition 3.5.9 Orthogonal projections. If U and V are orthogonal complements in a Euclidean space X, then they are also complementary subspaces in X. Hence Theorem 3.1.37 shows that they are the ranges of complementary projections P and Q. In this case, these projections are called the *orthogonal projections* on U and V, respectively. Hence, if P is the orthogonal projection on U, then $U = \text{Range}\ P = PX$ and $V = \text{Range}\ Q = (I - P)X$.

As mentioned in Remarks 3.1.40, a subspace U of X does not determine a projection on U. Such a projection is determined only after a complementary space V of U is specified. In a Euclidean space X, every subspace U has a distinguished complement $V = U^{\perp}$. Hence, in Euclidean spaces, every subspace U specifies a distinguished projection, namely, the orthogonal projection on U.

Theorem 3.5.10 *Let $E = \{ \mathbf{e}_1, \ldots, \mathbf{e}_k \}$ be an orthonormal basis for a subspace U of a Euclidean space X. Define $P : X \to X$ by*

$$P\mathbf{x} = \langle \mathbf{x}, \mathbf{e}_1 \rangle \mathbf{e}_1 + \cdots + \langle \mathbf{x}, \mathbf{e}_k \rangle \mathbf{e}_k.$$

Then $P : X \to X$ is the orthogonal projection on U. More explicitly, P is a projection with $U = PX$ and $V = (I - P)X$ is the orthogonal complement of U.

Proof. We see that $P\mathbf{x} \in U$ for all $\mathbf{x} \in X$ and that $P\mathbf{u} = \mathbf{u}$ for all $\mathbf{u} \in U$. Hence $P^2\mathbf{x} = P(P\mathbf{x}) = P\mathbf{x}$ for all $\mathbf{x} \in X$ and therefore $P^2 = P$. This shows that P is a projection and that $U = \text{Range } P$. Also, as in the proof of Theorem 3.5.6 above, we see that $\mathbf{x} - P\mathbf{x}$ is orthogonal to each $\mathbf{u} \in U$.

Let $Q = I - P$ be the complementary projection and $V = \text{Range } Q$. To complete the proof, we have to show that V is the orthogonal complement of U. If $\mathbf{v} \in V$, then $Q\mathbf{v} = (I - P)\mathbf{v} = \mathbf{v} - P\mathbf{v}$ is orthogonal to each $\mathbf{u} \in U$, as observed above. Therefore $V = \text{Range } (I - P)$ is indeed the orthogonal complement of U. $\square$

Remark 3.5.11 The projection $P : X \to X$ in Theorem 3.5.10 above is defined in terms of a basis E for U. The properties of P obtained in that theorem show that P is uniquely determined in terms of U. Hence P is independent of the choice of the basis E for U. Another important property is given in Theorem 3.5.12 below. This property also shows that P is determined by U alone. Geometrically, this property means that the point $P\mathbf{x}$ in U is the closest point to $\mathbf{x}$ among the points in U.

Theorem 3.5.12 *Let U be a subspace of a Euclidean space X. Let $P : X \to X$ be the orthogonal projection on U. Let $\mathbf{x} \in X$. Then*

$$\|\mathbf{x} - P\mathbf{x}\| \le \|\mathbf{x} - \mathbf{a}\| \textit{ for all } \mathbf{a} \in U.$$

Proof. Let $\mathbf{v} = \mathbf{x} - P\mathbf{x}$. Then $\langle \mathbf{v}, \mathbf{u} \rangle = 0$ for all $\mathbf{u} \in U$ by Theorem 3.5.10. Now $\mathbf{x} - \mathbf{a} = (\mathbf{x} - P\mathbf{x}) + (P\mathbf{x} - \mathbf{a}) = \mathbf{v} + \mathbf{u}$, where $\mathbf{u} = (P\mathbf{x} - \mathbf{a}) \in U$. Hence $\mathbf{v} \perp \mathbf{u}$ and $\|\mathbf{v} + \mathbf{u}\|^2 = \|\mathbf{v}\|^2 + \|\mathbf{u}\|^2$ by the Pythagorean Theorem 3.4.12. Consequently,

$$\|\mathbf{x} - \mathbf{a}\|^2 = \|\mathbf{v} + \mathbf{u}\|^2 = \|\mathbf{v}\|^2 + \|\mathbf{u}\|^2 \ge \|\mathbf{v}\|^2 = \|\mathbf{x} - P\mathbf{x}\|^2. \quad \square$$

Example 3.5.13 Let $X = \mathbb{R}^3$ be the xyz-space and

$$U = \{ (x, y, z) \in \mathbb{R}^3 \mid x + y + z = 0 \}.$$

Show that U is a subspace of X and find the orthogonal projection P on U.

First solution. It is obvious that U is a subspace. To find P, we need to find an orthonormal basis for U. One example of a nonzero vector in U is $\mathbf{a} = (1,\ -1,\ 0)$. Another vector in U not in the span of $\mathbf{a}$ is $\mathbf{c} = (1,\ 0,\ -1)$. But we want to find a vector in U orthogonal to $\mathbf{a}$, in order to facilitate the passage to an orthonormal basis. By inspection, we see that $\mathbf{b} = (1,\ 1,\ -2)$ is such a vector. The vectors $\mathbf{a}$ and $\mathbf{b}$ are nonzero and orthogonal to each other. Hence they are linearly independent. They must be a basis for U since $(\dim U) = 2$. In fact $(\dim U) \geq 2$ since U contains two linearly independent vectors. Also $(\dim U) < 3$, since it is clear that $U \neq \mathbb{R}^3$. Hence

$$\mathbf{e}_1 = (\|\mathbf{a}\|)^{-1}\mathbf{a} = 2^{-1/2}(1,\ -1,\ 0) \text{ and } \mathbf{e}_2 = (\|\mathbf{b}\|)^{-1}\mathbf{b} = 6^{-1/2}(1,\ 1,\ -2)$$

constitute an orthonormal basis for U. Therefore, for all $\mathbf{r} = (x,\ y,\ z) \in \mathbb{R}^3$,

$$\begin{aligned} P\mathbf{r} &= \langle \mathbf{r},\ \mathbf{e}_1\rangle\mathbf{e}_1 + \langle \mathbf{r},\ \mathbf{e}_2\rangle\mathbf{e}_2 \text{ and} \\ P(x,\ y,\ z) &= (1/2)(x-y)(1,\ -1,\ 0) + (1/6)(x+y-2z)(1,\ 1,\ -2) \\ &= (1/3)(2x-y-z,\ -x+2y-z,\ -x-y+2z). \end{aligned}$$

To verify that this is the correct answer, it is enough to show that $P\mathbf{r} \in U$ and $(\mathbf{r} - P\mathbf{r}) \perp U$, that is, $(\mathbf{r} - P\mathbf{r}) \perp \mathbf{u}$ for all $\mathbf{u} \in U$. First, we see that $P\mathbf{r} \in U$, since the sum of the coordinates of $P(x,\ y,\ z)$ is

$$(1/3)((2x-y-z) + (-x+2y-z) + (-x-y+2z)) = 0.$$

To see that $(\mathbf{r} - P\mathbf{r}) \perp U$, we compute $(\mathbf{r} - P\mathbf{r})$ as

$$((x,\ y,\ z) - P(x,\ y,\ z)) = (1/3)(x+y+z)(1,\ 1,\ 1).$$

We see that $(1,\ 1,\ 1) \perp U$. In fact, if $(u_1,\ u_2,\ u_3) \in U$, then $u_1 + u_2 + u_3 = 0$ and therefore $\langle (1,\ 1,\ 1),\ (u_1,\ u_2,\ u_3)\rangle = u_1 + u_2 + u_3 = 0$. Hence $(\mathbf{r} - P\mathbf{r}) \perp U$.

Second solution. Instead of P, we may find the complementary projection $Q = (I - P)$. This is the orthogonal projection on the orthogonal complement V of U. Since $(\dim U) = 2$, we have $(\dim V) = 1$. Hence Q should be easier to find. One example of a nonzero vector orthogonal to U is $\mathbf{g} = (1,\ 1,\ 1)$. In fact, U is the set of $\mathbf{r} \in \mathbb{R}^3$ such that $\langle \mathbf{r},\ \mathbf{g}\rangle = 0$. Hence V is the space spanned by $\mathbf{g}$. An orthonormal basis for V is

$$\mathbf{e} = (\|\mathbf{g}\|)^{-1}\mathbf{g} = 3^{-1/2}(1,\ 1,\ 1).$$

Therefore $Q\mathbf{r} = \langle \mathbf{r},\ \mathbf{e}\rangle\mathbf{e}$ is

$$Q(x,\ y,\ z) = (1/3)(x+y+z)(1,\ 1,\ 1).$$

Then we obtain

$$P(x,\ y,\ z) = (I - Q)((x,\ y,\ z) = ((x,\ y,\ z) - (1/3)(x+y+z)(1,\ 1,\ 1).$$

This agrees with the previous result, as already verified. △

Problems

3.78 Let X be a Euclidean space. Let $U_1, \ldots, U_m$ be subspaces of X such that $U_i \perp U_j$ for all $i \neq j$ and $W = U_1 + \cdots + U_m$. Let P_i be the orthogonal projection on U_i. Show that $I = P_1 + \cdots + P_m$ and

$$\|\mathbf{x}\|^2 = \|P_1\mathbf{x}\|^2 + \cdots + \|P_m\mathbf{x}\|^2 \quad \text{for all } \mathbf{x} \in X.$$

3.79 Let $(W, \langle\ ,\ \rangle)$ be a Euclidean space, and let $\mathbf{a}$ be a nonzero vector in W. Let $c \in \mathbb{R}$ and let $E = \{\, \mathbf{w} \in W \mid \langle \mathbf{w}, \mathbf{a} \rangle = c \,\}$. Show that

$$\min_{\mathbf{e} \in E} \|\mathbf{w}_0 - \mathbf{e}\| = \frac{|\langle \mathbf{w}_0, \mathbf{a} \rangle - c|}{\|\mathbf{a}\|} \quad \text{for all } \mathbf{w}_0 \in W.$$

3.80 Let U, V be subspaces of an inner product space. Show that if $U \subset V$, then $V^\perp \subset U^\perp$. Hence, deduce that $U^\perp \cap V^\perp \subset (U \cap V)^\perp$.

3.81 Let U and V be subspaces of a Euclidean space and assume that $U \cap V = \{\mathbf{0}\}$. Is it true that $U^\perp \cap V^\perp = \{\mathbf{0}\}$? Is it true that $(U + V)^\perp = U^\perp + V^\perp$?

3.82 Let U be a subspace of X with $\dim U = (\dim X) - 1$. Show that U is the kernel of a nonzero $f \in L(X, \mathbb{R})$.

3.83 Let $\mathbf{u}_1, \ldots, \mathbf{u}_m$ be distinct vectors in a Euclidean space X. Assume that $\mathbf{u}_i \perp \mathbf{u}_j$ for all $i \neq j$. Let $\mathbf{a} \in X$. Let S be the set of all numbers of the form

$$\left\| \mathbf{a} - \sum_{k=1}^{m} c_k \mathbf{u}_k \right\|,$$

where $c_1, \ldots, c_m$ range over all real numbers. Find $\inf S$.

3.84 Let U be a subspace in a Euclidean space X. Let $\mathbf{c} \in X$ and

$$E = \mathbf{c} + U = \{\, \mathbf{c} + \mathbf{u} \mid \mathbf{u} \in U \,\}.$$

Find $\inf_{\mathbf{e} \in E} \|\mathbf{b} - \mathbf{e}\|$ for each $\mathbf{b} \in X$.

3.85 Let $E = \{\, \mathbf{e}_1, \ldots, \mathbf{e}_k\}$ and $U = \{\, \mathbf{u}_1, \ldots, \mathbf{u}_k\}$ be orthonormal subsets of a Euclidean space X such that Span E = Span U. Is it true that

$$\max_{\mathbf{x} \in X, \|\mathbf{x}\|=1} |\langle \mathbf{x}, \mathbf{e}_1 \rangle|^2 + \cdots + |\langle \mathbf{x}, \mathbf{e}_k \rangle|^2 = \max_{\mathbf{x} \in X, \|\mathbf{x}\|=1} |\langle \mathbf{x}, \mathbf{u}_1 \rangle|^2 + \cdots + |\langle \mathbf{x}, \mathbf{u}_k \rangle|^2?$$

3.86 Let X be a Euclidean space with an orthonormal basis $\{\mathbf{u}_1, \ldots, \mathbf{u}_m\}$. Let $1 \leq k < m$. Show that for each $\mathbf{x} \in X$, there is a $\mathbf{y} \in X$ such that

$$\mathbf{x} = \langle \mathbf{x}, \mathbf{e}_1 \rangle \mathbf{e}_1 + \cdots + \langle \mathbf{x}, \mathbf{e}_k \rangle \mathbf{e}_k + \langle \mathbf{y}, \mathbf{e}_{k+1} \rangle \mathbf{e}_{k+1} + \cdots + \langle \mathbf{y}, \mathbf{e}_m \rangle \mathbf{e}_m.$$

3.87 Let $T : X \to X$ be an orthogonal projection. Is it true that

$$\langle T\mathbf{x}, \mathbf{y} \rangle = \langle \mathbf{x}, T\mathbf{y} \rangle \quad \text{for all } \mathbf{x}, \mathbf{y} \text{ in } X?$$

3.6 SPECTRAL THEOREM

The spectral theorem is of central importance in linear algebra. It summarizes many properties of linear transformations on Euclidean spaces. The proof of the spectral theorem that we shall develop in this section is simple, although not elementary. The usual route to the spectral theorem proceeds via the fundamental theorem of algebra, which is a familiar result but one that requires a good deal of preliminary work. Our alternative approach relies upon the Bolzano-Weierstrass theorem in Euclidean spaces, Theorem 4.2.9. (The proof of that result, in chapter 4, is entirely independent of the ideas developed in this section.)

We will use the spectral theorem mainly for two purposes. First, the theorem gives us a geometrical picture of orthogonal projections between the subspaces of a Euclidean space. Second, the theorem allows us to provide a direct geometrical proof of a basic fact about the effect of a linear transformation upon the volume of a set: when a linear transformation maps one Euclidean space into another, the volumes of the images of sets under that transformation are multiplied by the absolute value of its determinant. For both of these purposes, an alternate formulation of the spectral theorem is convenient. This formulation is stated below as Theorem 3.6.4. It is equivalent to the standard version, Theorem 3.6.12. The standard version of the spectral theorem is not used in this book.

Eigenbases

The property of linear maps that we want to show is the following. For any linear map $T : X \to Y$ between two Euclidean spaces, there is an orthonormal basis for X that is mapped to an orthogonal set of vectors in Y. We will call such a basis an *eigenbasis for* T. The key step in the proof of this result is Lemma 3.6.1, which states that any linear transformation between Euclidean spaces attains a maximum norm on the unit ball. The proof depends crucially upon Theorem 4.2.9, the Bolzano-Weierstrass theorem in Euclidean spaces. After convergence in Euclidean spaces is discussed, we will restate Lemma 3.6.1 as Lemma 4.5.45 with a short proof.

Lemma 3.6.1 *Let X and Y be Euclidean spaces. Let $T : X \to Y$ be a linear transformation. Then there is a unit vector $\mathbf{e} \in X$ such that $\|T\mathbf{x}\| \leq \|T\mathbf{e}\|$ for all unit vectors $\mathbf{x} \in X$.*

Proof. By the boundedness of linear transformations, Theorem 3.4.25, there is a number M such that $\|T\mathbf{x}\| \leq M\|\mathbf{x}\|$ for all $\mathbf{x} \in X$. Hence, $B = \{\, \|T\mathbf{x}\| \mid \|\mathbf{x}\| = 1 \,\}$ is a bounded set of numbers. If $\alpha = \sup B$, then there is a sequence $\mathbf{x}_n \in X$ such that $\|\mathbf{x}_n\| = 1$ and such that $\lim_n \|T\mathbf{x}_n\| = \alpha$. By the Bolzano-Weierstrass theorem, Theorem 4.2.9, there is a subsequence $\mathbf{x}_{n_k}$ and a vector $\mathbf{e} \in X$ such that $\lim_k \|\mathbf{x}_{n_k} - \mathbf{e}\| = 0$. Then $\lim_k |\, \|\mathbf{x}_{n_k}\| - \|\mathbf{e}\| \,| = |\, 1 - \|\mathbf{e}\| \,| = 0$ and

$$\lim_k |\, \|T\mathbf{x}_{n_k}\| - \|T\mathbf{e}\| \,| \leq \|T\mathbf{x}_{n_k} - T\mathbf{e}\| \leq M\|\mathbf{x}_{n_k} - \mathbf{e}\| = 0,$$

by Corollary 3.4.8. It follows that $\|\mathbf{e}\| = 1$ and $\|T\mathbf{e}\| = \alpha$. Therefore $\|T\mathbf{x}\| \leq \|T\mathbf{e}\|$ for all unit vectors $\mathbf{x} \in X$. □

Lemma 3.6.2 *Let $T : X \to Y$ be a linear map between two Euclidean spaces. Let $\mathbf{e} \in X$ be a unit vector such that $\|T\mathbf{x}\| \leq \|T\mathbf{e}\|$ for all unit vectors $\mathbf{x} \in X$. Then $T\mathbf{x} \perp T\mathbf{e}$ in Y whenever $\mathbf{x} \perp \mathbf{e}$ in X.*

Proof. Let $\mathbf{u}$ be a unit vector in X and $\mathbf{u} \perp \mathbf{e}$. Let

$$A = \|T\mathbf{e}\|^2,\ B = \langle T\mathbf{e},\, T\mathbf{u}\rangle, \text{ and } C = \|T\mathbf{u}\|^2.$$

An easy check shows that $\mathbf{v} = \cos t\,\mathbf{u} + \sin t\,\mathbf{e}$ is a unit vector in X. Hence, $\|T\mathbf{v}\|^2 \leq A$ by the hypothesis about $\mathbf{e}$. Therefore

$$\begin{aligned} \|T\mathbf{v}\|^2 &= \langle \cos t\,T\mathbf{u} + \sin t\,T\mathbf{e},\, \cos t\,T\mathbf{u} + \sin t\,T\mathbf{e}\rangle \\ &= C\cos^2 t + 2B\sin t\cos t + A\sin^2 t \leq A. \end{aligned}$$

It follows that

$$2B\tan t \quad \leq \quad (A - C)$$

for all $t \in (-\pi/2,\, \pi/2)$. This implies that $B = \langle T\mathbf{e},\, T\mathbf{u}\rangle = 0$. □

Definition 3.6.3 Eigenbases. Let $T : X \to Y$ be a linear transformation between two Euclidean spaces. An *eigenbasis for T* is any orthonormal basis

$$\mathbb{E} = \{\, \mathbf{e}_1,\, \ldots,\, \mathbf{e}_n \}$$

for X such that

$$\langle T\mathbf{e}_i,\, T\mathbf{e}_j\rangle_Y = 0 \text{ for } i \neq j.$$

Theorem 3.6.4 Spectral theorem (eigenbasis version). *Every linear transformation $T : X \to Y$ between two Euclidean spaces has an eigenbasis.*

Proof. Let $T : X \to Y$ be a linear map. We proceed by induction on $n = \dim X$. If $n = 1$, then either of the two unit vectors in X forms an eigenbasis for T. Assume that the result is true for n-dimensional spaces, and let X be an $n + 1$-dimensional space. Use Lemma 3.6.1 to find a unit vector $\mathbf{e}_0 \in X$ such that $\|T\mathbf{u}\| \leq \|T\mathbf{e}_0\|$ for all unit vectors $\mathbf{u} \in X$. Let X_0 be the orthogonal complement of the space spanned by $\mathbf{e}_0$. Lemma 3.6.2 shows that $T\mathbf{e} \perp TX_0$. Since X_0 is an n-dimensional space, we may use the inductive hypothesis to find an orthonormal basis $\mathbb{E}_0 = \{\mathbf{e}_1, \ldots, \mathbf{e}_n\}$ for X_0 such that $\langle T\mathbf{e}_i, T\mathbf{e}_j\rangle = 0$ whenever $i, j = 1, \ldots, n$ and $i \neq j$. But then $\mathbb{E} = \{\mathbf{e}_0, \mathbf{e}_1, \ldots, \mathbf{e}_n\}$ is an eigenbasis for T. □

Self-Adjoint Transformations

The standard version of the spectral theorem is about *self-adjoint transformations.* These are linear transformations $T : X \to X$ such that $\langle T\mathbf{x}, \mathbf{y}\rangle = \langle \mathbf{x}, T\mathbf{y}\rangle$ for all $\mathbf{x}, \mathbf{y} \in X$. They are also defined in terms of *the adjoint transformation.*

Theorem 3.6.5 *Let $T : X \to Y$ be a linear transformation between two Euclidean spaces. Then there is a unique linear transformation $T^* : Y \to X$ such that*

$$\langle T\mathbf{x}, \mathbf{y}\rangle_Y = \langle \mathbf{x}, T^*\mathbf{y}\rangle_X \textit{ for all } \mathbf{x} \in X \textit{ and } \mathbf{y} \in Y.$$

Proof. Let $\mathbf{y} \in Y$ be fixed. Then $f(\mathbf{x}) = \langle T\mathbf{x}, \mathbf{y}\rangle_Y$, $\mathbf{x} \in X$, defines a linear map $f : X \to \mathbb{R}$. By Theorem 3.4.21, there is a unique vector $\mathbf{u} \in X$ such that

$$f(\mathbf{x}) = \langle T\mathbf{x}, \mathbf{y}\rangle_Y = \langle \mathbf{x}, \mathbf{u}\rangle_X, \quad \text{for all } \mathbf{x} \in X.$$

This defines a map $S : Y \to X$ such that

$$\langle T\mathbf{x}, \mathbf{y}\rangle_Y = \langle \mathbf{x}, S(\mathbf{y})\rangle_X \text{ for all } \mathbf{x} \in X, \mathbf{y} \in Y.$$

It remains to show that $S : Y \to X$ is linear. Let $\alpha, \beta \in \mathbb{R}$, $\mathbf{u}, \mathbf{v} \in Y$. Then

$$\begin{aligned}
\langle \mathbf{x}, S(\alpha\mathbf{u} + \beta\mathbf{v})\rangle &= \langle T\mathbf{x}, \alpha\mathbf{u} + \beta\mathbf{v}\rangle_X = \langle T\mathbf{x}, \alpha\mathbf{u}\rangle_X + \langle T\mathbf{x}, \beta\mathbf{v}\rangle_X \\
&= \alpha\langle T\mathbf{x}, \mathbf{u}\rangle_X + \beta\langle T\mathbf{x}, \mathbf{v}\rangle_X \\
&= \alpha\langle \mathbf{x}, S(\mathbf{u})\rangle_Y + \beta\langle \mathbf{x}, S(\mathbf{v})\rangle_Y \\
&= \langle \mathbf{x}, \alpha S(\mathbf{u}) + \beta S(\mathbf{v})\rangle_Y
\end{aligned}$$

for all $\mathbf{x} \in X$. Hence $S(\alpha\mathbf{u} + \beta\mathbf{v}) = \alpha S(\mathbf{u}) + \beta S(\mathbf{v})$. □

Definition 3.6.6 The adjoint transformation. The linear transformation S defined by Theorem 3.6.5 is called the *adjoint transformation* of $T : X \to Y$. It is denoted by $T^* : Y \to X$.

Definition 3.6.7 Self-adjoint transformations. A linear transformation $T : X \to X$ is called a *self-adjoint transformation* if $T = T^*$.

Lemma 3.6.8 *Let $\mathbb{E}$ be an eigenbasis for $T : X \to Y$. If $\mathbf{e}_i \in \mathbb{E}$, then*

$$T^*T\mathbf{e}_i = \mu_i \mathbf{e}_i \text{ where } \mu_i = ||T\mathbf{e}_i||^2.$$

Proof. We have $\langle \mathbf{e}_j, T^*T\mathbf{e}_i \rangle_X = \langle T\mathbf{e}_j, T\mathbf{e}_i \rangle_Y = ||T\mathbf{e}_i||^2 \delta_{ij}$. $\square$

Definition 3.6.9 Eigenvectors and eigenvalues. Let X be a Euclidean space and let $S : X \to X$ be a linear transformation. A vector $\mathbf{e} \in X$ is called an *eigenvector* of S if $\mathbf{e} \neq \mathbf{0}$ and if $S\mathbf{e} = \lambda \mathbf{e}$ with $\lambda \in \mathbb{R}$. The number λ is called the *eigenvalue* of the eigenvector $\mathbf{e} \in X$.

Lemma 3.6.10 *Every self-adjoint transformation $S : X \to X$ has an eigenvector.*

Proof. Let $\mathbf{e}$ be a vector in an eigenbasis of S. Let $\alpha = ||S\mathbf{e}||$. If $S\mathbf{e} = -\alpha\mathbf{e}$ then $\mathbf{e}$ is an eigenvector with the eigenvalue $\lambda = -\alpha$. Otherwise, $\mathbf{u} = (\alpha\mathbf{e} + S\mathbf{e}) \neq \mathbf{0}$. Lemma 3.6.8 shows that $S^2\mathbf{e} = \alpha^2\mathbf{e}$ since $S^* = S$. Hence, $S(\alpha\mathbf{e} + S\mathbf{e}) = \alpha(\alpha\mathbf{e} + S\mathbf{e})$. Therefore, $\mathbf{u}$ is an eigenvector of S with the eigenvalue $\lambda = \alpha$. $\square$

Lemma 3.6.11 *Let $S : X \to X$ be a self-adjoint transformation. Let $\mathbf{u}$ be an eigenvector of S. If $\mathbf{v} \perp \mathbf{u}$, then $S\mathbf{v} \perp \mathbf{u}$.*

Proof. We have $\langle \mathbf{v}, \mathbf{u} \rangle = 0$. Let $S\mathbf{u} = \lambda\mathbf{u}$. Hence,

$$\langle S\mathbf{v}, \mathbf{u} \rangle = \langle \mathbf{v}, S\mathbf{u} \rangle = \lambda \langle \mathbf{v}, \mathbf{u} \rangle = 0$$

shows that $S\mathbf{v} \perp \mathbf{u}$. $\square$

Theorem 3.6.12 Spectral theorem. *Let $S : X \to X$ be a self-adjoint transformation. Then X has an orthonormal basis consisting of the eigenvectors of S.*

Proof. Proceed by induction on $n = \dim X$. If $n = 1$ then any nonzero vector is an eigenvector of S. The result is clear in this case. Assume the result for n-dimensional spaces, and let X be an $(n+1)$-dimensional space. Apply Lemma 3.6.10 to find a unit eigenvector $\mathbf{e}$. Let X_0 be the orthogonal complement of the one-dimensional space spanned by $\mathbf{e}$. Lemma 3.6.11 shows that the subspace X_0 is invariant under S. Since X_0 is an n-dimensional space, the induction hypothesis shows that there is an orthonormal basis $\mathbb{E}_0$ for X_0 consisting of the eigenvectors of S. Then $\mathbb{E} = (\mathbf{e}, \mathbb{E}_0)$ is an orthonormal basis for X consisting of the eigenvectors of S. □

Remarks 3.6.13 Two versions of the spectral theorem. Theorems 3.6.4 and 3.6.12, our two versions of the spectral theorem, are equivalent. We have seen that Theorem 3.6.12 follows from Theorem 3.6.4 through elementary arguments. The following theorem shows that the converse is also true.

Theorem 3.6.14 *Let $T : X \to Y$ be a linear map between Euclidean spaces and let $T^* : Y \to X$ be its adjoint. Then an orthonormal basis $\mathbb{E}$ of X is an eigenbasis of T if and only if each $\mathbf{e}_i \in \mathbb{E}$ is an eigenvector of $S = T^*T : X \to X$.*

Proof. Let $\mathbb{E}$ be an eigenbasis for T. If $\mathbf{e} \in \mathbb{E}$, then Lemma 3.6.8 shows that $T^*T\mathbf{e} = \|T\mathbf{e}\|^2\mathbf{e}$. Hence $\mathbf{e}$ is an eigenvector of S with the eigenvalue $\|T\mathbf{e}\|^2$.

Conversely, assume that $\mathbb{E}$ is an orthonormal basis of X such that each $\mathbf{e} \in \mathbb{E}$ is an eigenvector of $S = T^*T$. If $\mathbf{e}_i$ and $\mathbf{e}_j$ are two different vectors in $\mathbb{E}$, then

$$\langle T\mathbf{e}_i, T\mathbf{e}_j\rangle_Y = \langle S\mathbf{e}_i, \mathbf{e}_j\rangle_X = \lambda_i\langle \mathbf{e}_i, \mathbf{e}_j\rangle_X = 0.$$

Hence $\mathbb{E}$ is an eigenbasis for T. □

Eigenbases for the Adjoint Transformation

There is a natural correspondence between the eigenbases of a linear transformation $T : X \to Y$ and the eigenbases of the adjoint transformation $T^* : Y \to X$. The situation is simplest for invertible transformations. We consider this case separately.

Lemma 3.6.15 *Let $T : X \to Y$ be an invertible transformation. Let $\mathbb{E}$ be an eigenbasis for T. Let $\mathbb{U} = \{T\mathbf{e}/\|T\mathbf{e}\| \mid \mathbf{e} \in \mathbb{E}\}$. Then $\mathbb{U}$ is an eigenbasis for $T^* : Y \to X$. If $\mathbf{e} \in \mathbb{E}$ and if $T\mathbf{e} = \alpha\mathbf{u}$ with $\mathbf{u} \in \mathbb{U}$, then $T^*\mathbf{u} = \alpha\mathbf{e}$.*

Proof. Let $\mathbf{e} \in \mathbb{E}$ and let $\alpha = \|T\mathbf{e}\|$. Then $\alpha \neq 0$ since T is invertible. If $\mathbf{u} = (1/\alpha)T\mathbf{e}$, then $T\mathbf{e} = \alpha\,\mathbf{u}$ and $T^*T\mathbf{e} = \alpha\,T^*\mathbf{u} = \alpha^2\mathbf{e}$ by Lemma 3.6.8. Hence $T^*\mathbf{u} = \alpha\mathbf{e}$. This also shows that $\mathbb{U}$ is an eigenbasis for T^*. □

Remarks 3.6.16 Kernel of a transformation. The kernel of a linear transformation $T : X \to Y$ was defined in Definition 3.1.28 as the subspace $\text{Ker } T =$

$\{ \mathbf{x} \mid T\mathbf{x} = \mathbf{0} \} = X_1 \subset X$. Lemma 3.6.17 below shows that an eigenbasis for T separates X_1 from its orthogonal complement $X_0 = X_1{}^{\perp}$. Note that not every orthonormal basis has the property formulated in Lemma 3.6.17.

Lemma 3.6.17 *Let $T : X \to Y$ be a linear transformation. Let $X_1 = \mathrm{Ker}\, T$ and let $X_0 = X_1{}^{\perp}$ be the orthogonal complement of X_1. Then each eigenbasis $\mathbb{E}$ for T decomposes as $\mathbb{E} = \mathbb{E}_0 \cup \mathbb{E}_1$, $\mathbb{E}_0 \cap \mathbb{E}_1 = \emptyset$, such that $\mathbb{E}_0$ spans X_0 and $\mathbb{E}_1$ spans X_1. Furthermore, if $\mathbb{E}_1'$ is any orthonormal basis for X_1, then $\mathbb{E}' = \mathbb{E}_0 \cup \mathbb{E}_1'$ is also an eigenbasis for T.*

Proof. Given an eigenbasis $\mathbb{E}$, let $\mathbb{E}_0 = \{ \mathbf{e} \mid \mathbf{e} \in \mathbb{E},\ T\mathbf{e} \neq \mathbf{0} \}$. The rest of the proof is left as an exercise. □

Theorem 3.6.18 *Let $T : X \to Y$ be a linear transformation and let $T^* : Y \to X$ be the adjoint transformation. Let $X_1 = \mathrm{Ker}\, T$ and $Y_1 = \mathrm{Ker}\, T^*$. Let $X_0 = X_1{}^{\perp}$ and $Y_0 = Y_1{}^{\perp}$. Let $\mathbb{E}$ be an eigenbasis for T. Let $\mathbb{E} = \mathbb{E}_0 \cup \mathbb{E}_1$, $\mathbb{E}_0 \cap \mathbb{E}_1 = \emptyset$, be the decomposition of $\mathbb{E}$ as obtained in Lemma 3.6.17. Let*

$$\mathbb{U}_0 = \{ T\mathbf{e}/\|T\mathbf{e}\| \mid \mathbf{e} \in \mathbb{E}_0 \} .$$

Let $\mathbb{U}_1$ be an orthonormal basis for Y_1. Then $\mathbb{U} = \mathbb{U}_0 \cup \mathbb{U}_1$ is an eigenbasis for $T^* : Y \to X$. Also, if $\mathbf{e} \in E_0$ and $\alpha = \|T\mathbf{e}\|$, then there is a unique $\mathbf{u} \in \mathbb{U}_0$ such that $T\mathbf{e} = \alpha\mathbf{u}$ and $T^*\mathbf{u} = \alpha\mathbf{e}$.

Proof. This is summary of the results obtained above. □

Application to Orthogonal Projections

An application of Theorem 3.6.18 to orthogonal projections gives a simple geometrical picture for these projections.

Let A and B be two subspaces of a Euclidean space X. The orthogonal projection on B is a linear map defined on X. The restriction of this map to A defines a linear transformation $T : A \to B$. Similarly, take the orthogonal projection on A and restrict this to B. One gets another linear transformation $S : B \to A$.

The situation is very simple if A and B are both one-dimensional spaces. Let $\mathbf{u}$ be a unit vector of A and let $\mathbf{v}$ be a unit vector of B. Then

$$T\mathbf{x} = \langle \mathbf{x},\, \mathbf{v} \rangle_X\, \mathbf{v} \text{ for all } \mathbf{x} \in A \text{ and } S\mathbf{y} = \langle \mathbf{y},\, \mathbf{u} \rangle_X\, \mathbf{u} \text{ for all } \mathbf{y} \in B.$$

Theorem 3.6.20 shows that the general case can be understood in terms of this simple case.

Lemma 3.6.19 *Let A and B be two subspaces of a Euclidean space X. Let $T : A \to B$ and $S : B \to A$ be the corresponding orthogonal projections. Then $S = T^*$ is the adjoint transformation of T.*

Proof. Let $\mathbf{a} \in A$ and $\mathbf{b} \in B$. Then $\mathbf{a} = \mathbf{a}_1 + T\mathbf{a}$ where $\mathbf{a}_1 \perp B$. Hence,

$$\langle \mathbf{a}, \mathbf{b} \rangle_X = \langle \mathbf{a}_1 + T\mathbf{a}, \mathbf{b} \rangle_X = \langle \mathbf{a}_1, \mathbf{b} \rangle_X + \langle T\mathbf{a}, \mathbf{b} \rangle_X = \langle T\mathbf{a}, \mathbf{b} \rangle_B.$$

Similarly we obtain $\langle \mathbf{a}, \mathbf{b} \rangle_X = \langle \mathbf{a}, S\mathbf{b} \rangle_A$. Hence $T^* = S$. □

Theorem 3.6.20 *Let A and B be two subspaces of a Euclidean space X. Then there is a decomposition*

$$A = A_0 \oplus A_1 \text{ with } A_0 \perp A_1 \text{ and } B = B_0 \oplus B_1 \text{ with } B_0 \perp B_1$$

such that $A_1 \perp B$, $B_1 \perp A$, and $\dim A_0 = \dim B_0$. Also, there are orthonormal bases $\mathbb{U}_0$ and $\mathbb{V}_0$ for A_0 and B_0 such that $T\mathbf{u}_i = \lambda_i \mathbf{v}_i$ and $S\mathbf{v}_i = \lambda_i \mathbf{u}_i$ for all $\mathbf{u}_i \in A_0$ and $\mathbf{v}_i \in B_0$.

Proof. This follows directly by an application of Theorem 3.6.18 to the adjoint transformations T and S. □

Note that $T\mathbf{u}_i = \mathbf{u}_i = \mathbf{v}_i = S\mathbf{v}_i$ is possible for some i. These vectors would span $A \cap B$. Otherwise, the two-dimensional spaces spanned by $\mathbf{u}_i$ and $\mathbf{v}_i$ are invariant under both T and S. These spaces are mutually orthogonal to each other, and in each one of them the projections T and S are like the projections between two vectors.

A Summary of Determinants

Determinants are a particularly important class of multilinear functions. They are essential in integration and are also important in other applications. Appendix C contains a review of the most important results about determinants, together with complete proofs. Here we summarize the main features of determinants.

Definition 3.6.21 Alternating multilinear functions. Let X and Y be two vector spaces, and $k \in \mathbb{N}$. A multilinear function $F : X^k \to Y$ is called *an alternating multilinear function* if

$$F(\mathbf{x}_{\lambda(1)}, \ldots, \mathbf{x}_{\lambda(k)}) = (\operatorname{sign} \lambda)\, F(\mathbf{x}_1, \ldots, \mathbf{x}_k)$$

for all permutations $\lambda \in \mathbb{S}_k$ of $\mathbb{N}_k = \{1, \ldots, k\}$. The set of alternating multilinear functions $F : X^k \to Y$ is denoted by $AML(X^k, Y)$. We see that this is a subspace of $ML(X^k, Y)$.

Definition 3.6.22 Determinant functions. If $\dim X = n$, then any nonzero element ψ of $AML(X^n, \mathbb{R})$ is called a *determinant function*, or simply a *determinant*.

Theorem 3.6.23 below expresses a key fact about determinants.

Theorem 3.6.23 *If* $\dim X = n$, *then* $AML(X^n, \mathbb{R})$ *is a one-dimensional space.*

Proof. See Corollary C.3.10 in Appendix C. □

Theorem 3.6.23 implies that any determinant function is a nonzero multiple of any other determinant function. Hence, there is essentially only one determinant function $X^n \to \mathbb{R}$, up to a nonzero multiplicative constant. (It is essential here that $\dim X = n$.) The following example tells us what this single determinant function looks like.

Example 3.6.24 Basic example of a determinant function. Let X be an n-dimensional space. Let $\mathbb{E} = (\mathbf{e}_1, \ldots, \mathbf{e}_n) \in X^n$ be an ordered basis for X, that is, an n-tuple of vectors that constitute a basis for X. Given any $(\mathbf{x}_1, \ldots, \mathbf{x}_n) \in X^n$, let $\mathbf{M} = \{x_{ij}\}$ be the matrix obtained from the coordinate expansions

$$\mathbf{x}_i = \sum_j x_{ij}\mathbf{e}_j$$

and let

$$\psi_{\mathbb{E}}(\mathbf{x}_1, \ldots, \mathbf{x}_n) = \det \mathbf{M} = \det\{x_{ij}\}.$$

Here $\det \mathbf{M}$ is the usual determinant of an $n \times n$ matrix (see Appendix C for the definition and basic properties). This is an alternating multilinear function. It follows that $\psi_{\mathbb{E}} : X^n \to \mathbb{R}$ is a determinant function. Note that

$$\psi_{\mathbb{E}}(\mathbf{e}_1, \ldots, \mathbf{e}_n) = 1 \tag{3.22}$$

for any basis $\mathbb{E} = (\mathbf{e}_1, \ldots, \mathbf{e}_n)$, since the identity matrix has determinant 1.

In light of Theorem 3.6.23, any other determinant function is a nonzero multiple of $\psi_{\mathbb{E}}$.

Definition 3.6.25 Determinant of a linear operator. Lemma C.4.1 shows that if $T : X \to X$ is a linear transformation, ψ is any determinant function and $\mathbb{E} = (\mathbf{e}_1, \ldots, \mathbf{e}_n)$ is any basis for X, then the number $\psi(T\mathbf{e}_1, \ldots, T\mathbf{e}_n)/\psi(\mathbf{e}_1, \ldots, \mathbf{e}_n)$ is independent of the choice of the determinant function ψ and the basis $\mathbb{E}$. It is called the *determinant of* T and is denoted by $\det T$. In particular, $\det T = \psi_{\mathbb{E}}(T\mathbf{e}_1, \ldots, T\mathbf{e}_n)$ for any basis $\mathbb{E}$ (since $\psi_{\mathbb{E}}(\mathbf{e}_1, \ldots, \mathbf{e}_n) = 1$).

Definition 3.6.26 Euclidean determinants. There is a special situation in Euclidean spaces. If $\mathbb{E}$ and $\mathbb{U}$ are two orthonormal bases for a Euclidean space, then they define the same determinant up to a factor of ± 1. Such a determinant is called a *Euclidean determinant.* Hence there are exactly two Euclidean determinants in a Euclidean space.

Definition 3.6.27 Determinant of a linear transformation between two Euclidean spaces. In general, there is no way to define a unique determinant of a linear transformation between two different vector spaces (of equal dimension) independently of the choice of a basis and a determinant function on each vector space. If $T : X \to Y$ is a linear transformation between two Euclidean spaces of equal dimension, however, then there is a natural choice for the determinant for T, although it is defined only up to a factor of ± 1. Let $\mathbb{U}$ and $\mathbb{V}$ be orthonormal bases for X and Y. Then define

$$\det T = \psi_{\mathbb{V}}(T\mathbf{e}_1, \ldots, T\mathbf{e}_n)/\psi_{\mathbb{U}}(\mathbf{e}_1, \ldots, \mathbf{e}_n), \tag{3.23}$$

where $\mathbb{E} = (\mathbf{e}_1, \ldots, \mathbf{e}_n)$ is any basis for X. This definition is independent of the choice of $\mathbb{E}$, but it depends minimally on the choices of the Euclidean determinants $\psi_{\mathbb{U}}$ and $\psi_{\mathbb{V}}$. Nevertheless, different choices change the result only up to a factor of ± 1. Fortunately, this is not important for most of our work with determinants, which depends only upon the absolute value of the determinant. In particular, this absolute value is the *volume multiplier* associated with the linear transformation.

Determinants as Volume Multipliers

Theorem 3.6.28 *Let $T : X \to Y$ be a linear transformation between two Euclidean spaces of the same dimension n. If $\mathbb{E}$ is an eigenbasis for T, then*

$$\det T = \pm\lambda_1 \cdots \lambda_n, \quad \text{with } \lambda_i = \|T\mathbf{e}_i\|.$$

Proof. The vectors $T\mathbf{e}_i$, $\mathbf{e}_i \in \mathbb{E}$, are orthogonal to each other, since $\mathbb{E}$ is an eigenbasis for T. Some of these vectors may vanish, but in any case, there is an orthonormal basis $\mathbb{U}$ for Y such that $T\mathbf{e}_i = \lambda_i \mathbf{u}_i$. Then

$$\begin{aligned} \psi_{\mathbb{U}}(T\mathbf{e}_1, \ldots, T\mathbf{e}_n) &= \psi_{\mathbb{U}}(\lambda_1\mathbf{u}_1, \ldots, \lambda_n\mathbf{u}_n) && (3.24)\\ &= \lambda_1 \cdots \lambda_n\, \psi_{\mathbb{U}}(\mathbf{u}_1, \ldots, \mathbf{u}_n) && (3.25)\\ &= \lambda_1 \cdots \lambda_n. && (3.26) \end{aligned}$$

Here (3.25) follows from the multilinearity of determinant functions. The conclusion follows from definition (3.23), since $\psi_{\mathbb{E}}(\mathbf{e}_1, \ldots, \mathbf{e}_n) = 1$ by(3.22). □

Remarks 3.6.29 Comments about volume. An important part of the theory of integration (Chapter 8) is to define a notion of volume on Euclidean spaces. Our definition will agree with intuitive ideas about volume recognized by Archimedes more than two thousand years ago. "Cubic boxes" spanned by orthonormal bases will have unit volume. The volume of a general rectangular box, spanned by orthogonal vectors, will be the product of its side lengths. Combining these points with Theorem 3.6.28, we see that a linear transformation transforms a certain cubic box of unit volume into a rectangular box of volume $|\det T|$. For now, we know only that this is true for this one particular cubic box. But it is a very plausible fact, which we shall eventually prove, that a linear transformation changes all volumes by the same factor. This factor will be called the volume multiplier of the transformation. Therefore the *volume multiplier of* T is equal to $|\det T|$.

Examples

Example 3.6.30 Finding an eigenbasis. Let $A = \begin{bmatrix} \mathbf{u} & \mathbf{v} \end{bmatrix}$ be a 2×2 matrix considered as a linear map from $\mathbb{R}^2$ to $\mathbb{R}^2$. Assume that $\|\mathbf{u}\| = \|\mathbf{v}\|$. As a first step towards finding an eigenbasis for A, let us find a unit vector $\mathbf{e}$ in $\mathbb{R}^2$ such that $\|A\mathbf{x}\| \leq \|A\mathbf{e}\|$ for all unit vectors $\mathbf{x}$ in $\mathbb{R}^2$. Each unit vector in $\mathbb{R}^2$ is of the form $(\cos\theta, \sin\theta)$ for some $\theta \in [0, 2\pi]$. Thus, we look for θ for which

$$\left\| A \begin{bmatrix} \cos\theta \\ \sin\theta \end{bmatrix} \right\|^2$$

is maximized. Now,

$$\begin{aligned} \left\| A \begin{bmatrix} \cos\theta \\ \sin\theta \end{bmatrix} \right\|^2 &= (\cos\theta\mathbf{u} + \sin\theta\mathbf{v}) \cdot (\cos\theta\mathbf{u} + \sin\theta\mathbf{v}) \\ &= \sin^2\theta\|\mathbf{v}\|^2 + \cos^2\theta\|\mathbf{u}\|^2 + \sin(2\theta)(\mathbf{u}\cdot\mathbf{v}) \\ &= \|\mathbf{u}\|^2 + \sin(2\theta)(\mathbf{u}\cdot\mathbf{v}). \end{aligned}$$

So, if $\mathbf{u}\cdot\mathbf{v} \geq 0$, then we choose θ so that $\sin(2\theta) = 1$. If $\mathbf{u}\cdot\mathbf{v} < 0$, we choose θ so that $\sin(2\theta) = -1$. As an illustration, let $A = \begin{bmatrix} 1 & 5 \\ 7 & 5 \end{bmatrix}$. Then $\|\mathbf{u}\| = \|\mathbf{v}\|$ and $\mathbf{u}\cdot\mathbf{v} > 0$. Thus, with $\theta = \pi/4$, we have

$$\mathbf{e} = (\cos\theta, \sin\theta) = \frac{1}{\sqrt{2}}(1, 1).$$

Hence,

$$\|A\mathbf{e}\| = 3\sqrt{10}.$$

Let $\mathbf{u} \in \mathbb{R}^2$ with $\mathbf{u} \perp \mathbf{e}$. Then $\mathbf{u} = t(-1, 1)$ for some scalar t. Since

$$A\begin{bmatrix} 1 \\ 1 \end{bmatrix} = \begin{bmatrix} 6 \\ 12 \end{bmatrix} \quad \text{and} \quad A\begin{bmatrix} -1 \\ 1 \end{bmatrix} = \begin{bmatrix} 4 \\ -2 \end{bmatrix}$$

are orthogonal, it follows that $A\mathbf{u} \perp A\mathbf{e}$, as predicted by Lemma 3.6.2.

Example 3.6.31 Eigenbases of real-valued functions. Let $T : X \to Y$ be a linear transformation between Euclidean spaces. If $\dim Y = 1$, then one can easily exhibit an eigenbasis for T. Let $\mathbf{u}$ be a unit vector in Y and express T in terms of a real valued function $f : X \to \mathbb{R}$ as $T\mathbf{x} = f(\mathbf{x})\,\mathbf{u}$. Then, by Theorem 3.4.21, there is a vector $\mathbf{a} \in X$ such that $f(\mathbf{x}) = \langle \mathbf{a}, \mathbf{x} \rangle$. Let $\mathbf{e}_1$ be a unit vector in X such that $\mathbf{a} = \alpha\,\mathbf{e}_1$. If $\|\mathbf{a}\| \neq 0$, one can take $\mathbf{e}_1 = \mathbf{a}/\|\mathbf{a}\|$. Complete $\mathbf{e}_1$ to an orthonormal basis $\mathbb{E} = (\mathbf{e}_1, \ldots, \mathbf{e}_n)$ for X. It is clear that this is an eigenbasis for $T : X \to Y$ since $T\mathbf{e}_i = \mathbf{0}$ for all $i = 2, \ldots, n$.

Example 3.6.32 Let $\mathbb{R}^3$ be represented as the xyz-space. Let A be the subspace $x + 2y + 3z = 0$ and let B be the subspace $z = 0$. Let $P : \mathbb{R}^3 \to B$ be the orthogonal projection on B. Let $G : A \to B$ be the restriction of P to A. Find eigenbases for P and for G.

Solution. An easy verification shows that $P(x, y, z) = (x, y, 0)$. Hence we see that the orthonormal basis

$$\begin{aligned} \mathbb{E} &= \{\,(1, 0, 0), (0, 1, 0), (0, 0, 1)\,\} \text{ of } \mathbb{R}^3 \text{ is mapped to} \\ P\mathbb{E} &= \{\,(1, 0, 0), (0, 1, 0), (0, 0, 0)\,\}. \end{aligned}$$

We see that $P\mathbb{E}$ is an orthogonal set in B. Hence $\mathbb{E}$ is an eigenbasis for P.

To find an eigenbasis for $G : A \to B$, consider A as the graph of the function $z = F(x, y) = -(1/3)(x + 2y)$. This is defined on the xy-plane B and takes values on the z-axis. Since it is a real-valued function defined on a Euclidean space B, it can be represented as an inner product. In fact, $z = \langle \mathbf{a}, (x, y) \rangle_B$, where $\mathbf{a} = (-1/3, -2/3)$. Express this vector as $\mathbf{a} = \alpha\mathbf{e}_1$, where $\alpha = \sqrt{5}/3$ and $\mathbf{e}_1 = (-1, -2)/\sqrt{5}$ is a unit vector. Example 3.6.31 shows that the orthonormal basis consisting of

$$\mathbf{e}_1 = (-1, -2)/\sqrt{5} \text{ and } \mathbf{e}_2 = (2, -1)/\sqrt{5}$$

is an eigenbasis for $F : B \to \mathbb{R}$. Let $S : B \to A$ be the transformation that takes a point $(x, y) \in B$ to the corresponding point $S(x, y) = (x, y, -(1/3)(x + 2y))$ on A, the graph of $F : B \to \mathbb{R}$. Theorem 3.6.20 shows that the normalized vectors

$$\|S\mathbf{e}_1)\|^{-1} S(\mathbf{e}_1) = -(3, 6, 5)/\sqrt{70} \;\text{ and }\; \|S\mathbf{e}_2)\|^{-1} S(\mathbf{e}_2) = (2, -1, 0)/\sqrt{5}$$

form an eigenbasis for the orthogonal projection $G : A \to B$. $\triangle$

Problems

3.88 Let A and B be two subspaces of a Euclidean space. Let $T : A \to B$, $S : B \to A$ be the corresponding orthogonal projections as in Lemma 3.6.19. Show directly, without using eigenbases, that T and S are adjoint transformations.

3.89 Show that any orthogonal projection $T : X \to X$ to a subspace of X is a self-adjoint transformation.

3.90 Let $T : X \to X$ be a self-adjoint transformation. Let U be an invariant subspace for T. That is, let $T\mathbf{u} \in U$ for all $\mathbf{u} \in U$. Show that $V = U^{\perp}$ is also invariant under T.

3.91 Give an example of a self-adjoint transformation $T : X \to X$ and an eigenbasis $\mathbb{E}$ for T such that no vector in $\mathbb{E}$ is an eigenvector of T.

3.92 Let A and B be two subspaces of X with $\dim A = \dim B$. Let $T : A \to B$ and $S : B \to A$ be the corresponding orthogonal projections. Also assume that there is an eigenbasis $\mathbb{U}$ for T such that $T\mathbf{u}_i \neq \mathbf{0}$ for all $\mathbf{u}_i \in \mathbb{U}$. Show that $|\det T| = |\det S|$. Also, if $A' = A^{\perp}$, $B' = B^{\perp}$ with the corresponding orthogonal projections $T' : A' \to B'$ and $S' : B' \to A'$, then

$$|\det T| = |\det S| = |\det T'| = |\det S'|.$$

Show that in the two-dimensional case this is a familiar statement: the angle between two lines is the same as the angle between their normals.

PART II

DIFFERENTIATION

CHAPTER 4

NORMED VECTOR SPACES

The norm of a vector in a Euclidean space V has already been defined as $\|\mathbf{v}\| = \langle \mathbf{v}, \mathbf{v} \rangle^{1/2}$. The most important properties of the norm are the following:

(1) $\|\mathbf{v}\| \geq 0$ for all $\mathbf{v} \in V$ and $\|\mathbf{v}\| = 0$ if and only if $\mathbf{v} = \mathbf{0}$.

(2) $\|t\mathbf{v}\| = |t| \, \|\mathbf{v}\|$ for all vectors $\mathbf{v} \in V$ and for all scalars $t \in \mathbb{R}$.

(3) $\|\mathbf{u} + \mathbf{v}\| \leq \|\mathbf{u}\| + \|\mathbf{v}\|$ for all vectors $\mathbf{u}, \mathbf{v} \in V$.

Any function that has these three properties is a reasonably well-behaved candidate for the length of a vector. A well-behaved concept of length gives us a well-behaved notion of distance between two vectors. That is the basic tool for defining limits and convergence. Thus, analysis depends upon norms in an essential way.

It turns out, however, that many arguments in analysis apply not just to Euclidean spaces, but also to any vector space X that is equipped with a function $\|\cdot\| : X \to \mathbb{R}$ that has the three properties listed above. Such functions are called *(general) norms* on X, and X is called a *normed vector space*. The first part of the chapter establishes

Analysis in Vector Spaces.
By M. A. Akcoglu, P. F. A. Bartha and D. M. Ha

the basic results about norms and about when a sequence in a normed vector space converges to a limit.

It is natural to wonder whether facts about convergence and limits might vary, depending upon one's choice of norm. A key result in the early portion of the chapter is that any two norms (on a finite-dimensional vector space) are equivalent: they define exactly the same class of convergent sequences and exactly the same limits. This part of the chapter also shows that we can define a very natural (and useful) norm on the vector space of linear transformations between two normed spaces, and the same point applies to multilinear transformations.

The middle section of the chapter defines the continuity of functions on normed spaces. In this section, we establish analogues of familiar results about the continuity of functions of a real variable as well as some basic facts about Cartesian products. Finally, by exploiting properties of the norm defined on the space of linear transformations, we show that the operation that maps a linear transformation to its inverse is continuous.

The final section defines the most important concepts of general topology: open and closed sets, boundaries, compactness, and connectedness. Still working in the setting of normed vector spaces, we establish fundamental results about compactness and continuity that will be used throughout the rest of the book.

4.1 PRELIMINARIES

Definition 4.1.1 Norms. Let X be a vector space. A function $\|\cdot\| : X \to \mathbb{R}$ is called a *norm* on X if it satisfies the following three conditions.

1. **Positive definiteness:** If $\mathbf{x} \in X$ and $\mathbf{x} \neq \mathbf{0}$, then $\|\mathbf{x}\| > 0$.
2. **Homogeneity:** If $\mathbf{x} \in X$ and $t \in \mathbb{R}$, then $\|t\mathbf{x}\| = |t|\,\|\mathbf{x}\|$.
3. **Triangle inequality:** If $\mathbf{x},\ \mathbf{y} \in X$, then $\|\mathbf{x} + \mathbf{y}\| \leq \|\mathbf{x}\| + \|\mathbf{y}\|$.

Definition 4.1.2 Normed spaces. A *normed space* is a vector space X together with a norm $\|\cdot\| : X \to \mathbb{R}$. We use the same notation, $\|\cdot\|$, for norms on different normed spaces. Where the distinction between different spaces is not clear, we write $\|\cdot\|_X$ for the norm on X.

Examples 4.1.3 Norms on $\mathbb{R}^n$. The standard Euclidean norm on $\mathbb{R}^n$ is

$$\|\mathbf{x}\| = \|(x_1, \ldots, x_n)\| = (x_1^2 + \cdots + x_n^2)^{1/2}.$$

It is induced by the standard inner product on $\mathbb{R}^n$. When $n = 1$, the Euclidean norm is simply the absolute value function on $\mathbb{R}$. Two other useful norms on $\mathbb{R}^n$ are

$$\|\mathbf{x}\| = \|(x_1, \ldots, x_n)\| = |x_1| + \cdots + |x_n|$$

and

$$\|\mathbf{x}\| = \|(x_1, \ldots, x_n)\| = \max(|x_1|, \ldots, |x_n|).$$

It is easy to verify that these are indeed norms. In both cases, the triangle inequality follows immediately from the definition and the special case of the triangle inequality for $\mathbb{R}$.

Example 4.1.4 Norms on $\mathbb{R}^1 = \mathbb{R}$. The absolute value $|a|$ of $a \in \mathbb{R}$ defines a norm on the one-dimensional space $\mathbb{R}^1 = \mathbb{R}$. Any norm on $\mathbb{R}^1$ is a positive multiple of the absolute value. To see this, let $\|\cdot\|$ be a norm on $\mathbb{R}^1$ and let $\|1\| = M$. Then $M > 0$ and $\|a\| = \|a1\| = |a|\,\|1\| = M\,|a|$ for all $a \in \mathbb{R}^1 = \mathbb{R}$. When we consider $\mathbb{R}$ as a normed space, we will always assume that the norm is the absolute value.

Example 4.1.5 Norms on Cartesian products. Let U_i be a normed space for $1 \le i \le k$, and let $X = U_1 \times \cdots \times U_k$. There are several natural ways to define a norm on X. The three most common are:

$$\begin{aligned}
\|\mathbf{x}\| = \|(\mathbf{u}_1, \ldots, \mathbf{u}_k)\| &= \|\mathbf{u}_1\| + \cdots + \|\mathbf{u}_k\| \\
\|\mathbf{x}\| = \|(\mathbf{u}_1, \ldots, \mathbf{u}_k)\| &= (\|\mathbf{u}_1\|^2 + \cdots + \|\mathbf{u}_k\|^2)^{1/2} \\
\|\mathbf{x}\| = \|(\mathbf{u}_1, \ldots, \mathbf{u}_k)\| &= \max(\|\mathbf{u}_1\|, \ldots, \|\mathbf{u}_k\|).
\end{aligned}$$

It is easy to verify that these are indeed norms on X.

Remarks 4.1.6 Non-Euclidean norms. There are many advantages to using a Euclidean norm when one is available. In some cases, however, there are other natural choices for a norm. The most important case is $L(X, Y)$, the vector space of all linear maps $T : X \to Y$. We shall define a very useful non-Euclidean norm on this space.

Definition 4.1.7 Equivalent norms. Let $\|\cdot\|$ and $\|\cdot\|'$ be two norms on a vector space X. If there are two constants $L, M \in \mathbb{R}$ such that, for all $\mathbf{x} \in X$,

$$\|\mathbf{x}\| \le L\,\|\mathbf{x}\|' \text{ and } \|\mathbf{x}\|' \le M\,\|\mathbf{x}\|,$$

then these are *equivalent norms on* X. We write $\|\cdot\| \sim \|\cdot\|'$ to indicate the equivalence of these norms. We can see easily that this is an equivalence relation among the norms on the vector space$X$. We will show below that any two norms on a (finite dimensional) vector space are equivalent.

Remarks 4.1.8 Triangle and reverse triangle inequalities. Consider three points $\mathbf{a}$, $\mathbf{b}$, and $\mathbf{c}$ in a normed space. The inequalities

$$\begin{aligned} \|\mathbf{a}-\mathbf{c}\| &\leq \|\mathbf{a}-\mathbf{b}\| + \|\mathbf{b}-\mathbf{c}\| & (4.1) \\ \big|\, \|\mathbf{a}-\mathbf{b}\| - \|\mathbf{a}-\mathbf{c}\| \,\big| &\leq \|\mathbf{b}-\mathbf{c}\| & (4.2) \end{aligned}$$

are used frequently. The first inequality (4.1) is just the triangle inequality

$$\|\mathbf{x}+\mathbf{y}\| \leq \|\mathbf{x}\| + \|\mathbf{y}\|$$

applied to $\mathbf{x} = \mathbf{a} - \mathbf{b}$ and $\mathbf{y} = \mathbf{b} - \mathbf{c}$; it is also referred to as the *triangle inequality*. The second inequality (4.2) is equivalent to the two inequalities

$$\|\mathbf{a}-\mathbf{b}\| - \|\mathbf{a}-\mathbf{c}\| \leq \|\mathbf{b}-\mathbf{c}\| \quad \text{and} \quad \|\mathbf{a}-\mathbf{c}\| - \|\mathbf{a}-\mathbf{b}\| \leq \|\mathbf{b}-\mathbf{c}\|.$$

Each of these is a rearrangement of a triangle inequality. For easy reference, we will call the inequality (4.2) the *reverse triangle inequality*.

Remarks 4.1.9 Geometric formulations. The triangle and the reverse triangle inequalities can be stated in terms of familiar geometrical concepts. If $\mathbf{a}$ and $\mathbf{b}$ are two points in a normed space X, then the set

$$\{\, \mathbf{x} \in X \mid \mathbf{x} = t\mathbf{a} + (1-t)\mathbf{b},\ 0 \leq t \leq 1 \,\}$$

is the line segment joining these two points. The length of this segment, defined as $\|\mathbf{a}-\mathbf{b}\|$, is the distance between the end points $\mathbf{a}$ and $\mathbf{b}$. Hence the triangle inequality says that the length of one side of a triangle is dominated by the sum of the lengths of the other two sides. The reverse triangle inequality says that the length of one side of a triangle dominates the difference between the lengths of the other two sides.

Problems

4.1 Let X be a vector space. If $\mathbf{a}, \mathbf{b}$ are in X, the line segment joining $\mathbf{a}$ and $\mathbf{b}$ is the set

$$L[\mathbf{a}, \mathbf{b}] = \{\, \mathbf{x} \in X \mid \mathbf{x} = t\mathbf{a} + (1-t)\mathbf{b},\ 0 \leq t \leq 1 \,\}.$$

Show that $L[\mathbf{a}, \mathbf{b}]$ is convex: If $\mathbf{u}, \mathbf{v}$ are in $L[\mathbf{a}, \mathbf{b}]$, then $s\mathbf{u} + (1-s)\mathbf{v} \in L[\mathbf{a}, \mathbf{b}]$ for all $0 \leq s \leq 1$.

4.2 If N_1, N_2 are norms on a vector space X, show that $N_1 + N_2$ is also a norm on X. What about the product $N_1 N_2$?

4.3 Let $\|\ \|$ be the standard Euclidean norm on $\mathbb{R}^n$, and let $\|\mathbf{x}\|' = |x_1| + \cdots + |x_n|$ for all $\mathbf{x} \in \mathbb{R}^n$. Show that there are constants A, B such that

$$A\|\mathbf{x}\|' \leq \|\mathbf{x}\| \leq B\|\mathbf{x}\|' \quad \text{for all } \mathbf{x} \in \mathbb{R}^n.$$

(So these two norms are equivalent.)

4.4 Let X, Y be vector spaces and suppose that $f : X \to Y$ is an isomorphism. Let $\| \|_X$ be a norm on X and define

$$\|\mathbf{y}\| = \|\mathbf{x}\|_X$$

whenever $\mathbf{y} = f(\mathbf{x})$ for (unique) $\mathbf{x}$ in X. Show that $\| \|$ is a norm on Y.

4.5 Let $\| \|$ be a norm on X and let $\mathbf{x}, \mathbf{y}$ be in X. Assume that $\|3\mathbf{x} - \mathbf{y}\| \leq 1$ and $\|5\mathbf{x} - 4\mathbf{y}\| \leq 3$. Show that $\|\mathbf{y}\| \leq 2$.

4.6 Let X be a normed space. For $\mathbf{a} \in X$ and $r > 0$, show that $B_r(\mathbf{a}) = \{ \mathbf{x} \in X \mid \|\mathbf{x} - \mathbf{a}\| < r \}$ is convex.

4.7 Let M be a subspace of a normed space X. If there is some $\mathbf{a} \in X$ and some $r > 0$ such that $B_r(\mathbf{a}) \subset M$, show that $M = X$.

4.8 Is it true that for any nonzero normed space X, the set $B_r(\mathbf{a})$ is infinite for all $\mathbf{a} \in X$ and all $r > 0$?

4.9 Let X be a normed space. Let $\mathbf{a} \in X$ and $r > 0$. Show that

$$\frac{1}{r}(-\mathbf{a} + B_r(\mathbf{a})) = B_1(\mathbf{0}).$$

4.10 Let $\| \|$ be a norm on X. Define $f(\mathbf{x}) = \|\mathbf{x}\|^2$ for all $\mathbf{x} \in X$. Is f a norm on X?

4.11 Let X be a vector space with $\dim X = 1$. Given a nonzero $\mathbf{e} \in X$ and a norm $\| \|$ on X, there is some positive constant C such that $\|a\mathbf{e}\| = |a|C$ for all $a \in \mathbb{R}$. True or false?

4.12 Let $\| \|$ be the standard Euclidean norm on $\mathbb{R}^n$, and let $\|\mathbf{x}\|'' = max(|x_1|, \cdots, |x_n|)$ for all $\mathbf{x} \in \mathbb{R}^n$. Show that this defines a norm, and that it is equivalent to the standard Euclidean norm.

4.13 Show that, for any real x, y, and z,

$$\begin{aligned}\sqrt{(x+2y)^2 + (y+2z)^2 + (z+2x)^2} &\leq \sqrt{(x-y)^2 + (y-z)^2 + (z-x)^2} \\ &+ 3\sqrt{x^2 + y^2 + z^2}.\end{aligned}$$

4.2 CONVERGENCE IN NORMED SPACES

The definitions and notation for representing a sequence are as defined in previous chapters. Our immediate objective is to define the convergence of a sequence in a normed space.

Convergent Sequences

Recall that a sequence r_n in $\mathbb{R}$ is called a *zero sequence* if $r_n \to 0$. More explicitly, r_n is a zero sequence if for each $\varepsilon > 0$ there is an $N \in \mathbb{N}$ such that $|r_n| < \varepsilon$ whenever $n \geq N$.

Definition 4.2.1 Convergent sequences. A sequence $\mathbf{x}_n$ in a normed space X is *said to converge to a point* $\mathbf{a} \in X$ if

$$\text{for each } \varepsilon > 0 \text{ there is an } N \in \mathbb{N} \text{ such that } \|\mathbf{x}_n - \mathbf{a}\| < \varepsilon \text{ for all } n \geq N.$$

Equivalently, $r_n = \|\mathbf{x}_n - \mathbf{a}\|$ is a zero sequence in $\mathbb{R}$. This condition is expressed as $\mathbf{x}_n \to \mathbf{a}$, or as $\lim_n \mathbf{x}_n = \mathbf{a}$. If $\mathbf{x}_n \to \mathbf{a}$, then $\mathbf{x}_n$ is a *convergent sequence* and $\mathbf{a}$ is the *limit* of this convergent sequence. This terminology is justified by Lemma 4.2.2 below.

Lemma 4.2.2 *A sequence cannot converge to two different points.*

Proof. Assume that $\mathbf{x}_n \to \mathbf{a}$ and $\mathbf{x}_n \to \mathbf{b}$. The triangle inequality 4.1.8 shows that

$$\|\mathbf{a} - \mathbf{b}\| \leq \|\mathbf{a} - \mathbf{x}_n\| + \|\mathbf{b} - \mathbf{x}_n\| \to 0.$$

Hence $\|\mathbf{a} - \mathbf{b}\| = 0$, and so $\mathbf{a} = \mathbf{b}$. $\square$

Lemma 4.2.3 *If* $\mathbf{x}_n \to \mathbf{a}$ *in* X*, then* $\|\mathbf{x}_n\| \to \|\mathbf{a}\|$ *in* $\mathbb{R}$.

Proof. The reverse triangle inequality 4.1.8 shows that

$$\big|\, \|\mathbf{x}_n\| - \|\mathbf{a}\| \,\big| \leq \|\mathbf{x}_n - \mathbf{a}\|.$$

Since $\|\mathbf{x}_n - \mathbf{a}\| \to 0$, we see that $\|\mathbf{x}_n\| \to \|\mathbf{a}\|$ in $\mathbb{R}$. $\square$

Definition 4.2.4 Bounded sets. Let B be a set in a normed space X. Then B is called a *bounded set* if there is an $M \in \mathbb{R}$ such that $\|\mathbf{x}\| \leq M$ for all $\mathbf{x} \in B$.

Definition 4.2.5 Bounded sequences. Let X be a normed space. A sequence $\mathbf{x}_n$ in X is is called a *bounded sequence* if its range is a bounded set. More explicitly, $\mathbf{x}_n$ is called a bounded sequence if there is an $M \in \mathbb{R}$ such that $\|\mathbf{x}_n\| \leq M$ for all $n \in \mathbb{N}$.

Example 4.2.6 Let c_k be a bounded sequence in $\mathbb{R}$ and let $\mathbf{u}_k$ be a bounded sequence in $\mathbb{R}^n$. Define a sequence $\mathbf{x}_k$ in $\mathbb{R}^n$ by

$$\mathbf{x}_k = (1/k)(c_1\mathbf{u}_1 + \cdots + c_k\mathbf{u}_k), \quad k \in \mathbb{N}.$$

Then $\mathbf{x}_k$ is also bounded. Indeed, since c_k and $\mathbf{u}_k$ are bounded, there is a constant C such that $|c_j| \leq C$ and $\|\mathbf{u}_j\| \leq C$ for all $j \in \mathbb{N}$. Then for all $k \in \mathbb{N}$, the triangle inequality gives

$$\begin{aligned} \|\mathbf{x}_k\| &= \|(1/k)(c_1\mathbf{u}_1 + \cdots + c_k\mathbf{u}_k)\| \\ &\leq (1/k)(|c_1|\,\|\mathbf{u}_1\| + \cdots + |c_k|\,\|\mathbf{u}_k\|) \\ &\leq (1/k)(kC^2) = C^2. \end{aligned}$$

Theorem 4.2.7 *Every convergent sequence is bounded.*

Proof. Let $\mathbf{x}_n \to \mathbf{a}$. Then by Lemma 4.2.3, $\|\mathbf{x}_n\| \to \|\mathbf{a}\|$. Hence, by Lemma 2.3.4, $(\|\mathbf{x}_n\|)$ is a bounded sequence of real numbers. $\square$

Bolzano-Weierstrass Theorem

Remark 4.2.8 Bolzano-Weierstrass theorem. The converse of the last theorem is obviously false. As a simple counterexample, let $\mathbf{a}$ be any nonzero vector and define $\mathbf{x}_n = (-1)^n\mathbf{a}$. Then $\mathbf{x}_n$ is bounded but does not converge. However, it turns out that in any finite-dimensional normed space, every bounded sequence has a convergent subsequence. This is the Bolzano-Weierstrass theorem, a key result in analysis. The Bolzano-Weierstrass theorem has already been proved for sequences in R. To extend it to arbitrary normed spaces, it turns out that the best strategy is first to prove it for the special case of Euclidean spaces. The general result will follow from this special case once we prove two additional results: the equivalence of all norms on a finite-dimensional normed space, and the fact that a Euclidean norm can always be defined on any finite-dimensional vector space.

Theorem 4.2.9 Bolzano-Weierstrass theorem in Euclidean spaces. *A bounded sequence in a Euclidean space has a convergent subsequence.*

Proof. Let X be an n-dimensional Euclidean space. We use an induction argument on n. The result is true for $n = 1$ by Theorem 2.4.5. Now assume the result for $(n-1)$-dimensional spaces, $n \geq 2$. Let $\{\,\mathbf{e}_1, \ldots, \mathbf{e}_n\}$ be an orthonormal basis for X. Let $\mathbf{x}_k$

be a bounded sequence in X. Let $r_k = \langle \mathbf{e}_n, \mathbf{x}_k \rangle$. Then $|r_k| \leq \|\mathbf{e}_n\| \, \|\mathbf{x}_k\| = \|\mathbf{x}_k\|$. Hence, r_k is a bounded sequence in $\mathbb{R}$. Therefore, it has a convergent subsequence r_m, $m \in \mathbb{K}$. Here $\mathbb{K}$ is an unbounded subset of $\mathbb{N}$.

Now the sequence $\mathbf{u}_m = \mathbf{x}_m - r_m \mathbf{e}_n$ is a bounded sequence in the subspace spanned by the vectors $\{ \mathbf{e}_1, \ldots, \mathbf{e}_{n-1} \}$. This subspace is an $(n-1)$-dimensional Euclidean space. Therefore, by the induction hypothesis, $\mathbf{u}_m$ has a convergent subsequence $\mathbf{u}_\ell$, $\ell \in \mathbb{L}$. Here $\mathbb{L}$ is an unbounded subset of $\mathbb{K}$. The subsequence r_ℓ, $\ell \in \mathbb{L}$, is still convergent as a subsequence of a convergent sequence. Hence, $\mathbf{x}_\ell = \mathbf{u}_\ell + r_\ell \mathbf{e}_n$, $\ell \in \mathbb{L}$, is a convergent subsequence of $\mathbf{x}_k$. □

Equivalence of Norms

We will now show that any norm on a finite-dimensional vector space is equivalent to a Euclidean norm. This implies the stronger-sounding result that any two norms are equivalent.

First, recall that one can define a Euclidean norm on any finite-dimensional vector space X. Take any basis $E = \{ \mathbf{e}_1, \ldots, \mathbf{e}_n \}$ for X and let

$$\|\mathbf{x}\| = (x_1(\mathbf{x})^2 + \cdots + x_n(\mathbf{x})^2)^{1/2},$$

where $x_i : X \to \mathbb{R}$ are the coordinate functions for the basis E. This is a Euclidean norm induced by the inner product defined as

$$\langle \mathbf{u}, \mathbf{v} \rangle = x_1(\mathbf{u})\, x_1(\mathbf{v}) + \cdots + x_n(\mathbf{u})\, x_n(\mathbf{v})$$

for all $\mathbf{u}$, $\mathbf{v} \in X$. The basis E is an orthonormal basis in this inner product. We shall assume that one such Euclidean norm is fixed in the following discussion.

Notations 4.2.10 Suppose X has a Euclidean norm $\|\cdot\|_{\text{euc}} : X \to \mathbb{R}$ and also another norm $\|\cdot\| : X \to \mathbb{R}$. Formally, we then have two different normed spaces, which we denote as $(X, \|\cdot\|_{\text{euc}})$ and as $(X, \|\cdot\|)$. We write that $\mathbf{x}_n \to \mathbf{a}$ in $(X, \|\cdot\|)$ just in case $\|\mathbf{x}_n - \mathbf{a}\| \to 0$.

To show that any norm is equivalent to the Euclidean norm involves proving two inequalities, as defined in Definition 4.1.7. We prove these as two separate lemmas.

Lemma 4.2.11 *There is an $L \in \mathbb{R}$ such that $\|\mathbf{x}\| \leq L \, \|\mathbf{x}\|_{\text{euc}}$ for all $\mathbf{x} \in X$.*

Proof. Let $E = \{ \mathbf{e}_1, \ldots, \mathbf{e}_n \}$ be an orthonormal basis for X for the given Euclidean norm. Let $x_i : X \to \mathbb{R}$ be the coordinate functions for E. Hence

$$\mathbf{x} = x_1(\mathbf{x})\, \mathbf{e}_1 + \cdots + x_n(\mathbf{x})\, \mathbf{e}_n.$$

Here $|x_i(\mathbf{x})| \leq \|\mathbf{x}\|_{\text{euc}}$ for all $\mathbf{x} \in X$ and for all $i = 1, \ldots, n$. Therefore

$$\begin{aligned} \|\mathbf{x}\| &\leq |x_1(\mathbf{x})| \, \|\mathbf{e}_1\| + \cdots + |x_n(\mathbf{x})| \, \|\mathbf{e}_n\| \\ &\leq (\|\mathbf{e}_1\| + \cdots + \|\mathbf{e}_n\|)\|\mathbf{x}\|_{\text{euc}} = L \, \|\mathbf{x}\|_{\text{euc}} \end{aligned}$$

with $L = \|\mathbf{e}_1\| + \cdots + \|\mathbf{e}_n\|$. □

Corollary 4.2.12 *If* $\mathbf{x}_n \to \mathbf{a}$ *in* $(X, \| \cdot \|_{\text{euc}})$, *then also* $\mathbf{x}_n \to \mathbf{a}$ *in* $(X, \| \cdot \|)$.

Proof. We have $\|\mathbf{x}_n - \mathbf{a}\| \leq L \, \|\mathbf{x}_n - \mathbf{a}\|_{\text{euc}} \to 0$. □

Lemma 4.2.13 *There is an* $M \in \mathbb{R}$ *such that* $\|\mathbf{x}\|_{\text{euc}} \leq M \, \|\mathbf{x}\|$ *for all* $\mathbf{x} \in X$.

Proof. We first show that there is an $m > 0$ such that

$$m \leq \|\mathbf{x}\| \text{ whenever } \|\mathbf{x}\|_{\text{euc}} = 1. \tag{4.3}$$

If (4.3) is false, then there is a sequence $\mathbf{x}_n \in X$ such that $\|\mathbf{x}_n\|_{\text{euc}} = 1$ and $\|\mathbf{x}_n\| \to 0$. But $\mathbf{x}_n$ is a bounded sequence in $(X, \| \cdot \|_{\text{euc}})$. Hence by the Bolzano-Weierstrass theoremin Euclidean spaces (Theorem 4.2.9), $\mathbf{x}_n$ has a subsequence $\mathbf{x}_{k_n}$ such that

$$\mathbf{x}_{k_n} \to \mathbf{a} \text{ in } (X, \| \cdot \|_{\text{euc}}).$$

Therefore, by Corollary 4.2.12 above, also

$$\mathbf{x}_{k_n} \to \mathbf{a} \text{ in } (X, \| \cdot \|).$$

Hence, by Lemma 4.2.3, $\|\mathbf{x}_{k_n}\|_{\text{euc}} \to \|\mathbf{a}\|_{\text{euc}}$ and $\|\mathbf{x}_{k_n}\| \to \|\mathbf{a}\|$. Then $\|\mathbf{a}\|_{\text{euc}} = 1$ since each $\|\mathbf{x}_{k_n}\|_{\text{euc}} = 1$ and $\|\mathbf{a}\| = 0$ since $\|\mathbf{x}_n\| \to 0$. But $\|\mathbf{a}\| = 0$ means that $\mathbf{a} = \mathbf{0}$ and $\|\mathbf{a}\|_{\text{euc}} = 1$ means that $\mathbf{a} \neq \mathbf{0}$. This contradiction shows that (4.3) is true.

Now let $\mathbf{x} \in X$. If $\mathbf{x} = 0$, then the conclusion of the lemma is clear. Assume that $\|\mathbf{x}\|_{\text{euc}} = t > 0$. Then $\|(1/t)\mathbf{x}\|_{\text{euc}} = 1$. Hence $m \leq \|(1/t)\mathbf{x}\|$ by (4.3). This means that $t = \|\mathbf{x}\|_{\text{euc}} \leq (1/m)\|\mathbf{x}\| = M\|\mathbf{x}\|$, where $M = 1/m$. □

Theorem 4.2.14 *Any two norms on a vector space are equivalent.*

Proof. Let $\| \cdot \|_i$, $i = 1, 2$, be any two norms, and let $\| \cdot \|_{\text{euc}}$ be a Euclidean norm on a vector space X. Use Lemmas 4.2.11 and 4.2.13 to find L_i, $M_i \in \mathbb{R}$ such that

$$\|\mathbf{x}\|_i \leq L_i \, \|\mathbf{x}\|_{\text{euc}} \text{ and } \|\mathbf{x}\|_{\text{euc}} \leq M_i \, \|\mathbf{x}\|_i \text{ for } i = 1, 2 \text{ and for all } \mathbf{x} \in X.$$

Then $\|\mathbf{x}\|_1 \leq L_1 \, M_2 \|\mathbf{x}\|_2$ and $\|\mathbf{x}\|_2 \leq L_2 \, M_1 \|\mathbf{x}\|_1$ for all $\mathbf{x} \in X$. □

Theorem 4.2.15 *A set in a vector space is bounded with respect to one norm if and only if it is bounded with respect to any other norm. A sequence* $\mathbf{x}_n$ *in a vector space* X *converges to* $\mathbf{a}$ *with respect to one norm if and only if it also converges to* $\mathbf{a}$ *with respect to any other norm.*

Proof. Let $\|\cdot\|$ and $\|\cdot\|'$ be two norms on X. Let $\|\cdot\| \leq L\|\cdot\|'$ and $\|\cdot\|' \leq M\|\cdot\|$. If a set A in X is bounded in $\|\cdot\|$, then there is a $K \in \mathbb{R}$ such that $\|\mathbf{a}\| \leq K$ for all $\mathbf{a} \in A$. In this case $\|\mathbf{a}\|' \leq KM$ for all $\mathbf{a} \in A$. Hence, A is also bounded in $\|\cdot\|'$. The proof for convergence is similar. □

Theorem 4.2.16 **Bolzano-Weierstrass theorem in normed spaces.** *Every bounded sequence in a normed space has a convergent subsequence.*

Proof. Let $\|\cdot\|$ be the given norm on X. Let $\|\cdot\|'$ be a Euclidean norm on X. If $\mathbf{x}_n$ is bounded in $(X, \|\cdot\|)$, then by Theorem 4.2.15 it is also bounded in $(X, \|\cdot\|')$. By the Bolzano-Weierstrass theorem in Euclidean spaces, 4.2.9, it has a subsequence converging in $(X, \|\cdot\|')$. Applying Theorem 4.2.15 again, this subsequence also converges in $(X, \|\cdot\|)$. □

Sequences in Cartesian Product Spaces

In addition to the Bolzano-Weierstrass theorem, the following basic result is a second consequence of the equivalence of all norms on a finite-dimensional normed space. It tells us that convergence of a sequence in a product space is equivalent to the convergence of all of the component sequences.

Theorem 4.2.17 *Let* $U_1, \cdots U_n$ *be normed spaces and let* $X = U_1 \times \cdots \times U_n$. *Let* $\mathbf{x}_{k,i}$ *be a sequence in* U_i *and let*

$$\mathbf{x}_k = (\mathbf{x}_{k,1}, \ldots \mathbf{x}_{k,n})$$

be the corresponding sequence in X. *Then*

$$\lim_k \mathbf{x}_k = \mathbf{x} = (\mathbf{x}_1, \ldots, \mathbf{x}_n) \text{ in } X$$

if and only if each $\lim_k \mathbf{x}_{k,i} = \mathbf{x}_i$ *in* U_i *for each* $i = 1, \ldots, n$.

Proof. Convergence is independent of the norm used in a space. This follows from Theorem 4.2.15 above. A convenient norm to use in X is

$$\|\mathbf{x}\| = \|(\mathbf{x}_1, \ldots, \mathbf{x}_n)\| = \max(\|\mathbf{x}_1\|, \ldots, \|\mathbf{x}_n\|).$$

It is easy to see that this is indeed a norm on X. Then it is clear that

$$\|\mathbf{x}_k - \mathbf{x}\| = \max(\|\mathbf{x}_{k,1} - \mathbf{x}_1\|, \ldots, \|\mathbf{x}_{k,n} - \mathbf{x}_n\|) \to 0$$

if and only if $\|\mathbf{x}_{k,i} - \mathbf{x}_i\| \to 0$ for each $i = 1, \ldots, n$. □

Cauchy Sequences

Definition 4.2.18 Cauchy sequences in normed spaces. A sequence $\mathbf{x}_n$ in a normed space X is called a *Cauchy sequence* if for each $\varepsilon > 0$ there is an integer $N \in \mathbb{N}$ such that $\|\mathbf{x}_n - \mathbf{x}_m\| < \varepsilon$ for all $m, n \geq N$.

Theorem 4.2.19 *A sequence in a normed space is a Cauchy sequence if and only if it is a convergent sequence.*

Proof. This proof is almost the same as the proof of the analogous Theorem 2.4.7 for Cauchy sequences of real numbers. First, suppose that $\mathbf{x}_n$ is a convergent sequence and let $\mathbf{x}_n \to \mathbf{a}$. Given $\varepsilon > 0$, find $N \in \mathbb{N}$ such that $\|\mathbf{x}_n - \mathbf{a}\| < \varepsilon/2$ for all $n \geq N$. Then

$$\|\mathbf{x}_n - \mathbf{x}_m\| \leq \|\mathbf{x}_n - \mathbf{a}\| + \|\mathbf{a} - \mathbf{x}_m\| < (\varepsilon/2) + (\varepsilon/2) = \varepsilon$$

for all $m, n \geq N$. Hence $\mathbf{x}_n$ is a Cauchy sequence.

Conversely, assume that $\mathbf{x}_n$ is a Cauchy sequence. Since $\mathbf{x}_n$ must then be bounded, we can apply the Bolzano-Weierstrass theorem to find a convergent subsequence $\mathbf{x}_{n_k}$. Let $\mathbf{a} = \lim \mathbf{x}_{n_k}$. Then for each $\varepsilon > 0$ there is a K such that $\|\mathbf{x}_{n_k} - \mathbf{a}\| < \varepsilon/2$ for all $k \geq K$. Since $\mathbf{x}_n$ is Cauchy, we may take K large enough so that $\|\mathbf{x}_n - \mathbf{x}_m\| < \varepsilon/2$ for all $m, n \geq K$. Pick some n_k large enough that both $k, n_k \geq K$. If $n \geq K$, then

$$\|\mathbf{x}_n - \mathbf{a}\| \leq \|\mathbf{x}_n - \mathbf{x}_{n_k}\| + \|\mathbf{x}_{n_k} - \mathbf{a}\|.$$

But $\|\mathbf{x}_n - \mathbf{x}_{n_k}\| < \varepsilon/2$ since $n, n_k \geq K$. Also $\|\mathbf{x}_{n_k} - \mathbf{a}\| < \varepsilon/2$ since $k \geq K$. Therefore $\|\mathbf{x}_n - \mathbf{a}\| < \varepsilon$ for all $n \geq K$. Hence $\mathbf{x}_n \to \mathbf{a}$. □

Example 4.2.20 Let $\mathbf{a}$ be a nonzero vector in a normed space X. Let t_n be a sequence of real numbers. Then $t_n\mathbf{a}$ converges if and only if t_n converges. Indeed, since $\|\mathbf{a}\| > 0$ and $|t_n - t_m|\,\|\mathbf{a}\| = \|t_n\mathbf{a} - t_m\mathbf{a}\|$, we see that $t_n\mathbf{a}$ is Cauchy if and only if t_n is Cauchy. Hence, by Theorem 4.2.19, $t_n\mathbf{a}$ converges if and only if t_n converges.

Theorem 4.2.21 *Let $\mathbf{u}_n$ be a sequence in a normed space X. Let $\mathbf{x}_n = \sum_{i=1}^{n} \mathbf{u}_i$ be the sequence of partial sums. If $\sum_n \|\mathbf{u}_n\| < \infty$, then $\mathbf{x}_n$ is a Cauchy sequence in X. If $\lim_n \mathbf{x}_n = \mathbf{a}$, then $\|\mathbf{x}_n - \mathbf{a}\| \leq \sum_{i>n} \|\mathbf{u}_i\|$.*

Proof. The sequence $S_n = \sum_{i=1}^{n} \|\mathbf{u}_n\|$ is a monotone and bounded sequence in $\mathbb{R}$. Therefore it converges by the monotone convergence theorem, Theorem 2.3.13. It follows that S_n is a Cauchy sequence in $\mathbb{R}$. Hence, given $\varepsilon > 0$, there is an $N \in \mathbb{N}$ such that $0 \leq S_{n+k} - S_n < \varepsilon$ for all $n \geq N$ and for all $k \in \mathbb{N}$. Hence

$$\begin{aligned}
\|\mathbf{x}_n - \mathbf{x}_{n+k}\| &\leq \|\mathbf{x}_n - \mathbf{x}_{n+1}\| + \cdots + \|\mathbf{x}_{n+k-1} - \mathbf{x}_{n+k}\| \\
&= \|\mathbf{u}_{n+1}\| + \cdots + \|\mathbf{u}_{n+k}\| = S_{n+k} - S_n < \varepsilon
\end{aligned}$$

whenever $n \geq N$ and $k \in \mathbb{N}$. This proves that $\mathbf{x}_n$ is a Cauchy sequence.

For the last part we have

$$\lim_k \|\mathbf{x}_n - \mathbf{x}_{n+k}\| = \|\mathbf{x}_n - \mathbf{a}\| \leq \lim_k (S_{n+k} - S_n) = \sum_{i>n} \|\mathbf{u}_i\| .$$

Here the first equality follows from Lemma 4.2.3 and from the observation that $\lim_k(\mathbf{x}_n - \mathbf{x}_{n+k}) = \mathbf{x}_n - \mathbf{a}$. □

Problems

4.14 If A, B are bounded subsets of a normed space X, show that $rA + sB$ is also a bounded subset of X for any scalars r, s.

4.15 Let $\mathbf{x}_n$ be a sequence in a normed space. Let $f : \mathbb{N} \to \mathbb{N}$ be an increasing function. If $\mathbf{x}_n$ converges, show that the sequence $\mathbf{x}_{f(n)}$ also converges.

4.16 Let $\mathbf{x}_n$ and $\mathbf{y}_n$ be convergent sequences in a normed space X. Let t, s be scalars. Show that the sequence $s\mathbf{x}_n + t\mathbf{y}_n$ is convergent and

$$\lim_{n\to\infty} (s\mathbf{x}_n + t\mathbf{y}_n) = s \lim_{n\to\infty} \mathbf{x}_n + t \lim_{n\to\infty} \mathbf{y}_n .$$

Also, show that

$$\lim_{n\to\infty} \|\mathbf{x}_n\| = \left\| \lim_{n\to\infty} \mathbf{x}_n \right\| .$$

4.17 Define norms $\| \ \|$ and $\| \ \|'$ on $\mathbb{R}^2$ by

$$\|\mathbf{x}\| = |x_1| + |x_2|, \quad \|\mathbf{x}\|' = \max\{|x_1|, |x_2|\} \quad \text{for all } \mathbf{x} = (x_1, x_2) \in \mathbb{R}^2.$$

Show directly that $\| \ \|$ and $\| \ \|'$ are equivalent norms on $\mathbb{R}^2$.

4.18 Let $\mathbf{x}_n$ be a sequence in a normed space X. For each $k \in \mathbb{N}$, let $\mathbf{y}_k = \mathbf{x}_{2k}$ and let $\mathbf{z}_k = \mathbf{x}_{2k-1}$. Show that $\mathbf{x}_n$ converges to $\mathbf{a}$ if and only if both $\mathbf{y}_n$ and $\mathbf{z}_n$ converge to $\mathbf{a}$.

4.19 Let X be an inner product space, and let $\| \ \|$ be the norm induced by the inner product. Suppose that $\mathbf{x}_n$ is an orthogonal sequence in X that converges. Find $\lim_{n\to\infty} \mathbf{x}_n$.

4.20 Let $\mathbf{x}_k$ be a sequence in a normed space. For $n \in \mathbb{N}$, set

$$\mathbf{t}_n = \sum_{k=1}^{n} (\mathbf{x}_{k+1} - \mathbf{x}_k), \quad d_n = \sum_{k=1}^{n} \|\mathbf{x}_{k+1} - \mathbf{x}_k\|.$$

Show that $\mathbf{t}_n$ converges if and only if $\mathbf{x}_k$ converges. If $\mathbf{x}_k$ converges, must d_n converge?

4.21 Suppose that $\mathbf{a}_n$ is a sequence in a normed space that converges to some $\mathbf{a}$. Show that

$$\lim_{n\to\infty} \frac{1}{n}(\mathbf{a}_1 + \cdots + \mathbf{a}_n) = \mathbf{a}.$$

4.22 For $k = 1, \ldots, n$, let $\|\cdot\|_k$ be norm on X_k. Is it true that there is a constant C such that whenever $\mathbf{x}_1 \in X_1, \ldots, \mathbf{x}_k \in X_k$, then

$$\|\mathbf{x}_1\|_1^2 + \cdots + \|\mathbf{x}_n\|_n^2 \le C \max_{1\le k\le n} (\|\mathbf{x}_k\|_k)?$$

4.23 Let X be a vector space with bases $\{\mathbf{u}_1, \ldots, \mathbf{u}_m\}$ and $\{\mathbf{v}_1, \ldots, \mathbf{v}_m\}$. Suppose $S \subset X$ and assume that there is a constant M such that whenever $\mathbf{x} \in S$ with $\mathbf{x} = c_1\mathbf{u}_1 + \cdots + c_m\mathbf{u}_m$, then $\max_{1\le k\le m} |c_k| \le M$. Show that there is a constant C such that whenever $\mathbf{x} \in S$ and $\mathbf{x} = d_1\mathbf{v}_1 + \cdots + d_m\mathbf{v}_m$, then

$$|d_1| + \cdots + |d_m| \le C.$$

4.24 Show that there is a positive constant A such that

$$\frac{x_1^2 + \cdots + x_n^2}{(1 + |x_1| + \cdots + |x_n|)^2} \ge A \quad \text{for all } \mathbf{x} \in \mathbb{R}^n \text{ with } \max_{1\le k\le n} |x_k| = 1.$$

4.25 Let $\mathbf{a}_k = (a_{k,0}, a_{k,1}, \ldots, a_{k,n})$ be a bounded sequence in $\mathbb{R}^{n+1}$. Show that there is an increasing sequence m_k in $\mathbb{N}$, and a polynomial p of degree no more than n such that

$$\lim_{k\to\infty} \max_{t\in[0,1]} |a_{m_k,0} + a_{m_k,1}t + \cdots + a_{m_k,n}t^n - p(t)| = 0.$$

4.3 NORMS OF LINEAR AND MULTILINEAR TRANSFORMATIONS

A linear map $T : X \to Y$ between two normed spaces has an important property called *boundedness*. This property leads to the definition of a natural norm on the vector space $L(X, Y)$ of all linear maps $X \to Y$.

Theorem 4.3.1 Boundedness of linear transformations. *Let X and Y be two normed spaces. Let $T : X \to Y$ be a linear transformation. Then there is an $M \in \mathbb{R}$*

such that $\|T\mathbf{x}\|_Y \leq M\,\|\mathbf{x}\|_X$ *for all* $\mathbf{x} \in X$. *Here,* $\|\ \|_X$ *and* $\|\ \|_Y$ *denote the norms on* X *and* Y.

Proof. Define $\|\mathbf{x}\| = \|T\mathbf{x}\|_Y + \|\mathbf{x}\|_X$ for all $\mathbf{x} \in X$. Then $\|\cdot\|$ is a norm on X. (What goes wrong if we try $\|\mathbf{x}\| = \|T\mathbf{x}\|_Y$?) Hence, $\|\cdot\|$ and $\|\cdot\|_X$ are equivalent. Thus, there is a constant C such that

$$\|T\mathbf{x}\|_Y + \|\mathbf{x}\|_X \leq C\|\mathbf{x}\|_X \quad \text{for all } \mathbf{x} \in X.$$

Thus, with $M = C - 1$, the result follows. □

Example 4.3.2 Let X and Y be two normed spaces. Let $T : X \to Y$ be an isomorphism (Definition 3.1.45). Then here are positive constants A, B such that

$$A\,\|\mathbf{x}\| \leq \|T\mathbf{x}\| \leq B\,\|\mathbf{x}\| \quad \text{for all } \mathbf{x} \in X.$$

This follows immediately from the fact that $\|\mathbf{x}\|' = \|T\mathbf{x}\|_Y$ is a norm (since T is an isomorphism), and hence equivalent to $\|\mathbf{x}\|$.

Theorem 4.3.3 Norms of linear transformations. *Let* X *and* Y *be two normed spaces. Let* $T : X \to Y$ *be a linear function. Then*

$$\|T\| = \sup\{\, \|T\mathbf{x}\| \mid \|\mathbf{x}\| = 1,\ \mathbf{x} \in X \,\} \tag{4.4}$$

exists and defines a norm on $L(X, Y)$.

Proof. If $T \in L(X, Y)$, then Theorem 4.3.1 shows that there is an $M \in \mathbb{R}$ such that $\|T\mathbf{x}\| \leq M$ whenever $\|\mathbf{x}\| = 1$, $\mathbf{x} \in X$. Hence the set in (4.4) is contained in the interval $[\,0,\ M\,]$. Therefore $\|T\|$ exists.

We check that $\|T\|$ satisfies the defining conditions for norms stated in Definition 4.1.1. Clearly $\|T\| \geq 0$ for all $T \in L(X, Y)$ and $\|T\| = 0$ if and only if $T = \mathbf{0}_{L(X,Y)}$, that is, if and only if $T\mathbf{x} = \mathbf{0}_Y$ for all $\mathbf{x} \in X$. Hence the positive definiteness of $\|T\|$ follows. Now let $T \in L(X, Y)$ and $t \in \mathbb{R}$. Then

$$\begin{aligned} \|tT\| &= \sup\{\, \|tT\mathbf{x}\| \mid \mathbf{x} \in X,\ \ \|\mathbf{x}\| = 1 \,\} \\ &= \sup\{\, |t|\,\|T\mathbf{x}\| \mid \mathbf{x} \in X,\ \ \|\mathbf{x}\| = 1 \,\} \\ &= |t| \sup\{\, \|T\mathbf{x}\| \mid \mathbf{x} \in X,\ \ \|\mathbf{x}\| = 1 \,\} = |t|\,\|T\|\,. \end{aligned}$$

This shows that $\|T\|$ is homogeneous. Note that the third of these equalities depends upon the following observation: if B is a bounded set of numbers, $r \geq 0$, and $B' = \{\, rs \mid s \in B \,\}$, then $\sup B' = r \sup B$. Finally, to verify the triangle inequality, let $T,\ S \in L(X, Y)$ and $\mathbf{x} \in X$, $\|\mathbf{x}\| = 1$. Then

$$\|(T + S)\mathbf{x}\| = \|T\mathbf{x} + S\mathbf{x}\| \leq \|T\mathbf{x}\| + \|S\mathbf{x}\| \leq \|T\| + \|S\|.$$

Hence $\|T + S\| \leq \|T\| + \|S\|$. □

Definition 4.3.4 Standard norms of linear transformations. The *standard norm on* $L(X, Y)$ is the norm defined in the preceding theorem, in terms of the norms on X and Y. Hence, with this standard norm, $\|T\mathbf{x}\|_Y \leq \|T\|_{L(X,Y)} \|\mathbf{x}\|_X$ for all $T \in L(X, Y)$ and for all $\mathbf{x} \in X$, $\mathbf{y} \in Y$.

Example 4.3.5 Let $T : \mathbb{R}^2 \to \mathbb{R}^2$ be given by $T\mathbf{u} = (x + y, x - y)$ for all $\mathbf{u} = (x, y) \in \mathbb{R}^2$. Then with respect to the standard Euclidean norm on $\mathbb{R}^2$,

$$\|T\mathbf{u}\|^2 = (x + y)^2 + (x - y)^2 = 2(x^2 + y^2) = 2\|\mathbf{u}\|^2.$$

Hence, $\|T\| \leq \sqrt{2}$. Since $\|T\mathbf{e}_1\| = \sqrt{2}$, it follows that $\|T\| = \sqrt{2}$.

Example 4.3.6 Let X, Y be (finite-dimensional) vector spaces. Let $\| \; \|_X, \| \; \|'_X$ and $\| \; \|_Y, \| \; \|'_Y$ be pairs of norms on X and Y, respectively. Then there is a constant C such that if $T : X \to Y$ is linear and $\|T\|$ is the norm of T with respect to $\| \; \|_X, \| \; \|_Y$, then the norm $\|T\|'$ of T with respect to $\| \; \|'_X$ and $\| \; \|'_Y$ satisfies

$$\|T\|' \leq C\|T\|.$$

The reason is that $L(X, Y)$ is a finite-dimensional normed vector space under the two different norms $T \mapsto \|T\|$ and $T \mapsto \|T\|'$. Consequently, the norms $\| \; \|$ and $\| \; \|'$ on $L(X, Y)$ are equivalent, and the desired conclusion follows.

Theorem 4.3.7 *If* $T \in L(X, Y)$, *then* $\|T\mathbf{x}\| \leq \|T\| \, \|\mathbf{x}\|$ *for all* $\mathbf{x} \in X$.

Proof. If $\|\mathbf{x}\| = 1$, then the result follows from the definition of $\|T\|$ in Theorem 4.3.3 above. Any $\mathbf{x} \in X$ can be expressed as $\mathbf{x} = t\mathbf{x}_0$ with $t \in \mathbb{R}$ and $\|\mathbf{x}_0\| = 1$, and in this case $\|\mathbf{x}\| = |t| \, \|\mathbf{x}_0\| = |t|$. Therefore

$$\|T\mathbf{x}\| = \|T(t\mathbf{x}_0)\| = |t| \, \|T\mathbf{x}_0\| \leq |t| \, \|T\| = \|T\| \, \|\mathbf{x}\|$$

as claimed. □

Example 4.3.8 Let $T : X \to Y$ be a linear map between normed spaces. If x_n is a Cauchy sequence in X, then Tx_n is also a Cauchy sequence in Y. This is because $\|Tx_n - Tx_m\| = \|T(x_n - x_m)\| \leq \|T\| \, \|x_n - x_m\|$ for all n, m.

Theorem 4.3.9 *If* $R \in L(X, Y)$ *and* $S \in L(Y, Z)$, *then* $\|SR\| \leq \|S\| \, \|R\|$. *Here* $SR \in L(X, Z)$ *is the composition of* R *and* S.

Proof. If $\mathbf{x} \in X$, then

$$\|(SR)\mathbf{x}\| = \|S(R\mathbf{x})\| \leq \|S\| \, \|R\mathbf{x}\| \leq \|S\| \, \|R\| \, \|\mathbf{x}\|.$$

This shows that $\|(SR)\mathbf{x}\| \leq \|S\| \, \|R\|$ whenever $\|\mathbf{x}\| = 1$. □

Example 4.3.10 Let X, Y be nontrivial normed spaces. Let $T : X \to Y$ be an isomorphism. Then $I_Y = TT^{-1}$ is the identity mapping on Y. Clearly, $\|I_Y\| = 1$ (since Y is nontrivial, there is a non-zero vector $\mathbf{y}$ with $I_Y(\mathbf{y}) = \mathbf{y}$). Thus,

$$1 = \|I_Y\| = \|TT^{-1}\| \leq \|T\| \, \|T^{-1}\|.$$

Also, since T is not the 0 mapping, $\|T\| > 0$. Hence,

$$(1/\|T\|) \leq \|T^{-1}\|.$$

Norms on Multilinear Functions

Let U_i and Y be normed spaces. It will be convenient to write $X_k = U_k \times \cdots \times U_1$ and let $ML_k(X_k,\, Y)$ denote the corresponding vector space of k-linear functions. There is a standard way to define a norm on $ML_k(X_k,\, Y)$, using induction on k.

For the induction step, we use the natural isomorphism

$$\vartheta : ML_{k+1}(X_{k+1},\, Y) \to L(U_{k+1},\, ML_k(X_k,\, Y))$$

defined in Definition 3.3.7. If $F \in ML_{k+1}(X_{k+1},\, Y)$, then F is a function of $(k+1)$ variables $\mathbf{u}_{k+1},\, \mathbf{u}_k,\, \ldots,\, \mathbf{u}_1$. For each $\mathbf{u}_{k+1} \in U_{k+1}$, we obtain $\vartheta F(\mathbf{u}_{k+1})$ by holding $\mathbf{u}_{k+1}$ fixed and considering F as a function of the remaining k variables. That is:

$$\vartheta F(\mathbf{u}_{k+1})(\mathbf{u}_k,\, \ldots,\, \mathbf{u}_1) = F(\mathbf{u}_{k+1},\, \mathbf{u}_k,\, \ldots,\, \mathbf{u}_1).$$

This defines $\vartheta F(\mathbf{u}_{k+1}) \in ML_k(X_k,\, Y)$ for each $\mathbf{u}_{k+1} \in U_{k+1}$.

Definition 4.3.11 Norm on $ML_k(X_k,\, Y)$. For $k = 1$, the norm on

$$ML_1(X_1,\, Y) = L(U_1,\, Y)$$

is just the standard norm on $L(U_1,\, Y)$. For the inductive step, we may suppose that the norm on $ML_k(X_k,\, Y)$ is defined. We know that there is a standard norm on

$$L(U_{k+1},\, ML_k(X_k,\, Y)).$$

If $F \in ML_{k+1}(X_{k+1},\, Y)$, then we define

$$\|F\|_{ML_{k+1}(X_{k+1},\, Y)} = \|\vartheta F\|_{L(U_{k+1},\, ML_k(X_k,\, Y))}.$$

These norms on $ML_k(X_k, Y)$ are defined as the *standard norms* on these spaces; as in the case of linear transformations, they depend upon the norms on the underlying vector spaces.

Remarks 4.3.12 The following theorems refer to several normed spaces and several norms, all of which should be clear from the context. To illustrate, suppose $F \in ML_{k+1}(X_{k+1}, Y)$. Then

$\|F\|$	is the standard norm on	$ML_{k+1}(X_{k+1}, Y)$.
$\|F(\mathbf{u}_{k+1}, \mathbf{u}_k, \ldots, \mathbf{u}_1)\|$	is the norm on	Y.
$\|\vartheta F\|$	is the norm on	$L(U_{k+1}, ML_k(X_k, Y))$.
$\|\vartheta F(\mathbf{u}_{k+1})\|$	is the norm on	$ML_k(X_k, Y)$.
$\|\vartheta F(\mathbf{u}_{k+1})(\mathbf{u}_k, \ldots, \mathbf{u}_1)\|$	is the norm on	Y.

The following theorem shows that the standard norm for multilinear functions has a useful property analogous to that of the standard norm for linear functions.

Theorem 4.3.13 *If* $F \in ML_k(X_k, Y)$, *then*

$$\|F(\mathbf{x})\| = \|F(\mathbf{u}_k, \ldots, \mathbf{u}_1)\| \leq \|F\| \cdot \|\mathbf{u}_k\| \cdots \|\mathbf{u}_1\| \tag{4.5}$$

for all $\mathbf{x} = (\mathbf{u}_k, \ldots, \mathbf{u}_1) \in X_k = U_k \times \cdots \times U_1$.

Proof. If $k = 1$, then (4.5) is true by Theorem 4.3.7. Suppose, for induction, that (4.5) is true for $k \in \mathbb{N}$. Let $F \in ML_{k+1}(X_{k+1}, Y)$. Then for all $\mathbf{x} = (\mathbf{u}_{k+1}, \mathbf{u}_k, \ldots, \mathbf{u}_1) \in X_{k+1}$,

$$\begin{aligned}
\|F(\mathbf{x})\| &= \|F(\mathbf{u}_{k+1}, \mathbf{u}_k, \ldots, \mathbf{u}_1)\| \\
&= \|\vartheta F(\mathbf{u}_{k+1})(\mathbf{u}_k, \ldots, \mathbf{u}_1)\| \\
&\leq \|\vartheta F(\mathbf{u}_{k+1})\| \cdot \|\mathbf{u}_k\| \cdots \|\mathbf{u}_1\| \\
&\leq \|\vartheta F\| \cdot \|\mathbf{u}_{k+1}\| \cdot \|\mathbf{u}_k\| \cdots \|\mathbf{u}_1\| \\
&= \|F\| \cdot \|\mathbf{u}_{k+1}\| \cdot \|\mathbf{u}_k\| \cdots \|\mathbf{u}_1\|.
\end{aligned}$$

Here the first inequality follows from the inductive hypothesis and the second inequality from Theorem 4.3.7. The last equality follows from the definition of $\|F\|$ in Definition 4.3.11. □

The next theorem will be useful in our study of differentiation.

Theorem 4.3.14 Increments of multilinear maps. *Let* $U_1, \ldots, U_k$ *and* Z *be normed spaces, let* $X_k = U_k \times \cdots \times U_1$, *and suppose that* $F \in ML_k(X_k, Z)$. *Let*

$$\mathbf{x} = (\mathbf{u}_k, \ldots, \mathbf{u}_1) \in X_k \text{ and } \mathbf{y} = (\mathbf{v}_k, \ldots, \mathbf{v}_1) \in X_k.$$

If $\|\mathbf{u}_i\| \leq R$, $\|\mathbf{v}_i\| \leq R$, *and* $\|\mathbf{u}_i - \mathbf{v}_i\| \leq \alpha$ *for each* $i = 1, \ldots, k$, *then*

$$\|F(\mathbf{x}) - F(\mathbf{y})\| \leq \|F\| \cdot kR^{k-1}\,\alpha. \tag{4.6}$$

Proof. Proceed by induction on $k \in \mathbb{N}$. If $k = 1$, then (4.6) reduces to

$$\|F(\mathbf{u}_1) - F(\mathbf{v}_1)\| \leq \|F\|\,\alpha \text{ for linear } F : U_1 \to Z.$$

This follows from Theorem 4.3.7. Assume for induction that (4.6) holds for k. Let $F \in ML_{k+1}(X_{k+1}, Z)$. To simplify the notation, let's write

$$(\mathbf{u}_{k+1}, \mathbf{u}_k, \ldots, \mathbf{u}_1) = (\mathbf{u}_{k+1}, \mathbf{u}) \text{ and } (\mathbf{v}_{k+1}, \mathbf{v}_k, \ldots, \mathbf{v}_1) = (\mathbf{v}_{k+1}, \mathbf{v}).$$

Note that $\|F(\mathbf{u}_{k+1}, \mathbf{u}) - F(\mathbf{v}_{k+1}, \mathbf{v})\|$ is dominated by the sum

$$\|F(\mathbf{u}_{k+1}, \mathbf{u}) - F(\mathbf{u}_{k+1}, \mathbf{v})\| + \|F(\mathbf{u}_{k+1}, \mathbf{v}) - F(\mathbf{v}_{k+1}, \mathbf{v})\|.$$

We estimate these two terms separately. For the first term we have

$$\begin{aligned}
\|F(\mathbf{u}_{k+1}, \mathbf{u}) - F(\mathbf{u}_{k+1}, \mathbf{v})\| &= \|\vartheta F(\mathbf{u}_{k+1})(\mathbf{u}) - \vartheta F(\mathbf{u}_{k+1})(\mathbf{v})\| \\
&\leq \|\vartheta F(\mathbf{u}_{k+1})\| \cdot kR^{k-1}\,\alpha \\
&\leq \|\vartheta F\| \cdot \|\mathbf{u}_{k+1}\| \cdot kR^{k-1}\,\alpha \\
&\leq \|F\| \cdot kR^k\,\alpha.
\end{aligned}$$

Here the first inequality follows from the induction hypothesis. The second inequality follows from Theorem 4.3.7 applied to the linear map ϑF. For the second term,

$$\begin{aligned}
\|F(\mathbf{u}_{k+1}, \mathbf{v}) - F(\mathbf{v}_{k+1}, \mathbf{v})\| &= \|\vartheta F(\mathbf{u}_{k+1})(\mathbf{v}) - \vartheta F(\mathbf{v}_{k+1})(\mathbf{v})\| \\
&= \|(\vartheta F(\mathbf{u}_{k+1}) - \vartheta F(\mathbf{v}_{k+1}))(\mathbf{v})\| \\
&\leq \|\vartheta F(\mathbf{u}_{k+1} - \mathbf{v}_{k+1})\| \cdot R^k \\
&\leq \|\vartheta F\| \cdot \|\mathbf{u}_{k+1} - \mathbf{v}_{k+1}\| \cdot R^k \\
&\leq \|F\| \cdot R^k\,\alpha.
\end{aligned}$$

The first inequality follows from Theorem 4.3.13. In the second inequality, we again apply Theorem 4.3.7 to the linear map ϑF. Adding these two estimates, we obtain (4.6) for $(k + 1)$. □

Example 4.3.15 Theorem 4.3.14 generalizes a familiar situation. Let each $U_i = \mathbb{R}$ with the absolute value norm. Define $F \in ML_k(X_k, \mathbb{R})$ by

$$F(u_1, \ldots, u_k) = u_1 \cdots u_k.$$

We can easily show that $\|F\| = 1$ by induction on k. Hence, if $|u_i|$, $|v_i| \leq R$ and $|u_i - v_i| \leq \alpha$, then Theorem 4.3.14 gives

$$|(u_1 \cdots u_k) - (v_1 \cdots v_k)| \leq kR^{k-1}\alpha.$$

In particular, $|u^k - v^k| \leq kR^{k-1}\alpha$ if $|u|$, $|v| \leq R$, $|u - v| \leq \alpha$.

Problems

4.26 Let $T : \mathbb{R}^2 \to \mathbb{R}^2$ be given by $T\mathbf{u} = (x+2y, 3x-y)$ for all $\mathbf{u} = (x, y) \in \mathbb{R}^2$. With respect to the standard Euclidean norm on $\mathbb{R}^2$, show that $\|T\| \leq \sqrt{11}$.

4.27 Let X be a nonzero normed space and let $T \in L(X, Y)$ with $T \neq \mathbf{0}$. What is $\{ \|T\mathbf{x}\| \mid \mathbf{x} \in X \}$?

4.28 Let $T \in L(X, Y)$ with $\|T\| \leq 1$. Show that if $\mathbf{u}, \mathbf{v}$ are vectors in the unit ball $\{ \mathbf{x} \in X \mid \|\mathbf{x}\| \leq 1 \}$, then $\|T\mathbf{u} - T\mathbf{v}\| \leq 2$.

4.29 Let $T : (X, \| \, \|) \to Y$ be a linear map between normed spaces. Let $L : X \to X$ be an isomorphism such that $\|L\mathbf{x}\| = 1$ for all $\mathbf{x} \in X$ with $\|\mathbf{x}\| = 1$. Define a norm $\| \, \|'$ on X by $\|\mathbf{x}\|' = \|L\mathbf{x}\|$ for all $\mathbf{x} \in X$. If A is the norm of $T : (X, \| \, \|) \to Y$, show that the norm B of $T : (X, \| \, \|') \to Y$ satisfies $B \geq A$.

4.30 Consider $\mathbb{R}^n$ with the norm $\|\mathbf{x}\| = (x_1^2 + \cdots + x_n^2)^{1/2}$. Let $a_1, \ldots, a_n$ be constants and define $T : \mathbb{R}^n \to \mathbb{R}^n$ by $T\mathbf{x} = (a_1 x_1, \ldots, a_n x_n)$ for all $\mathbf{x} = (x_1, \ldots, x_n) \in \mathbb{R}^n$. Show that $\|T\| \geq \max_{1 \leq k \leq n} |a_k|$.

4.31 For $k = 1, \ldots, n$, let $T_k : Y \to X_k$ be linear maps between normed spaces. Let $\| \, \|_k$ be the norm on X_k. Consider the norm $\| \, \|'$ on $\mathbf{X} = X_1 \times \cdots \times X_k$ given by

$$\|(\mathbf{x}_1, \ldots, \mathbf{x}_n)\|' = (\|\mathbf{x}_1\|_1^2 + \cdots + \|\mathbf{x}_n\|_n^2)^{\frac{1}{2}}.$$

Let $T : Y \to \mathbf{X}$ by $T\mathbf{y} = (T_1\mathbf{y}, \ldots, T_n\mathbf{y})$ for all $\mathbf{y} \in Y$. Show that T is linear. Is it true that

$$\|T\|^2 \leq \|T_1\|^2 + \cdots + \|T_n\|^2?$$

4.32 Suppose that X, Y are normed spaces. Let $T_n \in L(X, Y)$ be such that $\|T_n\| \leq 1$ for all $n \in \mathbb{N}$. Show that there is some $T \in L(X, Y)$ with $\|T\| \leq 1$ and some subsequence T_{n_k} of T_n such that

$$\|T - T_{n_k}\| \to 0.$$

4.33 Let X, Y be normed spaces. For $\mathbf{x} \in X$, define $\widehat{\mathbf{x}} : L(X, Y) \to Y$ by $\widehat{\mathbf{x}}(T) = T(\mathbf{x})$ for all $T \in L(X, Y)$. Show that $\widehat{\mathbf{x}}$ is a bounded linear map with $\|\widehat{\mathbf{x}}\| \leq \|\mathbf{x}\|$.

4.34 Let $T \in L(X, X)$. Recall that λ is an eigenvalue of T if there is some $\mathbf{u} \in X$ with $\mathbf{u} \neq \mathbf{0}$ such that $T\mathbf{u} = \lambda\mathbf{u}$. Show that for all eigenvalues λ of T, we have

$$|\lambda| \leq \|T\|.$$

4.35 Let $T \in L(X, X)$ and suppose that there is some $S \in L(X, X)$ such that $S^2 = T$. Then $\sqrt{\|T\|} \leq \|S\|$. True or false?

4.36 Let X, Y be normed spaces and let $L : X \to Y$ be linear. Define $F : \mathbb{R} \times X \to Y$ by $F(r, \mathbf{x}) = rL\mathbf{x}$ for all $(r, \mathbf{x}) \in \mathbb{R} \times X$. Compute $\|F\|$.

4.4 CONTINUITY IN NORMED SPACES

There are two familiar definitions of continuity of a function at a point. We start with a lemma that proves the equivalence of these two conditions.

Lemma 4.4.1 *Let X and Y be two normed spaces, $\mathbf{a} \in A \subset X$, and $f : A \to Y$ a function. Then the following are equivalent.*

(1) For each $\varepsilon > 0$ there is a $\delta > 0$ such that $\|f(\mathbf{x}) - f(\mathbf{a})\| < \varepsilon$ whenever $\|\mathbf{x} - \mathbf{a}\| < \delta$ and $\mathbf{x} \in A$.

(2) If $\mathbf{x}_n$ is a sequence in A and if $\mathbf{x}_n \to \mathbf{a}$ in X, then $f(\mathbf{x}_n) \to f(\mathbf{a})$ in Y.

Proof. Assume (1). Let $\mathbf{x}_n$ be a sequence in A and assume that $\mathbf{x}_n \to \mathbf{a}$. Given $\varepsilon > 0$, find $\delta > 0$ as in (1). Since $\mathbf{x}_n \to \mathbf{a}$, there is an N such that $\|\mathbf{x}_n - \mathbf{a}\| < \delta$ for all $n \geq N$. Hence, $\|f(\mathbf{x}_n) - f(\mathbf{a})\| < \varepsilon$ for all $n \geq N$. Therefore $f(\mathbf{x}_n) \to f(\mathbf{a})$ and (2) follows.

Conversely, assume that (1) is not true. Then there is an $\alpha > 0$ with the following property. For each $\delta > 0$ there is an $\mathbf{x} \in A$ such that $\|\mathbf{x} - \mathbf{a}\| < \delta$ but $\|f(\mathbf{x}_n) - f(\mathbf{a})\| \geq \alpha$. In particular, for each $n \in \mathbb{N}$ there is $\mathbf{x}_n$ such that

$$\|\mathbf{x}_n - \mathbf{a}\| < (1/n) \text{ but } \|f(\mathbf{x}_n) - f(\mathbf{a})\| \geq \alpha.$$

Then $\mathbf{x}_n \to \mathbf{a}$ but $f(\mathbf{x}_n) \not\to f(\mathbf{a})$. Hence (2) is not true. Therefore (1) and (2) are equivalent. □

Definition 4.4.2 Continuity of functions. Let X and Y be two normed spaces, $\mathbf{a} \in A \subset X$, and $f : A \to Y$ a function. Then f is said to be *continuous at* $\mathbf{a}$ if f satisfies one of the equivalent conditions above in Lemma 4.4.1. If f is continuous at each $\mathbf{a} \in A$, then f is said to be *continuous on* A.

Remarks 4.4.3 Continuity and norms. As proved in Theorem 4.2.15, the convergence of a sequence in a vector space is independent of the choice of the norm on that space. As a result, the continuity of a function between the vector spaces is independent of the choice of norms on these spaces.

Remarks 4.4.4 General theorems on continuity. Most of the functions we shall consider in this course are linear, multilinear, or obtained from linear and multilinear functions by composition, taking limits, or taking inverses. First, we establish the continuity of linear and multilinear functions. Then we show that continuity is preserved under composition, taking limits, and taking inverses, with some mild restrictions. This will establish the continuity of almost all the functions we consider in this course. Even with these functions, however, there may be some special points at which continuity is not clear. This happens especially in taking inverses (which includes division, the inverse of multiplication). The following three special examples illustrate how to proceed in such cases. The first example is a natural one; the others are contrived to reveal possible difficulties.

Special Examples

Example 4.4.5 Define $f : \mathbb{R}^2 \to \mathbb{R}$ by

$$f(x,\, y) = \begin{cases} xy/(x^2 + y^2) & \text{if } (x,\, y) \neq (0,\, 0) \\ 0 & \text{if } (x,\, y) = (0,\, 0). \end{cases}$$

Is f continuous at $(0,\, 0)$? The restriction of f to the x-axis or to the y-axis is identically the zero function. That is continuous everywhere. But we can draw no general conclusion about the continuity of f from these restrictions. Suppose we restrict f to the line $y = mx$, $m \neq 0$. This restriction is the function $\varphi : \mathbb{R} \to \mathbb{R}$ defined as $\varphi(x) = f(x,\, mx)$ for all $x \in \mathbb{R}$. We see that $\varphi(x) = m/(1 + m^2)$ if $x \neq 0$ and $\varphi(0) = 0$. Hence φ is discontinuous at $x = 0$. Therefore f is also discontinuous at $(0,\, 0)$. △

Example 4.4.6 Define $f : \mathbb{R}^2 \to \mathbb{R}$ by

$$f(x,\, y) = \begin{cases} 1 & \text{if } x \neq 0 \text{ and } y = 2x^2 \\ 0 & \text{otherwise.} \end{cases}$$

See Figure 4.1. This function is discontinuous at $(0,\, 0)$. We can see this by restricting f to the parabola $y = 2x^2$. This restriction is the function $\varphi : \mathbb{R} \to \mathbb{R}$ defined as $\varphi(x) = f(x,\, 2x^2)$ for all $x \in \mathbb{R}$. We see that $\varphi(x) = 1$ if $x \neq 0$ and $\varphi(0) = 0$. Hence φ is discontinuous at $x = 0$. Therefore f is also discontinuous at $(0,\, 0)$.

Note that unlike the previous example, the restriction of f to any straight line passing through $(0,\, 0)$ is continuous at $(0,\, 0)$. Certainly, the restrictions of f to the coordinate axes are identically zero. Now consider a line $y = mx$, $m \neq 0$. Then $\varphi(x) = f(x,\, mx)$ is obtained as follows. First, $\varphi(0) = f(0,\, 0) = 0$. Let $x \neq 0$. Then $\varphi(x) = 1$ if $y = mx = 2x^2$, that is, if $x = m/2$, and $\varphi(x) = 0$ if $x \neq m/2$. This function is continuous at $x = 0$. (It is discontinuous at $x = m/2$, but this is not important.)

This example shows that a function may have continuous restrictions to all straight lines passing through a point but still be discontinuous at that point. △

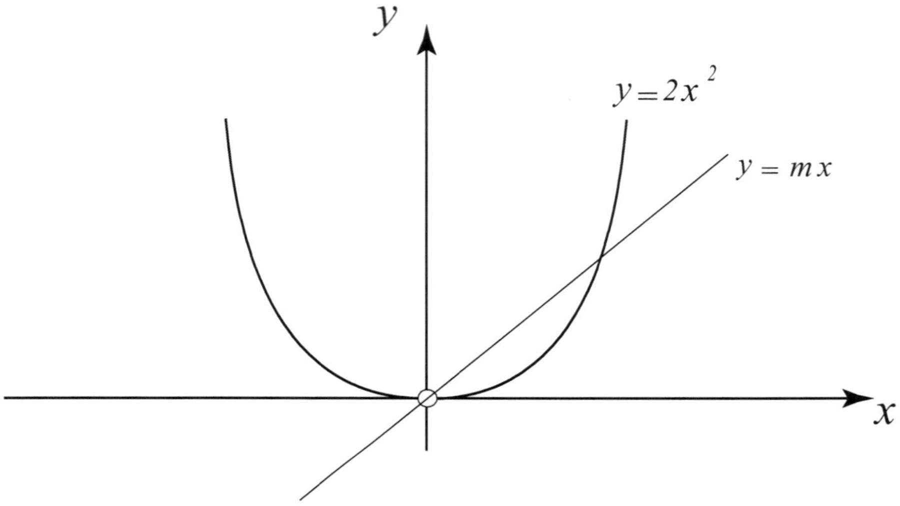

Figure 4.1. For Example 4.4.6.

Example 4.4.7 Define $f : \mathbb{R}^2 \to \mathbb{R}$ by

$$f(x, y) = \begin{cases} (y - x^2)(3x^2 - y)/x^4 & \text{if } x \neq 0 \text{ and } x^2 < y < 3x^2 \\ 0 & \text{otherwise.} \end{cases}$$

This function is even more pathological than the one in Example 4.4.6. It is discontinuous at $(0, 0)$ even though its restriction to any straight line is continuous everywhere. (Recall that the restriction of the previous function to $y = mx$ was continuous at $(0, 0)$ but discontinuous at $(m/2, m^2/2)$.) △

Continuity of Linear and Multilinear Functions

Theorem 4.4.8 Continuity of linear functions. *A linear map $T : X \to Y$ is continuous on X.*

Proof. Let $M \in \mathbb{R}$ be such that $\|T\mathbf{x}\| \leq M \|\mathbf{x}\|$, $\mathbf{x} \in X$. Let $\mathbf{x}_n \to \mathbf{a}$. Then

$$\|T\mathbf{x}_n - T\mathbf{a}\| = \|T(\mathbf{x}_n - \mathbf{a})\| \leq M \|\mathbf{x}_n - \mathbf{a}\| \to 0.$$

Hence $T\mathbf{x}_n \to T\mathbf{a}$. □

Theorem 4.4.9 Continuity of multilinear functions. *A multilinear map*

$$F : X_k = U_k \times \cdots \times U_1 \to Y$$

is continuous on X_k.

Proof. As continuity is independent of the choice of the norms, we may choose a convenient norm on X_k. We will let

$$\|\mathbf{x}\| = \|(\mathbf{u}_k, \ldots, \mathbf{u}_1)\| = \max(\|\mathbf{u}_k\|, \ldots, \|\mathbf{u}_1\|).$$

It is easy to check that this is a norm on X_k. Fix $\mathbf{a} \in X_k$. Choose an $R \in \mathbb{R}$ so that $\|\mathbf{a}\| < R$. Let $\varepsilon > 0$ be given. Choose $\delta > 0$ so that

$$\|F\| \cdot kR^{k-1}\delta < \varepsilon.$$

Without loss of generality, assume that $\|\mathbf{a}\| + \delta < R$. We now apply Theorem 4.3.14, which allows us to estimate the increment $\|F(\mathbf{x}) - F(\mathbf{a})\|$. If $\|\mathbf{x} - \mathbf{a}\| < \delta$, then we see that this theorem gives

$$\|F(\mathbf{x}) - F(\mathbf{a})\| \leq \|F\| \cdot kR^{k-1}\delta < \varepsilon.$$

Hence $F : X_k \to Y$ is continuous at an arbitrary point $\mathbf{a} \in X_k$. □

Continuity in Cartesian Products

Theorem 4.4.10 *Let X_is and Y_is be normed spaces and consider the Cartesian product spaces*

$$X = X_k \times \cdots \times X_1 \text{ and } Y = Y_k \times \cdots \times Y_1.$$

Let $A_i \subset X_i$ and $A = A_k \times \cdots \times A_1$. Let $f_i : A_i \to Y_i$ be a function for each $i = 1, \ldots, k$. Define $f : A \to Y$ as

$$f(\mathbf{x}) = (f_k(\mathbf{x}_k), \ldots, f_1(\mathbf{x}_1))$$

for all $\mathbf{x} = (\mathbf{x}_k, \ldots, \mathbf{x}_1) \in A$. Then f is continuous at $\mathbf{a} = (\mathbf{a}_k, \ldots, \mathbf{a}_1) \in A$ if and only if each f_i is continuous at $\mathbf{a}_i \in A_i$.

Proof. Let $\mathbf{x}_n = (\mathbf{x}_{kn}, \ldots, \mathbf{x}_{1n}) \in A$ be a sequence. Theorem 4.2.17 shows that this sequence converges to $\mathbf{a}$ in X if and only if each component $\mathbf{x}_{ni}$ converges to $\mathbf{a}_i$ in X_i. Similarly

$$f(\mathbf{x}_n) = (f_k(\mathbf{x}_{kn}), \ldots, f_1(\mathbf{x}_{1n})) \in Y$$

converges to $f(\mathbf{a})$ in Y if and only if each component $f_i(\mathbf{x}_{in})$ converges to $f_i(\mathbf{a}_i)$ in Y_i. Then the proof follows from the definition of continuity. □

Continuity Under Compositions

Let X, Y, Z, be normed spaces, and $A \subset X, B \subset Y$. Let $f : A \to Y$ and $g : B \to Z$ be two functions. Let $h = g \circ f$ be the composite function, $h(\mathbf{x}) = g(f(\mathbf{x}))$ for all $\mathbf{x} \in C$, where $C = A \cap f^{-1}(B)$ is the set of all $\mathbf{x} \in A$ such that $f(\mathbf{x}) \in B$.

Theorem 4.4.11 *If f is continuous at $\mathbf{a} \in C$ and g is continuous at $\mathbf{b} = f(\mathbf{a})$, then $h = g \circ f$ is continuous at $\mathbf{a}$.*

Proof. Let $\mathbf{x}_n$ be a sequence in C such that $\mathbf{x}_n \to \mathbf{a}$. Then $f(\mathbf{x}_n) = \mathbf{y}_n \to \mathbf{b} = f(\mathbf{a})$, since f is continuous at $\mathbf{a}$. But $\mathbf{y}_n$ is a sequence in B and $\mathbf{y}_n \to \mathbf{b}$. Hence, $h(\mathbf{x}_n) = g(\mathbf{y}_n) \to g(\mathbf{b}) = h(\mathbf{a})$, since g is continuous at $\mathbf{b}$. This shows that h is continuous at $\mathbf{a}$. □

Applications

Most of the following results could be proven directly from the definitions of convergence and continuity. We present them here as consequences of the preceding results about the continuity of linear and multilinear functions, the composition of continuous functions, and Cartesian products.

Lemma 4.4.12 *Let $a_n \to a$ in $\mathbb{R}$ and $\mathbf{x}_n \to \mathbf{x}$ in a normed space X. Then $a_n\mathbf{x}_n \to a\mathbf{x}$ in X.*

Proof. We apply Theorem 4.2.17 on sequences in Cartesian products. We see that $(a_n, \mathbf{x}_n) \in \mathbb{R} \times X$ converges to $(a, \mathbf{x})$ in $\mathbb{R} \times X$. We then use the fact that multiplication by scalars,

$$M : (\mathbb{R} \times X) \to X \text{ defined by } M(r, \mathbf{u}) = r\mathbf{u},$$

is multilinear, and hence continuous. Therefore, $M(a_n, \mathbf{x}_n) = a_n\mathbf{x}_n$ converges to $M(a, \mathbf{x}) = a\mathbf{x}$ in X. □

Lemma 4.4.13 *If $\mathbf{u}_n \to \mathbf{u}$ and $\mathbf{v}_n \to \mathbf{v}$ in X, then $(\mathbf{u}_n + \mathbf{v}_n) \to (\mathbf{u} + \mathbf{v})$ in X.*

Proof. Theorem 4.2.17 shows that $(\mathbf{u}_n, \mathbf{v}_n) \to (\mathbf{u}, \mathbf{v})$ in $X \times X$. But addition $F(\mathbf{a}, \mathbf{b}) = \mathbf{a} + \mathbf{b}$ is a linear operation $F : (X \times X) \to X$ and therefore continuous. Hence $F(\mathbf{u}_n, \mathbf{v}_n)$ converges to $F(\mathbf{u}, \mathbf{v})$ in X. This is the conclusion of the lemma. □

Theorem 4.4.14 Continuity of linear combinations. *Let X and Y be normed spaces. Let $r_i : A \to \mathbb{R}$ and $f_i : A \to Y$ be continuous functions, where $A \subset X$ and $i = 1, \ldots, k$. Let $f = \sum_i r_i f_i$. Then $f : A \to Y$ is continuous.*

Proof. Let $\mathbf{a} \in A$, $\mathbf{x}_n \in A$, and $\mathbf{x}_n \to \mathbf{a}$ in X. Then $r_i(\mathbf{x}_n) \to r_i(\mathbf{a})$ in $\mathbb{R}$ and $f_i(\mathbf{x}_n) \to f_i(\mathbf{a})$ in X. Then Lemmas 4.4.12 and 4.4.13 and an easy induction argument show that

$$f(\mathbf{x}_n) = \sum_i r_i(\mathbf{x}_n) f_i(\mathbf{x}_n) \to \sum_i r_i(\mathbf{a}) f_i(\mathbf{a}) = f(\mathbf{a})$$

in Y. Hence $f : A \to Y$ is continuous at each $\mathbf{a} \in A$. □

Theorem 4.4.15 Continuity of inner products. *Let X be a normed space and Y a Euclidean space. Let $f : A \to Y$ and $g : A \to Y$ be continuous, $A \subset X$. Then $\rho(\mathbf{x}) = \langle f(\mathbf{x}), g(\mathbf{x}) \rangle$ defines a continuous function $\rho : A \to \mathbb{R}$.*

Proof. Define $h : A \to (Y \times Y)$ by $h(\mathbf{x}) = (f(\mathbf{x}), g(\mathbf{x}))$, $\mathbf{x} \in A$. Then h is continuous by Theorem 4.4.10 on continuity in Cartesian products. Define the inner product function $P : (Y \times Y) \to \mathbb{R}$ by $P(\mathbf{u}, \mathbf{v}) = \langle \mathbf{u}, \mathbf{v} \rangle$, $(\mathbf{u}, \mathbf{v}) \in Y \times Y$. Since P is multilinear, it is continuous. Then $\rho = P \circ h : A \to \mathbb{R}$ is the composition of two continuous functions and hence is continuous. □

Example 4.4.16 Let $T_n \to T$ in $L(X, Y)$ and $\mathbf{x}_n \to \mathbf{x}$ in X. Then $T_n \mathbf{x}_n \to T\mathbf{x}$ in Y. To see this, note that $(T_n, \mathbf{x}_n) \to (T, \mathbf{x})$ in $L(X, Y) \times X$, again by Theorem 4.2.17. Also, the mapping

$$F : (L(X, Y) \times X) \to Y$$

defined by $F(T, \mathbf{x}) = T\mathbf{x} \in Y$, for $(T, \mathbf{x}) \in (L(X, Y) \times X)$, is multilinear, and hence continuous. Therefore $T_n \mathbf{x}_n = F(T_n, \mathbf{x}_n) \to F(T, \mathbf{x}) = T\mathbf{x}$ in Y. △

Continuity of the Inverse Function

Let X and Y be two normed spaces, $A \subset X$. If a function $f : A \to Y$ is one-to-one on A, then it has an inverse function $g : B \to X$, where $B = f(A)$. Assume that $f : A \to Y$ is continuous. In general, $g : B \to X$ is not continuous, as we show by means of two counterexamples. The first example is rather artificial; the second is more natural.

Example 4.4.17 Let $X = Y = \mathbb{R}$. Let

$$A = \{\, 1/k \mid k \in \mathbb{N} \,\} \subset [0, 1].$$

Define $f : A \to \mathbb{R}$ by $f(1) = 0$ and $f(1/k) = 1/k$ if $k \geq 2$. Then $f : A \to \mathbb{R}$ is continuous. Actually, any $F : A \to \mathbb{R}$ is continuous on A. The reason is that if $x_n \in A$ is a sequence in A and if $x_n \to a \in A$, then there is an $N \in \mathbb{N}$ such that $x_n = a$ for all $n \geq N$. (Why?) Any convergent sequence in A must eventually be a constant sequence. Hence $F(x_n) \to F(a)$ for any $F : A \to \mathbb{R}$.

Now the particular function f defined above is one-to-one on A. Let $B = f(A)$, and let the inverse inverse of f be $g : B \to \mathbb{R}$. We see that $0 \in B$ and the sequence $y_n = 1/n$, $n \geq 2$ is a sequence in B converging to $0 \in B$. But $g(y_n) = (1/n) \not\to g(0) = 1$. Therefore g is discontinuous at $0 \in B$. △

Example 4.4.18 Let $X = \mathbb{R}$, $Y = \mathbb{R}^2$, $A = [0, 1)$. We denote the points in X as $t \in \mathbb{R}$ and the points in Y as $(x, y) \in \mathbb{R}^2$. We define $f : A \to \mathbb{R}^2$ as

$$f(t) = (\cos 2\pi t, \sin 2\pi t), \quad 0 \leq t < 1\,.$$

Then $B = f(A)$ is the unit circle. We see that f is one-to-one and continuous on A. But the inverse function g is not continuous at $(1, 0) \in B$. To see this, define a sequence (x_n, y_n) as follows:

$$(x_n, y_n) = \begin{cases} (\cos 2\pi/n, \sin 2\pi/n) & \text{if } n \text{ is odd} \\ (\cos 2\pi(n-1)/n, \sin 2\pi(n-1)/n) & \text{if } n \text{ is even.} \end{cases}$$

Then we see that $(x_n, y_n) \in B$ converges to $(1, 0) \in B$. But

$$t_n = g(x_n, y_n) = \begin{cases} 1/n & \text{if } n \text{ is odd} \\ (n-1)/n & \text{if } n \text{ is even} \end{cases}$$

does not converge in $\mathbb{R}$. Therefore g is not continuous at $(1, 0)$. △

These counterexamples are due to a deficiency in the domain A of the original function $f : A \to Y$. If A satisfies one natural additional condition, then the inverse of a continuous and invertible function on A is, in fact, continuous. This additional condition is called *compactness.*

Definition 4.4.19 Compact sets. Let X be normed space and $C \subset X$. Then C is called a *compact set* if every sequence in C has a subsequence converging to a point in C.

Example 4.4.20 Let A, B be compact subsets of a normed space X. Then $A + B$ is also compact. To see this, let $\mathbf{c}_n$ be a sequence in $A + B$. Then there are sequences $\mathbf{a}_n$ in A and $\mathbf{b}_n$ in B such that $\mathbf{c}_n = \mathbf{a}_n + \mathbf{b}_n$ for all n. Since A is compact, there is a subsequence $\mathbf{a}_{n_k}$ of $\mathbf{a}_n$ that converges to $\mathbf{a} \in A$. Now $\mathbf{b}_{n_k}$ is a sequence in B, and by the compactness of B, it has a subsequence $\mathbf{b}_{m_j}$ that converges to $\mathbf{b} \in B$. Since

$\mathbf{a}_{m_j}$ is a subsequence of $\mathbf{a}_{n_k}$, it must also converge to $\mathbf{a}$. Thus, $\mathbf{c}_{m_j}$ is a subsequence of $\mathbf{c}_n$ and $\mathbf{c}_{m_j} = \mathbf{a}_{m_j} + \mathbf{c}_{m_j} \rightarrow \mathbf{a} + \mathbf{b} \in A + B$. Hence, $A + B$ is compact.

In the next section we will provide several equivalent formulations of compactness, including one that is easy to verify. Hence the following theorem is a useful and important result.

Theorem 4.4.21 *Let X and Y be two normed spaces. Let A be a compact subset of X. Let $f : A \rightarrow Y$ be a one-to-one and continuous function on A. Then the inverse function $g : B \rightarrow X$ is continuous on $B = f(A)$.*

Proof. Let $\mathbf{b} \in B$. Hence $\mathbf{b} = f(\mathbf{a})$ and $\mathbf{a} = g(\mathbf{b})$. We will show that $g : B \rightarrow X$ is continuous at $\mathbf{b}$. Let $\mathbf{y}_n$ be a sequence in B and $\mathbf{y}_n \rightarrow \mathbf{b}$. Let $\mathbf{x}_n = g(\mathbf{y}_n)$. We must show that $\mathbf{x}_n \rightarrow \mathbf{a}$.

If $\mathbf{x}_n \not\rightarrow \mathbf{a}$, then $\|\mathbf{x}_n - \mathbf{a}\| \not\rightarrow 0$. In this case, there is a number $\alpha > 0$ and a subsequence $\mathbf{x}_k$, $k \in \mathbb{K}$, such that $\|\mathbf{x}_k - \mathbf{a}\| \geq \alpha$ for all $k \in \mathbb{K}$. Now $\mathbf{x}_k$, $k \in \mathbb{K}$, is still a sequence in the compact set A. Hence it has a subsequence $\mathbf{x}_\ell$, $\ell \in \mathbb{L}$, that converges to a point $\mathbf{a}' \in A$. We see that $\|\mathbf{a} - \mathbf{a}'\| \geq \alpha > 0$. Hence $\mathbf{a}' \neq \mathbf{a}$ and therefore $f(\mathbf{a}) \neq f(\mathbf{a}')$, since f is one-to-one on A. But then $\mathbf{y}_\ell = f(\mathbf{x}_\ell) \rightarrow f(\mathbf{a}) \neq f(\mathbf{a}')$, which means that f is not continuous at $\mathbf{a}' \in A$. This contradiction shows that $\mathbf{x}_n \rightarrow \mathbf{a}$. □

Continuity Under Limits

Consider a sequence of functions $f_n : A \rightarrow Y$ defined on a set A in a normed space X and taking values in another normed space Y. Assume that the sequence $f_n(\mathbf{x})$ in Y converges for each $\mathbf{x} \in A$. In this case $\lim_n f_n(\mathbf{x}) = f(\mathbf{x})$ defines a new limit function $f : A \rightarrow Y$, and we say that f_n converges *pointwise* to f. Suppose that each $f_n : A \rightarrow Y$ is continuous. Can we conclude that f is also continuous on A? Not in general, but we do get continuity if we assume *uniformity of the convergence*. We prove this result below, but first we offer a counterexample for the general case where we have only pointwise convergence.

Example 4.4.22 A sequence of continuous functions with a discontinuous limit. Let $X = Y = \mathbb{R}$ and $A = [0, 1]$. Let $f_n(x) = x^n$ for each $n \in \mathbb{N}$ and for each $x \in A$. Each f_n is a polynomial and therefore is continuous on A. Also, $f(x) = \lim_n f_n(x) = \lim_n x^n$ exists for each $x \in A$. But $f(x) = 0$ if $0 \leq x < 1$ and $f(1) = 1$. Hence $f(x)$ is discontinuous at $x = 1 \in A$. △

Lemma 4.4.23 *Let X and Y be two normed spaces and $A \subset X$. Let $f : A \rightarrow Y$ be a function with the following property: for each $\varepsilon > 0$, there is a continuous function*

$g : A \to Y$ such that $\|f(\mathbf{x}) - g(\mathbf{x})\| < \varepsilon$ for all $\mathbf{x} \in A$. Then $f : A \to Y$ is a continuous function on A.

Proof. To show that $f : A \to Y$ is continuous at $\mathbf{a} \in A$, we will show that for each $\varepsilon > 0$ there is a $\delta > 0$ such that $\|f(\mathbf{x}) - f(\mathbf{a})\| < \varepsilon$ whenever $\mathbf{x} \in A$ and $\|\mathbf{x} - \mathbf{a}\| < \delta$. Given $\varepsilon > 0$, first find a continuous function $g : A \to Y$ such that

$$\|f(\mathbf{x}) - g(\mathbf{x})\| < \varepsilon/3 \text{ for all } \mathbf{x} \in A.$$

Then find a $\delta > 0$ such that $\|g(\mathbf{x}) - g(\mathbf{a})\| < \varepsilon/3$ whenever $\mathbf{x} \in A$ and $\|\mathbf{x} - \mathbf{a}\| < \delta$. This can be done since $g : A \to Y$ is continuous at $\mathbf{a} \in A$. Now if $\mathbf{x} \in A$ and $\|\mathbf{x} - \mathbf{a}\| < \delta$, then

$$\begin{aligned} \|f(\mathbf{x}) - f(\mathbf{a})\| &= \|f(\mathbf{x}) - g(\mathbf{x}) + g(\mathbf{x}) - g(\mathbf{a}) + g(\mathbf{a}) - f(\mathbf{a})\| \\ &\le \|f(\mathbf{x}) - g(\mathbf{x})\| + \|g(\mathbf{x}) - g(\mathbf{a})\| + \|g(\mathbf{a}) - f(\mathbf{a})\| \\ &< (\varepsilon/3) + (\varepsilon/3) + (\varepsilon/3) = \varepsilon. \end{aligned}$$

Hence $f : A \to Y$ is continuous at $\mathbf{a} \in A$. □

Definition 4.4.24 Uniform convergence. Let X and Y be two normed spaces and $A \subset X$. For each $n \in \mathbb{N}$, let $f_n : A \to Y$ be a function. Let $f : A \to Y$ be another function. Then f_n is said to *converge to f uniformly on A* if the following condition is satisfied: for each $\varepsilon > 0$ there is an $N \in \mathbb{N}$ such that $\|f_n(\mathbf{x}) - f(\mathbf{x})\| < \varepsilon$ for all $n \ge N$ and for all $\mathbf{x} \in A$.

Remarks 4.4.25 Convergence and uniform convergence. Uniform convergence on A is a stronger condition than convergence at each point in A. The sequence $f_n(x) = x^n$ considered in Example 4.4.22 provides a good illustration. We see that for each $x \in A = [0, 1]$ the sequence of numbers $f_n(x) = x^n$ converges to $f(x)$ in $\mathbb{R}$. Here $f(x) = 0$ if $0 \le x < 1$ and $f(1) = 1$. This sequence does not converge uniformly on A. In fact, given any $n \in \mathbb{N}$, we can find a point $a \in A$, $0 \le a < 1$, such that $|f_n(a) - f(a)| = a^n \ge (1/2)$. We simply take $a = (1/2)^{1/n}$. Hence there is no $n \in \mathbb{N}$ such that $|f_n(x) - f(x)| < (1/2)$ for all $x \in A$, and the convergence of f_n to f is not uniform. Note that this also follows from Theorem 4.4.29 below.

Definition 4.4.26 Uniform continuity. Let X and Y be two normed spaces and $A \subset X$. Suppose $f : A \to Y$. f is *uniformly continuous on A* if for each $\varepsilon > 0$ there is $\delta > 0$ such that for all $\mathbf{x}_1, \mathbf{x}_2 \in A$, $\|f(\mathbf{x}_1) - f(\mathbf{x}_2)\| < \varepsilon$ whenever $\|\mathbf{x}_1 - \mathbf{x}_2\| < \delta$.

It should be clear that uniform continuity of a function on A implies its continuity on A. The concept of uniform continuity is needed for the following example and in the proof of Theorem 4.4.29.

Example 4.4.27 Let $f_n : A \to Y$ and let $g : Y \to Z$, where $A \subset X$ and X, Y, Z are normed spaces. Assume that g is uniformly continuous and the sequence f_n converges uniformly on A. Then $g \circ f_n$ also converges uniformly on A. To see this, suppose that f_n converges uniformly on A to $f : A \to Y$. Let $\epsilon > 0$ be given. Then by the uniform continuity of g, there is some $\delta > 0$ such that whenever $\mathbf{y}_1, \mathbf{y}_2$ are in Y and $\|\mathbf{y}_1 - \mathbf{y}_2\| < \delta$, then $\|g(\mathbf{y}_1) - g(\mathbf{y}_2)\| < \epsilon$. Since f_n converges uniformly on A to f, there is some N such that

$$\|f_n(\mathbf{x}) - f(\mathbf{x})\| < \delta \quad \text{for all } n \geq N \text{ and all } \mathbf{x} \in A.$$

Hence, for all $n \geq N$ and all $\mathbf{x} \in A$,

$$\|g(f_n(\mathbf{x})) - g(f(\mathbf{x}))\| < \epsilon.$$

Thus, $g \circ f_n$ converges uniformly on A to $g \circ f$.

Example 4.4.28 Let $S_n(x) = \sum_{k=1}^{n} (1/k^2) \sin(kx)$ for all $x \in A$, where $A \subset R$. Then S_n converges uniformly on A. More generally, let f_k be a sequence of real-valued functions defined on a set A, and let M_k be a sequence of real numbers such that $|f_k(x)| \leq M_k$ for all $x \in A$. If $\sum_{k=1}^{\infty} M_k < \infty$, then $\sum_{k=1}^{\infty} f_k$ converges uniformly on A. To see this, let $\epsilon > 0$. Put $S_n(x) = \sum_{k=1}^{n} f_k(x)$ for all $x \in A$ and all $n \in \mathbb{N}$. Since $\sum_{k=1}^{\infty} M_k < \infty$, there is some N such that

$$\sum_{k=m+1}^{n} M_k < \epsilon \quad \text{for all } n > m \geq N.$$

Then for all $n > m \geq N$ and all $x \in A$,

$$|S_n(x) - S_m(x)| = \left| \sum_{k=m+1}^{n} f_k(x) \right| \leq \sum_{k=m+1}^{n} |f_k(x)| \leq \sum_{k=m+1}^{n} M_k < \epsilon.$$

Hence, $S_n(x)$ is a Cauchy sequence for all $x \in A$. Let $S(x) = \lim_{n \to \infty} S_n(x)$ for all $x \in A$. By letting $n \to \infty$ in the above inequalities, we get

$$|S(x) - S_m(x)| \leq \epsilon \quad \text{for all } m \geq N \text{ and all } x \in A.$$

This shows that S_n converges uniformly on A to S.

Theorem 4.4.29 Uniform limits of continuous functions. *Let X and Y be two normed spaces and $A \subset X$. Let $f_n : A \to Y$ be a sequence of continuous functions on A converging uniformly on A to a function $f : A \to Y$. Then $f : A \to Y$ is also a continuous function on A.*

Proof. We see that $f : A \to Y$ satisfies the hypothesis of Lemma 4.4.23 above. Indeed, given any $\varepsilon > 0$, there is a continuous function $f_n : A \to Y$ such that $\|f_n(\mathbf{x}) - f(\mathbf{x})\| < \varepsilon$ for all $\mathbf{x} \in A$. Then the same lemma shows that $f : A \to Y$ is continuous. □

Sequences of Polynomials

Almost all of the functions we consider in the course, apart from counterexamples, are the uniform limits of polynomials. We now give a sufficient condition for the uniform convergence of a sequence of polynomials.

Notations 4.4.30 Sequences of polynomials. Let X and Y be normed spaces. If $S \in ML_n(X^n, Y)$, then

$$f(\mathbf{x}) = S(\mathbf{x}, \ldots, \mathbf{x}), \quad \mathbf{x} \in X,$$

defines a corresponding homogeneous polynomial $f : X \to Y$. For each $n \in \mathbb{N}$, suppose $S_n \in ML_n(X^n, Y)$ and let $f_n : X \to Y$ be the corresponding homogeneous polynomial. Then $F_n = \sum_{i=1}^{n} f_i$, $n \in \mathbb{N}$, defines a sequence of polynomials. Finally, for each $r > 0$, let

$$B_r = \{\, \mathbf{x} \mid \mathbf{x} \in X, \ \ \|\mathbf{x}\| < r \,\}.$$

Theorem 4.4.31 Continuity of homogeneous polynomials. *With the notations in 4.4.30, $f : X \to Y$ is continuous. Also, $\|f(\mathbf{x})\| \le \|S\| \, \|\mathbf{x}\|^n$, $\mathbf{x} \in X$.*

Proof. Define $T : X \to X^n$ by $T\mathbf{x} = (\mathbf{x}, \ldots, \mathbf{x})$. Then T is linear and therefore continuous. Also, $S : X^n \to Y$ is multilinear and therefore continuous. We see that $f = S \cdot T : X \to Y$ is the composition of two continuous functions. Hence, $f : X \to Y$ is also continuous. Also,

$$\|S(\mathbf{x}_n, \ldots, \mathbf{x}_1)\| \le \|S\| \cdot \|\mathbf{x}_n\| \cdots \|\mathbf{x}_1\|$$

by Theorem 4.3.13. Hence $\|f(\mathbf{x})\| \le \|S\| \cdot \|\mathbf{x}\|^n$ follows. □

Theorem 4.4.32 Convergence of polynomials. *With the notations in 4.4.30, assume that there is an $R > 0$ such that $\sum_n \|S_n\| \, r^n < \infty$ whenever $0 \le r < R$. Then*

$$\lim_n F_n(\mathbf{x}) = F(\mathbf{x}) \ \ \textit{exists in } Y \textit{ for each } \mathbf{x} \in B_R$$

and defines a continuous function $F : B_R \to Y$.

Proof. If $\mathbf{x} \in B_R$, then $\|\mathbf{x}\| < R$. Let r be such that $\|x\| < r < R$. Hence

$$\|f_n(\mathbf{x})\| \le \|S_n\| \, \|\mathbf{x}\|^n \le \|S_n\| \, r^n.$$

Then Theorem 4.2.21 shows that $F_n(\mathbf{x})$ is a Cauchy sequence in X and that $F(\mathbf{x}) = \lim_n F_n(\mathbf{x})$ exists in Y. The same theorem also shows that

$$\begin{aligned} \|F(\mathbf{x}) - F_n(\mathbf{x})\| &\le \sum_{k>n} \|f_k(\mathbf{x})\| \le \sum_{k>n} \|S_k\| \, \|\mathbf{x}\|^k \\ &< \sum_{k>n} \|S_k\| \, r^k = L - L_n, \end{aligned}$$

where $L_n = \sum_{k=1}^{n} \|S_k\| r^k$ and $L = \lim_n L_n$ in $\mathbb{R}$. Hence, given $\varepsilon > 0$, we can find an $N \in \mathbb{N}$ such that $(L - L_n) < \varepsilon$ for all $n \geq N$. This shows that the convergence $F_n(\mathbf{x}) \to F(\mathbf{x})$ is uniform for $\mathbf{x} \in B_r$. Hence Theorem 4.4.29 shows that $F : B_r \to Y$ is continuous. This implies continuity on B_R, since each $\mathbf{x} \in B_R$ is contained in some B_r, $r < R$. □

The Inversion Operator

The passage from an invertible function to its inverse function is called the *inversion operation*. We will consider this operation only on the class of invertible linear transformations. The main result is that inversion is a continuous function (with respect to the standard norm). The proof of this fact relies upon an elegant analogy between series of real numbers and series of linear mappings.

Definition 4.4.33 The inversion operator. Let $L_0(X, Y)$ be the set of all invertible linear mappings $T : X \to Y$, a subset of $L(X, Y)$. Let

$$\mathrm{Inv} : L_0(X, Y) \to L(Y, X)$$

be the function defined as $\mathrm{Inv}(T) = T^{-1} \in L(Y, X)$. Then Inv is called the *inversion operator* (on $L_0(X, Y)$).

Remarks 4.4.34 Nonlinearity of the inversion. The set of invertible linear maps, $L_0(X, Y)$, is a subset of $L(X, Y)$ but not a *subspace* of $L(X, Y)$. Hence $L_0(X, Y)$ is not a vector space and Inv is not a linear operator. The following special case makes this point even more obvious.

Example 4.4.35 Inversion of linear mappings $\mathbb{R} \to \mathbb{R}$. Any linear mapping $\mathbb{R} \to \mathbb{R}$ is just multiplication by a constant. Hence $L(\mathbb{R}, \mathbb{R})$ can be identified with $\mathbb{R}$: if $a \in \mathbb{R}$, then $a \in L(\mathbb{R}, \mathbb{R})$ is the linear mapping $x \to ax$. This mapping is invertible if and only if $a \neq 0$. Hence

$$L_0(\mathbb{R}, \mathbb{R}) = \{\, a \mid a \in \mathbb{R},\ \ a \neq 0 \,\}.$$

The inversion operator $\mathrm{Inv} : L_0(\mathbb{R}, \mathbb{R}) \to L(\mathbb{R}, \mathbb{R})$ is $\mathrm{Inv}(a) = 1/a$ for all $a \neq 0$.

Example 4.4.36 A series expression for inversion in $L_0(\mathbb{R}, \mathbb{R})$. We continue with the previous example. If $|a| < 1$, then we know that the geometric series

$$q_n(a) = 1 + a + a^2 + \cdots + a^n$$

converges to $(1 - a)^{-1} = \mathrm{Inv}(1 - a)$. The result of the inversion can be expressed as the limit of a sequence of polynomials. More generally, if $p \neq 0$ and if $|t/p| < 1$, then the geometric series

$$q_n(t/p) = 1 + (t/p) + (t/p)^2 + \cdots + (t/p)^n$$

converges to $p\ (p-t)^{-1} = p\ \mathrm{Inv}(p-t)$. We shall show that an analogous result holds in general.

Notations 4.4.37 Let X be a normed space. Let $L(X,\ X)$ be the normed space of all linear transformations $X \to X$, using the standard norm for linear transformations as defined in Definition 4.3.4 and Theorem 4.3.3. Recall that $\|SR\| \leq \|S\|\ \|R\|$ for all $S,\ R \in L(X,\ X)$, as shown in Theorem 4.3.7. Write $T^2 = T \cdot T$ for the composition of T with itself and $T^{n+1} = T \cdot T^n$ for $n \in \mathbb{N}$. The result about composition implies that for all n, $\|T^n\| \leq \|T\|^n$. Finally, recall that I is the identity mapping.

The following theorem develops the analogy with Example 4.4.36.

Theorem 4.4.38 Inversion in $L_0(X,\ X)$. *Let $P \in L_0(X,\ X)$ and put $Q = P^{-1}$. If $T \in L(X,\ X)$ and if $\|QT\| < 1$, then $(P-T) \in L_0(X,\ X)$ (i.e., $(P-T$ is invertible) and the sequence*

$$Q_n(QT) = I + (QT) + (QT)^2 + \cdots + (QT)^n$$

converges to $(P-T)^{-1}P$.

This implies that the inversion operator, Inv : $L_0(X,\ X) \to L(X,\ X)$, *is continuous.*

Proof. First, consider the special case where $P = I$. In this case, $Q = I$ as well, our assumption is that $\|T\| < 1$, and our desired conclusion is that the sequence

$$Q_n(T) = I + T + T^2 + \cdots + T^n$$

converges to $(I-T)^{-1}$. This special case is easy to prove. First, observe that $Q_n(T)$ is a Cauchy sequence in $L(X,\ X)$ because

$$\|Q_m(T) - Q_n(T)\| \leq \|T^{n+1}\| + \cdots + \|T^m\| \leq \|T\|^{n+1} + \cdots + \|T\|^m$$

and $\|T\| < 1$. So, $Q_n(T)$ converges to some $Q \in L(X,\ X)$. Then by simple manipulations, $Q_n(T)(I-T) = (I - T^{n+1})$ converges to both $Q(I-T)$ and I, showing that $Q(I-T) = I$. Similarly, $(I-T)Q = I$. Hence, $Q = (I-T)^{-1} = \mathrm{Inv}(I-T)$.

The general case follows from the special case by substituting QT for T, where $Q = P^{-1}$ as before. Since $\|QT\| < 1$, we conclude that $(I - QT)$ is invertible and

$$Q_n(QT) = I + (QT) + (QT)^2 + \cdots + (QT)^n$$

converges to $(I - QT)^{-1}$. By simple manipulations,

$$(P - T) = P(I - P^{-1}T) = P(I - QT).$$

It follows that $(P - T)^{-1} = (I - QT)^{-1}Q$ exists whenever $(I - QT)^{-1}$ exists, and $(P - T)^{-1}P = (I - QT)^{-1}$ is the limit of the sequence $Q_n(QT)$. This shows that $(P - T)^{-1} \to P^{-1}$ as $T \to 0$; hence, Inv is continuous. □

Problems

4.37 Let $f : X \to X$ be defined by $f(\mathbf{x}) = a\mathbf{x} + \mathbf{b}$, where X is a normed space, a is a constant, and $\mathbf{b}$ is a fixed vector in X. Show that f is uniformly continuous on X.

4.38 If X is a normed space and $f : X \to \mathbb{R}$, $g : X \to \mathbb{R}$ are uniformly continuous on X, must the product $fg : X \to \mathbb{R}$ be uniformly continuous on X?

4.39 Is there a continuous function $f : \mathbb{R}^3 \to \mathbb{R}$ with the following properties:

For each $k \in \mathbb{N}$, there are $\mathbf{x}_k, \mathbf{y}_k$ with $\|\mathbf{x}_k\| = \|\mathbf{y}_k\| = 1$ such that

$$\|\mathbf{x}_k - \mathbf{y}_k\| < (1/k) \quad \text{and} \quad f(\mathbf{x}_k) - f(\mathbf{y}_k) > 1?$$

4.40 Consider $\mathbb{R}^2$ with the standard Euclidean norm. Suppose that $f : \mathbb{R}^2 \to \mathbb{R}$ is continuous and $f(r, s) = 0$ whenever r is rational and s is irrational. Show that $f(x, y) = 0$ for all $(x, y) \in \mathbb{R}^2$.

4.41 Let $\| \;\|_1, \| \;\|_2$ be norms on a vector space X. Let $f : (X, \| \;\|_1) \to \mathbb{R}$ be defined by

$$f(\mathbf{x}) = \|\mathbf{x}\|_2 \quad \text{for all } \mathbf{x} \in X.$$

Show that f is uniformly continuous.

4.42 Let X be a normed space and let $L : X \to X$ be an isomorphism. Show that for any linear map $T : X \to X$, there is a constant A such that

$$\|T\mathbf{x}\| \le A\,\|L\mathbf{x}\| \quad \text{for all } \mathbf{x} \in X.$$

4.43 If $I - S \in L_0(X, X)$, must it follow that $\|S\| < 1$?

4.44 Let M be a proper subspace of $L(X, X)$. Show that

$$L_0(X, X) \cap (L(X, X) \setminus M) \neq \emptyset.$$

4.45 Suppose that $T_n \in L_0(X, X)$ for all $n \in N$ and $T_n \to T$. Must it be true that $T \in L_0(X, X)$?

4.46 Let M be a nonempty subset of a normed space X. Define $d : X \to \mathbb{R}$ by

$$d(\mathbf{x}) = \inf \{ \|\mathbf{x} - \mathbf{m}\| \mid \mathbf{m} \in M \} \quad \text{for all } \mathbf{x} \in X.$$

Is d continuous on X? Is d uniformly continuous on X?

4.47 Let X and Y be normed spaces. Let M be a nonempty compact subset of X. Let $\mathcal{C}(M, Y)$ be the collection of all continuous functions from M into $(Y, \| \ \|_Y)$. For f, g in $\mathcal{C}(M, Y)$ and scalar t, define $f + g$ and tf in the usual way: for all $\mathbf{m} \in M$,

$$\begin{aligned}(f+g)(\mathbf{m}) &= f(\mathbf{m}) + g(\mathbf{m}) \\ (tf)(\mathbf{m}) &= t(f(\mathbf{m})).\end{aligned}$$

Show that $\mathcal{C}(M, Y)$ is a vector space, which may or may not be finite-dimensional. Furthermore, show that the function

$$\|f\| = \max_{\mathbf{m} \in M} \|f(\mathbf{m})\|_Y \quad \text{for all } f \in \mathcal{C}(M, Y)$$

defines a norm on $\mathcal{C}(M, Y)$.

4.48 Let $T \in L(X, X)$. Show that

$$e^T = \lim_n \left(I + \frac{1}{1!} T + \cdots + \frac{1}{n!} T^n \right)$$

exists and defines a continuous function $e^{\cdot} : L(X, X) \to L(X, X)$.

4.5 TOPOLOGY OF NORMED SPACES

In the previous section we defined compact subsets of a normed space, but only in order to prove the continuity of inverse functions. It turns out that compactness is an extremely important property of certain sets; indeed, it plays an essential role in many significant theorems. In this section, we provide an alternative characterization of compactness in terms of the *open* subsets of a normed space. The investigation of properties of open sets in a normed space, and more generally of concepts defined in terms of open sets, is referred to as the *topology* of a normed space. This section presents some basic results about the topology of a normed space.

Open Sets

Definition 4.5.1 Open balls. Let X be a normed space. *An open ball* in X is a set of the form

$$B_r(\mathbf{a}) = \{\, \mathbf{x} \in X \mid \|\mathbf{x} - \mathbf{a}\| < r \,\}.$$

Here $\mathbf{a} \in X$ is any point and $r > 0$. More explicitly, $B_r(\mathbf{a})$ is *the open ball of radius r about (or with its center at) point* $\mathbf{a}$.

Remark 4.5.2 The shape of a ball. If $X = \mathbb{R}^2$ or $\mathbb{R}^3$ with its standard Euclidean norm, then a ball looks like an ordinary ball. With other norms, a ball has a different shape. In our illustrations, we usually represent balls as discs, keeping in mind that this may not correspond to their actual shape. Figure 4.2 shows three balls of radius d in $\mathbb{R}^2$. These balls are with respect to the norms, from left to right,

$$(x^2 + y^2)^{1/2}, \quad (|x| + |y|), \quad \text{and} \quad \max(|x|, |y|).$$

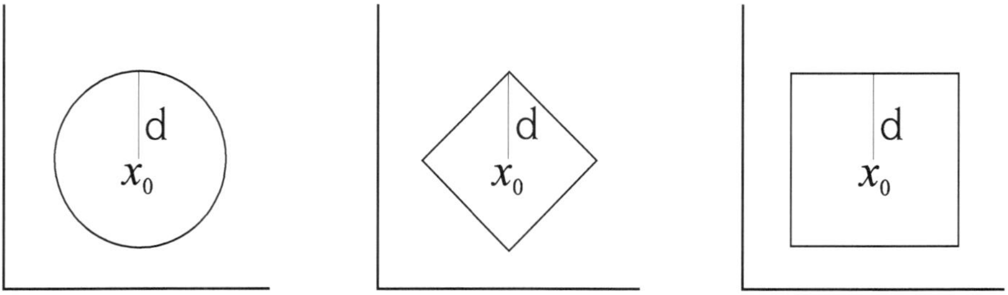

Figure 4.2. Balls of radius d in three different norms.

Definition 4.5.3 Open sets. A set G in a normed space X is called *open* if, whenever G contains a point, it also contains an open ball about that point. More explicitly, G is open if for every $\mathbf{a} \in G$ there is an $r > 0$ such that $B_r(\mathbf{a}) \subset G$. To justify the terminology of Definition 4.5.1, we show that open balls are also open sets.

Theorem 4.5.4 *An open ball is an open set.*

Proof. Let $B_r(\mathbf{a})$ be an open ball. Let $\mathbf{x} \in B_r(\mathbf{a})$. Then $p = \|\mathbf{x} - \mathbf{a}\| < r$. If $s = r - p$, then $s > 0$. We claim that $B_s(\mathbf{x}) \subset B_r(\mathbf{a})$. If $\mathbf{u} \in B_s(\mathbf{x})$, then $\|\mathbf{u} - \mathbf{x}\| < s$. Hence, by the triangle inequality 4.1.8,

$$\|\mathbf{u} - \mathbf{a}\| \le \|\mathbf{u} - \mathbf{x}\| + \|\mathbf{x} - \mathbf{a}\| < s + p = r.$$

Therefore $\mathbf{u} \in B_r(\mathbf{a})$ and, consequently, $B_s(\mathbf{x}) \subset B_r(\mathbf{a})$. □

Theorem 4.5.5 Properties of the collection of open sets. *The collection of open sets in a normed space has the following three properties.*

(1) The whole space X and the empty set $\emptyset$ are open.

(2) A finite intersection of open sets is open.

(3) Any union of open sets is open.

Proof. These conditions follow easily from the definitions. To prove (2), let $G_1, \ldots, G_n$ be open sets and let $G = \cap_{i=1}^{n} G_i$. Assume $\mathbf{x} \in G$. Since $\mathbf{x} \in G_i$ for each i, there is an $r_i > 0$ such that $B_{r_i}(\mathbf{x}) \subset G_i$. Let $r = \min_i r_i$. Then $r > 0$ and $B_r(\mathbf{x}) \subset G_i$ for all $i = 1, \ldots, n$. Hence $B_r(\mathbf{x}) \subset G$, and therefore G is also open. The proofs of parts (1) and (3) are left as exercises. □

Definition 4.5.6 Topological spaces. Let X be a nonempty set together with a collection $\mathcal{O}$ of subsets of X. If $\mathcal{O}$ contains X and $\emptyset$ and $\mathcal{O}$ is closed under finite intersections and arbitrary unions, then $\mathcal{O}$ is called a *topology on* X. A set X together with a topology is called a *topological space*. The subsets of a topological space X contained in the collection $\mathcal{O}$ are called the *open sets* in X. A concept is called a *topological concept* if it can be defined in terms of open sets only. We will show that convergence and continuity are topological concepts.

Theorem 4.5.5 above shows that a normed space becomes a topological space with the definition of open sets given in Definition 4.5.3. Note that the phrase "the topology of a normed space" sometimes refers to the collection of open sets (as in the preceding definition) and sometimes to the investigation of topological concepts (as in the opening remarks of this section). In practice, there should be no confusion.

Theorem 4.5.7 Topology of a vector space. *Any two norms on a (finite dimensional) vector space X define the same topology on X.*

Proof. Let $\|\cdot\|$ and $\|\cdot\|'$ be two norms on X. Let G be open in $\|\cdot\|$. Let $\mathbf{a} \in G$. Then there is an $r > 0$ such that $\mathbf{x} \in G$ whenever $\|\mathbf{x} - \mathbf{a}\| < r$. Theorem 4.2.14 shows that any two norms on X are equivalent. In particular, there is an $M > 0$ such that $\|\cdot\| \le M\|\cdot\|'$. Then we see that $\mathbf{x} \in G$ whenever $\|\mathbf{x} - \mathbf{a}\|' \le r/M$. Hence G is also open in $\|\cdot\|'$. The proof of the converse is the same. □

Definition 4.5.8 Neighborhoods of a point. Let X be a normed space and $\mathbf{a} \in X$. Any open set containing $\mathbf{a}$ is called a *neighborhood of* $\mathbf{a}$. Hence a nonempty open set is a neighborhood of each of the points that it contains.

Interiors, Exteriors, and Boundaries

Definition 4.5.9 Boundary points and boundaries. Let E be a set in a normed space X. A point $\mathbf{a} \in X$ is called *a boundary point of E* if every neighborhood of $\mathbf{a}$ intersects both E and its complement $E^c = X \setminus E$. The set of all boundary points of E is called *the boundary of E* and is denoted as ∂E.

Definition 4.5.10 Interiors and exteriors. Let E be a set in a normed space X. The set $E^0 = E \setminus \partial E$ is called the *interior of E*. The points in E^o are the *interior points of E*. The set $E^{\text{ext}} = X \setminus (E \cup \partial E)$ is called the *exterior of E*. The points in E^{ext} are *the exterior points of E*.

Remarks 4.5.11 Boundaries of complementary sets. If E is any set in a normed space, then $\partial E = \partial E^c$. This follows easily from the observations that the roles of E and E^c are interchangeable in Definition 4.5.10 and that $(E^c)^c = E$.

Remarks 4.5.12 Boundaries of subsets. If $A \subset B$ are subsets of a normed space, it does not follow that $\partial A \subset \partial B$. For example, take $B = \mathbb{R}$ and $A = \mathbb{Q}$. Then $\partial A = \mathbb{R}$ but $\partial B = \emptyset$.

Example 4.5.13 Let

$$\begin{aligned} D &= \left\{ (x, y, z) \in \mathbb{R}^3 \;\middle|\; x^2 + y^2 + z^2 = 1 \right\}, \\ E &= \left\{ (x, y, z) \in \mathbb{R}^3 \;\middle|\; x^2 + y^2 + z^2 < 1 \right\}, \\ F &= \left\{ (x, y, z) \in \mathbb{R}^3 \;\middle|\; x^2 + y^2 + z^2 \leq 1 \right\}. \end{aligned}$$

Then $\partial D = \partial E = \partial F = D = \left\{ (x, y, z) \in \mathbb{R}^3 \;\middle|\; x^2 + y^2 + z^2 = 1 \right\}$.

Theorem 4.5.14 Boundaries and open sets. *A set G in a normed space X is open if and only if it contains none of its boundary points. Equivalently, G is open if and only if $G \cap \partial G = \emptyset$.*

Proof. Assume that G is open. Let $\mathbf{a} \in G$. Then G is a neighborhood of $\mathbf{a}$. But G does not intersect G^c. Hence $\mathbf{a}$ has a neighborhood that does not intersect both G and G^c. Therefore $\mathbf{a}$ is not a boundary point of G.

Conversely, assume that G is such that $G \cap \partial G = \emptyset$. Let $\mathbf{a} \in G$. Then $\mathbf{a}$ is not in ∂G, so $\mathbf{a}$ has a neighborhood H that does not intersect both G and G^c. But H intersects G since $\mathbf{a}$ is common to these two sets. Therefore $H \cap G^c = \emptyset$. This means that $\mathbf{a} \in H \subset G$, where H is an open set. Hence there is an $r > 0$ such that $B_r(\mathbf{a}) \subset H \subset G$, proving that G is an open set. $\square$

Remarks 4.5.15 Empty boundaries. Note that the space X and the empty set $\emptyset$ have no boundary points. It turns out that these are the only two sets in a normed space that have no boundary points. It is a good exercise to show that if E and E^c are both nonempty, then their boundaries are also nonempty. (*Hint*: let $\mathbf{a} \in E$ and $\mathbf{b} \in E^c$. Let $S = \{ t \mid 0 \leq t \leq 1,\ (1-t)\mathbf{a} + t\mathbf{b} \in E \}$. Let $\gamma = \sup S$ and $\mathbf{c} = (1-\gamma)\mathbf{a} + \gamma\mathbf{b}$. Show that $\mathbf{c} \in \partial E$.)

Closed Sets

Definition 4.5.16 Closed sets. A set F in a normed space X is said to be a *closed set* if its complement $F^c = X \setminus F$ is an open set.

Example 4.5.17 Let $\mathbf{a} \in X$, where X is a normed space. Then $\{\mathbf{a}\}$ is closed in X. To see this, let $G = X \setminus \{\mathbf{a}\}$. If $\mathbf{x} \in G$ and $r = \|\mathbf{x} - \mathbf{a}\|$, then $r > 0$ and $B(\mathbf{x}, r/2)$ does not contain $\mathbf{a}$. Thus, $B(\mathbf{x}, r/2) \subset G$. Thus, G is open; hence, $\{\mathbf{a}\}$ is closed in X.

Example 4.5.18 The surface of the ball $B_r(\mathbf{a})$ is the set

$$S_r(\mathbf{a}) = \{ \mathbf{x} \in X \mid \|\mathbf{x} - \mathbf{a}\| = r \}.$$

This set is closed. To see this, suppose that $\mathbf{c} \notin S_r(\mathbf{a})$. Let $p = \|\mathbf{c} - \mathbf{a}\|$ and $s = |p - r|$. Then $s > 0$. We see that $B_s(\mathbf{c}) \cap S_r(\mathbf{a}) = \emptyset$, whether $p < r$ or $r < p$. This shows that the complement of $S_r(\mathbf{a})$ is open. $\triangle$

Example 4.5.19 A finite union of closed subsets of a normed space is closed. If $A_1, \ldots, A_m$ are closed subsets of a normed space X, then $\bigcap_{k=1}^{m}(X \setminus A_k)$ is open, being an intersection of finitely many open subsets of X. But

$$\bigcap_{k=1}^{m}(X \setminus A_k) = X \setminus \left(\bigcup_{k=1}^{m} A_k\right)$$

so that $\bigcup_{k=1}^{m} A_k$ is also closed in X.

Theorem 4.5.20 Boundaries and closed sets. *A set F in a normed space X is closed if and only if it contains all of its boundary points. Equivalently, F is closed if and only if $\partial F \subset F$.*

Proof. This follows from the definitions and from Theorem 4.5.14. $\square$

Remarks 4.5.21 Sets that are both open and closed. Note that the whole space X and the empty set $\emptyset$ are both open and closed. These are the only two sets in a normed space that are both open and closed. This follows from the remarks in 4.5.15 above.

Definition 4.5.22 Closure of a set. Let E be any subset of a normed space X. Then $\overline{E} = E \cup \partial E$ is called the *closure of* E.

Lemma 4.5.23 *If an open set G intersects $\overline{E}$, then it also intersects E.*

Proof. Let $\mathbf{a} \in G \cap \overline{E}$. If $\mathbf{a} \in E$, the conclusion follows. If $\mathbf{a} \in \partial E$, then G is a neighborhood of the boundary point $\mathbf{a}$. Hence G must intersect E. $\square$

Lemma 4.5.24 *The closure $\overline{E}$ of any set E is a closed set.*

Proof. Let $\mathbf{a} \notin \overline{E}$. Let G be a neighborhood of $\mathbf{a}$. Then G intersects E^c since $\mathbf{a} \in G \cap E^c$. If G intersects $\overline{E}$, then it also intersects E. This follows from Lemma 4.5.23. Therefore, if every neighborhood of $\mathbf{a}$ intersects $\overline{E}$, then $\mathbf{a} \in \partial E$, which is a contradiction. Hence, $\mathbf{a}$ has a neighborhood contained in $(\overline{E})^c$. Therefore $(\overline{E})^c$ is open. $\square$

Closed Sets and Convergent Sequences

Theorem 4.5.25 *Let F be a set in a normed space X. Then F is a closed set if and only if every convergent sequence in F converges to a point in F.*

Proof. Assume that F is a closed set. Hence F^c is open. If $\mathbf{a} \in F^c$, then there is an $r > 0$ such that $B_r(\mathbf{a}) \subset F^c$. Therefore, if $\mathbf{x}_n$ is a sequence in F, then $\|\mathbf{x}_n - \mathbf{a}\| \geq r$ for all $n \in \mathbb{N}$. Hence $\mathbf{x}_n \not\to \mathbf{a}$. This means that if $\mathbf{x}_n$ converges, it must converge to a point in F.

Conversely, assume that every convergent sequence in F converges to a point in F. Let $\mathbf{a} \in \partial F$. Then $B_{1/n}(\mathbf{a})$ intersects F for all $n \in \mathbb{N}$. Hence there is an $\mathbf{x}_n \in F$ such that $\|\mathbf{x}_n - \mathbf{a}\| < 1/n$. This shows that there is a sequence $\mathbf{x}_n$ in F converging to $\mathbf{a} \in \partial F$. Therefore $\partial F \subset F$ and F is closed. $\square$

Example 4.5.26 Any one-dimensional subspace of a normed space is closed. To see this, let $E = \text{span}\,\{\mathbf{a}\}$, where $\mathbf{a} \in X$. Let $\mathbf{x}_n$ be a sequence in E that converges to some $\mathbf{z} \in E$. Then for each n, we have $\mathbf{x}_n = t_n\mathbf{a}$ for some scalar t_n. Since $\mathbf{x}_n$ converges, the sequence t_n must also converge in $\mathbb{R}$ to some $s \in \mathbb{R}$ (as was shown in a previous example). Hence, $\mathbf{x}_n = t_n\mathbf{a} \to s\mathbf{a} \in E$. Thus, E is closed.

Theorem 4.5.27 Limits of convergent sequences in E. *A point $\mathbf{a} \in X$ can be the limit of a sequence in E if and only if $\mathbf{a} \in \overline{E}$.*

Proof. Note that $\overline{E}$ is a closed set by Lemma 4.5.24 above. Hence, if $\mathbf{x}_n \in E \subset \overline{E}$ and if $\mathbf{x}_n \to \mathbf{a}$, then Theorem 4.5.25 above shows that $\mathbf{a} \in \overline{E}$.

Conversely, assume that $\mathbf{a} \in \overline{E}$. If $\mathbf{a} \in E$, then the constant sequence $\mathbf{x}_n = \mathbf{a}$ is a sequence in E converging to $\mathbf{a}$. If $\mathbf{a} \notin E$, then $\mathbf{a} \in \partial E$. Therefore, every ball $B_{1/n}(\mathbf{a})$ contains a point $\mathbf{x}_n \in E$. Then we see that $\mathbf{x}_n \to \mathbf{a}$. □

Continuous Functions and Topology

There is a close connection between continuity and topology. The following theorem shows that continuity can be defined in terms of open sets.

Theorem 4.5.28 Continuity and open sets. *Let X and Y be two normed spaces and $A \subset X$. Then a function $f : A \to Y$ is continuous if and only if for each open set H in Y there is an open set G in X such that $f^{-1}(H) = A \cap G$.*

Proof. Assume that $f : A \to Y$ is continuous. Let H be an open set in Y. If $\mathbf{x} \in f^{-1}(H)$, then $f(\mathbf{x}) = \mathbf{y} \in H$. Since H is open, there is a $\varepsilon > 0$ such that $B_\varepsilon(\mathbf{y}) \subset H$. Since f is continuous at $\mathbf{x} \in f^{-1}(H)$, there is a $\delta > 0$ such that $\|f(\mathbf{x}') - f(\mathbf{x})\| < \varepsilon$ whenever $\|\mathbf{x}' - \mathbf{x}\| < \delta$ and $\mathbf{x}' \in A$. Therefore $f(A \cap B_\delta(\mathbf{x})) \subset B_\varepsilon(\mathbf{y}) \subset H$. Note that $\delta > 0$ depends on $\mathbf{x}$. To show this explicitly, we write $\delta(\mathbf{x})$ instead of δ. With this notation, let

$$G = \bigcup\nolimits_{\mathbf{x} \in f^{-1}(H)} B_{\delta(\mathbf{x})}(\mathbf{x}).$$

Then G is open in X, since it is a union of open balls. Also, $f^{-1}(H) = A \cap G$.

Conversely, assume that $f : A \to Y$ is such that for each open $H \subset Y$ there is an open $G \subset X$ such that $f^{-1}(H) = A \cap G$. Let $\mathbf{x} \in A$ and $\mathbf{y} = f(\mathbf{x})$. Given $\varepsilon > 0$, let $H = B_\varepsilon(\mathbf{y})$. This is an open set in Y. Hence, there is an open $G \subset X$ such that $f^{-1}(H) = A \cap G$; i.e., $f(A \cap G) \subset B_\varepsilon(\mathbf{y}) = H$. Since $\mathbf{x} \in f^{-1}(H) = A \cap G$, there is a $\delta > 0$ such that $B_\delta(\mathbf{x}) \subset G$. This means that $f(\mathbf{x}') \in B_\varepsilon(\mathbf{y})$ whenever $\mathbf{x}' \in A \cap B_\delta(\mathbf{x})$. Hence f is continuous at $\mathbf{x} \in A$. □

Corollary 4.5.29 Continuous functions with open domains. *Let X and Y be two normed spaces. Let $A \subset X$ be an open set. A function $f : A \to Y$ is continuous if and only if the inverse image $f^{-1}(G)$ of any open set G in Y is an open set in X.*

Proof. Apply Theorem 4.5.28. □

Corollary 4.5.30 Continuous functions defined over the whole space. *Let X and Y be two normed spaces. A function $f : X \to Y$ defined on the whole space X is continuous if and only if the inverse image of every open set is open. Equivalently, $f : X \to Y$ is continuous if and only if the inverse image of every closed set is closed.*

Proof. Apply Theorem 4.5.28. □

Example 4.5.31 All subspaces of a vector space are closed sets. In fact, every subspace K of X is the kernel of a linear transformation $T : X \to X$. An example is the coordinate projection on a complementary subspace, as in Definition 3.1.42. Then $K = T^{-1}(\{\mathbf{0}\})$ is the inverse image of the closed set $\{\mathbf{0}\}$ under the continuous function $T : X \to X$. △

Continuous Functions and Connected Sets

Definition 4.5.32 Connected sets. A set C in a normed space X is called a *connected set* if it satisfies the following condition: if P and Q are two nonempty sets in X such that $P \cup Q = C$, then $C \cap \overline{P} \cap \overline{Q}$ is nonempty.

Remarks 4.5.33 Connectedness in $\mathbb{R}$ and in normed spaces. Definition 4.5.32 is the same as Definition 2.6.12 for connected subsets of $\mathbb{R}$. But there is a difference between the intuitive ideas of connectedness in $\mathbb{R}$ and in general vector spaces. In $\mathbb{R}$ every connected set is an interval: if a connected set contains two points, then it also contains all the points in between. There is an analogous notion of connectedness for general spaces, called arcwise connectedness. These two notions, connectedness and arcwise connectedness, are different in general, but they coincide for subsets of $\mathbb{R}$, and for open subsets of a normed space (see problem 4.83).

Definition 4.5.34 Arcwise connected sets. A set C in a normed space X is called *arcwise connected* if it has the following property: given two points $\mathbf{a}, \mathbf{b} \in C$ there is a continuous function $f : I = [0, 1] \to X$ such that $f(0) = \mathbf{a}$, $f(1) = \mathbf{b}$ and such that $f(t) \in C$ for all $t \in I$. In this case we say that $\mathbf{a}$ *and* $\mathbf{b}$ *can be joined by an arc in* C.

Theorem 4.5.35 Continuity and connected sets. *Let X and Y be two normed spaces and $A \subset X$. Let $f : A \to Y$ be a continuous function. If $C \subset A$ is a connected subset of X, then $B = f(C)$ is a connected subset of Y.*

Proof. Let R and S be two nonempty sets in Y such that $R \cup S = B$. Let $P = A \cap f^{-1}(R)$ and $Q = A \cap f^{-1}(S)$. Then we see that P and Q are two nonempty sets in X such that $P \cup Q = C$. Since C is a connected set, there is a point $\mathbf{c} \in C \cap \overline{P} \cap \overline{Q}$. We will show that $f(\mathbf{c}) \in B \cap \overline{R} \cap \overline{S}$. This means that $B \cap \overline{R} \cap \overline{S}$ is nonempty and therefore B is connected.

Since $\mathbf{c} \in \overline{P}$, there is a sequence $\mathbf{u}_n \in P \subset C$ such that $\mathbf{u}_n \to \mathbf{c}$. This follows from Theorem 4.5.27. Then $f(\mathbf{u}_n) \in R$ and $f(\mathbf{u}_n) \to f(\mathbf{c})$ by the continuity of f. Hence, by the same theorem, $f(\mathbf{c}) \in \overline{R}$. Therefore $f(\mathbf{c}) \in B \cap \overline{R}$. Similarly, there is a sequence $\mathbf{v}_n \in Q \subset C$ such that $\mathbf{v}_n \to \mathbf{c}$. Then $f(\mathbf{v}_n) \in S$ and $f(\mathbf{v}_n) \to f(\mathbf{c}) \in \overline{S}$. This shows that $f(\mathbf{c}) \in B \cap \overline{R} \cap \overline{S}$. □

Theorem 4.5.36 *Let A be a connected set in a normed space X. Let $f : A \to \mathbb{R}$ be a real-valued continuous function. If f takes two values on A, then it also takes all the values in between. More explicitly, if $f(\mathbf{a}) = r \leq t \leq s = f(\mathbf{b})$, with $\mathbf{a}, \mathbf{b} \in A$, then there is a point $\mathbf{c} \in A$ such that $f(\mathbf{c}) = t$.*

Proof. Theorem 4.5.35 shows that $f(A)$ is a connected set in $\mathbb{R}$. But we know from Theorem 2.6.13 that any connected subset of $\mathbb{R}$ is an interval. Hence, if $f(A)$ contains r and s and if $r \leq t \leq s$, then $f(A)$ also contains t. □

Theorem 4.5.37 Intermediate value theorem. *Let $a, b \in \mathbb{R}$, $a < b$. Let $f : [a, b] \to \mathbb{R}$ be a continuous function. If $f(a) = r \leq t \leq s = f(b)$ then there is a $c \in \mathbb{R}$ such that $a \leq c \leq b$ and $f(c) = t$.*

Proof. Apply Theorem 4.5.36 with the connected set $A = [a, b] \subset \mathbb{R}$. □

Compact Sets

Compact sets were defined in Definition 4.4.19 as follows. A set C in a normed space is called *compact* if every sequence in C has a subsequence converging to a point in C. Compact sets play a central role in analysis. They have several equivalent definitions. One of these definitions is established in Theorem 4.5.38 below. Another defining property is obtained in Theorem 4.5.42. This property is called the *Heine-Borel property*. Usually it is taken as the principle definition of compactness in more general settings.

Theorem 4.5.38 *A subset of a (finite dimensional) normed space is compact if and only if it is bounded and closed.*

Proof. Assume that C is not bounded. Then for each $n \in \mathbb{N}$ there is an $\mathbf{x}_n \in C$ such that $\|\mathbf{x}_n\| > n$. Then every subsequence of $\mathbf{x}_n$ is an unbounded sequence and therefore cannot converge. Hence C is not compact.

Now assume that C is not closed. Then Theorem 4.5.25 shows that there is a sequence $\mathbf{x}_n$ in C converging to a point $\mathbf{a} \notin C$. Every subsequence of this sequence also converges to $\mathbf{a} \notin C$. Therefore $\mathbf{x}_n$ has no subsequence converging to a point in C. Hence C is not compact.

Conversely, assume that C is closed and bounded. Then every sequence in C is a bounded sequence. Therefore it has a convergent subsequence by the Bolzano-Weierstrass theorem, 4.2.16. Then Theorem 4.5.25 shows that this subsequence converges to a point in C, since C is closed. Therefore C is compact. □

Definition 4.5.39 Open covers. A collection of open sets $\mathcal{G} = \{ G \}$ is called an *open cover* for a set E if E is contained in the union of this collection. Equivalently, $\mathcal{G} = \{ G \}$ is an open cover for E if for each $\mathbf{x} \in E$ there is a $G \in \mathcal{G}$ such that $\mathbf{x} \in G$.

Theorem 4.5.40 *Let $\mathcal{G}$ be an open cover for a compact set C. Then there is a $\delta > 0$ with the following property: if $\mathbf{u}, \mathbf{v} \in C$ and $\|\mathbf{u} - \mathbf{v}\| < \delta$, then there is a $G \in \mathcal{G}$ that contains both $\mathbf{u}$ and $\mathbf{v}$.*

Proof. Assume that the conclusion of the theorem is false. Then for each $n \in \mathbb{N}$ there are $\mathbf{u}_n, \mathbf{v}_n \in C$ such that $\|\mathbf{u}_n - \mathbf{v}_n\| < 1/n$, but there is no set $G \in \mathcal{G}$ that contains both $\mathbf{u}_n$ and $\mathbf{v}_n$. Since C is compact, there is a subsequence $\mathbf{u}_k$, $k \in \mathbb{K}$ such that $\mathbf{u}_k \to \mathbf{a} \in C$. It follows easily that $\mathbf{v}_k \to \mathbf{a}$. Now there is a $G \in \mathcal{G}$ such that $\mathbf{a} \in G$. Since G is open, there is an $r > 0$ such that $B_r(\mathbf{a}) \subset G$. But $\mathbf{u}_k$ and $\mathbf{v}_k$ are both in $B_r(\mathbf{a})$ for all sufficiently large $k \in \mathbb{K}$. Then $\mathbf{u}_k$ and $\mathbf{v}_k$ are in the same $G \in \mathcal{G}$ for such k. This contradiction proves the theorem. □

Theorem 4.5.41 *Let $\mathcal{G}$ be an open cover for a compact set C. Then there is an $r > 0$ such that each ball $B_r(\mathbf{x})$, $\mathbf{x} \in C$, is contained in a $G \in \mathcal{G}$.*

Proof. Assume that the conclusion of the theorem is false. Then for each $n \in \mathbb{N}$ there is an $\mathbf{x}_n \in C$ such that $B_{1/n}(\mathbf{x}_n)$ is not contained in any $G \in \mathcal{G}$. Then there is a subsequence $\mathbf{x}_k$, $k \in \mathbb{K}$, such that $\mathbf{x}_k \to \mathbf{a} \in C$. Now we know that there is a $G \in \mathcal{G}$ such that $\mathbf{a} \in G$. In this case there is an $R > 0$ such that $B_R(\mathbf{a}) \subset G$. Choose $k \in \mathbb{K}$ sufficiently large such that $\|\mathbf{x}_k - \mathbf{a}\| < R/2$ and $(1/k) < R/2$. Then we see that $B_{1/k}(\mathbf{x}_k) \subset B_R(\mathbf{a}) \subset G \in \mathcal{G}$. This contradiction proves the theorem. □

Theorem 4.5.42 Heine-Borel Theorem. *A set C in a normed space is compact if and only if every open cover for C has a finite subcover.*

Proof. First, suppose that C is compact, and let $\mathcal{G}$ be an open cover for C. Find $r > 0$ as in Theorem 4.5.41 above. Then for each $\mathbf{x} \in C$, $B_r(\mathbf{x})$ is contained in a

$G \in \mathcal{G}$. We claim that there are finitely many points $\mathbf{x}_1, \ldots, \mathbf{x}_n \in C$ such that

$$C \subset E_n = \bigcup_{i=1}^{n} B_r(\mathbf{x}_i). \tag{4.7}$$

Let $\mathbf{x}_1 \in C$ be arbitrary. If $C \subset E_1 = B_r(\mathbf{x}_1)$, then we are done. Otherwise choose $\mathbf{x}_2 \in (C \setminus E_1)$. Continuing in this way, if $\mathbf{x}_1, \ldots, \mathbf{x}_n$ have been chosen and if $C \not\subset E_n$, then choose $\mathbf{x}_{n+1} \in (C \setminus E_n)$. If this process does not terminate, then we obtain a sequence $\mathbf{x}_n$ in C. The distance between any two points in this sequence is at least r. Any subset of this sequence has the same property. By compactness, there is a convergent subsequence $\mathbf{x}_k$, $k \in \mathbb{K}$, such that $\mathbf{x}_k \to \mathbf{a} \in C$. Hence there are $k_1, k_2 \in \mathbb{K}$ such that $\|\mathbf{x}_{k_i} - \mathbf{a}\| < r/2$, $i = 1, 2$, and $k_1 < k_2$. We see that this implies $\|\mathbf{x}_{k_2} - \mathbf{x}_{k_1}\| < r$. Hence $\mathbf{x}_{k_2} \in B_r(\mathbf{x}_{k_1}) \subset E_{k_1}$. This contradicts the choice of $\mathbf{x}_{k_2}$. Therefore there must be an $n \in \mathbb{N}$ such that $C \subset E_n$. Since each $B_r(\mathbf{x}_i)$ is contained in some $G_i \in \mathcal{G}$, we see that (4.7) gives a finite cover for E.

Conversely, assume that $C \subset X$ is such that every open cover for C has a finite subcover. Let $\mathbf{x}_n$ be a sequence in C. For each $\mathbf{a} \in X$ and $r > 0$, let

$$\mathbb{K}_r(\mathbf{a}) = \{\, k \in \mathbb{N} \mid \|\mathbf{x}_k - \mathbf{a}\| < r \,\}.$$

We see that $\mathbf{x}_n$ has a subsequence converging to $\mathbf{a}$ if and only if $\mathbb{K}_r(\mathbf{a})$ is an unbounded set for all $r > 0$. Hence, if no subsequence of $\mathbf{x}_n$ converges to a point $\mathbf{a} \in C$, then for each $\mathbf{a} \in C$ there is an $r(\mathbf{a}) > 0$ such that $B_{r(\mathbf{a})}(\mathbf{a})$ contains only finitely many $\mathbf{x}_n$ s. In this case $\{B_{r(\mathbf{a})}(\mathbf{a})\}$, $\mathbf{a} \in C$, is an open cover for C. Hence C is contained in a finite union of these balls. This means that C contains $\mathbf{x}_n$ for only finitely many n s. This contradiction shows that $\mathbf{x}_n$ must have a subsequence converging to a point in C. Hence C is compact. □

Compact Sets Under Continuous Functions

Theorem 4.5.43 *Let X and Y be two normed spaces. Let C be a compact subset in X and $f : C \to Y$ a continuous function. Then $B = f(C)$ is a compact subset of Y.*

Proof. We have to show that every sequence $\mathbf{y}_n$ in B must have a convergent subsequence $\mathbf{y}_{k_n}$ converging to a point $\mathbf{b}$ in B. Let $\mathbf{y}_n$ be a sequence in $B = f(C)$. Then each $\mathbf{y}_n = f(\mathbf{x}_n)$ for some $\mathbf{x}_n \in C$. Since $\mathbf{x}_n$ is a sequence in the compact set C, it has a convergent subsequence $\mathbf{x}_{k_n}$ converging to a point $\mathbf{a} \in C$. Let $\mathbf{b} = f(\mathbf{a})$. Then $\mathbf{b} \in B$, and the continuity of f at $\mathbf{a}$ implies that

$$\mathbf{y}_{k_n} = f(\mathbf{x}_{k_n}) \to f(\mathbf{a}) = \mathbf{b} \in B.$$

This shows that B is compact. □

Theorem 4.5.44 *Let X be a normed space. Let A be a nonempty compact subset of X. Then any real-valued continuous function $f : A \to \mathbb{R}$ is bounded and attains its maximum and minimum on A. More explicitly, if $f : A \to \mathbb{R}$ is a continuous function defined on a nonempty compact set A, then there are points $\mathbf{a}, \mathbf{b} \in A$ such that $f(\mathbf{a}) \leq f(\mathbf{x}) \leq f(\mathbf{b})$ for all $\mathbf{x} \in A$.*

Proof. The set $B = f(A) \subset \mathbb{R}$ is the image of the compact set $A \subset X$ under the continuous function $f : A \to \mathbb{R}$. Then Theorem 4.5.43 shows that B is a compact subset of $\mathbb{R}$. Hence B is bounded and closed. It is also nonempty, since A is nonempty. Therefore $m = \inf B$ and $M = \sup B$ exist. Both belong to ∂B. But $\partial B \subset B$ since B is closed. Hence $m, M \in B = f(A)$ and, therefore, there are points $\mathbf{a}, \mathbf{b} \in A$ such that $m = f(\mathbf{a})$ and $M = f(\mathbf{b})$. Then $b = f(\mathbf{x}) \in B$ satisfies $m \leq b \leq M$. □

Recall that the proof of the spectral theorem, Theorem 3.6.4, depended on Lemma 3.6.1. It was obtained by an application of the Bolzano-Weierstrass theorem. Theorem 4.5.44 allows us to give a short proof of this result, restated as Lemma 4.5.45.

Lemma 4.5.45 *Let X and Y be Euclidean spaces. Let $T : X \to Y$ be a linear transformation. Then there is a unit vector $\mathbf{e} \in X$ such that $\|T\mathbf{x}\| \leq \|T\mathbf{e}\|$ for all unit vectors $\mathbf{x} \in X$.*

Proof. The function that takes $\mathbf{x} \in X$ to $\|T\mathbf{x}\| \in \mathbb{R}$ is continuous. Therefore, it reaches its maximum value on the compact set $\{\, \mathbf{x} \mid \|\mathbf{x}\| = 1 \,\}$. □

Uniform Continuity and Compactness

Uniform continuity was defined in Definition 4.4.26. Recall that a function may be continuous but not uniformly continuous on a set. Theorem 4.5.46 below shows that the situation is different for compact sets.

Theorem 4.5.46 Uniform continuity on compact sets. *Let X and Y be two normed spaces, and let A be a compact set in X. Then any continuous function $f : A \to Y$ is also uniformly continuous on A.*

Proof. If $f : A \to Y$ is not uniformly continuous on A, then there is an $\alpha > 0$ such that for each $\delta > 0$ there are two points $\mathbf{p}, \mathbf{q} \in A$ with $\|\mathbf{p} - \mathbf{q}\| < \delta$ but $\|f(\mathbf{p}) - f(\mathbf{q})\| \geq \alpha$. Then there are two sequences $\mathbf{p}_n$ and $\mathbf{q}_n$ in A such that $\|\mathbf{p}_n - \mathbf{q}_n\| < (1/n)$ and $\|f(\mathbf{p}_n) - f(\mathbf{q}_n)\| \geq \alpha$ for all $n \in \mathbb{N}$. Since A is compact, there is a subsequence $\mathbf{p}_{k_n}$ that converges to a point $\mathbf{a} \in A$. Then

$$\|\mathbf{q}_{k_n} - \mathbf{a}\| \leq \|\mathbf{q}_{k_n} - \mathbf{p}_{k_n}\| + \|\mathbf{p}_{k_n} - \mathbf{a}\| < (1/n) + \|\mathbf{p}_{k_n} - \mathbf{a}\|$$

shows that $\mathbf{q}_{k_n} \to \mathbf{a}$. Hence $f(\mathbf{p}_{k_n}) \to f(\mathbf{a})$ and $f(\mathbf{q}_{k_n}) \to f(\mathbf{a})$ by the continuity of f at $\mathbf{a} \in A$. This shows that

$$\|f(\mathbf{p}_{k_n}) - f(\mathbf{q}_{k_n})\| \leq \|f(\mathbf{p}_{k_n}) - f(\mathbf{a})\| + \|f(\mathbf{q}_{k_n}) - f(\mathbf{a})\| \to 0.$$

This contradicts the fact that $\|f(\mathbf{p}_{k_n}) - f(\mathbf{q}_{k_n})\| \geq \alpha > 0$ for all $n \in \mathbb{N}$. Hence f must be uniformly continuous on A. □

Distance Between Sets

Definition 4.5.47 Distance between sets. Let A and B be two nonempty sets in a normed space. Then

$$\rho(A, B) = \inf \{ \|\mathbf{a} - \mathbf{b}\| \mid \mathbf{a} \in A, \mathbf{b} \in B \}$$

is called the *distance between A and B.* Note that, the infimum is taken over a nonempty set of nonnegative numbers. Hence, the distance $\rho(A, B)$ is well-defined for any two nonempty sets A and B.

Lemma 4.5.48 *Let K be a compact set and let F be a closed set in a normed space X. If $K \cap F = \emptyset$ then $\rho(K, F) > 0$. Also, there is a compact set E such that $K \subset E^o \subset E \subset G = F^c = X \setminus F$.*

Proof. Since $G = F^c$, by assumption we have $K \subset G$ where K is closed and G is open. Hence for each $\mathbf{x} \in K$ there is an $r(\mathbf{x}) > 0$ such that $B_{r(\mathbf{x})}(\mathbf{x}) \subset G$. Then $\{B_{r(\mathbf{x})/3}(\mathbf{x})\}$, $\mathbf{x} \in K$, is an open cover for K. By the Heine-Borel Theorem 4.5.42, it has a finite subcover. Hence there are finitely many $\mathbf{x}_i \in K$ such that $K \subset \cup_i B_{r(\mathbf{x}_i)/3}(\mathbf{x}_i)$. Let $r = \min r(\mathbf{x}_i)/3$. Then $r > 0$. If $\mathbf{x} \in K$ then there is an $\mathbf{x}_i$ such that $\mathbf{x} \in B_{r(\mathbf{x}_i)/3}(\mathbf{x}_i)$. This means that $B_r(\mathbf{x}) \subset B_{r(\mathbf{x}_i)}(\mathbf{x}_i) \subset G$. Hence $\|\mathbf{x} - \mathbf{y}\| \geq r$ for any $\mathbf{y} \in F$. Therefore $\rho(K, F) \geq r > 0$.

For the second part, let $E = \cup_i \overline{B}_{2r(\mathbf{x}_i)/3}$. This is a compact set, since it is a finite union of compact sets. We can easily verify that E satisfies the requirements of the lemma. □

Definition 4.5.49 Convex sets. A C be a set in a vector space. Assume that if C contains two points $\mathbf{a}$ and $\mathbf{b}$, then it also contains the line segment

$$L = \{ t\mathbf{b} + (1 - t)\mathbf{a} \mid 0 \leq t \leq 1 \}$$

joining these two points. In this case, C is called a *convex set.*

Problems

4.49 Show that if a closed set contains E then it also contains $\overline{E}$.

4.50 Given any set E in a normed space X, show that the sets E^o, E^{ext}, and ∂E are pairwise disjoint and their union is X. Give examples of sets E such that $\partial E = X$, or such that $E^o = X$, or such that $E^{\text{ext}} = X$.

4.51 Let A, B be subsets of a normed space. Show that

$$\partial(A \cup B) \subset \partial A \cup \partial B.$$

Give an example to show that it is possible to have $\partial(A \cup B) \neq \partial A \cup \partial B$. If $A \subset B$, must it follow that $\partial A \subset \partial B$?

4.52 Show that $\mathbf{x} \in E^o$ if and only if there is an $r > 0$ such that $B_r(\mathbf{x}) \subset E$.

4.53 Given a normed space X, find an example of a set E such that $\partial E = X$

4.54 Let $S = \{ (x, y, z) \in \mathbb{R}^3 \mid 2x - y + 5z = 0 \}$. Show that $\partial S = S$.

4.55 Let $\mathbf{x}_n$ be a sequence in a normed space X and assume that $\mathbf{x}_n \to \mathbf{a}$ for some $\mathbf{a} \in X$. Let $S = \{ \mathbf{x}_n \mid n \in \mathbb{N} \}$. Is it true that $\partial S = \{\mathbf{a}\}$?

4.56 Let $\mathbf{x}_n$ be a sequence in a normed space X and assume that $\mathbf{x}_n \to \mathbf{a}$ for some $\mathbf{a} \in X$. Show that $T = \{ \mathbf{x}_n \mid n \in \mathbb{N} \} \cup \{\mathbf{a}\}$ is a closed subset of X.

4.57 Let X be a normed space. Verify that $\partial\emptyset = \emptyset = \partial X$. If $E \subset X$, must it be true that $\partial(\partial E) \subset \partial E$?

4.58 Show that $\partial(\mathbf{a} + E) = \mathbf{a} + \partial E$ for all $\mathbf{a} \in X$ and for all $E \subset X$.

4.59 Let F be any finite subset of a nonzero normed space X. Show that $\partial F = F$. Hence, deduce that F is closed.

4.60 Let A be a bounded subset of a normed space. Show that $A \cup \partial A$ is also bounded.

4.61 Let F be a finite subset of an open set O in a normed space. Show that $O \setminus F$ is also open.

4.62 Let E be a subset of a normed space X. Let $\mathbf{a} \in X$. Show that E is open if and only if $\mathbf{a} + E$ is open. Similarly, if c is any nonzero scalar, show that E is open if and only if cE is open.

4.63 Show that an arbitrary intersection of closed sets is closed and a union of finitely many closed sets is also closed.

4.64 Let M, N be subsets of a Euclidean space such that $\mathbf{m} \perp \mathbf{n}$ for all $\mathbf{m} \in M$ and all $\mathbf{n} \in N$. Let $\mathbf{m}_k$ be a sequence in M and let $\mathbf{n}_k$ be a sequence in N. Assume that $\mathbf{m}_k + \mathbf{n}_k$ is a bounded sequence. Show that there exists a sequence ℓ_k in $\mathbb{N}$ such that both $\mathbf{m}_{\ell_k}$ and $\mathbf{n}_{\ell_k}$ converge.

4.65 Let $U = \{ (x, y, z) \in \mathbb{R}^3 \mid |2x + 3y + z| < 1, |x - y + 5z| < 3 \}$. Show that U is open in the Euclidean space $\mathbb{R}^3$.

4.66 Let X and Y be normed spaces. Let $T : X \to X$ be linear. Assume that T maps a basis of X to a basis of Y. Show that $G \subset X$ is open if and only if $T(G)$ is open.

4.67 Suppose that X and Y are normed spaces and let $f : X \to Y$ be continuous. Let $D \subset X$. If f is one-to-one on X, show that $f(\partial D) \subset \partial(f(D))$. What happens if f is not one-to-one?

4.68 Show that the set of all isomorphisms $X \to X$ is an open subset of $L(X, X)$.

4.69 Let E be a set in a normed space X. A point $\mathbf{a} \in X$ is an *accumulation point of E* if $E \cap B_r(\mathbf{a})$ is an infinite set (that is, contains infinitely many points) for all $r > 0$. Show that every bounded infinite set has an accumulation point. Is every point in ∂E an accumulation point of E?

4.70 Let E be a set in a normed space X. A sequence in E is *dense in E* if every point in E is an accumulation point (Problem 4.69) of this sequence. Show that every infinite set contains dense sequences.

Problems on Compact Sets

4.71 Let A, B be compact subsets of a normed space. Show that $A \cap B$ and $A \cup B$ are compact.

4.72 Let $S = \{ \mathbf{x} = (x_1, x_2, x_3) \mid 1 \leq x_1^2 + x_2^2 + x_3^2 \leq 2 \}$. Show that S is compact in the Euclidean space $\mathbb{R}^3$.

4.73 Let X be a normed space. Let A be a compact subset of X, and let B be a nonempty finite subset of X. Must $A + B = \{ \mathbf{a} + \mathbf{b} \mid \mathbf{a} \in A, \mathbf{b} \in B \}$ be compact?

4.74 Let $n \in \mathbb{N}$. Give an example of compact subsets E, F of $\mathbb{R}^n$ such that $E \setminus F$ is not compact.

4.75 Let E be a subset of a normed space X. If $\mathbf{a} \in X$ is an accumulation point (Problem 4.69) of E then $E \setminus \{\mathbf{a}\}$ is not compact. If $\mathbf{u} \in \partial E$, is it true that $E \setminus \{\mathbf{u}\}$ is not compact?

4.76 Let $U = \{ \mathbf{x} = (x_1, x_2, x_3) \in \mathbb{R}^3 \mid 1 \leq \max_{1 \leq k \leq 3} |x_k| \leq 3 \}$. Show that there is an $\mathbf{a} \in U$ such that

$$a_1^2 + a_2^2 + a_3^2 = \min \{ x_1^2 + x_2^2 + x_3^2 \mid \mathbf{x} \in U \}.$$

4.77 Let E_n be a sequence of nonempty compact subsets of a normed space X. If $E_{k+1} \subset E_k$ for all $k \in \mathbb{N}$, show that $\cap_n E_n \neq \emptyset$. Give an example of a sequence A_n of nonempty subsets of $\mathbb{R}$ with $A_{k+1} \subset A_k$ for all $k \in \mathbb{N}$, but such that $\cap_n E_n = \emptyset$.

4.78 Let X and Y be normed spaces. Let M be a compact set in X. Let $f_1, \ldots, f_m$ be continuous functions $M \to Y$. Let

$$\mathbf{x}_k = (x_{k1}, x_{k2}, \ldots, x_{km}) \in \mathbb{R}^m, \quad k \in \mathbb{N}.$$

be a bounded sequence in $\mathbb{R}^m$. For each $k \in \mathbb{N}$, define

$$g_k = x_{k1} f_1 + \cdots + x_{km} f_m.$$

Show that there is a continuous function $g : M \to Y$ such that a subsequence of g_k converges to g uniformly on M. Is g unique?

4.79 Let X and Y be normed spaces. Let M be a compact set in X. Let $f : M \to Y$ be a continuous and one-to-one function. Show that a sequence $\mathbf{m}_n \in M$ converges if and only if $f(\mathbf{m}_n) \in Y$ converges.

Problems on Connected Sets

4.80 Let A be a connected subset of a normed space X and assume that there is an $\mathbf{a} \in A$ such that $A \setminus \{\mathbf{a}\}$ is connected. Let B be any nonempty subset of normed space Y such that for each $\mathbf{y} \in B$, the set $B \setminus \{\mathbf{y}\}$ is not connected. Show that there is no one-to-one, onto continuous map $f : A \to B$. Deduce that if $c < d$ are real numbers and I is an interval of the form $[a, b), (a, b]$ or $[a, b]$ with real numbers $a < b$, then there is no one-one onto continuous function $f : I \to (c, d)$.

4.81 Let $C = A \cup B$, where A and B are two sets in the xy-plane defined by

$$A = \{ (0, y) \mid -1 < y < 1 \} \text{ and } B = \{ (x, \sin(1/x)) \mid 0 < x < 1 \}.$$

Show that C is connected but not arcwise connected.

4.82 Let G be an open and connected set. If A and B are open sets and if $G = A \cup B$, then show that A or B is empty.

4.83 Show that if an open set is connected, then it is also arcwise connected. (*Hint:* Let G be an open set and $\mathbf{a} \in G$. Let A be the set of all points in G that can be joined to $\mathbf{a}$ by an arc in C. Let $B = G \setminus A$. Show that A and B are both open.)

4.84 Show that any open set is the union of a sequence of pairwise disjoint connected open sets.

4.85 Show that any open set in $\mathbb{R}$ is the union of a sequence of pairwise disjoint open intervals. Give an example to show that this union need not be a finite union, even for bounded open sets.

Remarks on Problem 4.85. Try to give a simple solution for this problem. At the same, time keep in mind that there are also very complicated open sets. Here is one: Let r_k be a dense sequence (Problem 4.70) in $(0,\ 1)$. Let $\varepsilon_k = 2^{-k-4}$. Let $G_k = (0,\ 1) \cap (r_k - \varepsilon_k,\ r_k + \varepsilon_k)$. Then $G = \cup_k G_k$ is an open set. Hence G is a union of pairwise disjoint open intervals. Probably no one knows the explicit forms of these intervals. Show, however, that $G \neq (0,\ 1)$.

Problems on Distances Between Sets

4.86 Let X be a normed space. Let $B \subset X$ be a nonempty set and $\mathbf{a} \in X$. Then $\rho(\mathbf{a},\ B) = \inf_{\mathbf{x} \in B} \|\mathbf{x} - \mathbf{a}\|$ is the *distance between the point* $\mathbf{a}$ *and the set* B. Show that there is a point $\mathbf{b} \in \overline{B}$ such that $\rho(\mathbf{a},\ B) = \|\mathbf{a} - \mathbf{b}\|$. Is $\mathbf{b} \in \overline{B}$ unique?

4.87 For any nonempty set E and for any $r > 0$ let $E_r = \cup_{\mathbf{x} \in E} B_r(\mathbf{x})$ be the *enlargement of* E by $r > 0$. Show that $\mathbf{x} \in E_r$ if and only if $\rho(\mathbf{x},\ E) < r$.

4.88 Show that $\mathbf{x} \in \overline{B}$ if and only if $\rho(\mathbf{x},\ B) = 0$. Show that $\mathbf{x} \in \overline{B} = \cap_{r>0} B_r$.

4.89 Let A and B be two nonempty sets in a normed space X. Let $\rho(\mathbf{x},\ B)$, $\mathbf{x} \in X$, be as defined in Problem 4.86, and $\rho(A,\ B)$, the distance between A and B, as defined in Definition 4.5.47. Show that $\rho(A,\ B) = \inf_{\mathbf{x} \in A} \rho(\mathbf{x},\ B)$.

4.90 Let A and B be two nonempty sets in a normed space X. If A is compact, then show that there are points $\mathbf{a} \in A$ and $\mathbf{b} \in B$ such that $\rho(A,\ B) = \|\mathbf{a} - \mathbf{b}\|$.

4.91 Let A and B be two nonempty sets in a normed space X. Give an example to show that there may not be any points $\mathbf{a} \in \overline{A}$ and $\mathbf{b} \in \overline{B}$ such that $\rho(A, B) = \|\mathbf{a}-\mathbf{b}\|$.

Problems on Convex Sets

4.92 Show that a convex set is either contained in a lower dimensional subspace or it contains an interior point.

4.93 Show that if a convex set in a normed space X contains points from a set A and from its complement $A^c = X \setminus A$, then it also contains points from ∂A. Give an example to show that this is not necessarily true for non-convex sets.

4.94 Let $\mathbf{a}$ be an interior point of a convex set C. If $\mathbf{b} \in \overline{C}$ and $0 \leq t < 1$, then show that $(1-t)\mathbf{a} + t\mathbf{b}$ is also an interior point of C. Give an example to show that this is not necessarily true for non-convex sets.

Problems on Oscillations

4.95 Let X and Y be normed spaces. Let $E \subset X$ and let $f : E \to Y$ be a bounded function. If $G \subset E$, then show that

$$\Omega(f, G) = \sup\{\, \|f(\mathbf{u}) - f(\mathbf{v})\| \mid \mathbf{u}, \mathbf{v} \in G \,\}$$

exists. It is called the *oscillation of f over the set G*. Also show that if $A \subset B \subset E$ then $\Omega(f, A) \leq \Omega(f, B)$.

4.96 With the notations of Problem 4.95, let $\mathbf{a} \in \overline{E}$. Show that

$$\omega(f, \mathbf{a}) = \lim_{r \to 0+} \Omega(f, E \cap B_r(\mathbf{a}))$$

exists. It is called the *oscillation of f at the point* $\mathbf{a}$.

4.97 Show that f is continuous at $\mathbf{a}$ if and only if $\omega(f, \mathbf{a}) = 0$.

4.98 Given $\alpha \in \mathbb{R}$, let $E(\alpha)$ be the set of $\mathbf{x} \in \overline{E}$ such that $\omega(f, \mathbf{x}) \geq \alpha$. Show that $E(\alpha)$ is a closed set for all $\alpha \in \mathbb{R}$.

CHAPTER 5

DERIVATIVES

The properties of linear transformations are relatively easy to grasp. The key idea of differential calculus is to consider the broader class of functions that can be approximated locally by linear transformations. The approximating linear transformation, which typically changes from point to point, is the *derivative* of the function at a point, while any function that can be approximated in this way is said to be *differentiable* at such points. Our objective is to investigate the extent to which the properties of linear transformations are passed on to this larger class of differentiable functions. Note that this is a direct generalization of differential calculus in the one-variable case, where we study functions that can be locally approximated by straight lines.

To begin our study, we first have to define the nature of approximation by a linear transformation. Second, we have to develop a test to find out if a given function can be so approximated, i.e., whether it is differentiable at a given point. It turns out, however, that checking the differentiability of a function is, in general, not easy. We shall concentrate on the sub-class of *continuously differentiable functions*. It happens

Analysis in Vector Spaces.
By M. A. Akcoglu, P. F. A. Bartha and D. M. Ha

that functions belonging to this class can be recognized by a routine test, and we shall make such functions the focus of our study.

Section 5.1 discusses differentiability for an important special case: the class of vector-valued functions of a real variable. The general definitions of differentiability and continuous differentiability are given in Section 5.2. The routine test for continuous differentiability is developed in Section 5.3. This test makes use of *directional* or *partial* derivatives, which are usually easy to compute. Subsequent sections explore the many applications of partial derivatives. Most importantly, these sections tell us how partial derivatives may be used to represent the full derivative of a function and how they enable us to differentiate sums, products and compositions of differentiable functions.

5.1 FUNCTIONS OF A REAL VARIABLE

In this section, A is an open subset of $\mathbb{R}$ and $f : A \to Y$ is a function. The range space Y is an arbitrary normed space.

Definition 5.1.1 Derivative. Let $a \in A$. If

$$\lim_{r \to 0} \frac{f(a+r) - f(a)}{r} = f'(a) \in Y \tag{5.1}$$

exists, then it is called the *derivative* of f at $a \in A$.

The limit in (5.1) is taken in Y. Hence, the assertion that the limit exists means that for each $\varepsilon > 0$, there is a $\delta > 0$ such that if $0 < |r| < \delta$, then

$$\left\| \frac{f(a+r) - f(a)}{r} - f'(a) \right\| < \varepsilon. \tag{5.2}$$

An equivalent formulation is that for each $\varepsilon > 0$, there is a $\delta > 0$ such that if $|r| < \delta$, then

$$\|f(a+r) - f(a) - rf'(a)\| \leq \varepsilon\,|r|. \tag{5.3}$$

This slight re-arrangement helps to set the stage for the general definition of the derivative in Section 5.2.

Finally, note that if $f'(a)$ exists, then

$$\lim_{r \to 0} \left\| \frac{f(a+r) - f(a)}{r} \right\| = \lim_{r \to 0} \frac{\|f(a+r) - f(a)\|}{|r|} = \|f'(a)\|. \tag{5.4}$$

This follows from the continuity of the norm function.

Remarks 5.1.2 Independence of the norm on Y. The above statements are independent of the norm on Y. This follows easily from the equivalence of any two norms on the (finite-dimensional) vector space Y.

Example 5.1.3 Let Y be a normed space with basis $B = \{\mathbf{u}_0, \ldots, \mathbf{u}_m\}$. Let $f : \mathbb{R} \to Y$ be defined by

$$f(t) = \mathbf{u}_0 + t\mathbf{u}_1 + \cdots + t^m \mathbf{u}_m \quad \text{for all } t \in \mathbb{R}.$$

For $a \in \mathbb{R}$, we have

$$\lim_{r \to 0} \frac{(a+r)^n - a^n}{r} = na^{n-1} \quad \text{for all positive integers } n.$$

Hence,

$$\begin{aligned} f'(a) &= \lim_{r \to 0} \frac{f(a+r) - f(a)}{r} \\ &= \lim_{r \to 0} \frac{1}{r} \sum_{k=1}^{m} ((a+r)^k - a^k)\mathbf{u}_k \\ &= \sum_{k=1}^{m} ka^{k-1}\mathbf{u}_k. \end{aligned}$$

Remarks 5.1.4 Openness of A. The assumption that A is open is implicitly used in these statements. In fact, since $a \in A$, there is a $c > 0$ such that $(a + r) \in A$ whenever $|r| < c$. Hence in (5.2) and (5.3), the term $f(a + r)$ is always defined whenever r is sufficiently small. No further restrictions on r are necessary. This is an important point in the definition of the derivative. There should be no restrictions on the increments of the variable other than that they are sufficiently small.

Remarks 5.1.5 Computations of derivatives. Ordinary rules of differentiation. To compute the derivative $f'(x)$, we have to evaluate the limit

$$\frac{f(x+r) - f(x)}{r}$$

in Y. If $Y = \mathbb{R}$, that is, if f is a real-valued function, then these evaluations are routine for elementary functions. The computations of these ordinary derivatives are worked out in a basic calculus course. We know, for example, how to compute the derivatives of polynomials, and of functions such as $\cos x$ or e^x. We shall refer to these computations as the *ordinary rules of differentiation*, and we shall assume that all such results are familiar to the reader. These computations are used in our examples and problems, but not in the development of the main results.

If Y is a general vector space, then the ordinary rules of differentiation may be applied to the components of f. Let $W = \{\mathbf{w}_1, \ldots, \mathbf{w}_m\}$ be a basis for Y. The components of $f : A \to Y$ are the functions $f_i : A \to \mathbb{R}$ defined by $f_i = y_i \cdot f$, where the functions $y_i : Y \to \mathbb{R}$ are the coordinate functions of the basis W. Hence,

$$f = f_1\mathbf{w}_1 + \cdots + f_m\mathbf{w}_m .$$

We see easily that $f'(x) \in Y$ exists if and only if each $f_i'(x) \in \mathbb{R}$ exists. In this case

$$f'(x) = f_1'(x)\mathbf{w}_1 + \cdots + f_m'(x)\mathbf{w}_m .$$

Each $f_i'(x)$ can be computed by the ordinary rules of differentiation.

Example 5.1.6 Take $Y = \mathbb{R}^4$ with the standard basis $\mathbf{e}_1, \mathbf{e}_2, \mathbf{e}_3, \mathbf{e}_4$. Let

$$f(t) = \left(\frac{1-t}{t}, \sin t^2, e^{5\cos t}, (t^2+1)^3\right), \quad t \in (0,1).$$

Then for all $t \in (0,1)$, we have

$$f_1(t) = \frac{1-t}{t}, \quad f_2(t) = \sin(t^2), \quad f_3(t) = e^{5\cos t}, \quad f_4(t) = (t^2+1)^3.$$

By the ordinary rules for differentiation,

$$\begin{aligned} f'(t) &= f_1'(t)\mathbf{e}_1 + f_2'(t)\mathbf{e}_2 + f_3'(t)\mathbf{e}_3 + f_4'(t)\mathbf{e}_4 \\ &= \left(-\frac{1}{t^2}, 2t\cos(t^2), -5\sin te^{5\cos t}, 6t(t^2+1)^2\right), \quad t \in (0,1). \end{aligned}$$

Example 5.1.7 Let $Y = L(\mathbb{R}^4, \mathbb{R}^2)$, the normed space of all linear transformations from $\mathbb{R}^4$ to $\mathbb{R}^2$. For each $x \in \mathbb{R}$, let $f(x)$ be the linear map from $\mathbb{R}^4$ to $\mathbb{R}^2$ defined for all $\mathbf{u} = (u_1, u_2, u_3, u_4) \in \mathbb{R}^4$ by

$$f(x)\mathbf{u} = \begin{bmatrix} -1 & 2 & x^2-1 & \ln(x^2+1) \\ xe^x & 5 & 7x & \cos x \end{bmatrix} \begin{bmatrix} u_1 \\ u_2 \\ u_3 \\ u_4 \end{bmatrix}.$$

We will compute $f'(x)$. The standard basis for Y consists of the linear maps $L_{ij}, i = 1, 2, j = 1, , 3, 4$, where the standard matrix for each $L_{ij} : \mathbb{R}^4 \to \mathbb{R}^2$ is the 2×4 matrix whose every entry is 0 except the (i, j) entry, which is 1. Thus, the coordinate functions of f with respect to the standard basis of Y are the functions f_{ij} given by $f_{11}(x) = -1, f_{12}(x) = 2, f_{13}(x) = x^2 - 1, f_{14}(x) = \ln(x^2+1)$, and so on. Hence, for each $x \in \mathbb{R}$ and each $\mathbf{u} = (u_1, u_2, u_3, u_4) \in \mathbb{R}^4$, we have

$$f'(x)\mathbf{u} = \begin{bmatrix} 0 & 0 & 2x & (2x)/(x^2+1) \\ e^x(1+x) & 0 & 7 & -\sin x \end{bmatrix} \begin{bmatrix} u_1 \\ u_2 \\ u_3 \\ u_4 \end{bmatrix}.$$

Definition 5.1.8 Affine approximations. Let $f : A \to Y$ be differentiable at $a \in A$. The *affine (or first-order polynomial) approximation* of f at a is the function $T_a : \mathbb{R} \to Y$ defined by $T_a x = f(a) + f'(a)(x - a)$, $x \in \mathbb{R}$. If $Y = \mathbb{R}$, then $y = T_a x = f(a) + f'(a)(x - a)$ is the familiar equation of the tangent line of the curve $y = f(x)$.

Derivatives and Increments

In this course, derivatives are used mainly to estimate the increments of functions. The starting point is the following result.

Theorem 5.1.9 Increments and derivatives. *Assume that $f'(a)$ exists and that $\|f'(\mathbf{a})\| < M$. Then there is a $\delta > 0$ such that*

$$\|f(a + r) - f(a)\| < M\,|r| \text{ whenever } |r| < \delta. \tag{5.5}$$

Proof. Since $\lim_{r \to 0}(1/|r|)\|f(a + r) - f(a)\| = \|f'(a)\|$, there is a $\delta > 0$ such that

$$\frac{\|f(a + r) - f(a)\|}{|r|} < M \tag{5.6}$$

whenever $0 < |r| < \delta$. □

Corollary 5.1.10 Derivatives and continuity. *If $f'(a)$ exists, then f is continuous at $a \in A$.*

Proof. Assume that $\|f'(\mathbf{a})\| < M$. If $r_n \to 0$, then we see that

$$\|f(a + r_n) - f(a)\| < M\,|r_n| \tag{5.7}$$

for all sufficiently large n. Hence $f(a + r_n) \to f(a)$ in Y. □

Theorem 5.1.9 may be thought of as letting us compare the increments of f to the increments of $\varphi(x) = M\,x$. We shall need the following generalization of Theorem 5.1.9, which allows us to compare the increments of f to the increments of other functions $\varphi : A \to \mathbb{R}$.

Theorem 5.1.11 *Let $a \in A$. If $\|f'(a)\| < \varphi'(a)$, then there is a $\delta > 0$ such that*

$$\|f(a + r) - f(a)\| \leq \varphi(r + a) - \varphi(a) \tag{5.8}$$

whenever $0 \leq r < \delta$.

Proof. Let $M \in \mathbb{R}$ be such that $\|f'(a)\| < M < \varphi'(a)$. We have

$$\begin{aligned} \lim_{r \to 0}(1/|r|)\,\|f(a+r) - f(a)\| &= \|f'(a)\| < M \text{ and} \\ \lim_{r \to 0}(1/r)(\varphi(a+r) - \varphi(a)) &= \varphi'(a) > M. \end{aligned}$$

In this case there is a $\delta > 0$ such that

$$\begin{aligned} \|f(a+r) - f(a)\| &\leq M\,|r| \text{ and} \\ \varphi(a+r) - \varphi(a) &\geq M\,r \end{aligned}$$

whenever $|r| < \delta$. We see that (5.8) is satisfied with this $\delta > 0$. □

Mean Value Theorems

The main tools for estimating the increments of functions belong to an all-important collection of results that we shall refer to as the *mean value theorems*. Here, we present two of these results. In what follows, Y is a normed space and A is an open subset of $\mathbb{R}$.

Theorem 5.1.12 Mean Value Theorem (variable upper bound). *Let $f : A \to Y$ and let $\varphi : A \to \mathbb{R}$. Assume that $f'(x)$ and $\varphi'(x)$ both exist and that $\|f'(x)\| \leq \varphi'(x)$ for each $x \in I = [\,a,\,b\,] \subset A$. Then*

$$\|f(b) - f(a)\| \leq \varphi(b) - \varphi(a). \tag{5.9}$$

Proof. Let $\varepsilon > 0$ and put $\psi(x) = \varphi(x) + \varepsilon x$. Define

$$J = \{\, x \in I \mid \|f(x) - f(a)\| \leq \psi(x) - \psi(a) \,\}. \tag{5.10}$$

Then J is bounded since $J \subset I$ and nonempty since $a \in J$. Hence $c = \sup J$ exists and $c \in I$. We show first that $c \in J$. This is clear if $c = a$. Otherwise, we see easily that there is a sequence $x_n \in J$ such that $x_n \to c$ in $\mathbb{R}$. Hence

$$\|f(x_n) - f(a)\| \leq \psi(x_n) - \psi(a) \tag{5.11}$$

for all $n \in \mathbb{N}$. Now, since $f'(c)$ and $\psi'(c) = \varphi'(c) + \varepsilon$ both exist, Corollary 5.1.10 shows that both f and ψ are continuous at $c \in I$. Hence $f(x_n) \to f(c)$ in Y and $\psi(x_n) \to \psi(c)$ in $\mathbb{R}$. Then, from (5.11) and from the continuity of the norm,

$$\|f(c) - f(a)\| \leq \psi(c) - \psi(a).$$

Hence $c \in J$.

Next, $\|f'(c)\| \leq \varphi'(c) < \varphi'(c) + \varepsilon = \psi'(c)$. Therefore, Theorem 5.1.11 shows that there is a $\delta > 0$ such that

$$\|f(c+r) - f(c)\| \leq \psi(c+r) - \psi(c)$$

whenever $0 \leq r < \delta$. Now assume that $c < b$. Then there is an r such that $0 < r < \delta$ and such that $(c+r) = e \leq b$. Then

$$\begin{aligned} \|f(e) - f(a)\| &\leq \|f(e) - f(c)\| + \|f(c) - f(a)\| \\ &\leq (\psi(e) - \psi(c)) + (\psi(c) - \psi(a)) = \psi(e) - \psi(a). \end{aligned}$$

This shows that $e \in J$. This is a contradiction, since $c < e$ and $c = \sup J$. Hence $c = b$. It follows that

$$\|f(b) - f(a)\| \leq \psi(b) - \psi(a) = (\varphi(b) - \varphi(a)) + \varepsilon(b - a). \tag{5.12}$$

But (5.12) is true for all $\varepsilon > 0$. Then (5.9) follows. □

Theorem 5.1.13 Mean Value Theorem (fixed upper bound). *Assume that $f'(x)$ exists and that $\|f'(x)\| \leq M$ for $a \leq x \leq b$. Then $\|f(b) - f(a)\| \leq M\,(b - a)$.*

Proof. Apply Theorem 5.1.12, with $\varphi(x) = Mx$. □

Even though Theorem 5.1.13 is a special case of Theorem 5.1.12, it is the version of the mean value theorem that we shall use most widely (although we employ Theorem 5.1.12 in the discussion of Taylor polynomials in the next subsection). We freely use the name 'mean value theorem' for either of these two results.

Example 5.1.14 Suppose that f, g are functions from A into Y and $I = [a, b] \subset A$. If $\|f'(x) - g'(x)\| \leq M$ for all $x \in I$, then

$$\|f(b) - f(a)\| \leq \|g(b) - g(a)\| + M\,|b - a|.$$

To see this, set $h(x) = f(x) - g(x)$ for all $x \in A$. Then $h'(x) = f'(x) - g'(x)$ at each x where f and g are differentiable. Thus, by assumption, $\|h'(x)\| \leq M$ for all $x \in I$. Hence, by Theorem 5.1.13,

$$\|f(b) - f(a) - (g(b) - g(a))\| = \|h(b) - h(a)\| \leq M\,\|b - a|.$$

Thus, it follows from the triangle inequality that

$$|\,\|f(b) - f(a)\| - \|g(b) - g(a)\|\,| \leq M\,|b - a|.$$

Hence, $\|f(b) - f(a)\| \leq \|g(b) - g(a)\| + M\,|b - a|$.

Remarks 5.1.15 Relation to other mean value theorems. The classical mean value theorem states the following. Suppose $I = [a, b]$ and $f : I \to \mathbb{R}$ is a real-valued function. If f is continuous on I and differentiable in the interior of I, then there is a $c \in \mathbb{R}$ such that $a < c < b$ and such that $f(b) - f(a) = f'(c)(b - a)$. This is a stronger result than our mean value theorems, but it is valid only for real-valued functions. There is no comparable result for vector-valued functions. It turns out, however, that the classical mean value theorem can be replaced by 5.1.12 or 5.1.13 in all the applications considered in this course. The mean value theorem 5.1.13 can be obtained from the classical result, but the direct proof given above seems to be shorter and more instructive. Also, note that the mean value theorem 5.1.12 is related to the well-known Cauchy Mean Value Theorem. Since we do not need this latter result, we shall not discuss it here.

Taylor Polynomials

Definition 5.1.16 Higher-order derivatives. If $f : A \to Y$ has a derivative $f'(x) \in Y$ at every $x \in A$, then we have a well-defined *derivative function* $f' : A \to Y$. If this new function also has a derivative at $x \in A$, then it is called the *second derivative* of f at $x \in A$ and denoted by $f''(x) \in Y$. If the second derivatives exist at each $x \in A$, then they define the *second-derivative function* $f'' : A \to Y$. Higher-order derivatives are defined by induction. If the nth-order derivative function $f^{(n)} : A \to Y$ exists and has a derivative $f^{(n+1)}(x) \in Y$ at every $x \in A$, then these derivatives define the *$(n+1)$st-order derivative function* $f^{(n+1)} : A \to Y$. We also write $f^{(1)} = f'$, $f^{(2)} = f''$, and $f^{(0)} = f$.

Lemma 5.1.17 *Assume that $f^{(k)}(a) = 0$ for $k = 0, 1, \ldots, (n-1)$ at a certain point $a \in A$. Also assume that there is an $R > 0$ such that $\|f^{(n)}(a + x)\| \le M$ whenever $|x| < R$. Then*

$$\|f(a + x)\| \le (1/n!)\, M\, |x|^n \quad \textit{whenever} \quad |x| < R. \tag{5.13}$$

Proof. First, assume that $x > 0$. Proceed by induction on $n \in \mathbb{N}$. Let $n = 1$. Let x be fixed, $0 < x < R$. Then $\|f'(a + t)\| \le M$ for all $t \in [0, x]$. Therefore by the mean value theorem 5.1.13,

$$\|f(a + x)\| = \|f(a + x) - f(a)\| \le ((a + x) - a)M = x\, M.$$

A similar result holds if $-R < x < 0$. This proves the result for $n = 1$.

Now assume the result for $(n-1)$, $n \ge 2$. Given $f : A \to Y$ satisfying the hypotheses of the lemma, let $g = f'$. Then

$$\begin{aligned} g^{(k)}(a) &= f^{(k+1)}(a) = 0 \text{ for } k = 0, 1, \ldots, (n-2), \text{ and} \\ \|g^{(n-1)}(a + x)\| &= \|f^{(n)}(a + x)\| \le M \text{ whenever } |x| < R. \end{aligned}$$

Hence, by the induction hypothesis,

$$\|g(a+x)\| = \|f'(a+x)\| \leq (1/(n-1)!)\, M\, x^{(n-1)} \tag{5.14}$$

whenever $0 \leq x < R$. Define

$$\varphi(x) = \begin{cases} (1/n!)\, M\, x^n & \text{if } x \geq 0, \\ 0 & \text{if } x < 0. \end{cases}$$

Let x be fixed, $0 < x < R$. Then we see that (5.14) can be expressed as

$$\|f'(a+t)\| \leq \varphi'(t)$$

for all $t \in [0,\ x]$. Hence the mean value theorem 5.1.12 shows that

$$\begin{aligned} \|f(a+x)\| &= \|f(a+x) - f(a)\| \\ &\leq (\varphi(x) - \varphi(0)) = (1/n!)\, M\, x^n. \end{aligned}$$

This proves the result for $x > 0$. Arguments for $x < 0$ are similar. □

Example 5.1.18 Let $f : \mathbb{R} \to Y$ be such that $f^{(n)}$ exists on $\mathbb{R}$ for all n. Assume that $f^{(n)}(0) = 0$ for all n and $\|f^{(n)}(x)\| \leq nR$ for all n and all $x \in \mathbb{R}$ with $|x| < R$. Then $f(x) = \mathbf{0}$ for all $x \in \mathbb{R}$. To see this, let $n \in \mathbb{N}$ and let $R > 0$ be arbitrary. Apply Lemma 5.1.17 with $a = 0, M = nR$ to obtain

$$\|f(x)\| \leq (1/n!)\,(nR)\,|x|^n \leq \frac{1}{(n-1)!}\, R^{n+1} \text{ whenever } |x| < R.$$

Since the above holds for all $n \in \mathbb{N}$, we get

$$\|f(x)\| \leq \lim_{n\to\infty} \frac{1}{(n-1)!}\, R^{n+1} = 0 \quad \text{whenever } |x| < R.$$

Since this is true for any $R > 0$, it follows that $f(x) = \mathbf{0}$ for all $x \in \mathbb{R}$.

Definition 5.1.19 Taylor polynomials. Let $f : A \to Y$ be a function. Assume that $f^{(n)}(a)$ exists for a certain $n \in \mathbb{N}$ and $a \in A$. Then the *(nth-degree) Taylor polynomial* $P_n : \mathbb{R} \to Y$ *of* $f : A \to Y$ at $a \in A$ is defined as

$$P_n(a+x) = \sum_{k=0}^{n} \frac{1}{k!}\, f^{(k)}(a)\, x^k \tag{5.15}$$

for all $x \in \mathbb{R}$. Note that the value of $P_n : \mathbb{R} \to Y$ at each $x \in \mathbb{R}$ is a linear combination of $(n+1)$ fixed vectors $f^{(k)}(a) \in Y$, $k = 0, 1, \ldots, n$. The coefficients of these fixed vectors are the powers $(1/k!)x^k$.

Example 5.1.20 Let $f(x) = (\cos x, e^x, x - \sin x)$ for all $x \in \mathbb{R}$. Then

$$\begin{aligned} f^{(1)}(x) &= (-\sin x, e^x, 1 - \cos x) \\ f^{(2)}(x) &= (-\cos x, e^x, \sin x) \\ f^{(3)}(x) &= (\sin x, e^x, \cos x) \\ f^{(4)}(x) &= (\cos x, e^x, -\sin x) \\ f^{(5)}(x) &= (-\sin x, e^x, -\cos x) \end{aligned}$$

The 5th-degree Taylor polynomial of f at 0 is $P_5 : \mathbb{R} \to \mathbb{R}^3$ given by

$$P_5(x) = \sum_{k=0}^{5} \frac{1}{k!} f^{(k)}(0)\, x^k \quad \text{for all } x \in \mathbb{R}.$$

Hence, for all $x \in \mathbb{R}$,

$$\begin{aligned} P_5(x) &= (1,1,0) + x\,(0,1,0) + \frac{x^2}{2!}\,(-1,1,0) \\ &\quad + \frac{x^3}{3!}\,(0,1,1) + \frac{x^4}{4!}\,(1,1,0) + \frac{x^5}{5!}\,(0,1,-1). \end{aligned}$$

Lemma 5.1.21 Derivatives of Taylor polynomials. *Let $P_n : \mathbb{R} \to Y$ be the nth-degree Taylor polynomial of $f : A \to Y$ at $a \in A$. Then*

$$P_n^{(k)}(a) = \begin{cases} f^{(k)}(a) & \text{if } 0 \le k \le n, \\ 0 & \text{if } k > n. \end{cases} \tag{5.16}$$

Proof. This follows by an easy computation. □

Theorem 5.1.22 Approximation by Taylor polynomials. *Let $f : A \to Y$ be a function. Assume that there is an $R > 0$ such that $f^{(n)}(a + x)$ exists and*

$$\|f^{(n)}(a + x)\| \le M \text{ for all } |x| < R.$$

Then $\|f(a + x) - P_{n-1}(a + x)\| \le (1/n!) M\, |x|^n$ whenever $|x| < R$.

Proof. Define $g : A \to Y$ by $g(x) = f(x) - P_{n-1}(x)$. We see that $g^{(k)}(a) = 0$ for $0 \le k \le (n - 1)$ and $\|g^{(n)}(a + x)\| = \|f^{(n)}(a + x)\| \le M$ whenever $|x| < R$. Then the proof follows from Lemma 5.1.17. □

Example 5.1.23 Let $f(x) = (\cos x,\ e^x,\ \sin x)$. Then,

$$\|f^{(n)}(x)\|^2 = \sin^2 x + e^{2x} + \cos^2 x = 1 + e^{2x}$$

for all $n \in \mathbb{N}$ and $x \in \mathbb{R}$. If $R > 0$ and $M = (1 + e^{2R})^{1/2}$, then

$$\|f^{(n)}(x)\| \leq (1 + e^{2R})^{1/2} = M \quad \text{whenever } |x| < R.$$

Hence, by Theorem 5.1.22,

$$\|f(x) - P_n(x)\| \leq (1/(n+1)!)M\,|x|^{n+1} \leq (1/(n+1)!)M\,R^{n+1}$$

whenever $|x| < R$. Since $\lim_n M\,R^n/n! = 0$ for all $R > 0$, we see that

$$f(x) = \lim_{n \to \infty} P_n(x) \quad \text{for all } x \in \mathbb{R}.$$

Taylor Series

Assume that $f^{(n)}(a)$ exists for all $n \in \mathbb{N}$. Then $P_n(x)$, as defined in (5.15), exists for each $n \in \mathbb{N}$. Then $P_n(x)$ is an approximation of $f(x)$ for x close to a for each fixed $n \in \mathbb{N}$. The exact formulation of this approximation is given by Taylor's theorem, Theorem 5.1.22. Now we fix $x \in \mathbb{R}$ and want to know if $\lim_n P_n(x)$ exists and is equal to $f(x)$. The answer is easy; it follows from Theorem 5.1.22 above.

Theorem 5.1.24 Approximation by Taylor series. *For each $n \in \mathbb{N}$ and $r > 0$, let*

$$M_n(r) = \sup\left\{ \|f^{(n)}(a + x)\| \mid |x| \leq r \right\}$$

if it exists. Assume that there is an $R > 0$ such that $(1/n!)\,M_n(R)R^n$ is a bounded sequence. Then

$$f(a + x) = f(a) + f'(a)x + \cdots + \frac{1}{n!}f^{(n)}(a)x^n + \cdots \tag{5.17}$$

$$= \lim_n \sum_{k=0}^{n} \frac{1}{k!} f^{(k)}(a)x^k = \lim_n P_n(a + x) \tag{5.18}$$

whenever $|x| < R$. Also, the series in (5.17) can be differentiated term-by-term to obtain, for all $k \in \mathbb{N}$,

$$f^{(k)}(a + x) = f^{(k)}(a) + f^{(k+1)}(a)x + \cdots + \frac{1}{n!}f^{(k+n)}(a)x^n + \cdots. \tag{5.19}$$

Proof. Let $(1/n!)\,M_n(R)R^n$ be a bounded sequence. We see that

$$\lim_n (1/n!)\,M_n(R)r^n = 0$$

whenever $|r| < R$. Then (5.17) follows from Theorem 5.1.22. For the second part, note that $g(a + x) = f^{(k)}(a + x)$ satisfies the same hypotheses as $f(a + x)$. □

Motions of Euclidean Spaces

Motions of Euclidean spaces provide instructive examples of differentiation. We consider only the following type of motions.

Definition 5.1.25 Motions. Let X be a Euclidean space. Let $A \subset \mathbb{R}$ be an open interval containing the origin $0 \in \mathbb{R}$. Let $\mathbf{s} : A \to X$ and $S : A \to L(X, X)$ be two differentiable functions. Assume that $S(0) = I$ is the identity on X. These two functions define a mapping $M(t) : X \to X$ by

$$M(t)\mathbf{a} = \mathbf{s}(t) + S(t)\mathbf{a}, \quad \mathbf{a} \in X$$

for each $t \in A$. Each $M(t) : X \to X$ is an affine mapping of X into itself. The family $\{M(t)\}$, $t \in A$, is called a *motion of* X. The *trajectory* of $\mathbf{a} \in X$ is the function $\mathbf{r} : A \to X$ defined as $\mathbf{r}(t) = M(t)\mathbf{a}$, $t \in A$. Note that $\mathbf{r}(0) = \mathbf{a}$. Hence $\mathbf{r}(t) = M(t)\mathbf{r}(0)$ and $\mathbf{r}(t)$, $t \in A$, is the trajectory of $\mathbf{r}(0)$. Note that $\mathbf{s}(t)$, $t \in A$, is the trajectory of $\mathbf{s}(0) \in X$.

Definition 5.1.26 Velocities. Notations are the same as in Definition 5.1.25 above. Let $\mathbf{r} : A \to X$ be a trajectory. Then its derivative at $t \in A$

$$\mathbf{r}'(t) = \lim_{h \to 0}(1/h)(\mathbf{r}(t+h) - \mathbf{r}(t))$$

is defined as the *velocity on this trajectory* at the point $\mathbf{r}(t)$. We see that

$$\mathbf{r}'(t) = \mathbf{s}'(t) + S'(t)(\mathbf{r}(0)).$$

The proof of this is given as Problem 5.11.

Definition 5.1.27 Rigid motions. A motion $M(t)$ is called a *rigid motion* if it preserves the distances between any two points. More explicitly, $M(t)$ is a rigid motion if

$$\|M(t)\mathbf{a} - M(t)\mathbf{b}\| = \|\mathbf{a} - \mathbf{b}\|$$

for all $\mathbf{a}$, $\mathbf{b} \in X$ and for all $t \in A$. It is clear that $M(t)$ is a rigid motion if and only if $S(t)$ is an isometry for each $t \in A$. This means that $\|S(t)\mathbf{x}\| = \|\mathbf{x}\|$ for each $\mathbf{x} \in X$ and for each $t \in A$.

Examples and applications of these notions are given as problems.

Problems

In the following problems, the norm on $\mathbb{R}^n$ is the standard Euclidean norm. Also, A always denotes an open interval in $\mathbb{R}$.

5.1 Define $f : \mathbb{R} \to \mathbb{R}^2$ by $f(t) = (\cos 2\pi t,\ \sin 2\pi t)$. Find $f^{(n)}(t)$ and show that $\langle f^{(n)}(t),\ f^{(n+1)}(t)\rangle = 0$ for all $t \in \mathbb{R}$ and $n \in \mathbb{N}$.

5.2 Define $f : \mathbb{R} \to \mathbb{R}^3$ by $f(t) = (\cos 2\pi t,\ \sin 2\pi t,\ 2\pi t)$. Find $f^{(n)}(t)$ and show that $\langle f^{(n)}(t),\ f^{(n+1)}(t)\rangle = 0$ for all $t \in \mathbb{R}$ and $n \in \mathbb{N}, n \geq 1$.

5.3 Assume that $f : A \to \mathbb{R}^n$ is differentiable and that $\|f(t)\| > 0$ for all $t \in A$. Show that $u(t) = f(t)/\|f(t)\|$ is also differentiable and $\langle u(t),\ u'(t)\rangle = 0$ for all $t \in A$.

5.4 Define $f : R \to \mathbb{R}^2$ by $f(t) = (e^t \cos 2\pi t,\ e^t \sin 2\pi t)$. Show that the angle between the vectors $f(t)$ and $f'(t)$ is constant.

5.5 Let X be a normed space. Given $A \in L(X,\ X)$ and $t \in \mathbb{R}$, let

$$f(t) = e^{tA} = \lim_n \left(I + \frac{t}{1!} A + \frac{t^2}{2!} A^2 + \cdots + \frac{t^n}{n!} A^n \right)$$

as in Problem 4.48. Show that $f : \mathbb{R} \to L(X,\ X)$ is differentiable. Find $f'(t)$.

5.6 A disc of radius $2R$ in the xy-plane rolls on the x-axis without gliding. At time t the center of the disc is at the point $(vt,\ 2R)$, where $v > 0$ is a constant. Let P be the point on this disc which is at point $(0,\ R)$ at time $t = 0$. Find the equation $f : \mathbb{R} \to \mathbb{R}^2$ of the the trajectory of P. Find the points where this trajectory has horizontal tangents $f'(t)$.

5.7 Let $f,\ g : A \to \mathbb{R}^n$ be two differentiable functions. Then show that

$$F = \langle f,\ g\rangle : A \to \mathbb{R}$$

is also differentiable and $F'(t) = \langle f'(t),\ g(t)\rangle + \langle f(t),\ g'(t)\rangle$.

5.8 Let $f,\ g : A \to \mathbb{R}^3$ be two differentiable functions. Let

$$F = f \times g : A \to \mathbb{R}^3,$$

with the usual cross product in $\mathbb{R}^3$. Show that $F : A \to \mathbb{R}^3$ is also differentiable and $F(t) = (f'(t) \times g(t)) + (f(t) \times g'(t))$.

5.9 **Frenet formulas.** Consider a function $\mathbf{r} : A \to \mathbb{R}^3$ as the equation of a curve C in $\mathbb{R}^3$. Then

$$\begin{aligned} \mathbf{u}(t) &= \mathbf{r}'(t)/\|\mathbf{r}'(t)\| \text{ is called the } \textit{unit tangent vector}, \\ \mathbf{n}(t) &= \mathbf{u}'(t)/\|\mathbf{u}'(t)\| \text{ is called the } \textit{unit principal normal vector}, \text{ and} \\ \mathbf{b}(t) &= \mathbf{u}(t) \times \mathbf{n}(t) \text{ is called the } \textit{binormal vector} \end{aligned}$$

of C at the point $\mathbf{r}(t)$. We assume that C is such that all three vectors are well-defined at every point of C. We let $p(t) = \|\mathbf{r}'(t)\|^{-1}$ and define

$$\begin{array}{lll} \text{the } \textit{curvature of } C \textit{ at } \mathbf{r}(t) & \text{as} & \rho(t) = \|\mathbf{u}'(t)\|\, p(t) \text{ and} \\ \text{the } \textit{torsion of } C \textit{ at } \mathbf{r}(t) & \text{as} & \tau(t) = \pm\|\mathbf{b}'(t)\|\, p(t)\,. \end{array}$$

Show that $(\mathbf{u}(t),\, \mathbf{n}(t),\, \mathbf{b}(t))$ is an orthonormal basis for $\mathbb{R}^3$ and that

$$\begin{aligned} \mathbf{u}'(t)\, p(t) &= \rho(t)\, \mathbf{n}(t), \\ \mathbf{n}'(t)\, p(t) &= -\rho(t)\, \mathbf{u}(t) - \tau(t)\, \mathbf{b}(t), \text{ and} \\ \mathbf{b}'(t)\, p(t) &= \tau(t)\, \mathbf{n}(t). \end{aligned}$$

The sign of the torsion $\tau(t)$ is defined by the last formula. The formulas above are known as the *Frenet formulas*. We will refer to the set of three vectors $(\mathbf{u}(t),\, \mathbf{n}(t),\, \mathbf{b}(t))$ as the *Frenet vectors* of the curve C at the point $\mathbf{r}(t)$.

5.10 Compute the Frenet vectors for the *helix*

$$\mathbf{r}(t) = (R\, \cos\omega t,\; R\, \sin\omega t,\; \lambda t), \quad t \in \mathbb{R},$$

and verify the Frenet formulas. Here $R > 0$ and $\omega,\, \lambda \in \mathbb{R}$ are constants.

Problems on the Motions of Euclidean Spaces

5.11 Let X be a Euclidean space and $S : A \to L(X,\, X)$ a differentiable function. Show that for each $\mathbf{a} \in X$ the function $\mathbf{r}(t) = S(t)\,\mathbf{a}$, $t \in A$, is also differentiable and $\mathbf{r}'(t) = S'(t)\,\mathbf{a}$.

5.12 Let X be a Euclidean space. Let $S : \mathbb{R} \to L(X,\, X)$ be such that

$$S(a + b) = S(b) \cdot S(a)$$

for all $a,\, b \in \mathbb{R}$. If $S'(0) = A$ exists then show that $S(t) = e^{At}$ for all $t \in \mathbb{R}$. Here $e^T = \sum_{n=0}^{\infty}(1/n!)T^n$, $T \in L(X,\, X)$, as defined in Problem 4.48. What is $S'(t)$ in terms of $S(t)$ and $S'(0)$?

5.13 Let X be a Euclidean space. Let $S : A \to L(X,\, X)$ be such that $\|S(t)\mathbf{v}\| = \|\mathbf{v}\|$ for all $t \in A$ and $\mathbf{v} \in X$. If $S'(t)$ exists, then show that $\langle S'(t)\mathbf{v},\, S(t)\mathbf{v}\rangle = 0$ for all $t \in A$ and $\mathbf{v} \in X$.

5.14 **Rotations of $\mathbb{R}^3$.** Let $\mathbf{w} \in \mathbb{R}^3$ be a unit vector and let $\omega \in \mathbb{R}$. The *rotation of $\mathbb{R}^3$ about the axis $\mathbf{w}$ with the angular velocity of ω* is defined as follows. Complete

$\mathbf{w}$ to an orthonormal basis $(\mathbf{u}, \mathbf{v}, \mathbf{w})$ so that $\mathbf{w} = \mathbf{u} \times \mathbf{v}$. Then let

$$\begin{aligned} R(t)(\alpha\mathbf{u} + \beta\mathbf{v} + \gamma\mathbf{w}) &= \alpha R(t)\mathbf{u} + \beta R(t)\mathbf{v} + \gamma R(t)\mathbf{w}, \quad \text{where} \\ R(t)\mathbf{u} &= \cos\omega t\,\mathbf{u} + \sin\omega t\,\mathbf{v} \\ R(t)\mathbf{v} &= -\sin\omega t\,\mathbf{u} + \cos\omega t\,\mathbf{v} \\ R(t)\mathbf{w} &= \mathbf{w} \end{aligned}$$

for all $\mathbf{x} = (\alpha\mathbf{u} + \beta\mathbf{v} + \gamma\mathbf{w}) \in \mathbb{R}^3$ and for all $t \in \mathbb{R}$. Show that $R(t)$ is a rigid motion. Also show that

$$R'(t)\mathbf{x} = \omega\,\mathbf{w} \times R(t)\mathbf{x}$$

for all $\mathbf{x} \in \mathbb{R}^3$ and for all $t \in \mathbb{R}$.

5.15 Let $S : A \to L(\mathbb{R}^3, \mathbb{R}^3)$ be a differentiable function such that $\|S(t)\mathbf{x}\| = \|\mathbf{x}\|$ for all $t \in A$ and $\mathbf{x} \in \mathbb{R}^3$. Show that for each $t \in A$ there is a unique vector $\mathbf{m}(t) \in \mathbb{R}^3$ such that $S'(t)\mathbf{x} = \mathbf{m}(t) \times S(t)\mathbf{x}$ for all $t \in A$ and for all $\mathbf{x} \in \mathbb{R}^3$.

Remarks. Rigid motions about a fixed point. We see that $S(t)$ in the preceding problem is a rigid motion about the fixed origin. This problem shows that at each instant $t \in A$, the velocities for the rigid motion are the same as the velocities in a rotation. The angular velocity of this rotation is given by $\omega(t) = \|\mathbf{m}(t)\|$ and the axis is given by the unit vector $\mathbf{w}(t) = (1/\omega(t))\mathbf{m}(t)$. They are called the *instantaneous angular velocity* and *instantaneous axis of rotation* of this rigid motion $S(t)$ with a fixed point.

5.16 Let $\mathbf{c} \in \mathbb{R}^3$. Define $A : \mathbb{R}^3 \to \mathbb{R}^3$ by $A\mathbf{x} = \mathbf{c} \times \mathbf{x}$ for $\mathbf{x} \in \mathbb{R}^3$. Compute $e^{At}\mathbf{x}$ for all $t \in \mathbb{R}$ and $\mathbf{x} \in \mathbb{R}^3$.

5.17 Helicoidal motions. Let $R(t) : \mathbb{R}^3 \to \mathbb{R}^3$ be a rotation given in terms of $\mathbf{w} \in \mathbb{R}^3$ and $\omega \in \mathbb{R}$ as in Problem 5.14. Let $T(t) : \mathbb{R}^3 \to \mathbb{R}^3$ be a translation given by $T(t)\mathbf{x} = \mathbf{x} + t\mathbf{a}$, $\mathbf{x} \in \mathbb{R}^3$, $t \in \mathbb{R}$, where $\mathbf{a} \in \mathbb{R}^3$ is fixed. Then

$$H(t)\mathbf{x} = t\mathbf{c} + R(t)\mathbf{x}, \quad t \in \mathbb{R}, \quad \mathbf{x} \in \mathbb{R}^3$$

is called a *helicoidal motion*. Show that helicoidal motions are rigid motions. If $\mathbf{r}(t) = H(t)\mathbf{x}$, then show that the velocities are given as $\mathbf{r}'(t) = \mathbf{c} + \omega\,\mathbf{w} \times R(t)\mathbf{x}$.

5.18 General rigid motions. Let $M(t)\mathbf{x} = \mathbf{s}(t) + S(t)\mathbf{x}$, $t \in A$, be a general rigid motion. If $\mathbf{r}(t) = M(t)\mathbf{x}$, then show that for each instant $t \in A$ the velocities $\mathbf{r}'(t)$ are the same as the velocities in a helicoidal motion. The translational and rotational parts of this helicoidal motion depend on $t \in A$. They are called the *instantaneous translations and rotations* in a general rigid motion.

5.19 Instantaneous translations and rotations of Frenet vectors. The Frenet vectors $(\mathbf{u}(t), \mathbf{n}(t), \mathbf{b}(t))$ of a curve $\mathbf{r} : A \to \mathbb{R}^3$ form an orthonormal basis for $\mathbb{R}^3$

for all $t \in A$. Hence, they define a rigid motion as follows. For convenience, assume that $0 \in A$ and define

$$S(t)(\alpha\mathbf{u}(0) + \beta\mathbf{n}(0) + \gamma\mathbf{b}(0)) = \alpha\mathbf{u}(t) + \beta\mathbf{n}(t) + \gamma\mathbf{b}(t)$$

for all $\mathbf{x} = (\alpha, \beta, \gamma) \in \mathbb{R}^3$. Then $M(t)\mathbf{x} = \mathbf{r}(t) + S(t)\mathbf{x}$ defines a rigid motion of $\mathbb{R}^3$. Find the instantaneous translations and rotations of this rigid motion.

Problems on Plane Curves

Let $\mathbf{r} : A \to \mathbb{R}^2 \subset \mathbb{R}^3$ be a plane curve C with the corresponding Frenet vectors $(\mathbf{u}(t), \mathbf{n}(t), \mathbf{b}(t))$. Let $\mathbf{k}$ be a unit vector orthogonal to the subspace $\mathbb{R}^2$ in the vector space $\mathbb{R}^3$.

5.20 Show that $\mathbf{u}(t)$ and $\mathbf{n}(t)$ are orthogonal to $\mathbf{k}$ and $\mathbf{b}(t) = \pm\mathbf{k}$ and $\tau(t) = 0$ for all $t \in A$. Also show that $\mathbf{n}(t) = \pm\mathbf{k} \times \mathbf{u}(t)$ and $\mathbf{u}(t) = \pm\mathbf{k} \times \mathbf{n}(t)$ for all $t \in A$.

5.21 The point $\mathbf{e}(t) = \mathbf{r}(t) + (1/\rho(t))\,\mathbf{n}(t)$ is called the *center of curvature of* C *at the point* $\mathbf{r}(t)$. Then $\mathbf{e} : A \to \mathbb{R}^2 \subset \mathbb{R}^3$ defines another curve E. Show that the unit tangent vector of E at $\mathbf{e}(t)$ is $\pm\mathbf{n}(t)$.

5.22 Define $S : A \to L(\mathbb{R}^2, \mathbb{R}^2)$ as follows. Let $t_0 \in A$ be fixed. Any $\mathbf{v} \in \mathbb{R}^2$ has a unique expression as $\mathbf{v} = \mathbf{r}(t_0) + a\,\mathbf{u}(t_0) + b\,\mathbf{n}(t_0)$. Then

$$S(t)\mathbf{v} = \mathbf{v}(t) = \mathbf{r}(t) + a\,\mathbf{u}(t) + b\,\mathbf{n}(t).$$

Show that the velocity field $\mathbf{v}'(t)$ at any instant $t \in A$ is the same as the rotational velocities of the plane $\mathbb{R}^2$ about the point $\mathbf{e}(t) = \mathbf{r}(t) + (1/\rho(t))\mathbf{n}(t)$ with the angular velocity $\omega(t) = \rho(t)\,\|\mathbf{r}'(t)\|$.

5.23 Let L and L' be the lines in $\mathbb{R}^2$ passing through the points $\mathbf{r}(t)$ and $\mathbf{r}(t')$ and in the directions of $\mathbf{n}(t)$ and $\mathbf{n}(t')$, respectively. Show that the intersection points $P(t, t')$ of these lines converge to $\mathbf{e}(t)$ as $t \to t'$.

5.24 Verify the results of the last four problems for the parabola given by $\mathbf{r}(t) = (t, (1/2)t^2) \in \mathbb{R}^2$ for all $t \in \mathbb{R}$.

5.2 DIFFERENTIABLE FUNCTIONS

The main purpose of this section is to define the differentiability of a function f between two normed spaces X and Y. As stated in the chapter introduction, a function is differentiable at a point if it can be locally approximated by a linear transformation. This linear transformation is then the derivative (or *full derivative*) of the function at that point. Plainly, this way of understanding the derivative requires a subtle shift from the familiar definition of the derivative as a *real number* in the one-variable case, and indeed from its definition as a *vector* in the special case that we have just presented in Section 5.1.

Perhaps the easiest way to motivate this new way of thinking about the derivative is to note the equivalence, in the one-variable case, between the existence of the numerical limit $f'(a)$ and the fact that, in a neighborhood of a, the graph of $f(x)$ can be approximated by the tangent line whose slope is $f'(a)$. That is, $f(a+r)$ can be very well estimated by $f(a)+[f'(a)](r)$. Similarly, consider the situation of Section 5.1, where $f : A \to Y$ for a normed space Y, A is an open subset of $\mathbb{R}$, and $a \in A$. In this case, the existence of the vector derivative $f'(a)$ is equivalent to the fact that, in a neighborhood of a, $f(a+r)$ can be very well estimated by $f(a) + [f'(a)](r)$. The meaning of 'very well estimated' is given precisely by the formulation for the derivative that we pointed out in (5.3): for each $\varepsilon > 0$, there is a $\delta > 0$ such that if $|r| < \delta$, then

$$\|f(a+r) - f(a) - [f'(a)](r)\| \leq \varepsilon |r|.$$

The notation $[f'(a)]$ is meant to be suggestive. Any vector $\mathbf{y}$ in Y corresponds to the linear transformation $T_y : \mathbb{R} \to Y$ given by $T_y(r) = r\mathbf{y}$. Conversely, any linear transformation T in $L(\mathbb{R}, Y)$ corresponds to the vector $\mathbf{y} = T(1)$. It turns out that thinking of $[f'(a)]$ as a linear transformation in $L(\mathbb{R}, Y)$, rather than as a vector, is the key that lets us generalize the definition to cases where the domain space X is not $\mathbb{R}$. We make this precise in Definition 5.2.1: the derivative $f'(\mathbf{a})$, when it exists, is a linear transformation from X to Y. (Note that we shall write $f'(\mathbf{a})\mathbf{u}$, rather than $[f'(\mathbf{a})](\mathbf{u})$, for the application of the linear transformation to a vector $\mathbf{u} \in X$.)

The main difficulties with this novel definition are how to tell when a function has a derivative and how to picture the derivative. We get some help by introducing the notion of a partial or directional derivative. Suppose that $X = \mathbb{R}^2$ and $Y = \mathbb{R}$, and we have a function $f : \mathbb{R}^2 \to \mathbb{R}$. Given a point $\mathbf{a} = (a, b)$, we can watch how f behaves along straight lines through $\mathbf{a}$—for example, horizontal lines of the form $(a+t, b)$ and vertical lines of the form $(a, b+t)$. Either of these restrictions turns f into a function of one variable t, and the familiar (numerical) derivatives associated with these restricted functions are called *partial* or *directional* derivatives. Furthermore, there is no need to restrict ourselves to horizontal and vertical lines. Directional derivatives may be taken along any line through $\mathbf{a}$ (Definition 5.2.11).

At this point, there are two important questions about the relationship between the full derivative and the directional derivatives. First: given the full derivative, can we compute the directional derivatives? Second: given the directional derivatives, can we compute the full derivative? In this section, we answer the first question in the affirmative and explain the computation. We answer the second question in the negative: unfortunately, the existence of all directional derivatives is compatible with the non-existence of the full derivative. The good news, deferred to Section 5.3, is that the existence of continuously differentiable directional derivatives *does* guarantee the existence of the full derivative. In this case, the directional derivatives even give us a convenient matrix representation of the full derivative.

In what follows, X and Y are any two normed spaces and A is an open set in X.

Definition 5.2.1 Differentiable functions. A function $f : A \to Y$ is said to be *differentiable at* $\mathbf{a} \in A$ if there is a linear map $T \in L(X, Y)$, such that

$$\lim_{\mathbf{r}\to\mathbf{0}} \frac{\|f(\mathbf{a}+\mathbf{r}) - f(\mathbf{a}) - T\mathbf{r}\|}{\|\mathbf{r}\|} = 0. \tag{5.20}$$

Lemma 5.2.4 will show that if such a $T : X \to Y$ exists, then it is unique. It is called the *derivative* (or *full derivative*) *of* f *at* $\mathbf{a} \in A$ and is denoted by $f'(\mathbf{a})$. Note that $f'(\mathbf{a}) : X \to Y$ is a linear operator and its value at $\mathbf{x} \in X$ is written as $f'(\mathbf{a})(\mathbf{x}) \in Y$.

Remarks 5.2.2 Other formulations. An explicit formulation of (5.20) is that for each $\varepsilon > 0$ there is a $\delta > 0$ such that

$$\|f(\mathbf{a}+\mathbf{r}) - f(\mathbf{a}) - T\mathbf{r}\| \le \varepsilon \|\mathbf{r}\| \text{ whenever } \|\mathbf{r}\| < \delta. \tag{5.21}$$

Since A is open, there is a $c > 0$ such that $B_c(\mathbf{a}) \subset A$. We choose $\delta < c$. Therefore, $(\mathbf{a}+\mathbf{r}) \in A$ and consequently $f(\mathbf{a}+\mathbf{r})$ is defined whenever $\|\mathbf{r}\| < \delta$.

An equivalent form of (5.21) is the following. For every zero-sequence $\mathbf{r}_n$ in X there is a zero-sequence t_n in $\mathbb{R}$ such that

$$\|f(\mathbf{a}+\mathbf{r}_n) - f(\mathbf{a}) - T\mathbf{r}_n\| \le t_n \|\mathbf{r}_n\|. \tag{5.22}$$

This form avoids the ε-δ statements. Sometimes this may be an advantage. The equivalence of (5.21) and (5.22) follows easily.

The next two lemmas establish that there can be at most one linear transformation T that satisfies (5.20). If the derivative exists, then it is unique.

Lemma 5.2.3 *If* $S \in L(X, Y)$ *and if* $\lim_{\mathbf{r}\to\mathbf{0}}(\|S\mathbf{r}\| / \|\mathbf{r}\|) = 0$, *then* $S = \mathbf{0}$.

Proof. The condition $\lim_{\mathbf{r}\to\mathbf{0}}(\|S\mathbf{r}\| / \|\mathbf{r}\|) = 0$ means that for each $\varepsilon > 0$ there is a $\delta > 0$ such that

$$(\|S\mathbf{r}\| / \|\mathbf{r}\|) \le \varepsilon \text{ whenever } 0 < \|\mathbf{r}\| < \delta.$$

Let $\mathbf{u} \in X$ be a nonzero vector. Choose a nonzero $t \in \mathbb{R}$ so that $\|t\mathbf{u}\| < \delta$. Then

$$(\|S(t\mathbf{u})\| \,/\, \|t\mathbf{u}\|) = (\|S\mathbf{u}\| \,/\, \|\mathbf{u}\|) \leq \varepsilon$$

and therefore $\|S\mathbf{u}\| \leq \varepsilon\|\mathbf{u}\|$. This is true for any $\varepsilon > 0$. Hence $S\mathbf{u} = \mathbf{0}$ for all $\mathbf{u} \in X$. This means that $S = \mathbf{0}$. □

Lemma 5.2.4 *Let $T, T' \in L(X, Y)$ both satisfy (5.20) in the definition of derivatives, 5.2.1. Then $T = T'$.*

Proof. Let $S = T - T'$. We have

$$\begin{aligned} \|S\mathbf{r}\| &= \|T\mathbf{r} - T'\mathbf{r}\| \\ &\leq \|f(\mathbf{a}+\mathbf{r}) - f(\mathbf{a}) - T'\mathbf{r}\| + \|f(\mathbf{a}+\mathbf{r}) - f(\mathbf{a}) - T\mathbf{r}\|. \end{aligned}$$

Then $\lim_{\mathbf{r}\to\mathbf{0}}(\|S\mathbf{r}\| \,/\, \|\mathbf{r}\|) = 0$ by (5.20) in the definition of derivatives, Definition 5.2.1. Hence $S = \mathbf{0}$ by Lemma 5.2.3. □

In light of the foregoing lemmas, we can speak of *the* derivative $f'(a)$ and we can define an associated affine approximation function.

Definition 5.2.5 Affine approximations. Let $f : A \to Y$ be differentiable at $\mathbf{a} \in A$. The *affine (or first-order polynomial) approximation* of f at $\mathbf{a}$ is the function $T_{\mathbf{a}} : X \to Y$ defined by $T_{\mathbf{a}}\mathbf{x} = f(\mathbf{a}) + f'(\mathbf{a})(\mathbf{x} - \mathbf{a})$, $\mathbf{x} \in X$. Note that the linear transformation $f'(\mathbf{a})$ does not approximate the function, but rather its increments. The function itself is approximated by the affine approximation $T_{\mathbf{a}}$.

Since f' is a function from X to $L(X, Y)$ and both of these are normed spaces, it makes sense to ask whether f' is a continuous function.

Definition 5.2.6 Continuously differentiable functions. A function $f : A \to Y$ is said to be *continuously differentiable on A* if there is a continuous function $f' : A \to L(X, Y)$, such that

$$\lim_{\mathbf{r}\to\mathbf{0}} \frac{\|f(\mathbf{a}+\mathbf{r}) - f(\mathbf{a}) - f'(\mathbf{a})\mathbf{r}\|}{\|\mathbf{r}\|} = 0 \tag{5.23}$$

for all $\mathbf{a} \in A$.

Remarks 5.2.7 An important question. How can we decide if a function is differentiable? Unfortunately there is no easy answer in general. Lemma 5.2.8 formulates a necessary condition for differentiability, but it is not a sufficient condition. For continuous differentiability, however, there is a routine test. Lemma 5.2.9 formulates a necessary condition for continuous differentiability that turns out also to be sufficient. This condition provides the desired test for continuous differentiability.

Lemma 5.2.8 A necessary condition for differentiability. *Assume that $f : A \to Y$ is differentiable at $\mathbf{a} \in A$. Then*

$$\lim_{t\to 0} \frac{1}{t}\left(f(\mathbf{a}+t\mathbf{r}) - f(\mathbf{a})\right) \tag{5.24}$$

exists for each $\mathbf{r} \in X$.

Proof. Assume that the derivative $T = f'(\mathbf{a})$ exists. We will show that the limit in (5.24) converges to $T\mathbf{r}$. This is clear if $\mathbf{r} = \mathbf{0}$. Otherwise,

$$\left\| \frac{f(\mathbf{a}+t\mathbf{r}) - f(\mathbf{a})}{t} - T\mathbf{r} \right\| = \|\mathbf{r}\| \frac{\|f(\mathbf{a}+t\mathbf{r}) - f(\mathbf{a}) - Tt\mathbf{r}\|}{\|t\mathbf{r}\|}$$

shows that $\lim_{t\to 0}(1/t)(f(\mathbf{a}+t\mathbf{r}) - f(\mathbf{a})) = T\mathbf{r}$. □

Example 5.3.1 below shows that the condition in Lemma 5.2.8 is not sufficient for differentiability. Note, however, that if the domain space X is one-dimensional, then Lemma 5.2.8 gives a necessary and sufficient condition for differentiability. This follows by an easy argument and is left as an exercise. Hence for one-dimensional domains, the existence of the limit in (5.24) can be taken as the definition of differentiability.

Lemma 5.2.9 A necessary condition for continuous differentiability. *Assume that $f : A \to Y$ is continuously differentiable on A. Then, for each fixed $\mathbf{r} \in X$,*

$$\lim_{t\to 0} \frac{1}{t}\left(f(\mathbf{a}+t\mathbf{r}) - f(\mathbf{a})\right) = F(\mathbf{a}) \tag{5.25}$$

exists and defines a continuous function $F : G \to Y$.

Proof. Lemma 5.2.8 shows that $F(\mathbf{a})$ exists and is equal to $f'(\mathbf{a})\mathbf{r}$. But, by hypothesis, $f' : A \to L(X, Y)$ is continuous. Then $F(\cdot) = f'(\cdot)\mathbf{r} : A \to Y$ is also continuous for each fixed $\mathbf{r} \in X$. □

Theorem 5.3.4 below shows that the condition in Lemma 5.2.9 is also sufficient for continuous differentiability. Hence this condition provides an easy test for continuous differentiability. These arguments are left to the next section. The rest of this section contains miscellaneous remarks about full and directional derivatives, together with some computational examples.

Restricted and Directional Derivatives

We continue to assume that A is an open set in X and that $f : A \to Y$ is a function.

Definition 5.2.10 Restricted derivatives. Let U be a subspace of X. Then $f : A \to Y$ is said to have a *restricted derivative* at $\mathbf{a} \in A$ along U if there is a linear map $S : U \to Y$ such that

$$\lim_{\mathbf{u}\in U,\, \mathbf{u}\to\mathbf{0}} \frac{\|f(\mathbf{a}+\mathbf{u}) - f(\mathbf{a}) - S\mathbf{u}\|}{\|\mathbf{u}\|} = 0. \tag{5.26}$$

Simply by replacing X with U in the proof of Lemma 5.2.4, we can see that if such an $S : U \to Y$ exists, then it is unique. S is called the *restricted derivative of f at $\mathbf{a} \in A$ along U* and is denoted by $f'_U(\mathbf{a})$. $f'_U(\mathbf{a}) : U \to Y$ is a linear operator and its value at $\mathbf{u} \in U$ is $f'_U(\mathbf{a})(\mathbf{u}) \in Y$.

Definition 5.2.11 Directional derivatives. Let $\mathbf{u} \in X$. If

$$\lim_{t\to 0} \frac{f(\mathbf{a}+t\mathbf{u}) - f(\mathbf{a})}{t} \tag{5.27}$$

exists, then it is called the *directional derivative of f at $\mathbf{a} \in A$ along $\mathbf{u}$*. The directional derivative is denoted as $f'(\mathbf{a};\, \mathbf{u})$. It is a vector in Y.

Remarks 5.2.12 Relations between the derivatives. If $f'(\mathbf{a}) \in L(X, Y)$ exists, then $f'_U(\mathbf{a}) \in L(U, Y)$ also exists for all subspaces $U \subset X$. Also, $f'_U(\mathbf{a}) = f'(\mathbf{a})|_U$ is the restriction of $f'(\mathbf{a})$ to U. Furthermore, if the restricted derivative $f'_U(\mathbf{a}) \in L(U, Y)$ exists, then the directional derivatives $f'(\mathbf{a};\, \mathbf{u}) \in Y$ also exist for all $\mathbf{u} \in U$.

The most significant relationship here is that if $f'_U(\mathbf{a})$ exists, then $f'(\mathbf{a};\, \mathbf{u}) = f'_U(\mathbf{a})(\mathbf{u})$. That is, we can compute the directional derivative along $\mathbf{u}$ by applying $f'_U(\mathbf{a})$ to $\mathbf{u}$. (Lemma 5.2.8, which shows that $f'(\mathbf{a};\, \mathbf{u}) = f'(\mathbf{a})(\mathbf{u})$, establishes this result for the case where $U = X$.)

The proofs for all of these claims are easy. All are similar to the proof of Lemma 5.2.8 and all of them are left as exercises.

Example 5.2.13 Suppose that $f : \mathbb{R}^2 \to \mathbb{R}^3$ is differentiable at $\mathbf{0}$. Given that

$$\lim_{r\to 0} \frac{f(r, r) - f(\mathbf{0})}{r} = (2, -1, 1) \quad \text{and} \quad \lim_{r\to 0} \frac{f(3r, 2r) - f(\mathbf{0})}{r} = (1, 4, 3),$$

we show how to compute

$$\lim_{r\to 0} \frac{f(r, -r) - f(\mathbf{0})}{r}.$$

First, note that the given equations are equivalent to

$$f'(\mathbf{0};(1,1)) = (2,-1,1) \quad \text{and} \quad f'(\mathbf{0};(3,2)) = (1,4,3).$$

Hence, since f is differentiable at $\mathbf{0}$, Theorem 5.2.12 implies that $f'(\mathbf{0})(1,1) = (2,-1,1)$ and $f'(\mathbf{0})(3,2) = (1,4,3)$. Furthermore, the same theorem implies that

$$\begin{aligned}\lim_{r\to 0}\frac{f(r,-r)-f(\mathbf{0})}{r} &= \lim_{r\to 0}\frac{f(\mathbf{0}+r(1,-1))-f(\mathbf{0})}{r}\\ &= f'(\mathbf{0})(1,-1).\end{aligned}$$

Since $(1,-1) = -5(1,1)+2(3,2)$ and $f'(\mathbf{0})$ is linear, we have

$$\begin{aligned}f'(\mathbf{0})(1,-1) &= -5f'(\mathbf{0})(1,1)+2f'(\mathbf{0})(3,2)\\ &= -5(2,-1,1)+2(1,4,3)\\ &= (-8,13,1).\end{aligned}$$

Hence,

$$\lim_{r\to 0}\frac{f(r,-r)-f(\mathbf{0})}{r} = f'(\mathbf{0};(1,-1)) = f'(\mathbf{0})(1,-1) = (-8,13,1).$$

Example 5.2.14 Let $f:\mathbb{R}^2\to\mathbb{R}^3$ be differentiable at $\mathbf{0}$ and $f(\mathbf{0})=\mathbf{0}$ with

$$\lim_{r\to 0}\frac{f(r,2r)}{r} = (1,-2,1) \quad \text{and} \quad \lim_{r\to 0}\frac{f(2r,r)}{r} = (2,1,5). \tag{5.28}$$

We show how to compute $f'(\mathbf{0})$. Note that since $f(\mathbf{0})=\mathbf{0}$, Equations (5.28) say that

$$f'(\mathbf{0};(1,2)) = (1,-2,1) \quad f'(\mathbf{0};(2,1)) = (2,1,5).$$

So, by Theorem 5.2.12,

$$f'(\mathbf{0})(1,2) = (1,-2,1) \quad \text{and} \quad f'(\mathbf{0})(2,1) = (2,1,5).$$

Also, for any $(x,y)\in\mathbb{R}^2$, we have

$$(x,y) = \frac{2y-x}{3}(1,2)+\frac{2x-y}{3}(2,1).$$

Thus, since $f'(\mathbf{0})$ is linear,

$$\begin{aligned}f'(\mathbf{0})(x,y) &= \frac{2y-x}{3}f'(\mathbf{0})(1,2)+\frac{2x-y}{3}f'(\mathbf{0})(2,1)\\ &= \frac{2y-x}{3}(1,-2,1)+\frac{2x-y}{3}(2,1,5).\end{aligned}$$

Differentiability of the Restricted Function

Lemma 5.2.15 Open sets in subspaces. *Let A be an open set in a normed space X. Let U be a subspace of X. Then $B = A \cap U$ is an open set in U, considered as a normed space by itself.*

Proof. Let $\mathbf{b} \in B \subset A$. Since A is open in X, there is a $\delta > 0$ such that if $\|\mathbf{x}\| < \delta$, then $\mathbf{b}+\mathbf{x} \in A$. But if $\mathbf{x} = \mathbf{u} \in U$, then also $\mathbf{b}+\mathbf{u} \in U$ and therefore $\mathbf{b}+\mathbf{u} \in A \cap U$. Hence $\mathbf{b} + \mathbf{u} \in B$ whenever $\mathbf{u} \in U$ and $\|\mathbf{u}\| < \delta$. This shows that B is an open set in U. $\square$

Lemma 5.2.16 Derivatives of restricted functions. *Let A be an open set in X and U a subspace of X. Let $B = A \cap U$. Let $f : A \to Y$ be a function and let $\varphi = f|_B : B \to Y$ be the restriction of f to B. If $\mathbf{b} \in B = A \cap U$ and if $f'(\mathbf{b}) \in L(X, Y)$ exists, then $\varphi'(\mathbf{b}) \in L(U, Y)$ also exists and $\varphi'(\mathbf{b}) = f'(\mathbf{b})|_U$.*

Proof. Lemma 5.2.15 above shows that B is an open set in U. We claim that $T = f'(\mathbf{b})|_U \in L(U, Y)$ is the derivative of $\varphi : B \to Y$ at $\mathbf{b} \in B$. In fact,

$$\begin{aligned} &\lim_{\mathbf{u}\to\mathbf{0}} \frac{\|\varphi(\mathbf{a}+\mathbf{u}) - \varphi(\mathbf{a}) - T\mathbf{u}\|}{\|\mathbf{u}\|} \\ &= \lim_{\mathbf{u}\in U,\, \mathbf{u}\to\mathbf{0}} \frac{\|f(\mathbf{a}+\mathbf{u}) - f(\mathbf{a}) - f'(\mathbf{a})\mathbf{u}\|}{\|\mathbf{u}\|} = 0 \end{aligned}$$

shows that $\varphi'(\mathbf{b})$ exists and $\varphi'(\mathbf{b}) = f'(\mathbf{b})|_U \in L(U, Y)$. $\square$

Functions of a Real Variable

Remarks 5.2.17 Special notation for functions of a real variable. Let the domain space X be a one-dimensional space. We can assume that $X = \mathbb{R}$ without loss of generality. In this case, all three types of derivatives are essentially the same. Let I be an open interval in $\mathbb{R}$. The standard definition of the derivative of $f : I \to Y$ at $a \in I$, Definition 5.1.1, is

$$f'(a) = \lim_{t\to 0}(1/t)(f(a+t) - f(a)).$$

We see that this corresponds to the directional derivative of f in the direction of $1 \in \mathbb{R}$. Hence it could be denoted as $f'(a;\, 1)$ or as $f'(a)1$. Obviously, we never denote derivatives in this way for functions of one variable. As noted in the introduction to this section, we employ the notation $f'(a)$ differently for the one-variable case.

Computations of Directional Derivatives

The following lemma shows that any directional derivative can be computed as the ordinary derivative of a function of a real variable. Hence we can use the standard rules of differentiation to compute directional derivatives.

Lemma 5.2.18 *Define* $\varphi(s) = f(\mathbf{a}+s\mathbf{e})$, *a function of one variable defined in terms of* f. *Then*

$$\varphi'(s) = f'(\mathbf{a} + s\mathbf{e};\ \mathbf{e})$$

whenever $\varphi'(s)$ *exists.*

Proof. We have

$$\begin{aligned}\varphi'(s) &= \lim_{r\to 0}\frac{\varphi(s+r)-\varphi(s)}{r}\\ &= \lim_{r\to 0}\frac{f(\mathbf{a}+(s+r)\mathbf{e})-f(\mathbf{a}+s\mathbf{e})}{r}\\ &= f'(\mathbf{a}+s\mathbf{e};\ \mathbf{e})\end{aligned}$$

whenever $\varphi'(s)$ exists. □

Lemma 5.2.19 Homogeneity of the directional derivatives. *If* $f'(\mathbf{a};\ \mathbf{e})$ *exists, then* $f'(\mathbf{a};\ t\mathbf{e})$ *also exists for all* $t \in \mathbb{R}$ *and* $f'(\mathbf{a};\ t\mathbf{e}) = tf'(\mathbf{a};\ \mathbf{e})$.

Proof. The result is trivial if $t = 0$. Otherwise, we have

$$\begin{aligned}\lim_{r\to 0}\frac{1}{r}(f(\mathbf{a}+rt\mathbf{e})-f(\mathbf{a})) &= t\lim_{r\to 0}\frac{1}{rt}(f(\mathbf{a}+rt\mathbf{e})-f(\mathbf{a}))\\ &= t\lim_{s\to 0}\frac{1}{s}(f(\mathbf{a}+s\mathbf{e})-f(\mathbf{a})) = t\,f'(\mathbf{a};\ \mathbf{e}).\ \square\end{aligned}$$

Remarks 5.2.20 If we know that $f'(\mathbf{a}) \in L(X,\ Y)$ exists, then Lemma 5.2.8 shows that $f'(\mathbf{a};\ \mathbf{e}) = f'(\mathbf{a})\mathbf{e}$. In this case, lemma 5.2.19 becomes a triviality. If the existence of $f'(\mathbf{a})$ is not known, then a small argument may be necessary to show the homogeneity of the directional derivatives.

Increments and Derivatives

The basic use of derivatives is to estimate the increments of a function. We formulate several relations between derivatives and increments.

Theorem 5.2.21 . *Let $T = f'(\mathbf{a}) \in L(X, Y)$ be the derivative of $f : A \to Y$ at $\mathbf{a} \in A$. Let $M = \|T\|$, using the usual norm on linear transformations. Then for each $\varepsilon > 0$ there is a $\delta > 0$ such that*

$$\|f(\mathbf{a}+\mathbf{r}) - f(\mathbf{a})\| \leq (M+\varepsilon)\|\mathbf{r}\| \tag{5.29}$$

whenever $\|\mathbf{r}\| < \delta$.

Proof. Given $\varepsilon > 0$, find a $\delta > 0$ such that

$$\|f(\mathbf{a}+\mathbf{r}) - f(\mathbf{a}) - T\mathbf{r}\| \leq \varepsilon \|\mathbf{r}\|$$

whenever $\|\mathbf{r}\| < \delta$. Hence

$$\begin{aligned} \|f(\mathbf{a}+\mathbf{r}) - f(\mathbf{a})\| &\leq \|T\mathbf{r}\| + \|f(\mathbf{a}+\mathbf{r}) - f(\mathbf{a}) - T\mathbf{r}\| \\ &\leq M\|\mathbf{r}\| + \varepsilon\|\mathbf{r}\| = (M+\varepsilon)\|\mathbf{r}\| \end{aligned}$$

whenever $\|\mathbf{r}\| < \delta$. □

Corollary 5.2.22 Derivatives and continuity. *If f has a derivative at $\mathbf{x}$, then f is continuous at $\mathbf{x}$.*

Proof. Let $\mathbf{r}_n$ be a zero-sequence in X. Theorem 5.2.21 shows that

$$f(\mathbf{a}+\mathbf{r}_n) \to f(\mathbf{a}) \text{ in } Y.$$

This is the continuity of f at $\mathbf{a}$. □

Example 5.2.23 Let a be a constant. Define $f : \mathbb{R}^2 \to \mathbb{R}^2$ by

$$f(x,y) = \begin{cases} xy/(x^2+y^2) & \text{if } (x,y) \neq (0,0) \\ a & \text{if } (x,y) = (0,0) \end{cases}$$

Let us show that $\lim_{(x,y)\to(0,0)} f(x,y)$ does not exist. Thus, no matter what the value of a is, f is not continuous at $(0,0)$. It follows that for all values of a, the function f cannot be differentiable at $(0,0)$.

To show that $\lim_{(x,y)\to(0,0)} f(x,y)$ does not exist, we let (x,y) approach $\mathbf{0}$ through different paths. Along the line $y = x$, we have

$$\lim_{(x,y)\to(0,0)} f(x,y) = \lim_{x\to 0} \frac{x^2}{2x^2} = \frac{1}{2}.$$

Along the line $y = 0$, we have

$$\lim_{(x,y)\to(0,0)} f(x,y) = \lim_{x\to 0} \frac{x(0)}{x^2+(0)^2} = \lim_{x\to 0} \frac{0}{x^2} = 0.$$

Problems

5.25 Consider the xy-plane as a horizontal plane with the z-axis pointing directly upward. Assume that $z = f(x,\, y) = 10 - x^2 - 2y^2$ describes the surface of a hill.

1. Find the directional derivatives of $z = f(x,\, y)$ at the point $(x = 2,\, y = 1)$ and along an arbitrary vector $(u,\, v)$.
2. Take $(u,\, v)$ as a unit vector, $u^2 + v^2 = 1$. Explain why the directional derivative $f'((2,\, 1);\, (u,\, v))$ can be considered as the slope a hiker would experience if she starts at the point $(2,\, 1,\, 4)$ on the hill and moves in the $(u,\, v)$ direction determined by the horizontal coordinates.
3. Show that there is a plane passing through $(2,\, 1,\, 4)$ such that all the slopes on this plane are the same as the corresponding slopes on on the hill.
4. Find the equation of this plane.
5. What are the directions $(u,\, v)$ of the steepest ascent, the steepest descent, and of zero slope?

5.26 The temperature distribution in a certain region of the xyz-space at the time t is given as $f(x,\, y,\, z,\, t) = (100 - x^2 - y^2 - 2z^2)e^{-t/2}$. A fly is moving in this region according to the law of motion $x(t) = 2 - t$, $y(t) = 1 + t^2$, and $z(t) = t^3$. Find the rate of change in the temperature this fly experiences at the time $t = 2$. Show that this rate of change is the directional derivative of f at a certain point in $\mathbb{R}^4$ and along a certain vector in $\mathbb{R}^4$. Find this point and this vector.

5.27 Find the directional derivative of

$$f(x,y) = \begin{cases} \dfrac{x^2y^2}{x^4+y^4}, & (x,y) \neq (0,0) \\ 0, & (x,y) = (0,0) \end{cases}$$

at an arbitrary point and in an arbitrary direction, if it exists.

5.28 Same as Problem 5.27 for

$$f(x,y) = \begin{cases} \dfrac{x^3}{x^2+y^2}, & (x,y) \neq (0,0), \\ 0, & (x,y) = (0,0)\,. \end{cases}$$

5.29 Same as Problem 5.27 for

$$f(x,y) = \begin{cases} xy\, \sin \dfrac{1}{x^2+y^2}, & (x,y) \neq (0,0), \\ 0\,, & (x,y) = (0,0)\,. \end{cases}$$

5.30 Same as Problem 5.27 for

$$f(x,y) = \begin{cases} \dfrac{xy}{(x^2+y^2)^{1/2}} \sin \dfrac{1}{x^2+y^2}, & (x,y) \neq (0,0), \\ 0, & (x,y) = (0,0). \end{cases}$$

5.31 Let $f : \mathbb{R}^2 \to \mathbb{R}^2$ be the polar coordinates function

$$f(r,\, \theta) = (x,\, y) = (r\cos\theta,\, r\sin\theta).$$

Find the directional derivative of f at the point $(r = 1,\, \theta = \alpha)$ along an arbitrary vector $(a,\, \lambda)$ in the $r\theta$-plane. The result will be a vector in the xy-plane. In particular, find the directional derivatives along the vectors $(1,\, 0)$ and $(0,\, 1)$. Show that these two directional derivatives are orthogonal to each other.

5.32 Let $f : \mathbb{R}^3 \to \mathbb{R}^3$ be the cylindrical coordinates function

$$f(r,\, \theta,\, \zeta) = (x,\, y,\, z) = (r\cos\theta,\, r\sin\theta,\, \zeta).$$

Find the directional derivative of f at the point $(r = 1,\, \theta = \alpha,\, \zeta = 0)$ along an arbitrary vector $(a,\, \lambda,\, h)$ in the $r\theta\zeta$-space. The result will be a vector in the xyz-space. In particular, find the directional derivatives along the vectors $(1,\, 0,\, 0)$, $(0,\, 1,\, 0)$, and $(0,\, 0,\, 1)$. Show that these three directional derivatives are orthogonal to each other.

5.33 Let $f(\rho,\, \varphi,\, \theta) = (x,\, y,\, z)$ be the spherical coordinates function

$$x = \rho\sin\varphi\cos\theta, \quad y = \rho\sin\varphi\sin\theta, \quad z = \rho\cos\varphi.$$

Find the directional derivative of f at the point $(\rho = 1,\, \theta = \alpha,\, \varphi = \beta)$ along an arbitrary vector $(a,\, \lambda,\, \mu)$ in the $r\theta\varphi$-space. In particular, find the directional derivatives along the vectors $(1,\, 0,\, 0)$, $(0,\, 1,\, 0)$, and $(0,\, 0,\, 1)$. Show that these three directional derivatives are orthogonal to each other.

5.3 EXISTENCE OF DERIVATIVES

Lemma 5.2.8 gives a necessary condition for differentiability. This condition is the existence of the directional derivatives in all directions. The following example shows that this condition is not sufficient for differentiability.

Example 5.3.1 A nondifferentiable function with directional derivatives. Let

$$f(x,\, y) = \begin{cases} 1 & \text{if } x \neq 0 \text{ and } y = x^2 \\ 0 & \text{otherwise.} \end{cases}$$

This is the same as an earlier example, Example 4.4.6. We know that this function $f : \mathbb{R}^2 \to \mathbb{R}$ is discontinuous at the origin. Hence Corollary 5.2.22 shows that it cannot be differentiable at the origin. But its directional derivatives at the origin exist in all directions, and they are all zero. To see this, restrict f to a line passing through the origin. If this line is one of the coordinate axes, then this restriction is identically zero. Hence it is differentiable everywhere. Now consider the restriction of f to the line $y = mx$, $m \neq 0$. This restriction is zero everywhere except at the point (m, m^2), where it has the value of 1. Hence the restriction is zero in an open interval containing the origin. Therefore $f'((0, 0); \mathbf{e}) = 0$ for all $\mathbf{e} \in \mathbb{R}^2$. △

Lemma 5.2.9 shows that the existence and the continuity of directional derivatives in all directions are necessary for continuous differentiability. We will now show that this condition is also sufficient. In what follows, X and Y are two normed spaces, and A is an open set in X.

Lemma 5.3.2 *Let $\mathbf{e} \in X$ be a fixed vector. Assume that the directional derivative*

$$h(\mathbf{x}) = \lim_{t \to 0} \frac{f(\mathbf{x} + t\mathbf{e}) - f(\mathbf{x})}{t} \tag{5.30}$$

exists for all $\mathbf{x} \in A$ and defines a continuous function $h : A \to Y$. Then for each $\mathbf{a} \in A$ and for each $\varepsilon > 0$, there is a $\delta > 0$ such that

$$\|f(\mathbf{x} + t\mathbf{e}) - f(\mathbf{x}) - th(\mathbf{a})\| \leq \varepsilon\,|t| \tag{5.31}$$

whenever $\|\mathbf{x} - \mathbf{a}\| < \delta$ and $|t| < \delta$. In particular, if $\mathbf{v}_n$ is a zero-sequence in X and t_n is a zero-sequence in $\mathbb{R}$, then there is a zero-sequence $\varepsilon_n \in \mathbb{R}$ such that

$$\|f(\mathbf{a} + \mathbf{v}_n + t_n\mathbf{e}) - f(\mathbf{a} + \mathbf{v}_n) - t_n h(\mathbf{a})\| \leq \varepsilon_n\,|t_n|. \tag{5.32}$$

Proof. Given $\varepsilon > 0$, there is a $\delta > 0$ such that

$$\|h(\mathbf{x} + t\mathbf{e}) - h(\mathbf{a})\| < \varepsilon$$

whenever $\|\mathbf{x} - \mathbf{a}\| < \delta$ and $|t| < \delta$. This is possible because of the continuity of $h : A \to Y$. Let $\mathbf{x}$ be fixed and $\|\mathbf{x} - \mathbf{a}\| < \delta$. Define

$$\begin{aligned}
\varphi(s) &= f(\mathbf{x} + s\mathbf{e}) - f(\mathbf{x}) - sh(\mathbf{a}). \text{ Then} \\
\varphi'(s) &= \lim_{t \to 0} (1/t)(\varphi(s + t) - \varphi(s)) \\
&= \lim_{t \to 0} (1/t)(f(\mathbf{x} + s\mathbf{e} + t\mathbf{e}) - f(\mathbf{x} + s\mathbf{e}) - th(\mathbf{a})) \\
&= h(\mathbf{x} + s\mathbf{e}) - h(\mathbf{a}).
\end{aligned}$$

Hence we see that $\|\varphi'(s)\| < \varepsilon$ for all $|s| \leq |t| < \delta$. Then

$$\|\varphi(t) - \varphi(0)\| = \|\varphi(t)\| = \|f(\mathbf{x} + t\mathbf{e}) - f(\mathbf{x}) - th(\mathbf{a})\| \leq \varepsilon\,|t|.$$

The last step follows from the mean value theorem, Theorem 5.1.13. Finally, (5.32) follows directly from (5.31). □

Lemma 5.3.3 *Assume that X is spanned by a subspace V together with a vector $\mathbf{e} \in X \setminus V$. Assume that the directional derivative*

$$h(\mathbf{x}) = \lim_{t \to 0} \frac{f(\mathbf{x} + t\mathbf{e}) - f(\mathbf{x})}{t} \tag{5.33}$$

exists for all $\mathbf{x} \in A$ and defines a continuous function $h : A \to Y$. Assume that the restricted derivative $f'_V(\mathbf{a}) = T \in L(V, Y)$ exists at $\mathbf{a} \in A$. Then f is differentiable at $\mathbf{a}$. Its derivative is given as

$$f'(\mathbf{a})\mathbf{x} = f'(\mathbf{a})(\mathbf{v} + r\,\mathbf{e}) = T\mathbf{v} + r\,h(\mathbf{a}) \tag{5.34}$$

for all $\mathbf{x} = (\mathbf{v} + r\mathbf{e}) \in X$ with $\mathbf{v} \in V$ and $r \in \mathbb{R}$.

Proof. Each $\mathbf{x} \in X$ has a unique representation as $\mathbf{x} = \mathbf{v} + t\mathbf{e}$ with $\mathbf{v} \in V$ and $t \in \mathbb{R}$. Note that $\mathbf{v} = P\mathbf{x}$ and $t\mathbf{e} = Q\mathbf{x}$ with the coordinate projections P and Q associated with this decomposition of X. Hence, $\|\mathbf{v}\| \leq K\|\mathbf{x}\|$ and $\|t\mathbf{e}\| \leq K\|\mathbf{x}\|$ for some $K > 0$. Define $S \in L(X, Y)$ by

$$S\mathbf{x} = S(\mathbf{v} + t\mathbf{e}) = T\mathbf{v} + th(\mathbf{a}), \quad \mathbf{x} = (\mathbf{v} + t\mathbf{e}) \in X. \tag{5.35}$$

We see that the increment $f(\mathbf{a} + \mathbf{x}) - f(\mathbf{a})$ is the sum of two increments

$$f(\mathbf{a} + \mathbf{v}) - f(\mathbf{a}) \text{ and} \tag{5.36}$$

$$f(\mathbf{a} + \mathbf{v} + t\mathbf{e}) - f(\mathbf{a} + \mathbf{v}), \tag{5.37}$$

where $\mathbf{x} = \mathbf{v} + t(\mathbf{e})$. Let $\mathbf{x}_n = \mathbf{v}_n + t_n\mathbf{e}$ be a zero-sequence in X. Then we see that, $\mathbf{v}_n$ is a zero-sequence in V and t_n is a zero-sequence in $\mathbb{R}$. Recall that T is the restricted derivative of f at $\mathbf{a} \in A$ along V. Hence, by the observations in Remarks 5.2.2, there is a zero-sequence $\alpha_n \in \mathbb{R}$ such that

$$\|f(\mathbf{a} + \mathbf{v}_n) - f(\mathbf{a}) - T\mathbf{v}_n\| \leq \alpha_n \|\mathbf{v}_n\|. \tag{5.38}$$

Also, Lemma 5.3.2 shows that there is a zero-sequence $\beta_n \in \mathbb{R}$ such that

$$\|f(\mathbf{a} + \mathbf{v}_n + t_n\mathbf{e}) - f(\mathbf{a} + \mathbf{v}_n) - t_n h(\mathbf{a})\| \leq \beta_n \, \|t_n\mathbf{e}\|. \tag{5.39}$$

Then, (5.38) and (5.39) imply that, with $\mathbf{x}_n = \mathbf{v}_n + t_n\mathbf{e}$,

$$\|f(\mathbf{a} + \mathbf{x}_n) - f(\mathbf{a}) - S\mathbf{x}_n\| \leq \alpha_n \|\mathbf{v}_n\| + \beta_n \, |t_n| \leq \varepsilon_n \|\mathbf{x}_n\|, \tag{5.40}$$

where $\varepsilon_n = (\alpha_n + \beta_n)K$. This proves (5.34). □

Theorem 5.3.4 Existence of derivatives. *Let $f : A \to Y$, where A is an open subset of X. Let $E = \{\mathbf{e}_1, \ldots, \mathbf{e}_n\}$ be a basis for X. If all the directional derivatives along the vectors $\mathbf{e}_i$ exist, and if they are continuous functions on A, then f is continuously differentiable on A.*

Proof. Apply an induction argument on $n = \dim X$. The result is correct if $n = 1$. In fact, in this case directional derivatives and derivatives are the same, as already observed in Remarks 5.2.17. Assume that the result is true for $(n-1)$-dimensional spaces. This implies that f has restricted derivatives along the subspace spanned by $(\mathbf{e}_1, \ldots, \mathbf{e}_{n-1})$. Then Lemma 5.3.3 shows that f is differentiable. Continuous differentiability follows from the continuity of the directional derivatives. □

Problems

5.34 Define $f : \mathbb{R}^2 \to \mathbb{R}$ by $f(0, 0) = 0$ and

$$f(x, y) = (x^2 + y^2) \cos \frac{1}{x^2 + y^2} \quad \text{if } x^2 + y^2 > 0.$$

Show that $f' : \mathbb{R}^2 \to L(\mathbb{R}^2, \mathbb{R})$ exists but is not continuous at $(0, 0) \in \mathbb{R}^2$.

5.35 Let $f : \mathbb{R}^3 \to \mathbb{R}$ be defined as $f(x, y, z) = 1$ if $x = t$, $y = t^2$, and $z = t^3$ for some $t > 0$, and $f(x, y, z) = 0$ otherwise. Show that f has a restricted derivative at the origin along any one- or two-dimensional subspace. Show that f is not differentiable at the origin.

5.36 Let $n \in \mathbb{N}$, $n \geq 2$. Give an example of a function $f : \mathbb{R}^n \to \mathbb{R}$ that has restricted derivatives at the origin in all $(n-1)$-dimensional subspaces of $\mathbb{R}^n$ but is not differentiable at the origin.

5.37 Let $f(x, y) = 1$ if $y < x^2 < 2y$ and $f(x, y) = 0$ otherwise. Show that f has directional derivatives at the origin in any direction. Is f differentiable at the origin?

5.38 Let $f(x, y) = 1$ if $2x < x^2 + y^2 < 4x$ and $f(x, y) = 0$ otherwise. Show that f has directional derivatives at the origin in any direction. Is f differentiable at the origin?

5.39 Suppose that $f : \mathbb{R}^2 \to \mathbb{R}$ has a directional derivative at the origin along the vector $(1, 0)$. Assume that for all $x \in \mathbb{R}$ the directional derivative of f at the point $(x, 0)$ and in the direction of $(1, 1)$ exists and is equal to $p(x)$. Show that if $p : \mathbb{R} \to \mathbb{R}$ is continuous, then f is differentiable at $(0, 0)$.

5.4 PARTIAL DERIVATIVES

Definition 5.4.1 Partial derivatives. Let $f : A \to Y$ be a function and let $\mathbf{x} \in A$. Let $E = \{ \mathbf{e}_1, \ldots, \mathbf{e}_n \}$ be a basis for the domain space X. The directional derivatives $f'(\mathbf{x};\ \mathbf{e}_i)$ in the directions of the basis vectors $\mathbf{e}_i$ are called the *partial derivatives of* $f : A \to Y$ *at* $\mathbf{x} \in A$ *with respect to the basis* E.

Definition 5.4.2 Functions of several variables. Functions defined on subsets of $\mathbb{R}^n$ are called *functions of several variables.* In discussing partial derivatives, we shall ignore the difference between functions defined on subsets of a finite-dimensional normed space X and functions defined on subsets of $\mathbb{R}^n$. The following conventions justify this practice.

Fix a basis $E = \{ \mathbf{e}_1, \ldots, \mathbf{e}_n \}$ for X. Let $x_j : X \to \mathbb{R}$ be the coordinate functions with respect to E. There is an isomorphism $\Psi : \mathbb{R}^n \to X$ that takes

$$(x_1, \ldots, x_n) \in \mathbb{R}^n$$

to

$$\Psi(x_1, \ldots, x_n) = x_1\mathbf{e}_1 + \cdots + x_n\mathbf{e}_n \in X.$$

Set

$$\widetilde{A} = \Psi^{-1}(A) = \{ (x_1, \ldots, x_n) \in \mathbb{R}^n \mid x_1\mathbf{e}_1 + \cdots + x_n\mathbf{e}_n \in A \}.$$

Then we may define a function $\widetilde{f} = (f \cdot \Psi) : \widetilde{A} \to Y$ by

$$\widetilde{f}(x_1, \ldots, x_n) = f(x_1\mathbf{e}_1 + \cdots + x_n\mathbf{e}_n).$$

We will ignore the difference between f and $\widetilde{f}$, and between A and $\widetilde{A}$. Hence we write $\mathbf{x} = (x_1, \ldots, x_n) \in A$ and

$$f(\mathbf{x}) = f(x_1, \ldots, x_n).$$

Definition 5.4.3 Standard notation for partial derivatives. Fix a basis $E = \{ \mathbf{e}_1, \ldots, \mathbf{e}_n \}$ for X. The *standard notation* for $f'(\mathbf{x};\ \mathbf{e}_j)$ is

$$\frac{\partial f}{\partial x_j}(\mathbf{x}) = \frac{\partial f}{\partial x_j}(x_1, \ldots, x_n). \tag{5.41}$$

This notation indicates the way partial derivatives are computed. To find $f'(\mathbf{x};\ \mathbf{e}_j)$, proceed as follows. Consider $f(\mathbf{x})$ as a function $g(x)$ of a single variable $x = x_j$, keeping the other coordinates x_k, $k \neq j$, fixed. Then

$$\begin{aligned} \frac{\partial f}{\partial x_j}(\mathbf{x}) &= \lim_{r \to 0} \frac{f(\mathbf{x} + r\mathbf{e}_j) - f(\mathbf{x})}{r} \\ &= \lim_{r \to 0} \frac{g(x_j + r) - g(x_j)}{r} = g'(x_j) \end{aligned}$$

is the differentiation of f with respect to the variable $x = x_j$, regarding all other coordinates x_k, $k \neq j$, as constants. Therefore a partial derivative of a function of several real variables can be obtained by the basic differentiation rules applied to a single real variable.

Let $\mathbf{e}_1, \mathbf{e}_2$ be the standard basis vectors of $\mathbb{R}^2$. Since points in $\mathbb{R}^2$ are usually denoted as (x, y), alternative notation for $f'(\mathbf{x};\, \mathbf{e}_1)$ and $f'(\mathbf{x};\, \mathbf{e}_2)$ is, respectively,

$$\frac{\partial f}{\partial x}(\mathbf{x}) = \frac{\partial f}{\partial x}(x, y) \quad \text{and} \quad \frac{\partial f}{\partial y}(\mathbf{x}) = \frac{\partial f}{\partial y}(x, y).$$

Similar alternative notation is used for $\mathbb{R}^3$.

Definition 5.4.4 The Jacobian matrix. Partial derivatives provide a convenient way to represent the full derivatives of a vector-valued function $f : A \to Y$ in terms of real-valued functions. Choose a basis $W = \{\, \mathbf{w}_1, \ldots, \mathbf{w}_m\}$ for Y and consider the component functions $f_i : A \to \mathbb{R}$ defined by $f = f_1\mathbf{w}_1 + \cdots + f_m\mathbf{w}_m$. Then

$$\frac{\partial f}{\partial x_j}(\mathbf{x}) = \frac{\partial f_1}{\partial x_j}(\mathbf{x})\mathbf{w}_1 + \cdots + \frac{\partial f_m}{\partial x_j}(\mathbf{x})\mathbf{w}_m.$$

If the domain space X is an n-dimensional space, then each component has n partial derivatives. Hence there are mn real-valued partial derivatives

$$\frac{\partial f_i}{\partial x_j}(\mathbf{x}), \quad i = 1, \ldots, m, \quad j = 1, \ldots, n.$$

We organize these mn real numbers as an $m \times n$ matrix, referred to as the *Jacobian matrix*. It is denoted as

$$\mathbf{J}f(\mathbf{x}) = \frac{\partial(f_1, \ldots, f_m)}{\partial(x_1, \ldots, x_n)}(\mathbf{x}) = \begin{bmatrix} \frac{\partial f_1}{\partial x_1}(\mathbf{x}) & \cdots & \frac{\partial f_1}{\partial x_n}(\mathbf{x}) \\ \vdots & \ddots & \vdots \\ \frac{\partial f_m}{\partial x_1}(\mathbf{x}) & \cdots & \frac{\partial f_m}{\partial x_n}(\mathbf{x}) \end{bmatrix} = \begin{bmatrix} f_1'(\mathbf{x};\, \mathbf{e}_1) & \cdots & f_1'(\mathbf{x};\, \mathbf{e}_n) \\ \vdots & \ddots & \vdots \\ f_m'(\mathbf{x};\, \mathbf{e}_1) & \cdots & f_m'(\mathbf{x};\, \mathbf{e}_n) \end{bmatrix}.$$

The Jacobian matrix plainly depends upon the choice of the bases $E \subset X$ and $W \subset Y$.

If all the partial derivatives $f'(\mathbf{x};\, \mathbf{e}_j)$, $j = 1, \ldots, n$, exist for all $\mathbf{x} \in A$, then the Jacobian matrix defines a function

$$\mathbf{J}f = \frac{\partial(f_1, \ldots, f_m)}{\partial(x_1, \ldots, x_n)} : A \to \mathbb{M}_{mn}.$$

It takes each point $\mathbf{x} \in A$ to a real-valued matrix with the entries $f_i'(\mathbf{x};\, \mathbf{e}_j)$.

Remarks 5.4.5 Jacobian matrix and the derivative. Example 5.3.1 shows that the Jacobian matrix for f at $\mathbf{a} \in A$ can exist even if f is not differentiable at $\mathbf{a}$. But if f is differentiable at $\mathbf{a} \in A$, then its derivative $f'(\mathbf{a}) : X \to Y$ is represented by the Jacobian matrix $\mathbf{J}(\mathbf{a})$ with respect to the bases E and W. This means that if

$$\mathbf{y} = (y_1, \ldots, y_m) = f'(\mathbf{a})(x_1, \ldots, x_n) = f'(\mathbf{a})(\mathbf{x}), \tag{5.42}$$

then $\mathbf{y}$ can be computed by matrix multiplication:

$$\mathbf{y} = \begin{bmatrix} y_1 \\ \vdots \\ y_m \end{bmatrix} = \begin{bmatrix} f_1'(\mathbf{a};\, \mathbf{e}_1) & \ldots & f_1'(\mathbf{a};\, \mathbf{e}_n) \\ \vdots & \ddots & \vdots \\ f_m'(\mathbf{a};\, \mathbf{e}_1) & \ldots & f_m'(\mathbf{a};\, \mathbf{e}_n) \end{bmatrix} \begin{bmatrix} x_1 \\ \vdots \\ x_n \end{bmatrix} = \mathbf{J}f(\mathbf{a})\,\mathbf{x}. \tag{5.43}$$

Here, as in Definition 5.4.3, we ignore the difference between elements of $\mathbb{R}^n$ and $\mathbb{M}_{n1}$. That is, we identify

$$(x_1, \ldots, x_n) \in \mathbb{R}^n \quad \text{and} \quad \begin{bmatrix} x_1 \\ \vdots \\ x_n \end{bmatrix} \in \mathbb{M}_{n1},$$

and denote both objects by $\mathbf{x}$. To verify (5.43), it is enough to observe that it gives the correct result for each $\mathbf{x} = \mathbf{e}_j$. In fact, if $\mathbf{x} = \mathbf{e}_j$, then (5.43) becomes

$$\mathbf{J}f(\mathbf{a})\,\mathbf{e}_j = \begin{bmatrix} f_1'(\mathbf{a};\, \mathbf{e}_j) \\ \vdots \\ f_m'(\mathbf{a};\, \mathbf{e}_j) \end{bmatrix},$$

which represents $f'(\mathbf{a};\, \mathbf{e}_j) = f'(\mathbf{a})\mathbf{e}_j$.

Example 5.4.6 Define $f : (\mathbb{R}^2 \setminus \{\mathbf{0}\}) \to (0, \infty) \times [0, 2\pi)$ by $f(x, y) = (r, \theta)$ where $r = (x^2 + y^2)^{1/2}$ and $\theta \in [0, 2\pi)$ with $\sin\theta = y/r$, $\cos\theta = x/r$. Let us compute the Jacobian matrix of f at (x, y), where $x \neq 0, y \neq 0$. Here, the components of f are

$$f_1(x, y) = (x^2 + y^2)^{1/2} \quad \text{and} \quad f_2(x, y) = \arcsin(y(x^2 + y^2)^{-1/2}).$$

We see that, after some computations,

$$\begin{aligned} \frac{\partial f_1}{\partial x}(x, y) &= x(x^2 + y^2)^{-1/2} = \cos\theta, \\ \frac{\partial f_1}{\partial y}(x, y) &= y(x^2 + y^2)^{-1/2} = \sin\theta, \\ \frac{\partial f_2}{\partial x}(x, y) &= -y(x^2 + y^2)^{-1} = -(1/r)\sin\theta, \text{ and} \\ \frac{\partial f_2}{\partial y}(x, y) &= x(x^2 + y^2)^{-1} = (1/r)\cos\theta. \end{aligned}$$

Thus, the Jacobian matrix of f at (x, y) is

$$\mathbf{J}f(x,\, y) = \begin{bmatrix} \cos\theta & \sin\theta \\ -(1/r)\sin\theta & (1/r)\cos\theta \end{bmatrix}.$$

Here the variables r and θ are used to simplify the expressions. △

Example 5.4.7 Define $g : \mathbb{R}^2 \to \mathbb{R}^2$ by $g(r,\, \theta) = (r\cos\theta,\, r\sin\theta)$. An easy computation shows that

$$\mathbf{J}g(r,\, \theta) = \begin{bmatrix} \cos\theta & -r\sin\theta \\ \sin\theta & r\cos\theta \end{bmatrix}.$$

We see that $\mathbf{J}g(r,\, \theta)$ is the inverse of $\mathbf{J}f(x,\, y)$ in Example 5.4.6. This is not accidental. In fact, the restriction of g to $(0, \infty) \times [0, 2\pi)$ is the inverse of f. The chain rule, Theorem 5.5.6 below, will show that g' is the inverse of f'. △

Example 5.4.8 Let $A = \mathbb{R}^2 \setminus \{\mathbf{0}\}$ and define $f : A \to \mathbb{R}^2$ by

$$f(x,\, y) = ((x^2 + y^2)/(2x),\, (x^2 + y^2)/(2y)) \quad \text{for all } (x, y) \in A.$$

Let us find all $h \in \mathbb{R}$ so that $f'(1, 2)(h, 1 - h) = (-5, -1/2)$. Here, the components of f are $f_1(x, y) = (x^2 + y^2)/2x$ and $f_2(x, y) = (x^2 + y^2)/2y$. Thus,

$$\begin{aligned} \frac{\partial f_1}{\partial x}(x, y) &= \frac{x^2 - y^2}{2x^2}, \quad \frac{\partial f_1}{\partial y}(x, y) = \frac{y}{x} \\ \frac{\partial f_2}{\partial x}(x, y) &= \frac{x}{y}, \quad \frac{\partial f_2}{\partial y}(x, y) = \frac{y^2 - x^2}{2x^2}. \end{aligned}$$

Hence, $f'(1, 2)(x, y) = (1/2)(-3x + 4y,\, x + 3y)$ for all $(x, y) \in A$. Thus,

$$f'(1, 2)(h, 1 - h) = (1/2)(-7h + 4,\, -2h + 3).$$

Hence, we must have $h = 2$. △

The Gradient Vector

The derivative of a real-valued function at a point is a real-valued linear function. Hence it can be represented as an inner product with a unique vector. This vector is called the *gradient* of the function. We formulate this as Lemma 5.4.9.

Lemma 5.4.9 *Let A be an open set in a Euclidean space X. Let $f : A \to \mathbb{R}$ be a real-valued differentiable function. Then, for each $\mathbf{a} \in A$, there is a unique vector $\nabla f(\mathbf{a}) \in X$ such that $f'(\mathbf{a})\mathbf{x} = \langle \nabla f(\mathbf{a}),\, \mathbf{x} \rangle$ for all $\mathbf{x} \in X$.*

Proof. The derivative at $\mathbf{a} \in A$ is a linear function $f'(\mathbf{a}) : X \to \mathbb{R}$. In this case, Theorem 3.4.21 shows that $f'(\mathbf{a})$ can be represented as an inner product with a unique vector as stated in the lemma. □

Definition 5.4.10 The gradient vector. Let A be an open set in a Euclidean space X. Let $f : A \to \mathbb{R}$ be a real-valued differentiable function. The vector $\nabla f(\mathbf{a}) \in X$ obtained in Lemma 5.4.9 is called the *gradient vector* of f at $\mathbf{a} \in A$. The function $\nabla f : A \to X$ is the *gradient function* of f.

Remarks 5.4.11 Inner products of the gradient vector. The inner product of the gradient vector $\nabla f(\mathbf{a}) \in X$ with a vector $\mathbf{x}$ is the directional derivative of f along the vector $\mathbf{x}$. This follows from Definition 5.4.10:

$$\langle \nabla f(\mathbf{a}), \mathbf{x} \rangle = f'(\mathbf{a})\mathbf{x} = f'(\mathbf{a}; \mathbf{x}) = \lim_{t \to 0} \frac{1}{t} \left(f(\mathbf{a} + t\mathbf{x}) - f(\mathbf{a}) \right).$$

Remarks 5.4.12 Cartesian coordinates of the gradient vector. Let $X = \mathbb{R}^n$ with its standard inner product and with its standard basis $(\mathbf{e}_1, \ldots, \mathbf{e}_n)$. Then

$$\begin{aligned} \nabla f(\mathbf{a}) &= \langle \nabla f(\mathbf{a}), \mathbf{e}_1 \rangle \, \mathbf{e}_1 + \cdots + \langle \nabla f(\mathbf{a}), \mathbf{e}_n \rangle \, \mathbf{e}_n \\ &= \frac{\partial f}{\partial x_1}(\mathbf{a}) \, \mathbf{e}_1 + \cdots + \frac{\partial f}{\partial x_n}(\mathbf{a}) \, \mathbf{e}_n. \end{aligned}$$

In fact, $\langle \nabla f(\mathbf{a}), \mathbf{e}_i \rangle = f'(\mathbf{a})\mathbf{e}_i$ is the partial derivative of f with respect to the ith coordinate x_i, by Definition 5.4.1.

Remarks 5.4.13 The gradient vector and the Jacobian matrix. Let $A \subset \mathbb{R}^n$ be an open set and let $f : A \to \mathbb{R}$ be a differentiable function. The Jacobian matrix

$$\mathbf{J}f = \left[\frac{\partial f}{\partial x_1}, \ldots, \frac{\partial f}{\partial x_n} \right] : A \to \mathbb{M}_{1n}$$

and the gradient function

$$\nabla f = \left(\frac{\partial f}{\partial x_1}, \ldots, \frac{\partial f}{\partial x_n} \right) : A \to \mathbb{R}^n$$

are two different notations for the linear transformation determined by the n partial derivatives of this function.

Definition 5.4.14 Local extremal values. Let $f : A \to \mathbb{R}$ be a real-valued function defined on an open set A in a vector space X. Then $\mathbf{a} \in A$ is called a *local maximum* for f if there is a $\delta > 0$ such that $f(\mathbf{x}) \leq f(\mathbf{a})$ whenever $\mathbf{x} \in A$ and $\|\mathbf{x} - \mathbf{a}\| < \delta$. Local minimum values are defined similarly. It is easy to see that if $\nabla f(\mathbf{a}) \neq \mathbf{0}$, then

f cannot have a local extremal value at $\mathbf{a}$. Thus, local maxima or minima for A will always be found among the set of points $\mathbf{a}$ for which $\nabla f(\mathbf{a}) = \mathbf{0}$.

Remarks 5.4.15 Higher-order partial derivatives. Higher-order derivatives are discussed in Chapter 7 in a systematic way. Here, however, we may define higher-order partial derivatives as we did in the case of functions of a real variable. In fact, in dealing with partial derivatives, we effectively consider functions with only one real variable, since the other variables are kept fixed. For example, let $f : \mathbb{R}^3 \to Y$ be a function of three variables x, y, z taking values in a normed space Y. Then notations like

$$\frac{\partial}{\partial y}\frac{\partial f}{\partial x} = \frac{\partial^2 f}{\partial y \partial x}, \quad \frac{\partial}{\partial x}\frac{\partial f}{\partial x} = \frac{\partial^2 f}{\partial x^2}, \quad \frac{\partial}{\partial z}\frac{\partial^2 f}{\partial y^2} = \frac{\partial^3 f}{\partial z \partial y^2}, \quad \text{etc.}$$

have obvious meanings. For example, the first means to take the derivative of f with respect to x, and then to differentiate the resulting function with respect to y. We will prove that for higher-order derivatives, the order of differentiation is generally immaterial. For the time being, however, we must perform the differentiations in the given order.

Functions that are n times continuously differentiable are called $\mathcal{C}^n$ *functions*. They are defined precisely in Definition 7.1.5, and discussed in the sequel.

Problems

5.40 Let $(u, v) = (3x+2y, 6x-4y)$. Find the gradients ∇u, ∇v, and the Jacobian matrix $\partial(u, v)/\partial(x, y)$ at a general point (x, y), if they exist.

5.41 Same as Problem 5.40 for $(u, v) = (xy, y/x)$.

5.42 Same as Problem 5.40 for $(u, v) = ((x^2 + y^2)/(2x), (x^2 + y^2)/(2y))$.

5.43 Same as Problem 5.40 for

$$(u, v) = (p(x, y) + q(x, y), p(x, y) - q(x, y)),$$

where $p(x, y) = ((x + 1)^2 + y^2)^{1/2}$ and $q(x, y) = ((x - 1)^2 + y^2)^{1/2}$.

5.44 Let $\mathbf{r} : \mathbb{R}^2 \to \mathbb{R}^2$ be the polar coordinates function

$$(x, y) = \mathbf{r}(r, \theta) = (r \cos\theta, r \sin\theta).$$

Find $\mathbf{e}_1 = \partial\mathbf{r}/\partial r$ and $\mathbf{e}_2 = \partial\mathbf{r}/\partial\theta$ at all points $(r, \theta) \in \mathbb{R}^2$. Show that if $r \neq 0$, then $\mathbf{e}_1(r, \theta)$ and $\mathbf{e}_2(r, \theta)$ is an orthogonal basis for the xy-plane. Find the Jacobian matrix $\partial(x, y)/\partial(r, \theta)$.

5.45 A particle moves in the xy-plane according to the law of motion

$$\mathbf{s}(t) = (x(t),\ y(t)),\ \text{where } x(t) = e^t \cos 2t \text{ and } y(t) = e^t \sin 2t.$$

Find the coordinates of the velocity $\mathbf{s}'(t)$ and acceleration $\mathbf{s}''(t)$ of this particle with respect to the basis $\mathbf{e}_1(e^t,\ 2t)$ and $\mathbf{e}_2(e^t,\ 2t)$ defined in Problem (5.44).

5.46 Let $\mathbf{r} : \mathbb{R}^3 \to \mathbb{R}^3$ be the cylindrical coordinates function

$$(x,\ y,\ z) = \mathbf{r}(r,\ \theta,\ \zeta) = (r\cos\theta,\ r\sin\theta,\ \zeta).$$

Find $\mathbf{e}_1 = \partial\mathbf{r}/\partial r$, $\mathbf{e}_2 = \partial\mathbf{r}/\partial\theta$, and $\mathbf{e}_3 = \partial\mathbf{r}/\partial\zeta$ at all points $(r,\ \theta,\ \zeta) \in \mathbb{R}^3$. Show that if $r \neq 0$ then $\mathbf{e}_1(r,\ \theta,\ \zeta)$, $\mathbf{e}_2(r,\ \theta,\ \zeta)$, and $\mathbf{e}_3(r,\ \theta,\ \zeta)$ is an orthogonal basis for the xyz-space. Find the Jacobian matrix $\partial(x,\ y,\ z)/\partial(r,\ \theta,\ \zeta)$.

5.47 A particle moves in the xyz-space according to the law of motion

$$\mathbf{s}(t) = (x(t),\ y(t),\ z(t)) \ \text{ where}$$

$$x(t) = e^t\cos 2t,\quad y(t) = e^t\sin 2t,\quad z(t) = 3t.$$

Express the velocity $\mathbf{s}'(t)$ and acceleration $\mathbf{s}''(t)$ of this particle in terms of the basis $\mathbf{e}_1(e^t,\ 2t,\ 3t)$, $\mathbf{e}_2(e^t,\ 2t,\ 3t)$, and $\mathbf{e}_3(e^t,\ 2t,\ 3t)$ defined in Problem 5.46.

5.48 Let $\mathbf{r} : \mathbb{R}^3 \to \mathbb{R}^3$ be the spherical coordinates function

$$(x,\ y,\ z) = \mathbf{r}(\rho,\ \varphi,\ \theta) = (\rho\sin\varphi\cos\theta,\ \rho\sin\varphi\sin\theta,\ \rho cos\varphi).$$

Find $\mathbf{e}_1 = \partial\mathbf{r}/\partial\rho$, $\mathbf{e}_2 = \partial\mathbf{r}/\partial\varphi$, and $\mathbf{e}_3 = \partial\mathbf{r}/\partial\theta$ at all points $(\rho,\ \varphi,\ \theta) \in \mathbb{R}^3$. Show that if $\rho \neq 0$, then $\mathbf{e}_1(\rho,\ \varphi,\ \theta)$, $\mathbf{e}_2(\rho,\ \varphi,\ \theta)$, and $\mathbf{e}_3(\rho,\ \varphi,\ \theta)$ is an orthogonal basis for the xyz-space. Find the Jacobian matrix $\partial(x,\ y,\ z)/\partial(\rho,\ \varphi,\ \theta)$.

5.49 A particle moves in the xyz-space according to the law of motion

$$\mathbf{s}(t) = (x(t),\ y(t),\ z(t)),\ \text{ where}$$

$$x(t) = e^t\sin 3t\cos 2t,\quad y(t) = e^t\sin 3t\sin 2t,\quad z(t) = 3t\cos 3t.$$

Express the velocity $\mathbf{s}'(t)$ and acceleration $\mathbf{s}''(t)$ of this particle in terms of the basis $\mathbf{e}_1(\rho,\ \varphi,\ \theta)$, $\mathbf{e}_2(\rho,\ \varphi,\ \theta)$, and $\mathbf{e}_3(\rho,\ \varphi,\ \theta)$ defined in Problem 5.48.

5.5 RULES OF DIFFERENTIATION

In this section, we obtain some general rules for differentiation. As before, X and Y are two normed spaces, $A \subset X$ is an open set in X, and $f : A \to Y$ is a function.

Lemma 5.5.1 Derivative of a sum. *Let $f, g : A \to Y$ be two differentiable functions on A. Then $(f+g) : A \to Y$ is also differentiable on A and*

$$(f+g)' = f' + g'. \tag{5.44}$$

Proof. This is left as an exercise (a straightforward application of the definition of the derivative). □

Lemma 5.5.2 Derivative of a constant function. *Let $f : A \to Y$ be a constant function. Then $f' : A \to L(X, Y)$ is the zero-function. That is,*

$$f'(\mathbf{a})(\mathbf{x}) = \mathbf{0} \in Y \textit{ for all } \mathbf{a} \in A \textit{ and } \mathbf{x} \in X.$$

Proof. This is also left as an exercise. □

Lemma 5.5.3 Derivative of a linear function. *Let $R : X \to Y$ be a linear function. Then $R' : X \to L(X, Y)$ is the constant function with the constant value $R \in L(X, Y)$. That is,*

$$R'(\mathbf{a}) = R \textit{ for all } \mathbf{a} \in X \textit{ and } R'(\mathbf{a})\mathbf{x} = R\mathbf{x} \textit{ for all } \mathbf{a}, \mathbf{x} \in X.$$

Proof. If $R : X \to Y$ is linear, then $\Delta R(\mathbf{x})(\mathbf{r}) = R\mathbf{r}$ and $\|\Delta R(\mathbf{x})(\mathbf{r}) - R\mathbf{r}\| = 0$ for all $\mathbf{x}, \mathbf{r} \in X$. This shows that $DR(\mathbf{x}) = R$. □

Remarks 5.5.4 Constants and constant functions. Sometimes it may be awkward to distinguish between a constant and a function that takes the same constant value at all points in the domain of its definition. If this constant value is a point R in $L(X, Y)$, the derivative function is helpful to make this distinction. Hence, if $R \in L(X, Y)$, then $DR : X \to L(X, Y)$ is the function that has this constant value R at each $\mathbf{x} \in X$.

Example 5.5.5 Derivative of the identity function. Let $I : X \to X$ be the identity function. Then $I \in L(X, X)$. Hence $DI : X \to L(X, X)$ is the constant function $DI(\mathbf{x}) = I \in L(X, X)$ for all $\mathbf{x} \in X$. △

The Chain Rule

The chain rule states that the composition of two differentiable functions is differentiable. The derivative of the composition is the composition of the derivatives. This is a key result for differentiation in the multi-variable case, just as it is for one-variable calculus. In the following, X, Y, and Z are normed spaces, A is an open set in X

and B is an open set in Y. We consider two functions $f : A \to Y$, $g : B \to Z$, and assume that $f(A) \subset B$.

Theorem 5.5.6 Chain rule. *If $f : A \to Y$ is differentiable at $\mathbf{a} \in A$ and if $g : B \to Z$ is differentiable at $\mathbf{b} = f(\mathbf{a})$, then*

$h = g \cdot f : A \to Z$ is differentiable at $\mathbf{a}$ and $h'(\mathbf{a}) = g'(\mathbf{b}) \cdot f'(\mathbf{a}) = g'(f(\mathbf{a})) \cdot f'(\mathbf{a})$.

If f and g are continuously differentiable functions, then $h : g \cdot f$ is also a continuously differentiable function.

Proof. Let $R = f'(\mathbf{a})$ and $S = g'(\mathbf{b})$. Given a zero-sequence $\mathbf{x}_n$ in X, let

$$\begin{aligned} f(\mathbf{a} + \mathbf{x}_n) - f(\mathbf{a}) &= \mathbf{y}_n, && (5.45) \\ g(\mathbf{b} + \mathbf{y}_n) - g(\mathbf{b}) &= \mathbf{z}_n. && (5.46) \end{aligned}$$

Then we have

$$h(\mathbf{a} + \mathbf{x}_n) - h(\mathbf{a}) = \mathbf{z}_n. \tag{5.47}$$

The last equality follows from

$$g(f(\mathbf{a} + \mathbf{x}_n)) - g(f(\mathbf{a})) = g(\mathbf{b} + \mathbf{y}_n) - g(\mathbf{b}). \tag{5.48}$$

Recall that $R = f'(\mathbf{a})$. Hence there is a zero-sequence r_n in $\mathbb{R}$ such that

$$\|\mathbf{y}_n - R\mathbf{x}_n\| = \|f(\mathbf{a} + \mathbf{x}_n) - f(\mathbf{a}) - R\mathbf{x}_n\| \leq r_n \|\mathbf{x}_n\|. \tag{5.49}$$

This was observed in Remarks 5.2.2. Therefore

$$\|\mathbf{y}_n\| \leq \|R\mathbf{x}_n\| + \|\mathbf{y}_n - R\mathbf{x}_n\| \leq (\|R\| + r_n) \|\mathbf{x}_n\|. \tag{5.50}$$

Hence $\mathbf{y}_n$ is a zero-sequence in Y. Recall that $S = g'(\mathbf{b})$. Therefore, as in (5.49), there is a zero-sequence s_n in $\mathbb{R}$ such that

$$\begin{aligned} \|\mathbf{z}_n - S\mathbf{y}_n\| &= \|g(\mathbf{b} + \mathbf{y}_n) - g(\mathbf{b}) - S\mathbf{y}_n\| \\ &\leq s_n \|\mathbf{y}_n\| \leq s_n(\|R\| + r_n) \|\mathbf{x}_n\|. \end{aligned} \tag{5.51}$$

Here the last inequality follows from (5.50). Now (5.51), (5.49), and

$$\mathbf{z}_n - (SR)\mathbf{x}_n = (\mathbf{z}_n - S\mathbf{y}_n) + S(\mathbf{y}_n - R\mathbf{x}_n) \tag{5.52}$$

show that

$$\|\mathbf{z}_n - (SR)\mathbf{x}_n\| \leq (s_n(\|R\| + r_n) + \|S\| r_n) \|\mathbf{x}_n\| = t_n \|\mathbf{x}_n\|. \tag{5.53}$$

Here $t_n = s_n(\|R\| + r_n) + \|S\| \, r_n$ is a zero-sequence in $\mathbb{R}$. Hence, by (5.47),

$$\|h(\mathbf{a} + \mathbf{x}_n) - h(\mathbf{a}) - (SR)\mathbf{x}_n\| \le t_n \, \|\mathbf{x}_n\|. \tag{5.54}$$

This shows that $h'(\mathbf{a}) = SR$, as observed in Remarks 5.2.2.

For the continuity of $h' : A \to L(X, Z)$, let $\mathbf{a}_n \in A$ and $\mathbf{a}_n \to \mathbf{a} \in A$. Then $g'(f(\mathbf{a}_n)) = (g' \cdot f)(\mathbf{a}_n) \to g'(f(\mathbf{a}))$ in $L(Y, Z)$, since the composition of two continuous functions is continuous. Also, $f'(\mathbf{a}_n) \to f'(\mathbf{a})$ in $L(X, Y)$. Then we see that $h'(\mathbf{a}_n) = (g' \cdot f)(\mathbf{a}_n) \cdot f'(\mathbf{a}_n) \to h'(\mathbf{a}) = (g' \cdot f)(\mathbf{a}) \cdot f'(\mathbf{a})$ in $L(X, Z)$ by the continuity of products. Hence $h' : A \to L(X, Z)$ is continuous. □

Example 5.5.7 Chain rule in terms of Jacobian matrices. With the notation introduced above for the chain rule, Theorem 5.5.6, let U, V, and W be bases for X, Y, and Z, respectively. Let $x_i : X \to \mathbb{R}$, $y_j : Y \to \mathbb{R}$, and $z_k : Z \to \mathbb{R}$ be the corresponding coordinate functions. Assume that p, q, and r are the dimensions of X, Y, and Z, respectively. If the chain rule

$$h'(\mathbf{a}) = g'(\mathbf{b}) \cdot f'(\mathbf{a}) \tag{5.55}$$

is expressed in terms of the Jacobian matrices, then it becomes

$$\mathbf{J}h(\mathbf{a}) = \mathbf{J}g(\mathbf{b}) \cdot \mathbf{J}f(\mathbf{a}). \tag{5.56}$$

Note that h takes its values in Z. Hence it has $r = \dim Z$ components h_k. Each component h_k depends on $p = \dim X$ variables x_i. With similar notation for the other functions, we see that (5.56) becomes, in terms of the entries,

$$\frac{\partial h_k}{\partial x_i} = \sum_{j=1}^{q} \frac{\partial g_k}{\partial y_j} \frac{\partial f_j}{\partial x_i}. \tag{5.57}$$

A more instructive form of (5.57) is as follows. Denote $f(\mathbf{x})$ as $\mathbf{y}(\mathbf{x})$ and $g(\mathbf{y})$ as $\mathbf{z}(\mathbf{y})$. In fact, f expresses the y_j coordinates in terms of the x_i coordinates. Hence, f can be considered as $q = \dim Y$ functions y_j, each depending on $p = \dim X$ variables x_i as

$$y_j = y_j(x_1, \ldots, x_p), \tag{5.58}$$

with $j = 1, \ldots, q$. Similarly, g consists of $r = \dim Z$ functions z_k, each depending on $q = \dim Y$ variables y_j as

$$z_k = z_k(y_1, \ldots, y_q), \tag{5.59}$$

with $k = 1, \ldots, r$. The composite function $h = g \cdot f$ has again the same components z_k, but each argument y_j is expressed in terms of x_is by (5.58). Hence one can state (5.57) as

$$\frac{\partial z_k}{\partial x_i} = \sum_{j=1}^{q} \frac{\partial z_k}{\partial y_j} \frac{\partial y_j}{\partial x_i} = \sum\nolimits_j \frac{\partial z_k}{\partial y_j} \frac{\partial y_j}{\partial x_i}. \tag{5.60}$$

When one takes the partial derivative of z_k with respect to x_i, it is understood that z_k is considered as a function of x_i-variables. If z_k is differentiated with respect to y_j, then it is considered as a function of y_j-variables. Equations (5.60) are easier to remember than Equations (5.57). $\triangle$

Example 5.5.8 Let $f : \mathbb{R} \to \mathbb{R}$ be a differentiable function. Assume that $f'(x) = 0$ if and only if $x = 0$. Define $H : \mathbb{R}^2 \to \mathbb{R}^2$ by

$$h(x, y) = (f(x^2 - y^2), f(x^2 + y^2)) \quad \text{for all } (x, y) \in \mathbb{R}^2.$$

Let us see that h is differentiable. Here, the components of h are $h_1(x, y) = f(x^2 - y^2)$ and $h_2(x, y) = f(x^2 + y^2)$. Thus, since f is differentiable,

$$\begin{aligned}
\frac{\partial h_1}{\partial x}(x, y) &= 2xf'(x^2 - y^2), \quad \frac{\partial h_1}{\partial y}(x, y) = -2yf'(x^2 - y^2) \\
\frac{\partial h_2}{\partial x}(x, y) &= 2xf'(x^2 + y^2), \quad \frac{\partial h_2}{\partial y}(x, y) = 2yf'(x^2 + y^2).
\end{aligned}$$

Since f' is continuous, it follows that the above partial derivatives are all continuous on $\mathbb{R}^2$. So, f is differentiable on $\mathbb{R}^2$. Also, the matrix for $f'(x, y)$ is

$$\mathbf{J}f(x, y) = \begin{bmatrix} 2xf'(x^2 - y^2) & -2yf'(x^2 - y^2) \\ 2xf'(x^2 + y^2) & 2yf'(x^2 + y^2) \end{bmatrix}. \quad \triangle$$

Total Differentials

Notations 5.5.9 The d-symbol. If $f : A \to \mathbb{R}$ is a real-valued differentiable function, then one also writes df for f'. In basic calculus courses, the notation df usually suggests some kind of smallness. There is no such implication here. Hence $df : A \to L(X, \mathbb{R})$ is a function with values $df(\mathbf{a}) \in L(X, \mathbb{R})$, $\mathbf{a} \in A$. Also, $df(\mathbf{a})\mathbf{x}$ is a real number for all $\mathbf{a} \in A$ and $\mathbf{x} \in X$.

We should be careful to use this notation only for real-valued functions. In advanced calculus courses, the d-symbol is usually reserved for *exterior derivatives*. Exterior derivatives will be considered in chapter 10. For real-valued functions, they are the same as the derivatives discussed here.

Example 5.5.10 Derivatives of the coordinate functions. Let $x_j : X \to \mathbb{R}$ be the coordinate functions with respect to a basis in X. They are real-valued functions. The notation dx_j is the standard notation for their derivatives. Each $x_j : X \to \mathbb{R}$ is a linear transformation. Therefore, as in Lemma 5.5.3, $dx_j : X \to L(X, R)$ is a constant function. Indeed, $dx_j(\mathbf{a}) = x_j$ for all $\mathbf{a} \in X$. Hence $dx_j(\mathbf{a})(\mathbf{x}) = x_j(\mathbf{x})$ for all $\mathbf{a}$, $\mathbf{x} \in X$. $\triangle$

Definition 5.5.11 Total differentials. Let A be an open set in a vector space X. Let $\{\mathbf{e}_1, \ldots, \mathbf{e}_n\}$ be a basis for X. Let $f : A \to \mathbb{R}$ be a continuously differentiable function. Then

$$df = \frac{\partial f}{\partial x_1}\, dx_1 + \cdots + \frac{\partial f}{\partial x_n}\, dx_n$$

is called the *total differential* of f. This is just the derivative of a real-valued function expressed in terms of the derivatives of the coordinate functions $dx_i : X \to \mathbb{R}$.

Lemma 5.5.12 Derivative of the inverse function. *Let $f : A \to Y$ be a differentiable and invertible function. Assume that its range $f(A) = B$ is open in Y. If the inverse function $g : B \to X$ is differentiable, then $g'(\mathbf{b}) = (f'(\mathbf{a}))^{-1}$ for all $\mathbf{a} \in A$ and $\mathbf{b} = f(\mathbf{a}) \in B$.*

Proof. We see that the composition $g \cdot f : X \to X$ is the identity function $I : X \to X$. Hence, by the chain rule, Theorem 5.5.6, and by appealing to Example 5.5.5,

$$g'(\mathbf{b}) \cdot f'(\mathbf{a}) = I'(\mathbf{a}) = I\,.$$

This shows that $g'(\mathbf{b}) = (f'(\mathbf{a}))^{-1}$. $\square$

Theorem 5.5.13 Derivatives and Cartesian products. *Let X, Y and U be normed spaces. Let A be an open subset of U. Let $f : A \to X$ and $g : A \to Y$ be two continuously differentiable functions. Define*

$$\varphi : A \to X \times Y \text{ by } \varphi(\mathbf{a}) = (f(\mathbf{a}),\, g(\mathbf{a})) \in X \times Y, \quad \mathbf{a} \in A.$$

Then φ is a continuously differentiable function on A with

$$\varphi'(\mathbf{a})(\mathbf{u}) = (f'(\mathbf{a})(\mathbf{u}),\, g'(\mathbf{a})(\mathbf{u})) \in X \times Y$$

for all $\mathbf{a} \in A$ and $\mathbf{u} \in U$.

Proof. Define $R : X \to X \times Y$ and $S : Y \to X \times Y$ as $R\mathbf{x} = (\mathbf{x}, \mathbf{0})$ and $S\mathbf{y} = (\mathbf{0}, \mathbf{y})$ for all $\mathbf{x} \in X$ and $\mathbf{y} \in Y$. Then we see that

$$\varphi = R \cdot f + S \cdot g\,.$$

Now R and S are linear functions. Hence, they are continuously differentiable. Therefore, φ is continuously differentiable by the chain rule, Theorem 5.5.6. To find its derivative at $\mathbf{a} \in A$, let $f(\mathbf{a}) = \mathbf{b} \in X$ and $g(\mathbf{u}) = \mathbf{c} \in Y$. Hence, again by the chain rule, Theorem 5.5.6,

$$\varphi'(\mathbf{a}) = R'(\mathbf{b}) \cdot f'(\mathbf{a}) + S'(\mathbf{c}) \cdot g'(\mathbf{a}) = R \cdot f'(\mathbf{a}) + S \cdot g'(\mathbf{a})\,.$$

The last equality follows from Lemma 5.5.3, which gives the derivatives of linear functions. Hence

$$\begin{aligned}\varphi'(\mathbf{a})(\mathbf{u}) &= R\cdot f'(\mathbf{a})(\mathbf{u}) + S\cdot g'(\mathbf{a})(\mathbf{u})\\ &= (f'(\mathbf{a})(\mathbf{u}),\, \mathbf{0}) + (\mathbf{0},\, g'(\mathbf{a})(\mathbf{u}))\\ &= (f'(\mathbf{a})(\mathbf{u}),\, g'(\mathbf{a})(\mathbf{u}))\end{aligned}$$

for all $\mathbf{u} \in U$. □

Problems

5.50 Let $f : \mathbb{R}^3 \to \mathbb{R}^3$ be differentiable. Assume that

$$f(x, y, z) = f(x + y, 0, x + z) \quad \text{for all } (x, y, z) \in \mathbb{R}^3.$$

Show that the linear transformation $f'(\mathbf{x}) : \mathbb{R}^3 \to \mathbb{R}^3$ is never onto for any $\mathbf{x} \in \mathbb{R}^3$ of the form $(a, 0, c)$.

5.51 Let $\varphi : X \to \mathbb{R}$ be a differentiable function. If $\varphi(\mathbf{a}) \neq 0$, then show that $\nabla(1/\varphi)(\mathbf{a}) = -(1/\varphi(\mathbf{a})^2)\nabla\varphi(\mathbf{a})$.

5.52 Let f be a real-valued differentiable function of $\mathbf{y} = (y_1, \ldots, y_n)$. Let $\mathbf{y} = \mathbf{y}(\mathbf{x})$ be a differentiable function of $\mathbf{x} = (x_1, \ldots, x_m)$. Show that

$$\nabla f = \sum\nolimits_j \frac{\partial f}{\partial y_j} \nabla y_j.$$

5.53 Let A be an open set in the Euclidean space $\mathbb{R}^n$. Let $f : A \to \mathbb{R}$ be a twice continuously differentiable function. Then the *Laplacian* of f is defined as

$$\Delta f(\mathbf{x}) = \frac{\partial^2 f}{\partial x_1^2}(\mathbf{x}) + \cdots + \frac{\partial^2 f}{\partial x_n^2}(\mathbf{x}),$$

where $\mathbf{x} = (x_1, \ldots, x_n) \in A$. Let $n = 2$ and let $(x_1, x_2) = (x, y)$. Express $f(x, y)$ in terms of polar coordinates $x = r\cos\theta$ and $y = r\sin\theta$ as $f(x, y) = F(r, \theta)$. Show that

$$\Delta f = \frac{\partial^2 f}{\partial x^2} + \frac{\partial^2 f}{\partial y^2} = \frac{\partial^2 F}{\partial r^2} + \frac{1}{r}\frac{\partial F}{\partial \theta} + \frac{1}{r^2}\frac{\partial^2 F}{\partial \theta^2}.$$

5.54 Let $f(x, y, z) = \varphi(x^2 + y^2 + z^2)$, where $\varphi : \mathbb{R} \to \mathbb{R}$ is a differentiable function. Find ∇f and Δf. Here Δf is as in Problem 5.53.

5.55 Let $k \in \mathbb{N}$. Let $f : \mathbb{R}^n \to \mathbb{R}$ be a differentiable function. If $f(t\mathbf{x}) = t^k(\mathbf{x})$ for all $\mathbf{x} \in \mathbb{R}^n$ and for all $t \in \mathbb{R}$, then show that $\langle \nabla f(\mathbf{x}), \mathbf{x} \rangle = kf(\mathbf{x})$.

5.56 Let $g : Y \to \mathbb{R}$ be a differentiable function. Let $f(\mathbf{x}) = g(T\mathbf{x})$, where $T : X \to Y$ is a linear transformation. Find ∇f in terms of ∇g.

5.57 Let $g : \mathbb{R} \to \mathbb{R}$ be a differentiable function. Define $f : X \times Y \to \mathbb{R}$ as $f(\mathbf{x}, \mathbf{y}) = g(\langle T\mathbf{x}, \mathbf{y} \rangle)$, where $T : X \to Y$ is a linear transformation. Find ∇f.

5.6 DIFFERENTIATION OF PRODUCTS

The rule for the differentiation of products is familiar from the one-variable case. Consider

$$f(x) = x^2 e^x \cos x \tag{5.61}$$

as an example. To differentiate f, consider it as the product of three functions $f_1(x) = x^2$, $f_2(x) = e^x$, and $f_3(x) = \cos x$. Then

$$\begin{aligned} f'(x) &= f_1'(x) f_2(x) f_3(x) + f_1(x) f_2'(x) f_3(x) + f_1(x) f_2(x) f_3'(x) \\ &= 2x\, e^x \cos x + x^2 e^x \cos x - x^2 e^x \sin x. \end{aligned}$$

To generalize this method, consider $f(x)$ as the composition of two functions $F : \mathbb{R} \to \mathbb{R}^3$ and $M : \mathbb{R}^3 \to \mathbb{R}$ defined by

$$\begin{aligned} F(x) &= (f_1(x), f_2(x), f_3(x)) \\ &= (x^2, e^x, \cos x) \in \mathbb{R}^3 \end{aligned}$$

and

$$M(y_1, y_2, y_3) = y_1 \cdot y_2 \cdot y_3 \in \mathbb{R}.$$

Then $f(x) = M(F(x))$. Therefore, by the chain rule, Theorem 5.5.6,

$$f'(x) = M'(F(x)) \cdot F'(x).$$

The derivative of $F : \mathbb{R} \to \mathbb{R}^3$ is obtained as

$$F'(x) = (2x, e^x, \cos x).$$

This follows from Theorem 5.5.13. We will show below that $M'(y_1, y_2, y_3) : R^3 \to \mathbb{R}$ is given as

$$M'(y_1, y_2, y_3)(v_1, v_2, v_3) = v_1 y_2 y_3 + y_1 v_2 y_3 + y_1 y_2 v_3 \tag{5.62}$$

for all $(v_1, v_2, v_3) \in \mathbb{R}^3$. Then the differentiation rule for $f(x) = x^2 e^x \cos x$ in (5.61) follows. Equation (5.62) is true for all multilinear functions. This is a key result in the differentiation of product-like functions.

Notations 5.6.1 Multilinear functions. The notations and definitions below were introduced in Section 3.3. Let $U_1, \ldots, U_k$ be vector spaces. We ignore the difference between the spaces

$$U_1 \times \cdots \times U_k \text{ and } U_1 \oplus \cdots \oplus U_k$$

and denote both spaces as X. Let $P_i : X \to X$ be the coordinate projection on the ith component. Hence, if $\mathbf{u}_j \in U_j$, $j = 1, \ldots, k$, then

$$P_i(\mathbf{u}_1 + \cdots + \mathbf{u}_k) = \mathbf{u}_i \text{ for all } i = 1, \ldots, k.$$

Also, let $Q_i = I - P_i$. Then $M : X \to Y$ is called a *multi-linear function* if the equation

$$T_i(\mathbf{a})\mathbf{x} = M(Q_i\mathbf{a} + P_i\mathbf{x}) = M(\mathbf{a} + P_i\mathbf{x}) - M(\mathbf{a}), \quad \mathbf{x} \in X \tag{5.63}$$

defines a linear map $T_i(\mathbf{a}) : X \to Y$ for each fixed $\mathbf{a} \in X$ and for each fixed $i = 1, \ldots, k$. The second equality in (5.63) follows from Lemma 3.3.5. In particular, if $\mathbf{a} \in X$, $\mathbf{u}_i \in U_i$, and $t \in \mathbb{R}$, then

$$\begin{aligned} T_i(\mathbf{a})(t\mathbf{u}_i) &= M(Q_i\mathbf{a} + P_i(t\mathbf{u}_i)) = M(Q_i\mathbf{a} + t\mathbf{u}_i) && (5.64) \\ &= M(\mathbf{a} + P_i(t\mathbf{u}_i)) - M(\mathbf{a}) && (5.65) \\ &= M(\mathbf{a} + t\mathbf{u}_i) - M(\mathbf{a}). && (5.66) \end{aligned}$$

Here $P_i(t\mathbf{u}_i) = t\mathbf{u}_i$ since $\mathbf{u}_i \in U_i$. Equation (5.65) follows from Lemma 3.3.5.

Theorem 5.6.2 Derivatives of multilinear functions. *Multilinear functions are differentiable everywhere. If $X = U_1 \oplus \cdots \oplus U_k$ and if $M : X \to Y$ is multilinear, then*

$$M'(\mathbf{a})(\mathbf{x}) = \sum_{i=1}^{k} M(Q_i\mathbf{a} + P_i\mathbf{x}) \tag{5.67}$$

for all $\mathbf{a}$, $\mathbf{x} \in X$. The notations are as in Notations 5.6.1 above.

Proof. If $\mathbf{u}_i \in U_i$, then the directional derivative of $M : X \to Y$ at $\mathbf{a} \in X$ and in the direction of $\mathbf{u}_i$ is

$$M'(\mathbf{a};\, \mathbf{u}_i) = \lim_{t \to 0} \frac{M(\mathbf{a} + t\mathbf{u}_i) - M(\mathbf{a})}{t} = M(Q_i\mathbf{a} + \mathbf{u}_i).$$

In fact, (5.66) and the linearity of $T_i(\mathbf{a}) : X \to Y$ show that the ratio above has the constant value of

$$T_i(\mathbf{a})\mathbf{u}_i = M(\mathbf{a} + \mathbf{u}_i) - M(\mathbf{a}) = M(Q_i\mathbf{a} + \mathbf{u}_i).$$

Multilinear functions $M : X \to Y$ are continuous by Theorem 4.4.9. Hence $M'(\mathbf{a};\, \mathbf{u}_i)$ is a continuous function of $\mathbf{a} \in X$ for each fixed $\mathbf{u}_i \in U_i$.

If E_i is a basis for U_i, then we see that $E = \cup_i E_i$ is a basis for $X = \oplus_i U_i$. Hence, X has a basis consisting of vectors contained in U_i spaces. Therefore $M : X \to Y$ has continuous directional derivatives with respect to the vectors in a basis. Therefore, $M : X \to Y$ is also continuously differentiable by the existence theorem, Theorem 5.3.4. To obtain (5.67), note that each $\mathbf{x} \in X = U_1 \oplus \cdots \oplus U_k$ has a representation as

$$\mathbf{x} = P_1\mathbf{x} + \cdots + P_k\mathbf{x}, \quad P_i\mathbf{x} \in U_i.$$

Hence (5.67) follows from $M'(\mathbf{a})(P_i\mathbf{x}) = M(Q_i\mathbf{a} + P_i\mathbf{x})$ and from the linearity of $M'(\mathbf{a}) : X \to Y$. $\square$

Remarks 5.6.3 Explicit notations. Theorem 5.6.2 above is the key result in the differentiation of products. This theorem implies, for example, that if

$$\mathbf{a} = (\mathbf{a}_1,\, \mathbf{a}_2,\, \mathbf{a}_3) \text{ and } \mathbf{x} = (\mathbf{x}_1,\, \mathbf{x}_2,\, \mathbf{x}_3)$$

are two vectors in $X = U_1 \times U_2 \times U_3 \sim U_1 \oplus U_2 \oplus U_3$, then

$$\begin{aligned} M'(\mathbf{a})(\mathbf{x}) &= M'(\mathbf{a}_1,\, \mathbf{a}_2,\, \mathbf{a}_3)(\mathbf{x}_1,\, \mathbf{x}_2,\, \mathbf{x}_3) \\ &= M(\mathbf{x}_1,\, \mathbf{a}_2,\, \mathbf{a}_3) + M(\mathbf{a}_1,\, \mathbf{x}_2,\, \mathbf{a}_3) + M(\mathbf{a}_1,\, \mathbf{a}_2,\, \mathbf{x}_3). \end{aligned}$$

This is a direct generalization of the usual product rule (5.62) above.

Examples of Product Differentiations

The following examples of product differentiation are all obtained by the direct application of Theorem 5.6.2 above. In these examples, A is an open set in a normed space X.

Example 5.6.4 Product of a scalar function with a vector-valued function. Let $r : A \to \mathbb{R}$ and $g : A \to Y$ be two differentiable functions. Define

$$f = rg : A \to Y \text{ by } f(\mathbf{x}) = r(\mathbf{x})g(\mathbf{x}) \in Y \text{ for all } \mathbf{x} \in A.$$

Then $f : A \to X$ is differentiable and

$$f'(\mathbf{x}) = r'(\mathbf{x})\, g(\mathbf{x}) + r(\mathbf{x})\, g'(\mathbf{x}) \in L(X,\, Y),$$

or, more explicitly, for all $\mathbf{u} \in X$,

$$f'(\mathbf{x})(\mathbf{u}) = r'(\mathbf{x})(\mathbf{u})\, g(\mathbf{x}) + r(\mathbf{x})\, g'(\mathbf{x})(\mathbf{u}) \in Y.$$

To prove this result, define $F : A \to \mathbb{R} \times Y$ by

$$F(x) = (r(x),\, g(x)) \in \mathbb{R} \times Y,\ x \in A$$

and $Q : \mathbb{R} \times Y \to Y$ by

$$Q(a,\, \mathbf{y}) = a\mathbf{y} \in Y,\ (a,\, \mathbf{y}) \in \mathbb{R} \times Y,$$

and apply Theorem 5.6.2 above. $\triangle$

Example 5.6.5 Inner products. Let Y be a Euclidean space. Let $f : A \to Y$ and $g : A \to Y$ be two differentiable functions. Define

$$h = \langle f,\, g \rangle : A \to \mathbb{R} \text{ by } h(\mathbf{x}) = \langle f(\mathbf{x}),\, g(\mathbf{x}) \rangle \text{ for all } \mathbf{x} \in A.$$

Then $h : A \to \mathbb{R}$ is differentiable and

$$h' = \langle f,\, g' \rangle + \langle f',\, g \rangle : A \to L(X,\, \mathbb{R}).$$

More explicitly,

$$h'(\mathbf{x}) = \langle f(\mathbf{x}),\, g'(\mathbf{x}) \rangle + \langle f'(\mathbf{x}),\, g(\mathbf{x}) \rangle : X \to \mathbb{R}$$

for all $\mathbf{x} \in A$. Still more explicitly,

$$h'(\mathbf{x})(\mathbf{u}) = \langle f(\mathbf{x}),\, g'(\mathbf{x})(\mathbf{u}) \rangle + \langle f'(\mathbf{x})(\mathbf{u}),\, g(\mathbf{x}) \rangle \in \mathbb{R}$$

for all $\mathbf{x} \in A$ and for all $\mathbf{u} \in X$. To prove this result, define $F : A \to Y \times Y$ by

$$F(\mathbf{x}) = (f(\mathbf{x}),\, g(\mathbf{x})) \in Y \times Y,\ \mathbf{x} \in A,$$

and $Q = Y \times Y \to \mathbb{R}$ by

$$Q(\mathbf{p},\, \mathbf{q}) = \langle \mathbf{p},\, \mathbf{q} \rangle \in \mathbb{R},\ (\mathbf{p},\, \mathbf{q}) \in Y \times Y,$$

and apply Theorem 5.6.2 above. $\triangle$

Example 5.6.6 Cross products in $\mathbb{R}^3$. Let $f,\, g : A \to \mathbb{R}^3$ be two differentiable functions. Define $\varphi : A \to \mathbb{R}^3$ as the cross product

$$\varphi(\mathbf{x}) = f(\mathbf{x}) \times g(\mathbf{x}) \in \mathbb{R}^3,\ \mathbf{x} \in A.$$

Then $\varphi : A \to \mathbb{R}^3$ is differentiable and

$$\varphi'(\mathbf{x})(\mathbf{u}) = f'(\mathbf{x})(\mathbf{u}) \times g(\mathbf{x}) + f(\mathbf{x}) \times g'(\mathbf{x})(\mathbf{u}) \in \mathbb{R}^3$$

for all $\mathbf{x} \in A$ and $\mathbf{u} \in X$. To see this, note that $Q(\mathbf{p},\, \mathbf{q}) = \mathbf{p} \times \mathbf{q}$ defines a multilinear function $Q : \mathbb{R}^3 \times \mathbb{R}^3 \to \mathbb{R}^3$. $\triangle$

Example 5.6.7 Mixed products in $\mathbb{R}^3$. Let $f,\, g,\, h : A \to \mathbb{R}^3$ be three differentiable functions. Define $\varphi : A \to \mathbb{R}$ as the mixed product

$$\varphi(\mathbf{x}) = \langle f(\mathbf{x}),\, g(\mathbf{x}) \times h(\mathbf{x})\rangle \in \mathbb{R},\ \mathbf{x} \in A.$$

Then $\varphi : A \to \mathbb{R}$ is differentiable and

$$\begin{aligned}\varphi'(\mathbf{x})(\mathbf{u}) &= \langle f'(\mathbf{x})(\mathbf{u}),\, g(\mathbf{x}) \times h(\mathbf{x})\rangle + \langle f(\mathbf{x}),\, g'(\mathbf{x})(\mathbf{u}) \times h(\mathbf{x})\rangle \\ &+ \langle f(\mathbf{x}),\, g(\mathbf{x}) \times h'(\mathbf{x})(\mathbf{u})\rangle \in \mathbb{R}\end{aligned}$$

for all $\mathbf{x} \in A$ and $\mathbf{u} \in X$. To see this, note that $Q(\mathbf{p},\, \mathbf{q},\, \mathbf{r}) = \langle \mathbf{p},\, \mathbf{q} \times \mathbf{r}\rangle$ defines a multilinear function $Q : \mathbb{R}^3 \times \mathbb{R}^3 \times \mathbb{R}^3 \to \mathbb{R}$. $\triangle$

Example 5.6.8 Products of linear maps. Let $R,\, S : A \to L(Y,\, Y)$ be differentiable functions. Define

$$T = S \cdot R : A \to L(Y,\, Y) \text{ as } T(\mathbf{x}) = S(\mathbf{x}) \cdot R(\mathbf{x}) \text{ for all } \mathbf{x} \in A.$$

Then $T : A \to L(Y,\, Y)$ is differentiable and

$$T' = S \cdot R' + S' \cdot R : A \to L(X,\, L(Y,\, Y)).$$

More explicitly,

$$T'(\mathbf{x}) = S(\mathbf{x}) \cdot R'(\mathbf{x}) + S'(\mathbf{x}) \cdot R(\mathbf{x}) : X \to L(Y,\, Y)$$

for all $\mathbf{x} \in A$. Still more explicitly,

$$T'(\mathbf{x})(\mathbf{u}) = S(\mathbf{x}) \cdot R'(\mathbf{x})(\mathbf{u}) + S'(\mathbf{x})(\mathbf{u}) \cdot R(\mathbf{x}) \in L(Y,\, Y)$$

for all $\mathbf{x} \in A$ and for all $\mathbf{u} \in X$. $\triangle$

Problems

5.58 Let $h : \mathbb{R}^3 \to \mathbb{R}$ be defined by

$$h(\mathbf{x}) = \|(x - y, y - z, x - z)\|^2 \quad \text{for all } \mathbf{x} = (x, y, z) \in \mathbb{R}^3,$$

where the norm is the Euclidean norm. Compute $h'(\mathbf{a})(\mathbf{x})$ for $\mathbf{a},\, \mathbf{x} \in \mathbb{R}^3$.

5.59 Let $T : X \to X$ be a linear transformation. Let $f(\mathbf{x}) = \langle T\mathbf{x},\, \mathbf{x}\rangle$, $\mathbf{x} \in X$. Show that $\langle \nabla f(\mathbf{a}),\, \mathbf{x}\rangle = \langle T\mathbf{a},\, \mathbf{x}\rangle + \langle T\mathbf{x},\, \mathbf{a}\rangle$, $\mathbf{a},\, \mathbf{x} \in X$.

5.60 Let $T,\, S \in L(X,\, Y)$ and $f(\mathbf{x}) = \langle T\mathbf{x},\, S\mathbf{x}\rangle_Y$, $\mathbf{x} \in X$. Find $\nabla f(\mathbf{a})$, $\mathbf{a} \in X$.

5.61 Let $T \in L(X, Y)$ and let $\varphi : X \to \mathbb{R}$ be a differentiable function. Find ∇f for $f(\mathbf{x}) = \langle \varphi(\mathbf{x})T\mathbf{x}, \mathbf{x} \rangle$, $\mathbf{x} \in X$.

5.62 Let $T \in L(\mathbb{R}^3, \mathbb{R}^3)$ and let $\varphi : X \to \mathbb{R}$ be a differentiable function. Define $F : \mathbb{R}^3 \to \mathbb{R}^3$ by $F(\mathbf{x}) = \varphi(\mathbf{x})(\mathbf{x} \times T\mathbf{x})$. Find $F'(\mathbf{a})(\mathbf{x})$.

CHAPTER 6

DIFFEOMORPHISMS AND MANIFOLDS

Diffeomorphisms are invertible mappings between two open sets that are continuously differentiable in both directions. The simplest type of diffeomorphism is an invertible linear transformation. This is a basic concept in linear algebra. Similarly, the study of diffeomorphisms is a basic part of differential calculus.

Most of the results on diffeomorphisms depend on the inverse function theorem. This is one of the central results we obtain in this course. It states that if a continuously differentiable function has an invertible derivative at a point, then its restriction to a neighborhood of that point is a diffeomorphism. We also call a continuously differentiable function a $\mathcal{C}^1$ *function*.

Diffeomorphisms and the inverse function theorem are the gateway to the study of manifolds: surfaces, curves and other lower-dimensional structures that are embedded in a larger space. A good example is the upper half of the surface of the unit sphere in $\mathbb{R}^3$, which is a two-dimensional manifold. We can represent this set as the image of the unit disk $\{ (x, y) \mid x^2 + y^2 < 1 \}$ in $\mathbb{R}^2$ under the mapping

Analysis in Vector Spaces.
By M. A. Akcoglu, P. F. A. Bartha and D. M. Ha

$\varphi(x, y) = (x, y, (1 - (x^2 + y^2))^{1/2})$. We can also represent the entire surface of the unit sphere *implicitly* as $\{ (x, y, z) \mid F(x, y, z) = 1 - (x^2 + y^2 + z^2) = 0 \}$. It turns out that there are a number of equivalent ways to define a manifold. In each case, the best approach is to define what a manifold looks like locally (in the neighborhood of each of its points) rather than to attempt to represent the entire structure by a single function. These ideas, as well as the important topic of differentiation on manifolds, are explored in sections 6.2 - 6.5.

6.1 THE INVERSE FUNCTION THEOREM

We begin with a statement of the inverse function theorem for $\mathcal{C}^1$ functions. Let $A \subset X$ be an open set and $F : A \to X$ a $\mathcal{C}^1$ function. If $F'(\mathbf{a}) : X \to X$ is invertible at $\mathbf{a} \in A$, then $\mathbf{a}$ has a neighborhood $G \subset A$ such that the restriction of F to G is a $\mathcal{C}^1$ diffeomorphism $F|_G : G \to X$.

The hardest part of proving the inverse function theorem is to demonstrate the existence of a continuous inverse. After this is done, the second part of the theorem—continuous differentiability of the inverse function—follows by routine arguments. First, we are going to obtain this second part. In what follows, X and Y are two normed spaces and A and B are open sets in X and Y, respectively.

Lemma 6.1.1 *Let $f : A \to Y$ be a function and $B = f(A)$. Assume that B is open in Y and f has a continuous inverse $g : B \to X$. If f is differentiable at $\mathbf{a} \in A$ and if its derivative $T = f'(\mathbf{a}) : X \to Y$ at $\mathbf{a} \in A$ is invertible, then g is differentiable at $\mathbf{b} = f(\mathbf{a})$ and $g'(\mathbf{b}) = S = T^{-1} : Y \to X$. Furthermore, if f is a $\mathcal{C}^1$ function on A with an invertible derivative $f'(\mathbf{a}) : X \to Y$ at every point $\mathbf{a} \in A$, then g is a $\mathcal{C}^1$ function on B.*

Proof. Let $\mathbf{y}_n \in Y$ be a zero-sequence. Since $\mathbf{b} = f(\mathbf{a}) \in B$ and since B is open, assume that $(\mathbf{b} + \mathbf{y}_n) \in B$ without loss of generality. Let $\mathbf{x}_n = g(\mathbf{b} + \mathbf{y}_n) - g(\mathbf{b})$. Then $\mathbf{x}_n \in X$ is a zero-sequence because of the continuity of g. Since T is the derivative of f at $\mathbf{a}$, there is another zero-sequence $r_n \in \mathbb{R}$ such that

$$\|f(\mathbf{a} + \mathbf{x}_n) - f(\mathbf{a}) - T\mathbf{x}_n\| = \|\mathbf{y}_n - T\mathbf{x}_n\| \leq r_n \|\mathbf{x}_n\|.$$

This follows from the definition of the derivative. Hence

$$\begin{aligned}
\|\mathbf{x}_n\| &\leq \|S\mathbf{y}_n\| + \|\mathbf{x}_n - S\mathbf{y}_n\| \\
&= \|S\mathbf{y}_n\| + \|S(T\mathbf{x}_n - \mathbf{y}_n)\| \\
&\leq \|S\| \, \|\mathbf{y}_n\| + \|S\| \, \|T\mathbf{x}_n - \mathbf{y}_n\| \\
&\leq \|S\| \, \|\mathbf{y}_n\| + \|S\| \, r_n \, \|\mathbf{x}_n\|, \quad \text{and, therefore,} \\
(1 - r_n \, \|S\|) \, \|\mathbf{x}_n\| &\leq \|S\| \, \|\mathbf{y}_n\|.
\end{aligned}$$

Since $r_n \to 0$, we will assume that $(1 - r_n \|S\|) > 1/2$ for all n. In this case,

$$\begin{aligned} \|\mathbf{x}_n\| &\leq 2\|S\|\,\|\mathbf{y}_n\|. \text{ Therefore} \\ \|g(\mathbf{b}+\mathbf{y}_n) - g(\mathbf{b}) - S\mathbf{y}_n\| &= \|\mathbf{x}_n - S\mathbf{y}_n\| \\ &= \|S(T\mathbf{x}_n - \mathbf{y}_n)\| \\ &\leq \|S\|\,\|\mathbf{y}_n - T\mathbf{x}_n\| \\ &\leq \|S\|\,r_n\,\|\mathbf{x}_n\| \\ &\leq 2r_n\,\|S\|^2\|\mathbf{y}_n\|. \end{aligned}$$

Since $2r_n\,\|S\|^2 \to 0$, we see that g is differentiable at $\mathbf{b}$ and $g'(\mathbf{b}) = S$.

To obtain the second part, assume that f is a $\mathcal{C}^1$ function on A with an invertible derivative $f'(\mathbf{a}) : X \to Y$ at every point $\mathbf{a} \in A$. Hence $f' : A \to L(X, Y)$ is a continuous function and its range is contained in $L_{\text{Inv}}(X, Y)$, the set of invertible linear operators. The first part of the proof shows that $g' = \text{Inv} \cdot f' \cdot g$ is the composition of $g : Y \to X$ with $f' : X \to L_{\text{Inv}}(X, Y)$ and with the inversion $\text{Inv} : L_{\text{Inv}}(X, Y) \to L_{\text{Inv}}(Y, X)$. The first two mappings are continuous by assumption. The inversion mapping is continuous by Theorem 4.4.38. Hence $g' : Y \to L(Y, X)$ is continuous as the composition of continuous functions. □

The next step is to prove the inverse function theorem for an important special case (roughly, the case where $\|Df(\mathbf{x}) - I\| < 1$, i.e., the derivative is "close" to the identity map). Subsequently, we show that the general case can be reduced to this special case.

Lemma 6.1.2 *Let $f : X \to X$ be a $\mathcal{C}^1$ function with an invertible derivative $f'(\mathbf{a}) : X \to X$ at every point $\mathbf{a} \in X$. Assume that there is a $\lambda < 1$ such that*

$$\|f(\mathbf{v}) - f(\mathbf{u}) - (\mathbf{v} - \mathbf{u})\| \leq \lambda\|\mathbf{v} - \mathbf{u}\|$$

for all $\mathbf{u}, \mathbf{v} \in X$. Then $f(X) = X$ and f has a $\mathcal{C}^1$ inverse $g : X \to X$.

Proof. If $f(\mathbf{u}) = f(\mathbf{v})$, then $\|\mathbf{v} - \mathbf{u}\| \leq \lambda\|\mathbf{v} - \mathbf{u}\|$. This implies $\|\mathbf{v} - \mathbf{u}\| = 0$, since $\lambda < 1$. Hence f is a one-to-one function. To show that it maps X onto X, we will show that the equation $f(\mathbf{x}) = \mathbf{c}$ has a solution $\mathbf{a} \in X$ for any given $\mathbf{c} \in X$. We obtain $\mathbf{a}$ as the limit of a sequence of approximate solutions $\mathbf{x}_n$. This is essentially Newton's iteration method, as explained below in Remarks 6.1.7.

Let $\mathbf{x}_0 = \mathbf{c}$ and $\mathbf{x}_{n+1} = \mathbf{x}_n + (\mathbf{c} - f(\mathbf{x}_n))$ for $n \geq 0$. Then we have, for $n \geq 1$,

$$\begin{aligned} \mathbf{x}_{n+1} - \mathbf{x}_n &= \mathbf{x}_n - \mathbf{x}_{n-1} - f(\mathbf{x}_n) + f(\mathbf{x}_{n-1}) \text{ and} && (6.1) \\ \|\mathbf{x}_{n+1} - \mathbf{x}_n\| &= \|\mathbf{x}_n - \mathbf{x}_{n-1} - f(\mathbf{x}_n) + f(\mathbf{x}_{n-1})\| && (6.2) \\ &\leq \lambda\|\mathbf{x}_n - \mathbf{x}_{n-1}\|. && (6.3) \end{aligned}$$

Then $\|\mathbf{x}_{n+1} - \mathbf{x}_n\| \leq \lambda^n \|\mathbf{x}_1 - \mathbf{x}_0\|$ for all $n \in \mathbb{N}$. This follows from (6.3) by an easy induction argument. Hence the sums

$$\sum_n \|\mathbf{x}_{n+1} - \mathbf{x}_n\| \leq \left(\sum_n \lambda^n\right) \|\mathbf{x}_1 - \mathbf{x}_0\| \leq (1-\lambda)^{-1}\|\mathbf{x}_1 - \mathbf{x}_0\|$$

remain bounded. Therefore $\mathbf{x}_n$ is a Cauchy sequence in X and $\lim_n \mathbf{x}_n = \mathbf{a}$ exists. Then $f(\mathbf{x}_n) \to f(\mathbf{a})$ by the continuity of $f : X \to X$. Hence, by taking limits in the inductive step $\mathbf{x}_{n+1} = \mathbf{x}_n + \mathbf{c} - f(\mathbf{x}_n)$, we obtain $f(\mathbf{a}) = \mathbf{c}$. Hence f is a one-to-one function that maps X onto X.

It follows that f has an inverse function $g : X \to X$. We claim that g is continuous. Let $\mathbf{p}, \mathbf{q} \in X$. Let $g(\mathbf{p}) = \mathbf{u}$, $g(\mathbf{q}) = \mathbf{v}$. Hence $\mathbf{p} = f(\mathbf{u})$ and $\mathbf{q} = f(\mathbf{v})$. We have

$$\begin{aligned} \|\mathbf{v} - \mathbf{u}\| &\leq \|\mathbf{q} - \mathbf{p}\| + \|\mathbf{q} - \mathbf{p} - (\mathbf{v} - \mathbf{u})\| \\ &= \|\mathbf{q} - \mathbf{p}\| + \|f(\mathbf{v}) - f(\mathbf{u}) - (\mathbf{v} - \mathbf{u})\| \\ &\leq \|\mathbf{q} - \mathbf{p}\| + \lambda \|\mathbf{v} - \mathbf{u}\|. \end{aligned}$$

Hence $\|g(\mathbf{q}) - g(\mathbf{p})\| = \|\mathbf{v} - \mathbf{u}\| \leq (1-\lambda)^{-1}\|\mathbf{q} - \mathbf{p}\|$. This gives the continuity (actually the uniform continuity) of g on X. Then Lemma 6.1.1 shows that $g' : X \to L(X, X)$ exists and is continuous. □

Lemma 6.1.2 generalizes to the case where $\|Df(\mathbf{x}) - T\| < \|T^{-1}\|$ for some invertible linear mapping T.

Corollary 6.1.3 *Let $f : X \to X$ be a $\mathcal{C}^1$ function with an invertible derivative $f'(\mathbf{a}) : X \to X$ at every point $\mathbf{a} \in X$. Let $T \in L(X, X)$ be an invertible transformation with inverse S. Assume that there is a λ such that $\lambda\|S\| < 1$ and such that*

$$\|f(\mathbf{v}) - f(\mathbf{u}) - T(\mathbf{v} - \mathbf{u})\| \leq \lambda\|\mathbf{v} - \mathbf{u}\|$$

for all $\mathbf{u}, \mathbf{v} \in X$. Then $f(X) = X$ and f has a $\mathcal{C}^1$ inverse $g : X \to X$.

Proof. Note that $\|\mathbf{x}\| = \|ST\mathbf{x}\| \leq \|S\|\,\|T\mathbf{x}\|$ for all $\mathbf{x} \in X$. Hence

$$\begin{aligned} \|Sf(\mathbf{u}) - Sf(\mathbf{v}) - (\mathbf{u} - \mathbf{v})\| &\leq \|S\|\|f(\mathbf{u}) - f(\mathbf{v}) - T(\mathbf{u} - \mathbf{v})\| \\ &\leq \lambda\|S\|\|\mathbf{u} - \mathbf{v}\|. \end{aligned}$$

If $\lambda\|S\| < 1$, then Lemma 6.1.2 shows that $h = Sf : X \to X$ is invertible. Then $Th = f$ is also invertible. □

The general case follows from Corollary 6.1.3. The key idea is that if F is $\mathcal{C}^1$ at $\mathbf{a}$, then the requirement of the corollary is met locally (with $T = DF(\mathbf{a})$). We construct a function f for which the requirement of the corollary is met over the whole space X, ensuring that f and F are identical in a neighborhood of $\mathbf{a}$.

Theorem 6.1.4 The inverse function theorem for $\mathcal{C}^1$ functions. *Let $A \subset X$ be an open set and $F : A \to X$ a $\mathcal{C}^1$ function. If $F'(\mathbf{a}) : X \to X$ is invertible at $\mathbf{a} \in A$, then $\mathbf{a}$ has a neighborhood $G \subset A$ such that the restriction of $F : A \to X$ to G is a $\mathcal{C}^1$ diffeomorphism $F|_G : G \to X$.*

Proof. Let $\mathbf{a} \in A$. Assume that $T = F'(\mathbf{a}) : X \to X$ is invertible. Given $\varepsilon > 0$, find an $r > 0$ such that $B_{3r}(\mathbf{a}) \subset A$ and (since F is $\mathcal{C}^1$)

$$\|F'(\mathbf{x}) - T\|_{L(X,X)} < \varepsilon \text{ for all } \mathbf{x} \in B_{3r}(\mathbf{a}) \subset A. \tag{6.4}$$

Let $\varphi : X \to \mathbb{R}$ be a $\mathcal{C}^1$ function such that $\varphi(\mathbf{x}) = 1$ if $\|\mathbf{x}\| \leq 1$ and $\varphi(\mathbf{x}) = 0$ if $\|\mathbf{x}\| \geq 2$. An example of φ is given in Lemma 6.1.6 below. Define, for all $\mathbf{x} \in X$,

$$f(\mathbf{a}+\mathbf{x}) = F(\mathbf{a}) + T\mathbf{x} + \varphi(\mathbf{x}/r)(F(\mathbf{a}+\mathbf{x}) - F(\mathbf{a}) - T\mathbf{x}). \tag{6.5}$$

Then

$$f(\mathbf{a}+\mathbf{x}) = F(\mathbf{a}+\mathbf{x}) \text{ if } \|\mathbf{x}\| \leq r \text{ and} \tag{6.6}$$

$$f(\mathbf{a}+\mathbf{x}) = F(\mathbf{a}) + T\mathbf{x} \text{ if } \|\mathbf{x}\| \geq 2r. \tag{6.7}$$

The function f is thus identical to F inside a ball of radius r around $\mathbf{a}$, and identical to an affine function outside a ball of radius $2r$ around $\mathbf{a}$. We claim that $f : X \to X$ is a $\mathcal{C}^1$ function. In fact, f is a sum of the products of $\mathcal{C}^1$ functions in the set $\|\mathbf{x} - \mathbf{a}\| < 3r$, so clearly f is a $\mathcal{C}^1$ function in this set. And if $\|\mathbf{x} - \mathbf{a}\| > 2r$, then $f(\mathbf{x}) = F(\mathbf{a}) + T(\mathbf{x} - \mathbf{a})$, so that f is also a $\mathcal{C}^1$ function in this region. Therefore f is a $\mathcal{C}^1$ function on X.

We want to estimate $\|f'(\mathbf{x})\|_{L(X,X)}$. We first claim that

$$\|F(\mathbf{a}+\mathbf{x}) - F(\mathbf{a}) - T\mathbf{x}\|_X \leq 2\varepsilon r \text{ and} \tag{6.8}$$

$$\|F'(\mathbf{a}+\mathbf{x}) - T\|_{L(X,X)} \leq \varepsilon \tag{6.9}$$

for all $\|\mathbf{x}\| \leq 2r$. Here (6.9) follows immediately from (6.4). To obtain (6.8), let $\lambda(\mathbf{x}) = F(\mathbf{a}+\mathbf{x}) - F(\mathbf{a}) - T\mathbf{x}$. Then $\|\lambda'(\mathbf{x})\| \leq \varepsilon$ by (6.9). Hence, by the Mean Value Theorem,

$$\|\lambda(\mathbf{x}) - \lambda(\mathbf{0})\| = \|F(\mathbf{a}+\mathbf{x}) - F(\mathbf{a}) - T\mathbf{x}\| \leq \varepsilon\|\mathbf{x}\| \leq 2\varepsilon r$$

whenever $\|\mathbf{x}\| \leq 2r$.

The next step, with (6.5) in mind, is to differentiate

$$\vartheta(\mathbf{x}) = \varphi(\mathbf{x}/r)(F(\mathbf{a}+\mathbf{x}) - F(\mathbf{a}) - T\mathbf{x})$$

to obtain

$$\begin{aligned}\vartheta'(\mathbf{x}) &= (1/r)(F(\mathbf{a}+\mathbf{x}) - F(\mathbf{a}) - T\mathbf{x})\varphi'(\mathbf{x}/r) \\ &\quad + \varphi(\mathbf{x}/r)(F'(\mathbf{a}+\mathbf{x}) - T).\end{aligned}$$

Note that $\vartheta'(\mathbf{x}) = 0$ for $\|x\| > 2r$, since then $\varphi'(\mathbf{x}/r)$ and $\varphi(\mathbf{x}/r)$ are both 0. To derive a uniform bound on $\vartheta(\mathbf{x})$, let

$$p = \sup_{\mathbf{x}} \|\varphi(\mathbf{x})\| \text{ and } q = \sup_{\mathbf{x}} \|\varphi'(\mathbf{x})\|. \tag{6.10}$$

They both exist since they are the upper bounds of continuous functions of compact support. Then if $\|x\| \leq 2r$, it follows from (6.8) and (6.9) that

$$\begin{aligned} \|\vartheta'(\mathbf{x})\| &\leq (1/r)\|F(\mathbf{a}+\mathbf{x}) - F(\mathbf{a}) - T\mathbf{x}\| \, \|\varphi'(\mathbf{x}/r)\| \\ &\quad + \|\varphi(\mathbf{x}/r)\| \, \|F'(\mathbf{a}+\mathbf{x}) - T\| \\ &\leq (1/r)2\varepsilon r q + p\varepsilon = (2q+p)\varepsilon. \end{aligned}$$

But this bound actually holds for all $\mathbf{x}$, since $\vartheta'(\mathbf{x}) = 0$ for $\|x\| > 2r$. By definition (6.5), we also have

$$\vartheta(\mathbf{x}) = f(\mathbf{a}+\mathbf{x}) - F(\mathbf{a}) - T\mathbf{x} \text{ and } \vartheta'(\mathbf{x}) = f'(\mathbf{a}+\mathbf{x}) - T.$$

Then, again by the Mean Value Theorem, we obtain

$$\|f(\mathbf{u}) - f(\mathbf{v}) - T(\mathbf{u}-\mathbf{v})\| \leq (2q+p)\varepsilon \, \|\mathbf{u}-\mathbf{v}\| \tag{6.11}$$

Choose $\varepsilon > 0$ so that $(2q+p)\varepsilon\|S\| < 1$, where $S = T^{-1}$. Then Corollary 6.1.3 shows that $f : X \to X$ is a $\mathcal{C}^1$ diffeomorphism. Hence f maps any open $G \subset X$ to an open $f(G)$, and its restriction to G is a diffeomorphism between G and $f(G)$. But if we set $G = B_r(\mathbf{a})$, then $f(\mathbf{x}) = F(\mathbf{x})$ for $\mathbf{x} \in G$. Hence F is a diffeomorphism between G and $F(G)$. □

To complete the proof of the inverse function theorem, the following two lemmas establish the existence of a function φ with the properties exploited in the above argument.

Lemma 6.1.5 *Define $\psi : \mathbb{R} \to \mathbb{R}$ as*

$$\psi(t) = \begin{cases} 1 & \text{if } t < 1 \\ 1 - (2/9)(t-1)^2 & \text{if } 1 \leq t < 5/2 \\ (2/9)(4-t)^2 & \text{if } 5/2 \leq t < 4 \\ 0 & \text{if } 4 \leq t. \end{cases}$$

Then ψ is a $\mathcal{C}^1$ function.

Proof. This is left as an exercise. See Figure 6.1. □

Lemma 6.1.6 *Given a Euclidean space X, there is a $\mathcal{C}^1$ function $\varphi : X \to \mathbb{R}$ such that $\varphi(\mathbf{x}) = 1$ if $\|\mathbf{x}\| \leq 1$ and $\varphi(\mathbf{x}) = 0$ if $\|\mathbf{x}\| \geq 2$.*

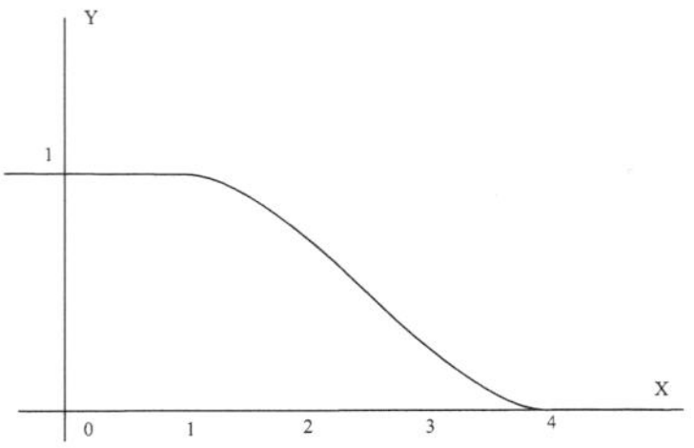

Figure 6.1. Graph of the function in Lemma 6.1.5.

Proof. Let $\varphi(\mathbf{x}) = \psi(\|\mathbf{x}\|^2)$, where $\psi : \mathbb{R} \to \mathbb{R}$ is the function obtained in Lemma 6.1.5. Note that $\|\mathbf{x}\|^2 = \langle \mathbf{x}, \mathbf{x} \rangle$ is a $\mathcal{C}^1$ function, $X \to \mathbb{R}$, as it is a polynomial. Hence φ is the composition of two $\mathcal{C}^1$ functions, and therefore a $\mathcal{C}^1$ function. □

Remarks 6.1.7 Relation to Newton's iteration method. Lemma 6.1.2 should be considered the core of the inverse function theorem. The fact that the equation $f(\mathbf{x}) = \mathbf{c}$ can be solved for all $\mathbf{c} \in X$ is the essential component of this theorem. The other arguments are more or less cleaning up the details. The solution of this equation is obtained by a simplified form of Newton's iteration process. Originally this is used to solve $f(x) = 0$, where $f : I \to \mathbb{R}$ is a continuously differentiable function defined on an open interval I. Let $a_i \in I$. If $f(a_1) = 0$, then we are done. Otherwise, replace the original function f by its affine approximation

$$f_1(x) = f(a_1) + f'(a_1)(x - a_1)$$

and solve $f_1(x) = 0$. This is a linear equation. Let a_2 be its solution. Under some reasonable assumptions, a_2 is a better approximation for a solution. Continue this process to obtain a sequence a_n. If it converges, the limit is a solution for $f(x) = 0$.

We follow this method in the proof of Lemma 6.1.2. Actually, things are even simpler. We start with $\mathbf{x}_1 \in X$. The affine approximation of f at $\mathbf{x}_1$ is

$$f_1(\mathbf{x}) = f(\mathbf{x}_1) + f'(\mathbf{x}_1)(\mathbf{x} - \mathbf{x}_1).$$

Instead of this function, we replace the derivative $f'(\mathbf{x}_1) : X \to X$ by the identity $I : X \to X$ and use

$$g_1(\mathbf{x}) = f(\mathbf{x}_1) + (\mathbf{x} - \mathbf{x}_1).$$

This is reasonable, since our hypotheses imply that all the derivatives of f are close to the identity. In fact, we see that if $f : X \to X$ is a differentiable function such that

$$\|f(\mathbf{v}) - f(\mathbf{u}) - (\mathbf{v} - \mathbf{u})\| \leq \lambda \|\mathbf{v} - \mathbf{u}\|$$

for all $\mathbf{u}, \mathbf{v} \in X$, then $\|f'(\mathbf{x}) - I\|_{L(X,\,X)} \leq \lambda$ for all $\mathbf{x} \in X$. The solutions of

$$g_n(\mathbf{x}) = f(\mathbf{x}_n) + (\mathbf{x} - \mathbf{x}_n) = \mathbf{c}$$

give a sequence that converges to the solution of $f(\mathbf{x}) = \mathbf{c}$.

Lemma 6.1.8 Graph diffeomorphisms. *Let X and Y be Euclidean spaces, A an open subset of X, and $f : A \to Y$ a $\mathcal{C}^1$ function. Define F as*

$$F(\mathbf{x}, \mathbf{y}) = (\mathbf{x}, \mathbf{y} + f(\mathbf{x})), \quad (\mathbf{x}, \mathbf{y}) \in A \times Y. \tag{6.12}$$

Then $F : A \times Y \to A \times Y$ is a diffeomorphism.

Proof. Let $(\mathbf{a}, \mathbf{b}) \in A \times Y$ and $(\mathbf{x}, \mathbf{y}) \in X \times Y$. We see that

$$F'(\mathbf{a}, \mathbf{b})(\mathbf{x}, \mathbf{y}) = (\mathbf{x}, \mathbf{y} + f'(\mathbf{a})\mathbf{x}). \tag{6.13}$$

We see that $F'(\mathbf{a}, \mathbf{b})(\mathbf{x}, \mathbf{y}) = (\mathbf{0}, \mathbf{0})$ implies that $(\mathbf{x}, \mathbf{y}) = (\mathbf{0}, \mathbf{0})$. Hence,

$$F'(\mathbf{a}, \mathbf{b}) : X \times Y \to X \times Y \tag{6.14}$$

is an isomorphism. Also, $F : A \times Y \to A \times Y$ is invertible with the inverse

$$F^{-1}(\mathbf{x}, \mathbf{y}) = (\mathbf{x}, \mathbf{y} - f(\mathbf{x}). \tag{6.15}$$

Hence $F : A \times Y \to A \times Y$ is a diffeomorphism. □

Definition 6.1.9 Graph diffeomorphisms. The diffeomorphism

$$F : A \times Y \to A \times Y \tag{6.16}$$

defined in Lemma 6.1.8 is called the *graph diffeomorphism* induced by $f : A \to Y$.

Examples of Diffeomorphisms

Let X and Y be two Euclidean spaces, A an open set in X, and $f : A \to Y$ a $\mathcal{C}^1$ function. To find out if f is a diffeomorphism, we start with the routine part of the test. Compute the derivative $f'(\mathbf{x}) \in L(X, Y)$ at a general point $\mathbf{x} \in A$ and see if this is an invertible linear transformation for all $\mathbf{x} \in A$. If f passes this test, then it may be a diffeomorphism. At this stage, we know that the range $B = f(A)$ is also an open set. (Why?) We do not know, however, if f has an inverse function $g : B \to A$. The inverse function exists if and only if f is a one-to-one function on A. The verification of this point may not be easy. Being a one-to-one function is not a local property, and here there is no help from calculus.

Nevertheless, the inverse function theorem tells us that if $f'(\mathbf{a}) : X \to Y$ is an invertible linear transformation for a certain $\mathbf{a} \in A$, then the restriction of f to a small enough neighborhood of $\mathbf{a}$ is a diffeomorphism. In general, this is all we need to know in differential calculus, as this study involves the properties of functions only in small neighborhoods.

It is useful to consider the part of the domain A where $f : A \to Y$ has an invertible derivative. We refer to this part as the *regular* part of the domain.

Definition 6.1.10 Regular part of the domain. Let $f : A \to Y$ be a $\mathcal{C}^1$ function. Then its *regular domain* (or the *regular part of its domain*) is

$$A_0 = \{\, \mathbf{a} \mid \mathbf{a} \in A, \;\; f'(\mathbf{a}) : X \to Y \;\text{ is invertible }\}. \tag{6.17}$$

Note that A_0 is also an open set. (Why?)

Example 6.1.11 Define $f : \mathbb{R} \to \mathbb{R}$ by $f(x) = x^2$. What is the regular part of its domain? Give examples of the subsets of the regular domain on which the restrictions of f are diffeomorphisms.

Solution. We have $f'(a) = 2a$ for all $a \in \mathbb{R}$. This defines an invertible transformation $x \to 2ax$ for all $a \neq 0$. Hence, $A_0 = \mathbb{R} \setminus \{0\} = (-\infty, 0) \cup (0, \infty)$. For all $a \in A_0$, there is a neighborhood of a where f is a diffeomorphism. In particular, there is a neighborhood of $a \in A_0$ in which f is one-to-one. This is not true at $0 \notin A_0$. There is no neighborhood of 0 in which f is one-to-one. We see that the restrictions of f to $(-\infty, 0)$ and to $(0, \infty)$ are diffeomorphisms. The restriction of f to $(-\infty, -3) \cup (1, 3)$ is also a diffeomorphism. △

Example 6.1.12 Define $f : \mathbb{R} \to \mathbb{R}$ by $f(x) = x^3$. What is the regular part of its domain? Give examples of the subsets of the regular domain on which the restrictions of f are diffeomorphisms.

Solution. We have $f'(a) = 3a^2$. Hence $A_0 = \mathbb{R} \setminus \{0\} = (-\infty, 0) \cup (0, \infty)$. We see that $f : \mathbb{R} \to \mathbb{R}$ is one-to-one and onto $\mathbb{R}$. Hence it has an inverse $g(x) = x^{1/3}$. The restriction of f to A_0 is a diffeomorphism $f_0 : A_0 \to \mathbb{R}$, but f itself is not a diffeomorphism, since its inverse g is not differentiable at 0. △

What about functions from $\mathbb{R}^2 \to \mathbb{R}^2$ or from $R^3 \to \mathbb{R}^3$? Chapter 1 provides many examples, and we can ask about their invertibility or the invertibility of their various restrictions. Some of the problems in this section require finding the regular parts of their domains. Here, we shall consider only the spherical coordinates mapping defined in Example 1.3.14.

Example 6.1.13 Spherical coordinates. Define $F : \mathbb{R}^3 \to \mathbb{R}^3$ as

$$F(\rho, \theta, \varphi) = (\rho \sin\varphi \cos\theta, \; \rho \sin\varphi \sin\theta, \; \rho\cos\varphi). \tag{6.18}$$

What is the regular part of its domain? Give examples of subsets of the regular domain on which the restrictions of F are diffeomorphisms.

Solution. The standard coordinates of the domain space $\mathbb{R}^3$ are denoted as $(\rho,\, \theta,\, \varphi)$ in this example. The derivative $F'(\rho,\, \theta,\, \varphi) : \mathbb{R}^3 \to \mathbb{R}^3$ transforms the vectors in the standard orthonormal basis of the domain space to the vectors

$$\frac{\partial F}{\partial \rho}(\rho,\, \theta,\, \varphi) = (\sin\varphi\cos\theta,\, \sin\varphi\sin\theta,\, \cos\varphi) \tag{6.19}$$

$$\frac{\partial F}{\partial \theta}(\rho,\, \theta,\, \varphi) = (-\rho\sin\varphi\sin\theta,\, \rho\sin\varphi\cos\theta,\, 0) \tag{6.20}$$

$$\frac{\partial F}{\partial \varphi}(\rho,\, \theta,\, \varphi) = (\rho\cos\varphi\cos\theta,\, \rho\cos\varphi\sin\theta,\, -\rho\sin\varphi) \tag{6.21}$$

in the range space $\mathbb{R}^3$. An easy check shows that these vectors are orthogonal to each other and their norms are, respectively, 1, $\rho\sin\varphi$, and ρ. Hence the standard orthonormal basis of the domain is an eigenbasis (Definition 3.6.3) for the derivative. The derivative is invertible unless $\rho\sin\varphi = 0$. Hence the regular domain of the spherical coordinates is obtained by removing the planes

$$\rho = 0, \quad \varphi = k\pi \ \ (k \in \mathbb{Z}) \tag{6.22}$$

from the $(\rho,\, \theta,\, \varphi)$-space. We see that all the values taken by F on its regular domain are also taken on the part of this domain defined by $0 < \rho$ and $0 < \varphi < \pi$. The restriction of F to this part is still not a diffeomorphism, since the points $\theta = \theta_0 + 2k\pi$, $k \in \mathbb{Z}$, with a fixed $\theta_0 \in \mathbb{R}$, are all mapped to the same point in the xyz-space. To obtain a diffeomorphism, we have to restrict the domain of θ to a convenient interval like $(0,\, 2\pi)$ or $(-\pi,\, \pi)$. Hence the restrictions of F to

$$A = \{\, (\rho,\, \theta,\, \varphi) \mid 0 < \rho,\ 0 < \theta < 2\pi,\ 0 < \varphi < \pi \,\} \text{ or to} \tag{6.23}$$

$$A' = \{\, (\rho,\, \theta,\, \varphi) \mid 0 < \rho,\ -\pi < \theta < \pi,\ 0 < \varphi < \pi \,\} \tag{6.24}$$

are diffeomorphisms. Again, an exact choice of these regions is not too important. Note that $\rho\sin\varphi = 0$ corresponds to the points on the z-axis in the xyz-space. Hence as long as a point $(x,\, y,\, z)$ is away from the z-axis, then a neighborhood of that point will be the diffeomorphic image of a region in the $\rho\theta\varphi$-space under the function F defined in (6.18). $\triangle$

Remarks 6.1.14 A visualization of spherical coordinates. One usually attaches the values of ρ, θ, and φ to the image point $F(\rho,\, \theta,\, \varphi)$ in the xyz-space to obtain a visualization of this coordinate system, as in Figure 6.2.

Example 6.1.15 For any open bounded interval I of $\mathbb{R}$, there is a diffeomorphism $f : I \to R$. We obtain this diffeomorphism as follows. There is a diffeomorphism

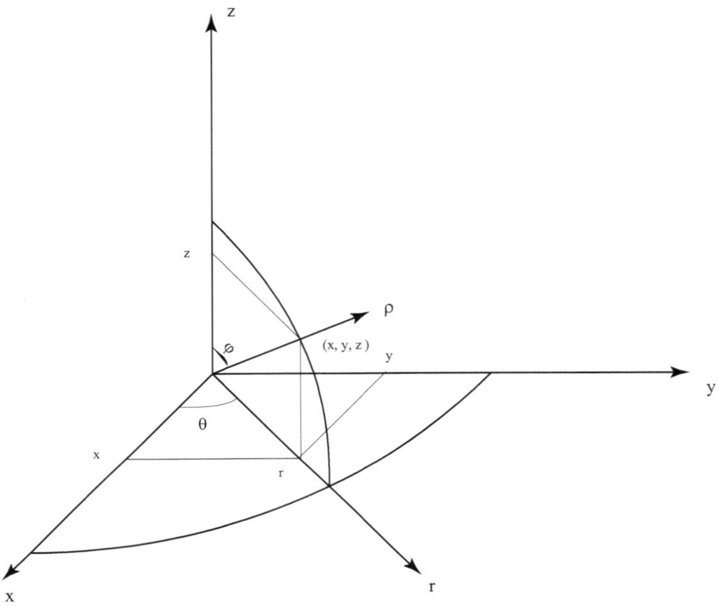

Figure 6.2. Spherical coordinates.

$g : I \to (-\pi/2, \pi/2)$. Also, the function $h(x) = \tan x$ is a diffeomorphism from $(-\pi/2, \pi/2)$ onto $\mathbb{R}$. Thus, $f = h \cdot g : I \to \mathbb{R}$ is a diffeomorphism.

Example 6.1.16 Let $g : \mathbb{R}^3 \to \mathbb{R}^3$ be defined by $g(\mathbf{x}) = (x + e^y, y + e^z, z + e^x)$ for all $\mathbf{x} = (x, y, z) \in \mathbb{R}^3$. Then

$$f'(\mathbf{x}) = \begin{bmatrix} 1 & e^y & 0 \\ 0 & 1 & e^z \\ e^x & 0 & 1 \end{bmatrix}.$$

Hence, $\det f'(\mathbf{x}) = 1 + e^{x+y+z} \neq 0$ for all $\mathbf{x} \in \mathbb{R}^3$. So, at each point $\mathbf{x} \in \mathbb{R}^3$, there is an open set U of $\mathbb{R}^3$ containing $\mathbf{x}$ such that the restriction of f to U is a diffeomorphism.

Example 6.1.17 There are $r > 0$ and $s > 0$ such that if $|a - 8| < r$ and $|b| < r$, then the system of equations

$$\begin{aligned} x^3 + 2x^2y + y^2 &= a \\ yx^2 - 3x^4y + y^3 &= b \end{aligned}$$

has a unique solution (x, y) such that $|x - 2| < s$ and $|y| < s$. To see this, let

$$f(x, y) = (x^3 + 2x^2y + y^2, yx^2 - 3x^4y + y^3) \quad \text{for all } (x, y) \in \mathbb{R}^2.$$

Then

$$f'(x, y) = \begin{bmatrix} 3x^2 + 4xy & 2x^2 + 2y \\ 2xy - 12x^3y & x^2 - 3x^4 + 3y^2 \end{bmatrix}.$$

Hence,

$$f'(2, 0) = \begin{bmatrix} 12 & 8 \\ 0 & -44 \end{bmatrix}$$

is invertible. Also, $f(2, 0) = (8, 0)$. Hence, by the inverse function theorem, there is an open set U containing $(2, 0)$ and an open set V containing $(8, 0)$ such that f maps U onto V in a one-to-one way. In particular, for all $(a, b) \in V$, there is a unique $(x, y) \in U$ such that $f(x, y) = (a, b)$.

Problems

6.1 Let

$$\begin{aligned} A &= \{\, \mathbf{x} = (x, y) \in \mathbb{R}^2 \mid x > 0, y > 0 \,\}, \\ B &= \{\, \mathbf{x} = (x, y) \in \mathbb{R}^2 \mid y > 0 \,\}. \end{aligned}$$

Let

$$f(\mathbf{x}) = (x^2 - y^2, 2xy) \quad \text{for all } \mathbf{x} \in A.$$

Show that f is a diffeomorphism from A onto B.

6.2 Let $D = \{\, \mathbf{x} = (x, y, z) \in \mathbb{R}^3 \mid x > 0, y > 0, z > 0 \,\}$. Let

$$f(\mathbf{x}) = (xy, yz, xz) \quad \text{for all } \mathbf{x} \in D.$$

Show that every point in D has a neighborhood on which f is a diffeomorphism.

6.3 Let D be an open subset of W and $f : D \to Z$ a continuously differentiable function. Suppose that $f'(\mathbf{x}) : W \to Z$ is an isomorphism for all $\mathbf{x} \in D$. Show that $f(D)$ is an open subset of Z. In addition, if f is one-to-one on D, then show that f is a diffeomorphism on D.

6.4 Let $f(\mathbf{x}) = (x^3 + x, y^3 + y, z^3 + z)$ for all $\mathbf{x} = (x, y, z) \in \mathbb{R}^3$. Show that $f : \mathbb{R}^3 \to \mathbb{R}^3$ is a diffeomorphism.

6.5 Let

$$\begin{aligned} A &= \{ (x,y) \in \mathbb{R}^2 \mid x \in \mathbb{R}, 0 < y < \pi \}, \\ B &= \{ (x,y) \in \mathbb{R}^2 \mid y > 0 \}. \end{aligned}$$

Define $f : A \to \mathbb{R}^2$ by

$$f(\mathbf{x}) = (e^x \cos y, e^x \sin y) \quad \text{for all } \mathbf{x} = (x,y) \in \mathbb{R}^2.$$

Show that f is a diffeomorphism from A onto B.

6.6 Let $F : \mathbb{R}^4 \to \mathbb{R}^4$ be defined by

$$F(\mathbf{x}) = (x_1, x_2, x_1^2 x_2^2 + x_3, x_1 - x_2 + x_4) \quad \text{for all } \mathbf{x} = (x_1, x_2, x_3, x_4).$$

Show that F is a diffeomorphism.

6.7 Define $f : \mathbb{R}^2 \to \mathbb{R}^2$ by $f(x, y) = (e^x + e^y, e^x - e^y)$. Show that f is a diffeomorphism. What is the range of f?

6.8 Define f from the xy-plane to the uv-plane by

$$(u, v) = f(x, y) = (3x + 2y, 6x + 4y).$$

Give some examples, if exist, of open sets in the xy-plane such that the restrictions of f to these sets are diffeomorphisms. (Cf. Problem 1.30.)

6.9 Repeat Problem 6.8 for

$$(u, v) = f(x, y) = (3x + 2y, 6x - 4y).$$

(Cf. Problem 1.31.)

6.10 Repeat Problem 6.8 for

$$(u, v) = f(x, y) = (xy, y/x).$$

(Cf. Problem 1.32.)

6.11 Repeat Problem 6.8 for

$$(u, v) = f(x, y) = ((x^2 + y^2)/(2x), (x^2 + y^2)/(2y)).$$

(Cf. Problem 1.33.)

6.12 Repeat Problem 6.8 for

$$(u, v) = f(x, y) = ((x^2 + y^2 + 1)/(2x), (x^2 + y^2 - 1)/(2y)).$$

(Cf. Problem 1.34.)

6.13 Repeat Problem 6.8 for

$$(u,\, v) = f(x,\, y) = (p(x,\, y) + q(x,\, y),\, p(x,\, y) - q(x,\, y)),$$

where $p(x,\, y) = ((x+1)^2 + y^2)^{1/2}$ and $q(x,\, y) = ((x-1)^2 + y^2)^{1/2}$.

(Cf. Problem 1.35.)

6.2 GRAPHS

Graphs are among the most basic examples of manifolds in Euclidean spaces. A manifold (to be defined formally in the next section) is a set that coincides with a graph in a neighborhood of each one of its points. Hence, a manifold can also be called a *local graph*.

Graphs were defined in Definition 1.3.1. From now on, we will use the term 'graph' in a slightly restricted sense. The intended meaning will be clear from the context.

Definition 6.2.1 Graphs. A set Γ in a Euclidean space Z is called a *graph* if there is a coordinate system $(X,\, Y)$ for Z, an open set A in X, and a $\mathcal{C}^1$ function $f : A \to Y$ such that

$$\Gamma = \{\, (\mathbf{x},\, \mathbf{y}) \in Z \mid \mathbf{x} \in A,\ \mathbf{y} = f(\mathbf{x}) \,\}\,.$$

Usually we assume that $(X,\, Y)$ is an orthogonal coordinate system. This is not a restriction of generality. If $\dim X = k$ and $\dim Z = n$, then Γ is a k-dimensional graph in an n-dimensional space. A function $f : A \to Y$ is also denoted as $\mathbf{y} = f(\mathbf{x})$, $\mathbf{x} \in A$, or even more simply as $\mathbf{y} = f(\mathbf{x})$. The function $\mathbf{y} = f(\mathbf{x})$ is called an *equation* or an *explicit equation* for Γ.

Definition 6.2.2 Tangent spaces. Let Γ be the graph of $\mathbf{y} = f(\mathbf{x})$. The *linear tangent space* $T_\mathbf{c}$ of Γ at $\mathbf{c} = (\mathbf{a},\, f(\mathbf{a})) \in \Gamma$ is the graph of the linear function $\mathbf{y} = f'(\mathbf{a})\mathbf{x}$. The *affine tangent space* $AT_\mathbf{c}$ of Γ at the same point is the graph of the affine function

$$\mathbf{y} = f(\mathbf{a}) + f'(\mathbf{a})(\mathbf{x} - \mathbf{a}). \tag{6.25}$$

This is an explicit equation for the affine tangent space at $\mathbf{c} = (\mathbf{a},\, f(\mathbf{a})) \in \Gamma$. When the meaning is clear from the context, we will omit the distinction between these two tangent spaces and call them both the *tangent space* at $\mathbf{c} \in \Gamma$. Note that $AT_\mathbf{c} = \mathbf{c} + T_\mathbf{c}$.

Remarks 6.2.3 Graph and tangent space as diffeomorphic images of A and X. Let Γ be the graph of a function $f : A \to Y$. Let $F(\mathbf{x},\, \mathbf{y}) = (\mathbf{x},\, \mathbf{y} + f(\mathbf{x}))$ be the graph diffeomorphism induced by f as defined in Definition 6.1.9. Consider A as a subset of $X \times Y \simeq X \oplus Y$. Then we see that $\Gamma = F(A)$ is the image of A under the diffeomorphism F. Similarly, the tangent space $T_{(\mathbf{a},\, f(\mathbf{a}))}$ is the image of X under the graph diffeomorphism $X \times Y \to X \times Y$ induced by $f'(\mathbf{a}) : X \to Y$. The affine tangent space $AT_{(\mathbf{a},\, f(\mathbf{a}))}$ is the image of X under the graph diffeomorphism induced by the affine function that takes $\mathbf{x}$ to $f(\mathbf{a}) + f'(\mathbf{a})(\mathbf{x} - \mathbf{a})$.

Remarks 6.2.4 Dependence on the coordinate system. The definition of the tangent space given above is in terms of a coordinate system. Different coordinate systems for a graph actually give the same tangent space. This is noted in Remarks 6.3.23 below. A direct proof of this independence is now possible, but later arguments give a more systematic approach.

Definition 6.2.5 Normal spaces. The orthogonal complement $N_{\mathbf{c}}$ of the tangent space $T_{\mathbf{c}}$ is called the *normal space*. Again, one may distinguish between the linear normal space $N_{\mathbf{c}}$ and the affine normal space $AN_{\mathbf{c}} = \mathbf{c} + N_{\mathbf{c}}$.

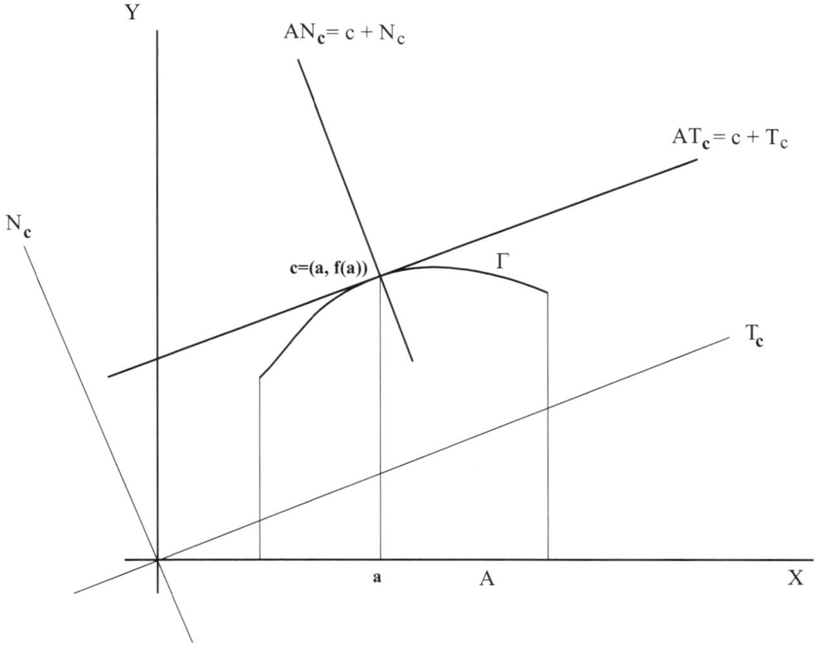

Figure 6.3. Graph with tangent and normal spaces.

Theorem 6.2.6 Equations of normal spaces. *Let Γ be the graph of a $\mathcal{C}^1$ function $\mathbf{y} = f(\mathbf{x})$ in the coordinate system (X, Y). Then*

$$\mathbf{x} = -f'(\mathbf{a})^*\mathbf{y} \text{ and } \mathbf{x} = \mathbf{a} - f'(\mathbf{a})^*(\mathbf{y} - f(\mathbf{a}))$$

are explicit equations for the linear normal space and for the affine normal space of Γ at the point $\mathbf{c} = (\mathbf{a}, f(\mathbf{a})) \in \Gamma$ and in the coordinate system (Y, X). Here $f'(\mathbf{a})^* : Y \to X$ is the adjoint of $f'(\mathbf{a}) : X \to Y$.

Proof. We have $(\mathbf{x}, \mathbf{y}) \in N_{\mathbf{a}}$ if and only if $(\mathbf{x}, \mathbf{y}) \perp T_{\mathbf{a}}$. Vectors in $T_{\mathbf{a}}$ are of the form $(\mathbf{u}, f'(\mathbf{a})\mathbf{u})$, $\mathbf{u} \in X$. Hence $(\mathbf{x}, \mathbf{y}) \in N_{\mathbf{a}}$ if and only if

$$\begin{aligned} \langle(\mathbf{x}, \mathbf{y}), (\mathbf{u}, f'(\mathbf{a})\mathbf{u})\rangle_Z &= \langle \mathbf{x}, \mathbf{u}\rangle_X + \langle \mathbf{y}, f'(\mathbf{a})\mathbf{u}\rangle_Y && (6.26)\\ &= \langle \mathbf{x}, \mathbf{u}\rangle_X + \langle f'(\mathbf{a})^*\mathbf{y}, \mathbf{u}\rangle_X && (6.27)\\ &= \langle \mathbf{x} + f'(\mathbf{a})^*\mathbf{y}, \mathbf{u}\rangle_X = 0 && (6.28)\end{aligned}$$

for all $\mathbf{u} \in X$. This happens if and only if $\mathbf{x} = -f'(\mathbf{a})^*\mathbf{y}$. This is an explicit equation of $N_{\mathbf{c}}$ in the coordinate system (Y, X). Then the equation of the affine normal space also follows. □

Scalar Equations

Let (X, Y) be a coordinate system in Z with $\dim X = k$, $\dim Y = \ell$, and $\dim Z = k + \ell = n$. Let $x_i : X \to \mathbb{R}$ and $y_j : Y \to \mathbb{R}$ be the coordinate functions with respect to some bases in X and in Y. Then a vectorial function $\mathbf{y} = f(\mathbf{x})$ is expressed in terms of ℓ scalar functions of k variables as

$$y_j = f_j(x_1, \ldots, x_k), \; j = 1, \ldots, \ell.$$

These are the explicit scalar equations of a k-dimensional graph Γ in an $n = (k + \ell)$-dimensional space. The scalar equations of the affine tangent space at $(a_1, \ldots, a_k; b_1, \ldots, b_\ell)$, where $b_j = f_j(a_1, \ldots, a_k)$, are

$$(y_j - b_j) = \sum_{i=1}^{k} (\partial f_j/\partial x_i)(a_1, \ldots, a_k)(x_i - a_i).$$

The affine normal space at the same point is given as

$$(x_i - a_i) = -\sum_{j=1}^{\ell} (\partial f_j/\partial x_i)(a_1, \ldots, a_k)(y_j - b_j).$$

Examples of Graphs

Example 6.2.7 Let $\Gamma = \{ ((x, y, z), (x^2 - yz, y^2 - xz)) \mid (x, y, z) \in \mathbb{R}^3 \}$. Then Γ is a three-dimensional graph in $\mathbb{R}^5$. Here, $Z = \mathbb{R}^5, X = \mathbb{R}^3, Y = \mathbb{R}^2, A = X,$

and $f : A \to Y$ is defined by $f(\mathbf{x}) = (x^2 - yz, y^2 - xz)$ for all $\mathbf{x} = (x, y, z) \in \mathbb{R}^3$. Note that $f : \mathbb{R}^3 \to \mathbb{R}^2$. Also,

$$f'(\mathbf{x}) = \begin{bmatrix} 2x & -z & -y \\ -z & 2y & -x \end{bmatrix}.$$

Let $\mathbf{a} = (1, -1, 2)$. Then the linear tangent space $\Gamma_{\mathbf{c}}$ to Γ at $\mathbf{c} = (\mathbf{a}, f(\mathbf{a}))$ is the graph of the function $L(\mathbf{x}) = f'(\mathbf{a})\mathbf{x}$. Now, $f'(\mathbf{a}) = \begin{bmatrix} 2 & -2 & 1 \\ -2 & -2 & -1 \end{bmatrix}$, so that

$$\begin{aligned} \Gamma_{\mathbf{c}} &= \left\{ (\mathbf{x}, f'(\mathbf{a})\mathbf{x}) \mid \mathbf{x} \in \mathbb{R}^3 \right\} \\ &= \left\{ (\mathbf{x}, (2x - 2y + z, -2x - 2y - z)) \mid \mathbf{x} = (x, y, z) \in \mathbb{R}^3 \right\}. \end{aligned}$$

Since $f(\mathbf{a}) = f(1, -1, 2) = (3, -1)$, the affine tangent space to Γ at $\mathbf{c}$ is

$$\begin{aligned} A_{\mathbf{c}} &= \left\{ (\mathbf{x}, f(\mathbf{a}) + f'(\mathbf{a})(\mathbf{x} - \mathbf{a}) \mid \mathbf{x} \in \mathbb{R}^3 \right\} \\ &= \left\{ (\mathbf{x}, (2x - 2y + z - 3, -2x - 2y - z + 1)) \mid \mathbf{x} \in \mathbb{R}^3 \right\}. \end{aligned}$$

The equations for the linear normal space and affine normal space in the coordinate system $(\mathbb{R}^2, \mathbb{R}^3)$ of $\mathbb{R}^5$ are, respectively,

$$\begin{aligned} \mathbf{x} &= -f'(a)^*\mathbf{y} \quad \text{and} \\ \mathbf{x} &= \mathbf{a} - f'(\mathbf{a})^*(\mathbf{y} - f(\mathbf{a})). \end{aligned}$$

Here,

$$f'(\mathbf{a})^* = \begin{bmatrix} 2 & -2 \\ -2 & -2 \\ 1 & -1 \end{bmatrix}.$$

In particular, the linear normal space to Γ at $\mathbf{c}$ is

$$N_{\mathbf{c}} = \left\{ (\mathbf{u}, (2v - 2u, 2u + 2v, v - u)) \mid \mathbf{u} = (u, v) \in \mathbb{R}^2 \right\}.$$

Similarly, the affine normal space to Γ at $\mathbf{c}$ is

$$\begin{aligned} A_{\mathbf{c}}^{\perp} &= \left\{ (\mathbf{u}, \mathbf{a} - f'(\mathbf{a})^*(\mathbf{u} - f(\mathbf{a}))) \mid \mathbf{u} \in \mathbb{R}^2 \right\} \\ &= \left\{ (\mathbf{u}, (-7 - 2v + 2u, 7 - 2u - 2v, -v + u)) \mid \mathbf{u} \in \mathbb{R}^2 \right\}. \end{aligned}$$

Example 6.2.8 The set

$$S = \left\{ (\mathbf{x}, x^2 + y + 2z) \mid \mathbf{x} = (x, y, z) \in \mathbb{R}^3 \right\}$$

is a surface in $\mathbb{R}^4$. The affine tangent space to S at $P((1, 0, 1), 3)$ is the set of all $(\mathbf{x}, u)$ in $\mathbb{R}^4$ given by

$$u - 3 = \langle \nabla f(1, 0, 1), \mathbf{x} - (1, 0, 1) \rangle = \langle (2, 1, 2), (x - 1, y, z - 1) \rangle$$

or

$$u - 3 = 2(x - 1) + y + 2(z - 1).$$

The equation of the normal line to S at P is

$$(x - 1, y, z - 1) = -(y - 3)\nabla f(1, 0, 1) = -(y - 3)(2, 1, 2).$$

Problems

6.14 Let Γ be the graph of $f : \mathbb{R}^3 \to \mathbb{R}^3$, where $f(\mathbf{x}) = (x^2y, y^2z, z^2x)$ for all $\mathbf{x} = (x, y, z) \in \mathbb{R}^3$. Let $\mathbf{a} = (1, -1, 1)$. Find the linear tangent space, the affine tangent space, the linear normal space, and the affine normal space to Γ at the point $(\mathbf{a}, f(\mathbf{a}))$.

6.15 Show that the linear tangent space to the graph of a function $f : \mathbb{R}^m \to \mathbb{R}^n$ at $(\mathbf{0}, f(\mathbf{0}))$ is the same as the linear tangent space to the graph of $y = f(\mathbf{x} - \mathbf{a})$ at the point where $\mathbf{x} = \mathbf{a}$.

6.16 Let Γ be the graph of some $f : \mathbb{R}^2 \to \mathbb{R}^2$. Suppose that the linear tangent space Γ at the point where $\mathbf{x} = \mathbf{0}$ is

$$\{ ((x, y), (2x - y, x + y)) \mid (x, y) \in \mathbb{R}^2 \}.$$

If $T : \mathbb{R}^2 \to \mathbb{R}^2$ is defined by $T(x, y) = (x^2 + 2x, y^2 + yx)$, find the linear tangent space to the graph of $g = f \cdot T$ at $(\mathbf{0}, g(\mathbf{0}))$.

6.17 Let Γ be the graph of $f : \mathbb{R}^m \to \mathbb{R}^n$. Let $\mathbf{a}, \mathbf{b}$ be in $\mathbb{R}^m$ such that $f'(\mathbf{b})^* f'(\mathbf{a}) = -I$. Show that the linear tangent space to Γ at $(\mathbf{a}, f(\mathbf{a}))$ and the linear tangent space to Γ at $(\mathbf{b}, f(\mathbf{b})$ are orthogonal subspaces of $\mathbb{R}^{m+n}$.

6.18 Find the affine tangent space to the surface

$$S = \{ (\mathbf{x}, x + e^y - z^2) \mid \mathbf{x} = (x, y, z) \in \mathbb{R}^3 \}$$

at the point $P((1, 0, 2), -2)$.

6.19 The linear tangent space to the graph of a function $f : \mathbb{R}^n \to \mathbb{R}$ at any point $(\mathbf{a}, f(\mathbf{a}))$ is a subspace of $\mathbb{R}^{n+1}$. Show that this subspace has dimension n.

6.20 If Γ is the graph of $f : \mathbb{R}^m \to \mathbb{R}^k$, is it true that the dimension of the linear tangent space to Γ at a point $(\mathbf{a}, f(\mathbf{a}))$ is always m?

6.3 MANIFOLDS IN PARAMETRIC REPRESENTATIONS

We would like to use graphs to investigate surfaces and curves and other similar structures in Euclidean spaces. Our definition of graphs is too restrictive for this purpose. The unit circle

$$\{\, (x,\, y) \in \mathbb{R}^2 \;\big|\; x^2 + y^2 = 1 \,\},$$

for example, is not a graph. In fact, there is no single equation that expresses one of these variables in terms of the others. One possibility is $y = \pm(1 - x^2)^{1/2}$. This is not a function, however, as it does not determine y uniquely. One may try $y = +(1 - x^2)^{1/2}$ to represent the upper half of the circle, but this does not give the points below the x-axis. Hence we must modify our definitions to include this and other important cases. Manifolds are defined for this purpose. They are the sets that agree with a graph in small neighborhoods.

Definition 6.3.1 Manifolds. A set M in a Euclidean space Z is called a *manifold* if for each $\mathbf{m} \in M$ there is an open set G and a graph Γ such that $\mathbf{m} \in G$ and $G \cap M = G \cap \Gamma$. If the graphs associated with the points of M are all of the same dimension k, then M is called a *k-dimensional manifold.* It is clear that a graph is also a manifold.

In a more explicit form, a set M in Z is a manifold if for each point $\mathbf{m} \in M$ there is an open set G, an orthogonal coordinate system $(X,\, Y)$ for Z, an open set $A \subset X$, and a $\mathcal{C}^1$ function $f : A \to Y$ with its graph Γ such that $\mathbf{m} \in G$ and $G \cap M = G \cap \Gamma$. If this can be done at each $\mathbf{m} \in M$ with a k-dimensional X, then M is called a k-dimensional manifold.

Example 6.3.2 Let M be a manifold in a Euclidean space Z. Let A be an open set in Z such that $B = M \cap A \neq \emptyset$. Then B is a manifold. To see this, let $\mathbf{m} \in B = M \cap A$. Then $\mathbf{m} \in M$, so there is some open set G and a graph Γ such that $\mathbf{m} \in G$ and $G \cap M = G \cap \Gamma$. Hence, $G_0 = A \cap G$ is an open set containing $\mathbf{m}$ and

$$G_0 \cap B = G_0 \cap (M \cap A) = (A \cap G) \cap (M \cap A) = (G \cap M) \cap A = \Gamma \cap A.$$

Since Γ is a graph and A is open, it is easy to verify that $\Gamma \cap A$ is also a graph. Hence, B is a manifold.

Example 6.3.3 A sphere is a manifold. Let

$$M_1 = \{\, (x,\, y,\, z) \in \mathbb{R}^3 \;\big|\; x^2 + y^2 + z^2 = 1,\; z > 0 \,\}$$

be the upper half of the unit sphere without the equator. Then M_1 is the graph of $z = +(1 - x^2 - y^2)^{1/2}$ defined on the open disc $x^2 + y^2 < 1$. The lower hemisphere is also a graph. To include the points on the equator, consider the graphs of $x = +(1 - y^2 - z^2)^{1/2}$ and $x = -(1 - y^2 - z^2)^{1/2}$ defined on $y^2 + z^2 < 1$. These four graphs still miss the two points $(0, \pm 1, 0)$. They can be covered by the graphs of $y = +(1 - x^2 - z^2)^{1/2}$ and $y = -(1 - x^2 - z^2)^{1/2}$. Hence we use six graphs to include all points of the unit sphere. Implicit representations of manifolds will provide an easier way of showing that a sphere is a manifold.

Definition 6.3.4 Curves and surfaces. A one-dimensional manifold is called a *curve* and an $(n-1)$-dimensional manifold in an n-dimensional space is called a *surface*. We see that a circle is a curve in $\mathbb{R}^2$ and a sphere is a surface in $\mathbb{R}^3$.

Example 6.3.5 Let $C = \{ (t, t^2, e^t, 1-t) \mid t \in \mathbb{R} \}$. Then C is a one-dimensional graph in $\mathbb{R}^4$. In particular, C is a curve in $\mathbb{R}^4$.

Let $S = \{ (x, y, z, x^2yz) \mid x > 0, y > 1, z > 2 \}$. Then S is a three-dimensional manifold in $\mathbb{R}^4$. Hence, S is a surface in $\mathbb{R}^4$.

Parametric Equations of Manifolds

An explicit equation for a graph is a function that defines this graph. In general, there are three ways to characterize a manifold: by graphs, by parametric equations, and by implicit representation. Having dealt with graphs, we now turn to parametric equations; implicit representations are discussed in the next section.

Parametric equations are familiar from cases such as the parametric equations of the unit circle: $x = \cos u,\ \ y = \sin u$. The omission of the domain of u is not accidental, as there are some difficulties to be resolved in connection with this domain. See the remarks in Remarks 6.3.10 below.

We will assume without loss of generality that all spaces considered below are Euclidean spaces even if this is not explicitly stated. In particular, U and V denote two subspaces of W and X and Y two subspaces of Z.

Lemma 6.3.6 *Let H be an open set in W and $\Phi : H \to Z$ a diffeomorphism. Let U be a subspace of W and $C = U \cap H$. Then each $\mathbf{c} \in C$ has neighborhood E in U such that $\Phi(E)$ is a graph in the coordinate system (X, Y), where $X = \Phi'(\mathbf{c})U$, and Y is any subspace of Z complementary to X. Also, there is an open set $G \subset Z$ such that $\Phi(E) = G \cap \Phi(C)$.*

Proof. Let $P : Z \to X$ and $Q : Z \to Y$ be the coordinate projections. Then $P \cdot \Phi : C \to X$ is a $\mathcal{C}^1$ function, as a composition of $\mathcal{C}^1$ functions. Also,

$$(P \cdot \Phi)'(\mathbf{c}) = P\Phi'(\mathbf{c}) : U \to X \tag{6.29}$$

by the chain rule. Note that $P\Phi'(\mathbf{c}) : U \to X$ is an isomorphism. In fact, $T = \Phi'(\mathbf{c}) : W \to Z$ is an isomorphism as the derivative of a diffeomorphism. Then the restriction of T to a subspace U is an isomorphism between U and $X = TU$. Therefore, the inverse function theorem gives an open set E in U such that $\mathbf{c} \in E \subset C$ and such that the restriction of $P \cdot \Phi : C \to X$ to E is a diffeomorphism $\eta : E \to X$. We have

$$\Phi(\mathbf{u}) = (P\Phi(\mathbf{u}),\, Q\Phi(\mathbf{u})) = (\eta(\mathbf{u}),\, Q\Phi(\mathbf{u})), \quad \mathbf{u} \in E.$$

Let $\eta(E) = A$ be the range of η and $\vartheta : A \to E$ the reverse diffeomorphism. If $\eta(\mathbf{u}) = \mathbf{x}$, then let $\mathbf{u} = \vartheta(\mathbf{x})$. Hence, $\Phi(\mathbf{u}) = (\mathbf{x},\, Q\Phi(\vartheta(\mathbf{x})))$ for all $\mathbf{u} \in E$. We see that $f = Q \cdot \Phi \cdot \vartheta : A \to Y$ is a composition of $\mathcal{C}^1$ functions. Hence, $f : A \to Y$ is also a $\mathcal{C}^1$ function. Therefore $\Phi(E)$ is the graph of this function in the coordinate system (X, Y). Also, $H \cap (E \times V)$ is an open set in H. Hence $G = \Phi(H \cap (E \times V))$ is an open set in Z containing $\mathbf{m} = \Phi(\mathbf{c})$. We see that $G \cap \Phi(C) = \Phi(E)$, since $U \cap H \cap (E \times V) = E$. □

Theorem 6.3.7 Parametric representations of manifolds. *Let H be an open set in W and $\Phi : H \to Z$ a diffeomorphism. Let U be a subspace of W and $C = U \cap H$. Then $\Phi(C) = M$ is a manifold in Z.*

Proof. Lemma 6.3.6 above shows that each $\mathbf{m} \in \Phi(C)$ has a neighborhood G in Z such that $G \cap \Phi(C)$ is a graph. Hence M is a manifold. □

Example 6.3.8 Let $f(\mathbf{x}) = (x^5 - y^5, x^5 + y^5)$ for all $\mathbf{x} = (x, y) \in \mathbb{R}^2$. Then

$$\det f'(\mathbf{x}) = \det \begin{bmatrix} 5x^4 & -5y^4 \\ 5x^4 & 5y^4 \end{bmatrix} = 50x^4y^4.$$

Let H be any open subset of $\mathbb{R}^2$ that does not contain any point $\mathbf{x}$ for which $x = 0$ or $y = 0$. Then $f'(\mathbf{x})$ is invertible for all $\mathbf{x} \in H$. Also, f is one-to-one on all of $\mathbb{R}^2$. In fact, $x^5 - y^5 = x_1^5 - y_1^5$ and $x^5 + y^5 = x_1^5 + y_1^5$ imply that $x^5 = x_1^5$. Then $x = x_1$ and hence, $y = y_1$. Therefore, $f : H \to \mathbb{R}^2$ is a diffeomorphism. Hence, by Theorem 6.3.7, $f(H)$ is a manifold in $\mathbb{R}^2$.

Definition 6.3.9 Parametric equations. Let C be an open set in a Euclidean space U. A function $\mathbf{z} = \varphi(\mathbf{u})$, $\mathbf{u} \in C$, is called a *parametric equation* for a manifold if $\varphi : C \to Z$ is the restriction of a diffeomorphism $\Phi : H \to Z$ to $C = H \cap U$. Here H is an open set in a space W that contains U as a subspace. The diffeomorphism $\Phi : H \to Z$ is referred to as an *underlying diffeomorphism*.

Remarks 6.3.10 A difficulty with parametric equations. If $\mathbf{z} = \varphi(\mathbf{u})$, $\mathbf{u} \in C$, is a parametric equation, then we see that $\varphi : C \to Z$ is a $\mathcal{C}^1$ function and its derivative $\varphi'(\mathbf{c}) : U \to Z$ is a one-to-one linear map at every point $\mathbf{c} \in C$. The converse is not true. A function with these properties is not necessarily a parametric equation. Let $\varphi : C \to Z$ be such a function. It may not be easy to find out if $\mathbf{z} = \varphi(\mathbf{u})$, $\mathbf{u} \in C$, is a parametric equation. Fortunately, this is not too important for our purposes. In differential calculus one is interested in the behavior of a function in the neighborhoods of a point. In this case, Theorem 6.3.12 shows that any point $\mathbf{c} \in C$ has a neighborhood $C_1 \subset C$ such that φ restricted to C_1 is a parametric equation $\mathbf{z} = \varphi(\mathbf{u})$, $\mathbf{u} \in C_1$. Hence $\varphi(C_1)$ is a manifold even though $\varphi(C)$ might not be a manifold. To refer to this type of case, it is useful to define local parametric equations.

Definition 6.3.11 Local parametric equations. Let C be an open set in a Euclidean space U. A function $\mathbf{z} = \varphi(\mathbf{u})$, $\mathbf{u} \in C$, is called a *local parametric equation* if every $\mathbf{c} \in C$ has a neighborhood $C_1 \subset C$ such that $\mathbf{z} = \varphi(\mathbf{u})$, $\mathbf{u} \in C_1$, is a parametric equation for a manifold M_1. There may not be a single manifold M that contains all such M_1 as subsets. The following is an important result.

Theorem 6.3.12 *Let C be an open set in U and $\varphi : C \to Z$ a $\mathcal{C}^1$ function such that $\varphi'(\mathbf{c}) : U \to Z$ is a one-to-one linear map at every point $\mathbf{c} \in C$. Then $\mathbf{z} = \varphi(\mathbf{u})$, $\mathbf{z} \in C$, is a local parametric equation.*

Proof. Let V be a space with $\dim V = \dim Z - \dim U$ and $W = U \times V$. Let $\mathbf{c} \in C$ and $X = \varphi'(\mathbf{c})U$. Since $\varphi'(\mathbf{c}) : U \to Z$ one-to-one, we see that $\dim X = \dim U$. Let $Y = X^{\perp}$. Then $\dim Y = \dim V$. Let $S : V \to Y$ be an isomorphism, and as usual let P and Q be orthogonal projections onto X and Y. For each $(\mathbf{u}, \mathbf{v}) \in C \times V \subset W$, let

$$\Phi(\mathbf{u}, \mathbf{v}) = (P\varphi(\mathbf{u}), Q\varphi(\mathbf{u}) + S\mathbf{v}) \in X \times Y = Z.$$

This defines a $\mathcal{C}^1$ function $\Phi : H \to Z$, where $H = C \times V$ is an open set in W. Its derivative $\Phi'(\mathbf{c}, \mathbf{0}) : W \to Z$ at $(\mathbf{c}, \mathbf{0}) \in H$ is given as

$$\begin{aligned} \Phi'(\mathbf{c}, \mathbf{0})(\mathbf{u}, \mathbf{v}) &= (P\varphi'(\mathbf{c})\mathbf{u}, Q\varphi'(\mathbf{c})(\mathbf{u}) + S\mathbf{v}) \\ &= (P\varphi'(\mathbf{c})\mathbf{u}, S\mathbf{v}), \quad \text{for all } (\mathbf{u}, \mathbf{v}) \in W. \end{aligned}$$

To obtain the second equality, note that $\varphi'(\mathbf{c})\mathbf{u} \in X$ by the definition of X. Hence $Q\varphi'(\mathbf{c})(\mathbf{u}) = \mathbf{0}$. We see that $\Phi'(\mathbf{c}, \mathbf{0}) : W \to Z$ is invertible. Hence, by the inverse function theorem, there is an open set $H_1 \subset W$ such that $(\mathbf{c}, \mathbf{0}) \in H_1 \subset H$ and such that the restriction of Φ to H_1 is a diffeomorphism $\Phi_1 : H \to Z$. Let $C_1 = U \cap H_1$. Then, by Theorem 6.3.7, $\varphi : C_1 \to Z$ is a parametric equation for the manifold $M_1 = \Phi_1(C_1)$ $\square$

Example 6.3.13 Let $C = (-2,\ 2 + (3/2)\pi) \subset \mathbb{R}$ and let $Z = \mathbb{R}^2$ be the xy-plane. Define $\varphi : C \to Z$ by

$$\varphi(u) = \begin{cases} (u,\ 0) & \text{if } -2 < u < 0, \\ (\sin u,\ 1 - \cos u) & \text{if } 0 \leq u < (3/2)\pi, \\ (-1,\ 1 + (3/2)\pi - u) & \text{if } (3/2)\pi \leq u < 2 + (3/2)\pi. \end{cases}$$

Then $\varphi : C \to \mathbb{R}^2$ is a $\mathcal{C}^1$ function with a one-to-one derivative at every point. Hence, φ is a local parametric equation. But $\varphi(C)$ is not a manifold and φ is not the parametric equation of any manifold. This is clear from Figure 6.4, as $\varphi(C)$ cannot be a manifold in any neighborhood of $(-1,\ 0)$. One may think that this is due to the fact that φ is a not a one-to-one function. Also consider $C_0 = (-2,\ 1+(3/2)\pi) \subset \mathbb{R}$ and

$$\varphi_0(u) = \begin{cases} (u,\ 0) & \text{if } -2 < u < 0, \\ (\sin u,\ 1 - \cos u) & \text{if } 0 \leq u < (3/2)\pi, \\ (-1,\ 1 + (3/2)\pi - u) & \text{if } (3/2)\pi \leq a < 1 + (3/2)\pi. \end{cases}$$

We see that $\varphi_0 : C_0 \to \mathbb{R}^2$ is a one-to-one local parametric equation but still is not a parametric equation. Problem 6.27 gives a sufficient condition for a local parametric equation to be also a parametric equation.

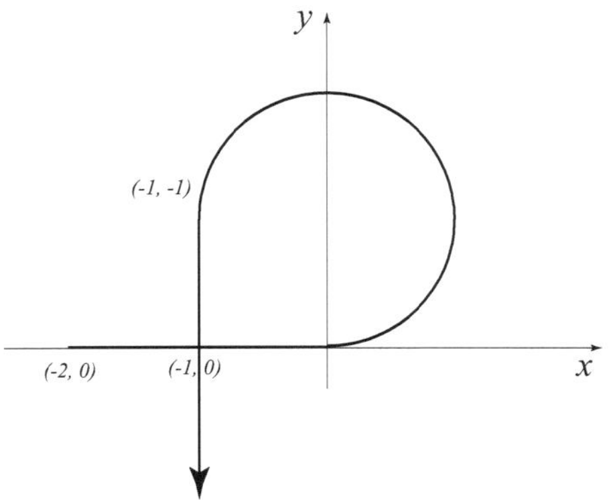

Figure 6.4. Local parametric equations in Example 6.3.13.

Example 6.3.14 Let $g : C \to \mathbb{R}^3$ be defined by $g(\mathbf{u}) = (u^2 + vw, v^4 - w, w^3)$, for all $\mathbf{u} = (u, v, w) \in C$, where

$$C = \left\{ \mathbf{u} \in \mathbb{R}^3 \mid u \neq 0, v \neq 0, w \neq 0 \right\}.$$

Then $\mathbf{z} = g(\mathbf{u}), \mathbf{u} \in C$ is a local parametric equation.

Example 6.3.15 Explicit and parametric equation. Let Γ be the graph of a $\mathcal{C}^1$ function $f : A \to Y$ in the coordinate system (X, Y). Then $\Gamma = M$ is a manifold. The explicit equation $\mathbf{y} = f(\mathbf{x})$ for Γ defines a natural parametric equation for M. In fact, let $H = A \times Y$ and let $\Phi : H \to Z$ be the graph diffeomorphism induced by φ, as defined in Definition 6.1.9. Hence

$$\Phi(\mathbf{z}) = \Phi(\mathbf{x}, \mathbf{y}) = (\mathbf{x}, \mathbf{y} + f(\mathbf{x})), \quad \mathbf{z} = (\mathbf{x}, \mathbf{y}) \in H.$$

The restriction of Φ to $A = H \cap X$ is a parametric equation

$$\varphi(\mathbf{x}) = (\mathbf{x}, f(\mathbf{x})), \mathbf{x} \in A, \text{ for } \Gamma.$$

Compatibility of Two Parametric Representations

Many concepts and operations related to manifolds are defined in terms of parametric equations. Clearly, we must show that such definitions are independent of the parametric equation used. Hence we develop some notations and relations about different parametric representations that agree on a part of a manifold.

Notations 6.3.16 Different parametric equations. Let H_i be open sets in W_i, $i = 1, 2$, and $\Phi_i : H_i \to Z$ diffeomorphisms, $G_i = \Phi_i(H_i)$. Let U_i be a subspace in W_i, and $C_i = H_i \cap U_i$. Assume that $\Phi(C_1) \cap \Phi_2(C_2) = M$ is not empty. Then $G_1 \cap G_2$ is not empty either, as it contains M. Without loss of generality, we will assume that $G_1 = G_2$. If originally this is not the case, then we let $G = G_1 \cap G_2$ and replace H_i by $H_i^1 = \Phi_i^{-1}(G) \subset H_i$ and Φ_i by the restriction of Φ_i to H_i^1. Hence we assume that

$$\Phi_i : H_i \to Z \text{ are two diffeomorphisms;} \tag{6.30}$$
$$\Phi_1(H_1) = \Phi_2(H_2) = G; \text{ and} \tag{6.31}$$
$$\Phi_1(C_1) = \Phi_2(C_2) = M, \text{ where} \tag{6.32}$$
$$C_i = U_i \cap H_i. \tag{6.33}$$

Definition 6.3.17 Equivalent parametric equations. Let $\varphi_i : C_i \to Z$ be two parametric equations. Let $\Phi_i : H_i \to Z$ be the underlying diffeomorphisms. Call φ_i s equivalent if Φ_i s are equivalent in the sense that they satisfy the conditions (6.30) - (6.33) in Notations 6.3.16. Arguments above in Notations 6.3.16 show that if $\varphi_1(C_1) \cap \varphi_2(C_2) = M$ is not empty, then there is no loss of generality in assuming that they are equivalent.

Lemma 6.3.18 Diffeomorphisms for equivalent parametric equations. *Let $\varphi_i : C_i \to Z$ be two equivalent parametric equations. Then there is a diffeomorphism $\vartheta : C_1 \to C_2$ such that $\varphi_1 = \varphi_2 \cdot \vartheta$.*

Proof. Use the underlying diffeomorphisms $\Phi_i : H_i \to G$ to define

$$\Theta = \Phi_2^{-1} \cdot \Phi_1 : H_1 \to H_2.$$

We see that $\Theta(C_1) = C_2$ and that $\Phi_1'(\mathbf{c}_1)U_1 = U_2$ for all $\mathbf{c}_1 \in C_1$. Let

$$\vartheta = \Theta|_{C_1} : C_1 \to H_2$$

be the restriction of Θ to C_1, We verify easily that $\vartheta(C_1) = C_2$ and $\vartheta : C_1 \to C_2$ is a diffeomorphism. It is clear that $\varphi_1 = \varphi_2 \cdot \vartheta$. □

Remarks 6.3.19 One can define $\vartheta : C_1 \to C_2$ directly as $\vartheta(\mathbf{u}_1) = \varphi_2^{-1}(\varphi_1(\mathbf{u}_1))$, without any reference to the underlying diffeomorphisms. This expression is defined since $\varphi_i : C_i \to M$ are both one-to-one functions that map C_i onto M. It is not clear, however, if $\vartheta : C_1 \to C_2$ is a diffeomorphism of C_1 onto C_2.

Lemma 6.3.20 *Let $\varphi_i : C_i \to Z$ be two equivalent parametric equations. If $\varphi_1(\mathbf{c}_1) = \varphi_2(\mathbf{c}_2) = \mathbf{m}$, then $\varphi_1'(\mathbf{c}_1)U_1 = \varphi_2'(\mathbf{c}_2)U_2$.*

Proof. Let $\vartheta : C_1 \to C_2$ be the diffeomorphism obtained in Lemma 6.3.18. Then $\varphi(\mathbf{c}_1) = \varphi_2(\vartheta(\mathbf{c}_1)) = \mathbf{m}$ shows that $\vartheta(\mathbf{c}_1) = \mathbf{c}_2$. Therefore

$$\varphi'(\mathbf{c}_1)U_1 = \varphi_2'(\mathbf{c}_2)\vartheta'(\mathbf{c}_1)U_1 = \varphi_2'(\mathbf{c}_2)U_2.$$

The first step follows from the chain rule. To obtain the second step, one observes that $\vartheta'(\mathbf{c}_1)U_1 = U_2$, since $\vartheta'(\mathbf{c}_1) : U_1 \to U_2$ is invertible. □

Remarks 6.3.21 A diffeomorphism for equivalent equations. In the proof of Lemma 6.3.20 above, one essentially uses only $\vartheta = \Theta|_{C_1} : C_1 \to C_2$. This can be defined without any reference to the underlying diffeomorphisms, as $\vartheta(\mathbf{u}_1) = \varphi_2^{-1}(\varphi_1(\mathbf{c}_1))$. This is defined since both $\varphi_i : C_i \to M$ are one-to-one functions that map C_i onto M. Hence, $\mathbf{m} = \varphi_1(\mathbf{u}_1) \in M$ and $\varphi_2^{-1}(\mathbf{m}) \in C_2$ exists. It is not clear, however, if $\vartheta : C_1 \to C_2$ is a diffeomorphism of C_1 onto C_2. By referring to the diffeomorphisms Φ_i, we see that this is indeed the case.

Tangent Spaces of Manifolds

Definition 6.3.22 Tangent spaces. Let M be a manifold in Z and $\mathbf{m} \in M$. Let $\varphi : C \to Z$, $C \subset U$, be a parametric equation for M with $\mathbf{m} \in \varphi(C)$. Then the linear tangent space of M at $\mathbf{m}$ is the subspace $T(\mathbf{m}) = \varphi'(\mathbf{c})U$ in Z. Note that if $\Phi : H \to Z$ is the underlying diffeomorphism for $\varphi : C \to Z$, then also $\Phi'(\mathbf{c})U = T(\mathbf{m})$. A parametric equation for $T(\mathbf{m})$ is $\mathbf{z} = \varphi'(\mathbf{c})\mathbf{u}$, $\mathbf{u} \in U$. Lemma 6.3.20 shows that $T(\mathbf{m})$ is independent of the parametric equation used.

Remarks 6.3.23 Agreement with the earlier definition. The tangent space of a graph was defined earlier in Definition 6.2.2. In fact, if a graph Γ is given by the explicit equation $\mathbf{y} = f(\mathbf{x})$, $\mathbf{x} \in A$, then

$$\mathbf{z} = \varphi(\mathbf{x}) = (\mathbf{x},\, f(\mathbf{x})), \quad \mathbf{x} \in A$$

is a parametric equation for Γ, as observed in Example 6.3.15. Then the tangent space $T(\mathbf{m})$ at $\mathbf{m} = \varphi(\mathbf{a})$ has the parametric equations

$$\mathbf{z} = \varphi'(\mathbf{a})\mathbf{x} = (\mathbf{x},\, f'(\mathbf{a})\mathbf{x}),\ \mathbf{x} \in X.$$

Therefore an explicit equation for $T(\mathbf{m})$ is $\mathbf{y} = f'(\mathbf{a})\mathbf{x}$.

Definition 6.3.24 Normal spaces. Let M be a manifold in Z and $\mathbf{m} \in M$. The *normal space* $N(\mathbf{m})$ at $\mathbf{m}$ is the orthogonal complement of the tangent space $T(\mathbf{m})$. Hence $N(\mathbf{m}) = T(\mathbf{m})^{\perp}$.

Some Properties of Tangent Spaces

Theorem 6.3.25 Tangent spaces and graphs. *Let M be a manifold and let $X = T(\mathbf{m})$ be the tangent space of M at $\mathbf{m} \in M$. Then $\mathbf{m}$ has a neighborhood G such that $M \cap G$ is a graph in the coordinate system $(X,\, Y)$, where Y is any subspace of Z complementary to X.*

Proof. Let $\Phi : H \to Z$ be a diffeomorphism such that

$$\mathbf{m} = \Phi(\mathbf{c}) \in \Phi(H \cap U) \subset M.$$

Lemma 6.3.6 shows that $\mathbf{m} \in M$ has a neighborhood G in Z such that $M \cap G$ is a graph in $(X,\, Y)$ where $X = \Phi'(\mathbf{c})U = T(\mathbf{m})$, and Y is any complementary space to X. □

Definition 6.3.26 Curves and their tangent vectors. Let Z be a vector space and I an open interval. A *curve* C in Z is a continuously differentiable function $\mathbf{r} : I \to Z$ with a nonzero derivative at every point $t \in I$. If M is a manifold in Z and if $\mathbf{r}(I) \subset M$, then C is a *curve on the manifold* M. If $\mathbf{z} \in Z$ and if $\mathbf{r}(t) = \mathbf{z}$ for some $t \in I$, then C is *a curve passing through* $\mathbf{z}$. The vector

$$\mathbf{r}'(t) = \lim_{s \to 0}(1/s)(\mathbf{r}(t+s) - \mathbf{r}(t)) \in Z$$

is the *tangent vector of* C at the point $\mathbf{r}(t) = \mathbf{z}$.

Theorem 6.3.27 Tangent vectors of curves on a manifold. *The tangent space* $T(\mathbf{m})$ *of a manifold* M *is the set of tangent vectors of the curves on* M *passing through* $\mathbf{m}$. *More explicitly,* $\mathbf{p} \in T(\mathbf{m})$ *if and only if there is a curve* $\mathbf{r} : I \longrightarrow M$ *on* M *and a* $t \in I$ *such that* $\mathbf{r}(t) = \mathbf{m}$ *and* $\mathbf{r}'(t) = \mathbf{p}$.

Proof. Let $\varphi : C \longrightarrow Z$ be a parametric equation with the underlying diffeomorphism $\Phi : H \longrightarrow Z$, $C = H \cap U$, and the inverse diffeomorphism $\Psi : G \longrightarrow W$, where $G = \Phi(H)$. Let $\mathbf{m} = \varphi(\mathbf{c}) \in M$. Let $\mathbf{z} = \mathbf{r}(t)$, $t \in I$, be a curve on M passing through $\mathbf{m} = \mathbf{r}(t_0)$. Then $\mathbf{s}(t) = \Psi(\mathbf{r}(t))$ is a curve in $C \subset U$ passing through $\mathbf{c} = \Psi(\mathbf{r}(t_0)) \in C$. Then $\mathbf{s}'(t_0) \in U$ since $\mathbf{s}(t) \in U$ for all $t \in I$, and (since $\varphi \cdot \mathbf{s}(t) = \mathbf{r}(t)$ for $t \in I$)

$$\mathbf{r}'(t_0) = (\varphi \cdot \mathbf{s})'(t_0)) = \varphi'(\mathbf{s}(t_0))\mathbf{s}'(t_0) = \varphi'(\mathbf{c})\mathbf{s}'(t_0) \in T(\mathbf{m}).$$

Conversely, if $\mathbf{p} \in T(\mathbf{m})$, then $\mathbf{q} = \Psi'(\mathbf{m})\mathbf{p} \in U$. Hence

$$\mathbf{s}(t) = t\mathbf{q},\, t \in I = (-\varepsilon, \varepsilon),$$

is a curve in C for some $\varepsilon > 0$. Then $\mathbf{r}(t) = \varphi(\mathbf{s}(t))$, $t \in I$, is a curve on M passing through $\mathbf{m} = \mathbf{r}(0)$. Also,

$$\mathbf{r}'(0) = (\varphi \cdot \mathbf{s})'(0) = \varphi'(\mathbf{c})\mathbf{s}'(0) = \varphi'(\mathbf{c})\mathbf{q} = \mathbf{p}. \quad \square$$

Problems

6.21 Let $M = \{\, (\mathbf{x}, (x^2 + y - z, y^2 + x + z)) \mid \mathbf{x} = (x, y, z) \in \mathbb{R}^3 \,\}$. Is M a manifold in $\mathbb{R}^5$? If it is, find a parametric equation for M and an underlying diffeomorphism.

6.22 Let $k \neq 0$ be a real number. Let $C = \{\, \mathbf{x} = (x, y) \in \mathbb{R}^2 \mid xy \neq 0 \,\}$. Define $f : C \longrightarrow \mathbb{R}^3$ by $f(\mathbf{x}) = (xy, x^2 + y^2, kx)$ for all $\mathbf{x} \in \mathbb{R}^2$. Show that f is a local parametric equation for a manifold.

6.23 Let $M \subset \mathbb{R}^n$ be an r-manifold and let $N \subset \mathbb{R}^m$ be an s-manifold. Show that $M \times N \subset \mathbb{R}^m \times \mathbb{R}^n = \mathbb{R}^{m+n}$ is an $(r + s)$-manifold.

6.24 Show that a compact manifold cannot be represented by a (single) parametric equation.

6.25 Let C be an open subset of U and $\varphi : C \longrightarrow Z$ a one-to-one $\mathcal{C}^1$ function with a one-to-one derivative at every point. Example 6.3.13 shows that $\varphi(C)$ does not have to be a manifold. Let C_0 be an open subset of C with closure $\overline{C_0} \subset C$. Show that $\varphi(C_0)$ is a manifold.

6.26 Let C be an open subset of U and $\varphi : C \to Z$ a one-to-one $\mathcal{C}^1$ function with a one-to-one derivative at every point. Let $\mathbf{a} \in C$ and $\mathbf{u} \in U$. Let $L \subset \mathbb{R}$ be the set of $t \in \mathbb{R}$ such that $\mathbf{a} + t\mathbf{u} \in C$. Define $\alpha : L \to Z$ by $\alpha(t) = \varphi(\mathbf{a} + t\mathbf{u})$, $t \in L$. Show that if $\varphi(C)$ is a manifold in Z then $\alpha(L)$ is also a manifold in Z.

6.27 Let C be an open subset of U and $\varphi : C \to Z$ a one-to-one $\mathcal{C}^1$ function with a one-to-one derivative at every point. Assume that the inverse function $\psi : \varphi(C) \to C$ is continuous. Show that $\varphi(C)$ is a manifold in Z.

6.4 MANIFOLDS IN IMPLICIT REPRESENTATIONS

Implicit representations of manifolds may be considered as the reverse of parametric representations. In the proof of Theorem 6.4.1 below, the inverse function theorem is used to reduce implicit representation to parametric representations.

Theorem 6.4.1 Implicit representations of manifolds. *Let* $\dim V < \dim Z$. *Let* E *be an open set in* Z *and let* $F : E \to V$ *be a* $\mathcal{C}^1$ *function. Then*

$$M = \{\, \mathbf{z} \mid \mathbf{z} \in E,\ \ F(\mathbf{z}) = \mathbf{0},\ \ F'(\mathbf{z})Z = V \,\}$$

is a manifold in Z. *The tangent space of* M *at* $\mathbf{m} \in M$ *is*

$$T(\mathbf{m}) = \operatorname{Ker} F'(\mathbf{m}).$$

Proof. Let $\mathbf{m} \in M$. Let $X = \operatorname{Ker} F'(\mathbf{m})$ and $Y = X^{\perp}$. Let $P : Z \to X$ and $Q : Z \to Y$ be the coordinate projections. We see that $\dim V = \dim Y$, since $F'(\mathbf{m}) : Z \to V$is onto V. Let U be a space with $\dim U = \dim X$ and $S : X \to U$ an isomorphism. Define $\vartheta : E \to W = U \times V$ by

$$\begin{aligned} \vartheta(\mathbf{z}) &= (SP\mathbf{z},\ F(\mathbf{z})) = (S\mathbf{x},\ F(\mathbf{x},\ \mathbf{y})).\ \text{Hence} \\ \vartheta'(\mathbf{m})(\mathbf{x},\ \mathbf{y}) &= (S\mathbf{x},\ F'(\mathbf{m})(\mathbf{x},\ \mathbf{y})) = (S\mathbf{x},\ F'(\mathbf{m})\mathbf{y}). \end{aligned}$$

Then $\vartheta'(\mathbf{m}) : Z \to W$ is an isomorphism since $SX = U$ and $F'(\mathbf{m})Y = V$. Also, $\vartheta : E \to W$ is a $\mathcal{C}^1$ function because both components are $\mathcal{C}^1$ functions. The inverse function theorem shows that there is an open set G in Z such that $\mathbf{m} \in G \subset E$ and such that the restriction of ϑ to G is a diffeomorphism

$$\vartheta|_G = \Psi : G \to W.$$

Let $\Psi(G) = H$. Note that $\Psi : G \to W$ maps $M \cap G$ to $U \cap H$, since $F(\mathbf{z}) = \mathbf{0}$ for $\mathbf{z} \in M$. Hence the reverse diffeomorphism $\Phi : H \to G$ maps $C = U \cap H$ to $M \cap G$. Therefore $M \cap G$ is a manifold with a parametric equation $\Phi|_C = \varphi : C \to Z$. Then

M is a manifold by Definition 6.3.1. We see that Φ maps U to $X = \text{Ker } F'(\mathbf{m})$. Hence X is the tangent space of M at $\mathbf{m} \in M$. □

Example 6.4.2 Let $E = \{\, \mathbf{x} = (x, y, u, v) \in \mathbb{R}^4 \mid x \neq v \,\}$. Let M be the set of all $\mathbf{x} \in E$ such that

$$\begin{aligned} x &= yu - v \\ u &= xv. \end{aligned}$$

Then M is a manifold in $\mathbb{R}^4$. To see this, define

$$F(\mathbf{x}) = (x - yu + v, xv - u) \quad \text{for all } \mathbf{x} = (x, y, u, v) \in E.$$

Then

$$F'(\mathbf{x}) = \begin{bmatrix} 1 & -u & -y & 1 \\ v & 0 & -1 & x \end{bmatrix}.$$

Since $x \neq v$, it follows that the matrix for $F'(\mathbf{x})$ has two linearly independent vectors. Hence, $F'(\mathbf{x}) : \mathbb{R}^4 \to \mathbb{R}^2$ is onto for all $\mathbf{x} \in E$. So, by Theorem 6.4.1,

$$M = \{\, \mathbf{x} \in E \mid F(\mathbf{x}) = \mathbf{0} \,\} = \left\{\, \mathbf{x} \in E \;\middle|\; F(\mathbf{x}) = \mathbf{0}, \quad F'(\mathbf{x})\mathbb{R}^2 = \mathbb{R}^2 \,\right\}$$

is a manifold in $\mathbb{R}^4$.

Also, the tangent space to M at $\mathbf{m} = (0, 1, 1, 1) \in M$ is Ker $F'(\mathbf{m})$, which is the set of all $(a, b, c, d) \in \mathbb{R}^4$ such that $F'(0, 1, 1, 1)(a, b, c, d) = \mathbf{0}$. That is, it is the set of all solutions (a, b, c, d) to the system

$$\begin{aligned} a - b - c + d &= 0 \\ a - c &= 0. \end{aligned}$$

Hence,

$$T'(0, 1, 1, 1) = \{\, (t, s, t, s) \mid t \in \mathbb{R}, s \in \mathbb{R} \,\}.$$

Definition 6.4.3 Implicit equations. Let E be a set and V a vector space. Let $F : E \to V$ be a function. Then $F(e) = \mathbf{0}$ is called the *implicit equation* of the set $S = F^{-1}(\{\, \mathbf{0} \,\})$. Hence $e \in S$ if and only if $F(e) = \mathbf{0}$.

Theorem 6.4.4 Implicit equations of manifolds. *Let E be an open set in Z and $F : E \to V$ a $\mathcal{C}^1$ function. Assume that $F'(\mathbf{z})Z = V$ whenever $F(\mathbf{z}) = \mathbf{0}$. Then $F(\mathbf{z}) = \mathbf{0}$ is an implicit equation of a manifold $M \subset E$.*

Proof. This is a reformulation of the first part of Theorem 6.4.1. □

Theorem 6.4.5 Implicit equations of tangent spaces. *Let $F(\mathbf{z}) = 0$ be the implicit equation of a manifold M. Then $F'(\mathbf{m})\mathbf{z} = \mathbf{0}$ is an implicit equation of the tangent space of M at $\mathbf{m} \in M$.*

Proof. This is a reformulation of the second part of Theorem 6.4.1. □

Another version of these results is known as the *implicit function theorem.*

Theorem 6.4.6 Implicit function theorem. *Let $E \subset Z = X \times Y$ be an open set in Z. Let $(\mathbf{a}, \mathbf{b}) \in E$. Let V be a space with $\dim V = \dim Y$. Let $F : E \to V$ be a $\mathcal{C}^1$ function with $F(\mathbf{a}, \mathbf{b}) = \mathbf{0}$ and $F'(\mathbf{a}, \mathbf{b})Y = V$. Then there is an open set A in X and a $\mathcal{C}^1$ function $f : A \to Y$ such that $\mathbf{a} \in A$, $f(\mathbf{a}) = \mathbf{b}$, and $F(\mathbf{x}, f(\mathbf{x})) = \mathbf{0}$ for all $\mathbf{x} \in A$.*

Proof. Theorems 6.4.4 and 6.4.5 show that $F(\mathbf{x}, \mathbf{y}) = \mathbf{0}$ is an implicit equation for a manifold M and that the tangent space of M at $(\mathbf{a}, \mathbf{b}) \in M$ is

$$\widetilde{X} = T(\mathbf{m}) = \operatorname{Ker} F'(\mathbf{a}, \mathbf{b}).$$

Define $\psi : E \to L(Y, V)$ by $\psi(\mathbf{e}) = F'(\mathbf{e})|_Y$, the linear map $F'(\mathbf{e})$ restricted to Y. That is, $\psi(\mathbf{e}) : Y \to V$ is defined by $\psi(\mathbf{e})(\mathbf{y}) = F'(\mathbf{e})(\mathbf{y})$ for all $\mathbf{y} \in Y$. Since F is a $\mathcal{C}^1$ function, ψ is continuous. Also, $\psi(\mathbf{a}, \mathbf{b})$ is invertible because $\dim Y = \dim V$ and $\psi(\mathbf{a}, \mathbf{b})Y = F'(\mathbf{a}, \mathbf{b})Y = V$. Because inversion is continuous, it follows that there is an open set U containing $(\mathbf{a}, \mathbf{b})$ with $U \subset G$ and such that $\psi(\mathbf{z})$ is invertible for all $\mathbf{z} \in U$. Thus, $F'(\mathbf{z})Y = V$ for all $\mathbf{z} \in U$. This implies that Y is complementary to $\widetilde{X}$. Then Theorem 6.3.25 shows that $(\mathbf{a}, \mathbf{b})$ has a neighborhood G such that $M \cap G$ is a graph in $(\widetilde{X}, Y)$. But X is also complementary to Y since (X, Y) is a coordinate system in Z. Then, we see easily that $M \cap G$ is also a graph in (X, Y). (cf. Problem 3.42). This is the conclusion of the theorem. □

Example 6.4.7 Let $F : \mathbb{R}^2 \times \mathbb{R}^2 \to \mathbb{R}^2$ be defined by

$$F(\mathbf{x}, \mathbf{u}) = (xv + yu - 1, xy - uv) \quad \text{for all } \mathbf{x} = (x, y) \in \mathbb{R}^2, \mathbf{u} = (u, v) \in \mathbb{R}^2.$$

Then

$$F'(x, y, u, v) = \begin{bmatrix} v & u & y & x \\ y & x & -v & -u \end{bmatrix}.$$

Let $\mathbf{c} = (1, 0, 0, 1) \in \mathbb{R}^2 \times \mathbb{R}^2$. Then $F(\mathbf{c}) = (0, 0)$ and

$$F'(\mathbf{c}) = \begin{bmatrix} 1 & 0 & 0 & 1 \\ 0 & 1 & -1 & 0 \end{bmatrix}.$$

Hence,

$$F'(\mathbf{c})(\{ (0, 0, u, v) \mid (u, v) \in \mathbb{R}^2 \} = \{ (v, -u) \mid (u, v) \in \mathbb{R}^2 \} = \mathbb{R}^2.$$

In the notation of Theorem 6.4.6

$$F'(\mathbf{c})(Y) = V.$$

Hence, by that theorem, there is an open set A in $\mathbb{R}^2$ containing $(1, 0)$ and a continuously differentiable function $f : A \to \mathbb{R}^2$ such that

$$F(\mathbf{x}, f(\mathbf{x})) = \mathbf{0} \quad \text{for all } \mathbf{x} \in A.$$

That is, the equation $F(\mathbf{x}, \mathbf{u}) = \mathbf{0}$ defines $\mathbf{u}$ implicitly as a differentiable function of $\mathbf{x}$ in the neighborhood A of $(1, 0)$. In particular, the system

$$\begin{aligned} xv + yu - 1 &= 0 \\ xy - uv &= 0 \end{aligned}$$

can be solved for (u, v) as a differentiable function of x and y for (x, y) in a neighborhood of $(1, 0)$.

Example 6.4.8 Let $E \subset \mathbb{R}^{n+m}$ be open and let $F : E \to \mathbb{R}^m$ have continuously differentiable component functions $F_1, \ldots, F_m$. Suppose that we denote a point $\mathbf{x} \in \mathbb{R}^n$ by $(x_1, \ldots, x_n)$ and a point $\mathbf{y} \in \mathbb{R}^m$ by $(y_1, \ldots, y_m)$ and a point $\mathbf{z} \in \mathbb{R}^{n+m}$ by $\mathbf{z} = (\mathbf{a}, \mathbf{b})$, where $\mathbf{a} \in \mathbb{R}^n, \mathbf{b} \in \mathbb{R}^m$. Assume that $(\mathbf{a}, \mathbf{b}) \in E$. If $F(\mathbf{a}, \mathbf{b}) = \mathbf{0}$ and

$$\det \begin{bmatrix} \dfrac{\partial F_1}{\partial y_1}(\mathbf{a}, \mathbf{b}) & \cdots & \dfrac{\partial F_1}{\partial y_m}(\mathbf{a}, \mathbf{b}) \\ \vdots & \cdots & \vdots \\ \dfrac{\partial F_m}{\partial y_1}(\mathbf{a}, \mathbf{b}) & \cdots & \dfrac{\partial F_m}{\partial y_m}(\mathbf{a}, \mathbf{b}) \end{bmatrix} \neq 0, \tag{6.34}$$

then the conclusion of Theorem 6.4.6 holds: there is an open set A in $\mathbb{R}^n$ and a $\mathcal{C}^1$ function $f : A \to \mathbb{R}^m$ such that $\mathbf{a} \in A$, $f(\mathbf{a}) = \mathbf{b}$, and $F(\mathbf{x}, f(\mathbf{x})) = \mathbf{0}$ for all $\mathbf{x} \in A$. This is because the last m columns of $F'(\mathbf{a}, \mathbf{b})$ are

$$\begin{bmatrix} \dfrac{\partial F_1}{\partial y_1}(\mathbf{a}, \mathbf{b}) \\ \vdots \\ \dfrac{\partial F_m}{\partial y_1}(\mathbf{a}, \mathbf{b}) \end{bmatrix}, \ldots, \begin{bmatrix} \dfrac{\partial F_1}{\partial y_m}(\mathbf{a}, \mathbf{b}) \\ \vdots \\ \dfrac{\partial F_m}{\partial y_m}(\mathbf{a}, \mathbf{b}) \end{bmatrix}$$

and (6.34) implies that these columns form a linearly independent set of m vectors, whence $F'(\mathbf{a}, \mathbf{b})\mathbb{R}^m = \mathbb{R}^m$.

As an illustration, consider the system

$$\begin{aligned} x^2u + yv &= 7 \\ xu^3 - y^2v &= 11. \end{aligned}$$

Let $f_1(\mathbf{x}, \mathbf{u}) = x^2u + yv - 7, f_2(\mathbf{x}, \mathbf{u}) = xu^3 - y^2v - 11$ for all $\mathbf{x} = (x, y), \mathbf{u} = (u, v)$. Then at a point (x, y, u, v), we have

$$\det \begin{bmatrix} \partial f_1/\partial u & \partial f_1/\partial v \\ \partial f_2/\partial u & \partial f_2/\partial v \end{bmatrix} = \det \begin{bmatrix} x^2 & y \\ 3u^2x & -y^2 \end{bmatrix} = -x^2y^2 - 3u^2xy.$$

Thus, at $(1, 1, -1, -1)$, the above determinant is $-4 \neq 0$. So, whenever (u, v) is near the point $(-1, -1)$, it can be solved uniquely in terms of x, y in a neighborhood of $(1, 1)$.

Example 6.4.9 The conclusion of the implicit function theorem may hold even if the hypotheses of the theorem are not satisfied. Let $F(x, y, z) = (x^4 - 16z^4, y - 2z)$ for all $(x, y, z) \in \mathbb{R}^3$ so that $F : \mathbb{R} \times \mathbb{R}^2 \to \mathbb{R}^2$. Then

$$F'(x, y, z) = \begin{bmatrix} 4x^3 & 0 & -64z^3 \\ 0 & 1 & -2 \end{bmatrix}.$$

Let $\mathbf{c} = (0, 0, 0)$. Then $F(\mathbf{c}) = \mathbf{0}$ and $F'(\mathbf{c}) = \begin{bmatrix} 0 & 0 & 0 \\ 0 & 1 & -2 \end{bmatrix}$. Hence, $F'(\mathbf{c})$ does not map $Y = \mathbb{R}^2$ onto $V = \mathbb{R}^2$. Let $f(x) = (x, x/2)$ for all $x \in \mathbb{R}$. Then $f : \mathbb{R} \to \mathbb{R}^2$ is continuously differentiable and $F(x, f(x)) = F(x, x, x/2) = \mathbf{0}$ for all $x \in \mathbb{R}$.

Implicit Functions and Jacobian Matrices

The implicit function theorem states a very plausible fact, even though it has a rather involved proof. Let $X = \mathbb{R}^m$ and $Y = V = \mathbb{R}^n$. Then the equation $F(\mathbf{z}) = \mathbf{0}$ can be expressed as

$$\begin{aligned} F_1(x_1, \cdots, x_m; y_1, \ldots, y_n) &= 0 \\ &\cdots \\ F_n(x_1, \cdots, x_m; y_1, \ldots, y_n) &= 0. \end{aligned}$$

These are n equations. We would like to solve them for the n unknowns y_j and express each y_j in terms of the m variables x_i. If these equations were linear equations, then they could be expressed as

$$\mathbf{A}\,\mathbf{x} + \mathbf{B}\,\mathbf{y} = \mathbf{0}.$$

Here $\mathbf{A}$ is an $n \times m$ and $\mathbf{B}$ is an $n \times n$ matrix, and $\mathbf{x}$, $\mathbf{y}$ are, respectively, $m \times 1$ and $n \times 1$ matrices, or column-vectors. We know that if $\mathbf{B}$ is an invertible matrix, then these equations can be solved for y_j. In fact, $\mathbf{y} = -\mathbf{B}^{-1}\,\mathbf{A}\,\mathbf{x}$.

In the general case, we have an equation of the form $F(\mathbf{x}, \mathbf{y}) = \mathbf{0}$. We are given that $\mathbf{c} = (\mathbf{a}, \mathbf{b})$ is a solution. We would like to solve this equation for all the values of $\mathbf{x}$ close to $\mathbf{a}$ and obtain $\mathbf{y} = f(\mathbf{x})$ as a function of $\mathbf{x}$. First, we try to do this approximately. We replace $F(\mathbf{x}, \mathbf{y})$ by its first-order Taylor polynomial

$$P(\mathbf{c})(\mathbf{z}) = F(\mathbf{c}) + F'(\mathbf{c})(\mathbf{z} - \mathbf{c}) = F'(\mathbf{c})(\mathbf{z} - \mathbf{c})$$

at $\mathbf{c} = (\mathbf{a}, \mathbf{b})$. Then we obtain the linear equation $F'(\mathbf{c})(\mathbf{z} - \mathbf{c}) = \mathbf{0}$ or

$$F'(\mathbf{a}, \mathbf{b})(\mathbf{x} - \mathbf{a}) + F'(\mathbf{a}, \mathbf{b})(\mathbf{y} - \mathbf{b}) = \mathbf{0}.$$

To express this equation in terms of scalar equations, we form the Jacobian matrices $\mathbf{A} = \{A_{ji}\}$ and $\mathbf{B} = \{B_{jk}\}$ with the components

$$A_{ji} = \frac{\partial F_j}{\partial x_i}(\mathbf{a}, \mathbf{b}) \text{ and } B_{jk} = \frac{\partial F_j}{\partial y_k}(\mathbf{a}, \mathbf{b}).$$

Then we obtain $\mathbf{A}\,(\mathbf{x} - \mathbf{a}) + \mathbf{B}\,(\mathbf{y} - \mathbf{b}) = \mathbf{0}$. This equation can be solved if $\mathbf{B}$ is invertible. We see that the invertibility of $\mathbf{B}$ means that if the derivative

$$F'(\mathbf{a}, \mathbf{b}) : (X \times Y) \to V$$

is restricted to vectors in Y, then it becomes an invertible mapping $Y \to V$. The implicit function theorem shows that if the approximate linear system of equations can be solved, then the original system of equations can be also solved.

Component Forms

Remarks 6.4.10 A summary of component forms. Let $\dim Z = n$ and $\dim V = m \leq n$. Choose an orthonormal basis $(\mathbf{e}_1, \ldots, \mathbf{e}_m)$ for V and express $F : Z \to V$ in terms of its components $F_i : Z \to \mathbb{R}$ so that

$$F(\mathbf{z}) = F_1(\mathbf{z})\mathbf{e}_1 + \cdots + F_m(\mathbf{z})\mathbf{e}_m.$$

This amounts to replacing V by $\mathbb{R}^m$. Hence we assume that $V = \mathbb{R}^m$ and let

$$F(\mathbf{z}) = (F_1(\mathbf{z}), \ldots, F_m(\mathbf{z})) \in \mathbb{R}^m.$$

Then the derivative $F'(\mathbf{a}) : Z \to \mathbb{R}^m$ is expressed as

$$F'(\mathbf{a})\mathbf{z} = (\langle \nabla F_1(\mathbf{a}), \mathbf{z} \rangle, \ldots, \langle \nabla F_m(\mathbf{a}), \mathbf{z} \rangle)$$

for all $\mathbf{z} \in Z$. The condition that $F'(\mathbf{a})Z = V$ is equivalent to saying that

$$\{\, \nabla F_1(\mathbf{a}), \ldots, \nabla F_m(\mathbf{a}) \,\}$$

is linearly independent and has m distinct elements. The vectorial equation $F'(\mathbf{a})\mathbf{z} = \mathbf{0}$ for the tangent space $T(\mathbf{a})$ becomes m scalar equations

$$\langle \nabla F_1(\mathbf{a}),\, \mathbf{z}\rangle = 0,\ \ldots,\ \langle \nabla F_m(\mathbf{a}),\, \mathbf{z}\rangle = 0.$$

Hence $T(\mathbf{a})$ is the orthogonal complement of the space spanned by the vectors

$$\{\, \nabla F_1(\mathbf{a}),\ \ldots,\ \nabla F_m(\mathbf{a})\,\}.$$

The space spanned by these vectors is then the normal space $N(\mathbf{a}) = T(\mathbf{a})^{\perp}$.

Example 6.4.11 The unit sphere in $\mathbb{R}^n$. The unit sphere S

$$x_1^2 + \cdots + x_n^2 = 1$$

is a manifold in $\mathbb{R}^n$. In fact, define $h : \mathbb{R}^n \to \mathbb{R} = V$ by

$$h(\mathbf{x}) = x_1^2 + \cdots + x_n^2 - 1,\ \mathbf{x} = (x_1,\ \ldots,\ x_n) \in \mathbb{R}^n.$$

Then S is given by a single scalar equation $h(\mathbf{x}) = 0$. This is the equation of a manifold if the gradient vectors of the components of h are linearly independent. In this case, there is only one component. The linear independence of a set of one vector ∇h means that $\nabla h \neq 0$. This is indeed the case, since

$$\nabla h(\mathbf{a}) = 2(a_1,\ \ldots,\ a_n) \neq \mathbf{0}$$

for all $\mathbf{a} \in E$ (where we may take $E = \mathbb{R}^n \setminus \{\mathbf{0}\}$). Therefore $h(\mathbf{x}) = 0$ defines a manifold S in $\mathbb{R}^n$. The normal space $N(\mathbf{a})$ is the one-dimensional space spanned by $\nabla h(\mathbf{a}) = 2(a_1,\ \ldots,\ a_n)$. The tangent space $T(\mathbf{a}) = N(\mathbf{a})^{\perp}$ is the set of all $\mathbf{x} \in \mathbb{R}^n$ such that

$$a_1 x_1 + \cdots + a_n x_n = 0.$$

Similarly, the equation of the affine tangent space $AT(\mathbf{a})$ is

$$\begin{aligned} a_1(x_1 - a_1) + \cdots + a_n(x_n - a_n) &= 0, \text{ or, equivalently,} \\ a_1 x_1 + \cdots + a_n x_n &= 1. \quad \triangle \end{aligned}$$

Problems

6.28 Let $F : A \to \mathbb{R}^m$, where A is an open set in $\mathbb{R}^n$ and F is of class $\mathcal{C}^1$ on A. Let M be the set of all $\mathbf{x} \in A$ such that $\dim(\text{Range } F'(\mathbf{x})) = m$. Show that if $M \neq \emptyset$, then M is an $(n - m)$-dimensional manifold.

6.29 Let $F : \mathbb{R}^3 \to \mathbb{R}^2$ be given by $F(x, y, z) = (y^2 + z, z + x)$ for all $\mathbf{x} = (x, y, z) \in \mathbb{R}^3$. Show that

$$M = \{ (x, y, z) \in \mathbb{R}^3 \mid F(\mathbf{x}) = (0, 0) \}$$

is a manifold in $\mathbb{R}^3$. Find the tangent space of M at $\mathbf{m} = (1, 1, -1)$. Also, find the normal space of M at the same point $\mathbf{m}$.

6.30 Let $F : \mathbb{R}^n \to \mathbb{R}$ be of class $\mathcal{C}^1$. Suppose that $c \in \mathbb{R}$ and

$$B_c = \{ \mathbf{x} \in \mathbb{R}^n \mid F(\mathbf{x}) = c \} \neq \emptyset.$$

Assume further that whenever $\mathbf{x} \in B_c$, we have $F'(\mathbf{x}) \neq \mathbf{0}$. Show that B_c is an $(n-1)$-manifold in $\mathbb{R}^n$.

6.31 Consider the system

$$\begin{aligned} x^2 + \frac{1}{2}y^2 + z^3 - z^2 &= \frac{3}{2} \\ x^3 + y^3 - 3y + z &= -3. \end{aligned}$$

Can we solve for y and z as a function of x for (x, y, z) in a neighborhood of $(-1, 1, 0)$?

6.32 Suppose that $F : \mathbb{R}^{n+m} \to \mathbb{R}^m$ is a continuously differentiable function with $F(\mathbf{a}, \mathbf{b}) = \mathbf{0}$ and

$$\det \left(\frac{\partial(F_1, \ldots, F_m)}{\partial(y_1, \ldots, y_m)}(\mathbf{a}, \mathbf{b}) \right) \neq 0.$$

Let A be a neighborhood of $\mathbf{a}$ in $\mathbb{R}^n$ and let f be a $\mathcal{C}^1$ function $f : A \to \mathbb{R}^m$ such that $f(\mathbf{a}) = \mathbf{b}$, and $F(\mathbf{x}, f(\mathbf{x})) = \mathbf{0}$ for all $\mathbf{x} \in A$. Show that

$$\mathbf{J}f(\mathbf{a}) = - \left(\frac{\partial(F_1, \ldots, F_m)}{\partial(y_1, \ldots, y_m)}(\mathbf{a}, \mathbf{b}) \right)^{-1} \cdot \frac{\partial(F_1, \ldots, F_m)}{\partial(x_1, \ldots, x_n)}(\mathbf{a}, \mathbf{b}).$$

6.33 Show that the system

$$\begin{aligned} xv + yu &= 1 \\ xy &= uv \end{aligned}$$

defines $(u, v) = h(x, y)$ implicitly as a function of (x, y) for (x, y) in a neighborhood of $(1, 0)$. Compute $h'(1, 0)$.

6.5 DIFFERENTIATION ON MANIFOLDS

Let M be a manifold in a Euclidean space Z. Let Y be another Euclidean space and $f : M \to Y$ a function. We shall define a concept of differentiation for such functions. The earlier definition of differentiation given in Definition 5.2.1 does not apply, as the domain of f may not be an open set in any vector space.

Definition 6.5.1 Derivatives of functions on manifolds. Let $\varphi : C \to Z$ be a parametric equation for a manifold M and $\mathbf{m} = \varphi(\mathbf{c})$. Then a function $f : M \to Y$ is said to be *differentiable at* $\mathbf{m} \in M$ if $g = f \cdot \varphi : C \to Y$ is differentiable at $\mathbf{c} \in C$. In this case, the derivative of $f : M \to Y$ at $\mathbf{m} \in M$ is defined as $f'(\mathbf{m}) = g'(\mathbf{c}) \cdot \varphi'(\mathbf{c})^{-1} \in L(T(\mathbf{m}), Y)$.

Note that $\varphi'(\mathbf{c}) : U \to Z$ is not invertible. But $\varphi'(\mathbf{c})$ is invertible as a linear map between U and the tangent space $T(\mathbf{m}) = \varphi'(\mathbf{c})U$. Hence $f'(\mathbf{m}) : T(\mathbf{m}) \to Y$ is well-defined. Nevertheless, our definition still has to be justified. Specifically, we need to show that the differentiability of f is independent of the parametric equation φ used.

Lemma 6.5.2 *Let $\varphi_i : C_i \to M$ be two equivalent parametric equations for M. Let $\varphi_i(\mathbf{c}_i) = \mathbf{m} \in M$. Then*

$$g_1 = f \cdot \varphi_1 : C_1 \to Y \text{ is differentiable at } \mathbf{c}_1 \text{ if and only if}$$
$$g_2 = f \cdot \varphi_2 : C_2 \to Y \text{ is differentiable at } \mathbf{c}_2.$$

If both are differentiable, then $f'(\mathbf{m}) = g_1'(\mathbf{c}_1) \cdot \varphi_1'(\mathbf{c}_1)^{-1} = g_2'(\mathbf{c}_2) \cdot \varphi_2'(\mathbf{c}_2)^{-1}$.

Proof. Let $\vartheta : C_1 \to C_2$ be the diffeomorphism obtained in Lemma 6.3.18 so that $\varphi_1 = \varphi_2 \cdot \vartheta$. Note that $\mathbf{c}_2 = \vartheta(\mathbf{c}_1)$. Assume that $g_2'(\mathbf{c}_2)$ exists. We have

$$\begin{aligned} g_1 &= f \cdot \varphi_1 = f \cdot \varphi_2 \cdot \vartheta = g_2 \cdot \vartheta \text{ and, therefore,} \\ g_1'(\mathbf{c}_1) &= (g_2 \cdot \vartheta)'(\mathbf{c}_1) = g_2'(\vartheta(\mathbf{c}_1))\vartheta'(\mathbf{c}_1) = g_2'(\mathbf{c}_2)\vartheta'(\mathbf{c}_1) \end{aligned}$$

by the chain rule. Hence $g_1'(\mathbf{c}_1)$ also exists and $g_1'(\mathbf{c}_1) = g_2'(\mathbf{c}_2)\vartheta'(\mathbf{c}_1)$. The other direction is similar. Now assume that the derivatives exist. Given $\mathbf{p} \in T(\mathbf{m})$, let $\mathbf{u}_i = \varphi_i'(\mathbf{c}_i)^{-1}\mathbf{p}$. Then $\mathbf{p} = \varphi_1'(\mathbf{c}_1)\mathbf{u}_1 = (\varphi_2 \cdot \vartheta)'(\mathbf{c}_1)\mathbf{u}_1 = \varphi_2'(\mathbf{c}_2)\vartheta'(\mathbf{c}_1)\mathbf{u}_1$ shows that $\mathbf{u}_2 = \vartheta'(\mathbf{c}_1)\mathbf{u}_1$. Therefore

$$\begin{aligned} f'(\mathbf{m})\mathbf{p} &= g_i'(\mathbf{c}_i) \cdot \varphi_i'(\mathbf{c}_i)^{-1}\mathbf{p} = g_i'(\mathbf{c}_i)\mathbf{u}_i, \text{ and} \\ g_1'(\mathbf{c}_1)\mathbf{u}_1 &= (g_2 \cdot \vartheta)'(\mathbf{c}_1)\mathbf{u}_1 = g_2'(\mathbf{c}_2)\vartheta'(\mathbf{c}_1)\mathbf{c}_1 = g_2'(\mathbf{c}_2)\mathbf{u}_2. \end{aligned}$$

Hence $f'(\mathbf{m}) : T(\mathbf{m}) \to Y$ is independent of the parametrization. □

A Special Case

Let M be a manifold in Z. In many cases of interest, a function f on M is the restriction of a differentiable function $F : G \to V$ defined on an open set $G \subset Z$ containing M. In this case, differentiation on M is simple. The derivative $f'(\mathbf{m}) : T(\mathbf{m}) \to V$ on M is obtained as the restriction of $F'(\mathbf{m}) : Z \to V$ to the tangent space $T(\mathbf{m})$. Hence $f'(\mathbf{m}) = F'(\mathbf{m})|_M$ if $f = F|_M$. We state this also as a lemma.

Lemma 6.5.3 *Let M be a manifold in Z and $\mathbf{m} \in M$. Let $F : G \to V$ be a differentiable function defined on an open set containing* $\mathbf{m}$. *Define*

$$f : (M \cap G) \to V \textit{ by } f(\mathbf{z}) = F(\mathbf{z}).$$

Then $f'(\mathbf{m}) : T(\mathbf{m}) \to Y$ exists and $f'(\mathbf{m}) : T(\mathbf{m}) \to V$ is the restriction of $F'(\mathbf{m}) : Z \to V$ to $T(\mathbf{m})$.

Proof. Let $\varphi : C \to Z$ be a parametric equation for M such that $\varphi(\mathbf{c}) = \mathbf{m}$ and $\varphi(C) \subset G$. Then $g = f \cdot \varphi = F \cdot \varphi$ shows that $g : A \to Y$ is differentiable. Hence $f'(\mathbf{m}) : T(\mathbf{m}) \to Y$ exists. Also, if $\mathbf{z} \in T(\mathbf{m})$, then

$$f'(\mathbf{m})\mathbf{z} = g'(\mathbf{a})\varphi'(\mathbf{c})^{-1}\mathbf{z} = F'(\varphi(\mathbf{c}))\varphi'(\mathbf{c})\varphi'(\mathbf{c})^{-1}\mathbf{z} = F'(\mathbf{m})\mathbf{z}.$$

This shows that $f'(\mathbf{m})$ is the restriction of $F'(\mathbf{m})$ to $T(\mathbf{m})$. □

Derivatives Along Curves on Manifolds

Let $F : G \to V$ be a differentiable function defined on an open set G. Then $F'(\mathbf{a})\mathbf{z}$ is the directional derivative of F at $\mathbf{a} \in G$ in the direction of $\mathbf{z} \in Z$. On manifolds one cannot, in general, take directional derivatives because the manifold will not contain straight lines running in every direction. Directional derivatives have to be replaced with derivatives along curves. Recall that a curve C on M is a $\mathcal{C}^1$ function $\mathbf{r} : I \to M$ defined on an open interval $I \subset \mathbb{R}$. The derivative $\mathbf{r}'(a)$ at $a \in I$ is the tangent vector of C at $\mathbf{r}(a) = \mathbf{m} \in M$, a vector in the space Z in which M is embedded. Theorem 6.3.27 shows that $\mathbf{r}'(a) \in T(\mathbf{r}(a))$. So the tangent vector of a curve in M at $\mathbf{m} \in M$ is in the tangent space $T(\mathbf{m})$. If $f'(\mathbf{m}) : T(\mathbf{m}) \to V$ exists for all $\mathbf{m} \in M$, then $f'(\mathbf{r}(a))\mathbf{r}'(a)$ is defined for all curves $\mathbf{r} : I \to M$. These are the derivatives along curves.

Theorem 6.5.4 Derivatives and derivatives along curves. *Let M be a manifold in Z and $f : M \to V$ a differentiable function. Let $\mathbf{r} : I \to M$ be a curve on M and $a \in I$. Let $\lambda = f \cdot \mathbf{r} : I \to V$. Then*

$$\lambda'(a) = \lim_{t \to 0} \frac{f(\mathbf{r}(a+t)) - f(\mathbf{r}(a))}{t} \tag{6.35}$$

exists in V *and* $\lambda'(a) = f'(\mathbf{r}(a))\mathbf{r}'(a)$.

Proof. Let $\varphi : C \to Z$ be a parametric equation for M and $\varphi(\mathbf{u}) = \mathbf{r}(a) = \mathbf{m}$. Let $\mathbf{s}(a) = \varphi^{-1}(\mathbf{r}(a))$ so that $\varphi(\mathbf{s}(a)) = \mathbf{r}(a)$. Then

$$\mathbf{r}'(a) = \varphi'(\mathbf{s}(a))\mathbf{s}'(a) \quad \text{and} \quad \lambda = f \cdot \mathbf{r} = f \cdot \varphi \cdot \mathbf{s} = g \cdot \mathbf{s}.$$

Now $f : M \to V$ is assumed to be differentiable. Hence $g = f \cdot \varphi$ is also differentiable by Definition 6.5.1. Therefore

$$\begin{aligned} \lambda'(a) &= g'(\mathbf{s}(a))\mathbf{s}'(a) = f'(\varphi(\mathbf{s}(a)))[\varphi'(\mathbf{s}(a))\mathbf{s}'(a)] \\ &= f'(\mathbf{r}(a))\mathbf{r}'(a). \quad \square \end{aligned}$$

Local Extremal Values on Manifolds

Let M be a manifold in X. Let $p : M \to \mathbb{R}$ be a real-valued function defined on M. A point $\mathbf{m}_0 \in M$ is called a *local maximum point* for p if there is a $\delta > 0$ such that $p(\mathbf{m}) \leq p(\mathbf{m}_0)$ whenever $\|\mathbf{m} - \mathbf{m}_0\| < \delta$ and $\mathbf{m} \in M$. Local minimum points are defined similarly.

Lemma 6.5.5 *Let* $\mathbf{m} \in M$ *be a local extremal point for* $p : M \to \mathbb{R}$. *Assume that* $p'(\mathbf{m}) : T(\mathbf{m}) \to \mathbb{R}$ *exists. Then* $p'(\mathbf{m}) = \mathbf{0}$.

Proof. Assume that there is a $\mathbf{u} \in T(\mathbf{m})$ such that $p'(\mathbf{m})\mathbf{u} \in \mathbb{R}$ is not zero. Use Theorem 6.3.27 to find a curve $\mathbf{r} : I \to M$ on M such that $\mathbf{r}(a_0) = \mathbf{m}$ for an $a_0 \in I$ and such that $\mathbf{r}'(a_0) = \mathbf{u}$. Define $s : I \to \mathbb{R}$ by $s(t) = p(\mathbf{r}(t))$. Then Theorem 6.5.4 shows that $s'(a_0) = p'(\mathbf{m})\mathbf{u} \neq 0$. Therefore $s : I \to \mathbb{R}$ cannot have a local extremal point at $a_0 \in I$. Then $\mathbf{m} = \mathbf{r}(a_0)$ cannot be a local extremal point for p. $\square$

Recall that $N(\mathbf{m}) = T(\mathbf{m})^{\perp}$ is the normal space at $\mathbf{m}$ of M, the orthogonal complement of the tangent space.

Lemma 6.5.6 *Let* $M \subset G \subset Z$, *where* H *is an open set. Let* $p : M \to \mathbb{R}$ *be the restriction of a differentiable function* $q : G \to \mathbb{R}$ *to* M. *Then* $p'(\mathbf{m}) = \mathbf{0}$ *if and only if* $\nabla q(\mathbf{m}) \in N(\mathbf{m})$.

Proof. In this case, $p'(\mathbf{m}) : T(\mathbf{m}) \to \mathbb{R}$ is just the restriction of $q'(\mathbf{m}) : \mathbb{R}^n \to \mathbb{R}$ to $T(\mathbf{m})$, as we proved in Lemma 6.5.3. Hence $p'(\mathbf{m}) = \mathbf{0}$ if and only if

$$p'(\mathbf{m})\mathbf{u} = q'(\mathbf{m})\mathbf{u} = \langle \nabla q(\mathbf{m}), \mathbf{u} \rangle = 0$$

for all $\mathbf{u} \in T(\mathbf{m})$. In this case $\nabla q(\mathbf{m}) \perp T(\mathbf{m})$ or $\nabla q(\mathbf{m}) \in N(\mathbf{m})$. $\square$

Finally, assume that $V = \mathbb{R}^m$ and $F : G \to \mathbb{R}^m$ is given as m scalar functions $F_i : G \to \mathbb{R}$. Hence $F = (F_1, \ldots, F_m) : G \to \mathbb{R}^m$.

Theorem 6.5.7 Lagrange multipliers. *Let G be an open set in Z. Let M be a manifold in G given by the equations $F_i(\mathbf{z}) = 0$, $i = 1, \ldots m$. Let $q : G \to \mathbb{R}$ be a differentiable function. Then the restriction of q to M can have a local extremal point at $\mathbf{m} \in M$ only if $\nabla q(\mathbf{m})$ is a linear combination of the gradient vectors $\nabla F_i(\mathbf{m})$.*

Proof. Lemmas 6.5.5 and 6.5.6 show that the restriction of $q : H \to \mathbb{R}$ to M can have an extremal point at $\mathbf{m}$ only if $\nabla q(\mathbf{m}) \in N(\mathbf{m})$. Then the result follows from the fact that $N(\mathbf{m})$ is spanned by the gradient vectors $\nabla F_i(\mathbf{m})$, as observed in Remarks 6.4.10. □

Example 6.5.8 Let us compute the maximum and minimum values of $x^3 + y^3 + z^3$, where $x^2 + y^2 + z^2 = 9$. Let $f(\mathbf{x}) = x^3 + y^3 + z^3$ and let $g(\mathbf{x}) = x^2 + y^2 + z^2$. We want to maximize f on $S = \{\, \mathbf{x} \in \mathbb{R}^3 \mid g(\mathbf{x}) = 9 \,\}$. Thus, we seek $\mathbf{x} \in S$ and $\lambda \in \mathbb{R}$ such that $\nabla f(\mathbf{x}) = \lambda \nabla g(\mathbf{x})$. Thus,

$$(3x^2, 3y^2, 3z^2) = \lambda(2x, 2y, 2z).$$

Since $x^2 + y^2 + z^2 = 9$, we have $\lambda \neq 0$. Assume first that $x \neq 0, y \neq 0, z \neq 0$. Then we obtain $x = y = z$ and, therefore, $x^2 = 3$. Thus, $\mathbf{x} = \pm(\sqrt{3}, \sqrt{3}, \sqrt{3})$. In this case, $f(\mathbf{x}) = \pm 9\sqrt{3}$.

Now, suppose that $x = 0 = y$. Then $z = \pm 3$ and $\lambda = (3z)/2$. Similarly, if $x = z = 0$, then $y = \pm 3, \lambda = (3y)/2$, and if $y = z = 0$, then $x = \pm 3, \lambda = (3x)/2$. In each of these three cases, $f(\mathbf{x}) = \pm 27$.

Finally, if only one of x, y, or z is zero, then the remaining two unknowns are equal to one another and each equals $\pm\sqrt{3}/2$. It is clear that the values of f in these cases are between -27 and 27. Hence, the maximum value of f is 27, which occurs at $(3, 0, 0)$, $(0, 3, 0)$, and $(0, 0, 3)$, and the minimum value of f is -27, which occurs at $(-3, 0, 0)$, $(0, -3, 0)$, and $(0, 0, -3)$.

Example 6.5.9 Let us find the maximum value of $u_1^2 \cdots u_n^2$, where

$$u_1^2 + \cdots + u_n^2 = 1.$$

Let $f(\mathbf{u}) = u_1^2 \cdots u_n^2$ and let $S = \{\, \mathbf{u} \in \mathbb{R}^n \mid u_1^2 + \cdots + u_n^2 = 1 \,\}$. Since f is continuous on the compact set S, we know that f must have a maximum value on S. Thus, there is some $\lambda \in \mathbb{R}$ and some $\mathbf{x} \in S$ such that $\nabla f(\mathbf{x}) = \lambda \nabla g(\mathbf{x})$. Thus,

$$\begin{aligned} x_1 x_2^2 \cdots x_n^2 &= \lambda x_1 \\ x_1^2 x_2 x_3^2 \cdots x_n^2 &= \lambda x_2 \\ &\cdots \\ x_1^2 \cdots x_{n-1}^2 x_n &= \lambda x_n. \end{aligned}$$

Multiplying the first equation by x_1, the second equation by x_2, ..., and the last equation by x_n, we deduce that

$$\lambda x_1^2 = \lambda x_2^2 = \cdots = \lambda x_n^2.$$

Since f takes on some positive values on S, the maximum value of f on S must also be positive. Hence, none of x_j can be zero. In particular, $\lambda \neq 0$. Therefore,

$$x_1^2 = x_2^2 = \cdots = x_n^2.$$

Thus, since $x_1^2 + \cdots + x_n^2 = 1$, we get

$$x_k^2 = \frac{1}{n} \quad \text{for all } k = 1, \ldots, n.$$

Therefore, the maximum value of f on S occurs when $x_k^2 = 1/n$ for all $k = 1, \ldots, n$. In particular,

$$u_1^2 \cdots u_n^2 = f(\mathbf{u}) \leq f(\mathbf{x}) = \left(\frac{1}{n}\right)^n = \left(\frac{u_1^2 + \cdots + u_n^2}{n}\right)^n \quad \text{for all } \mathbf{u} \in S.$$

That is,

$$\left(u_1^2 \cdots u_n^2\right)^{\frac{1}{n}} \leq \frac{u_1^2 + \cdots + u_n^2}{n} \quad \text{whenever } u_1^2 + \cdots + u_n^2 = 1. \tag{6.36}$$

From the above, we obtain the *arithmetic-geometric mean inequality*:

Let $t_1, \ldots, t_n$ be nonnegative numbers. Then

$$(t_1 \cdots t_n)^{\frac{1}{n}} \leq \frac{t_1 + \cdots + t_n}{n}. \tag{6.37}$$

To prove (6.37), let $t_1, \ldots, t_n$ be any nonnegative numbers. Put $S = t_1 + \cdots + t_n$. If $S = 0$, then each $t_k = 0$ and (6.37) is obvious. Assume that $S > 0$. Since each t_k/S is nonnegative, there is some $u_k \in \mathbb{R}$ with $t_k/S = u_k^2$. Then

$$u_1^2 + \cdots + u_n^2 = 1.$$

Thus, (6.37) follows from (6.36) because

$$\frac{1}{S}(t_1 \cdots t_n)^{\frac{1}{n}} = \left(u_1^2 \cdots + u_n^2\right)^{\frac{1}{n}} \leq \frac{u_1^2 + \cdots + u_n^2}{n} = \frac{t_1 + \cdots + t_n}{Sn}.$$

Example 6.5.10 Suppose that $f : \mathbb{R}^n \to \mathbb{R}$ is a continuously differentiable function. Assume that for any $\alpha \in \mathbb{R}$, the system of equations

$$\frac{\partial f}{\partial x_k}(\mathbf{u}) = \alpha u_k, \quad k = 1, \ldots, n$$

has at most one solution $\mathbf{u}$ with $u_1^2 + \cdots + u_n^2 = 1$. Then f is constant on $\{\, \mathbf{x} \in \mathbb{R}^n \mid x_1^2 + \cdots + x_n^2 = 1 \,\}$. To see this, let $g(\mathbf{x}) = x_1^2 + \cdots + x_n^2$. Put $S = \{\, \mathbf{x} \in \mathbb{R}^n \mid g(\mathbf{x}) = 1 \,\}$, a compact set in $\mathbb{R}^n$. Hence, f must have a maximum value and a minimum value on S. But these extreme values of f must occur at points $\mathbf{u} \in S$ for which there is some $\lambda \in \mathbb{R}$ with $\nabla f(\mathbf{u}) = \lambda \nabla g(\mathbf{u})$. Hence, with $\alpha = 2\lambda$, we must have

$$\frac{\partial f}{\partial x_k}(\mathbf{u}) = \alpha u_k \quad \text{for all } k = 1, \ldots, n.$$

By assumption, there is at most one such $\mathbf{u} \in S$. This implies that the maximum and minimum values of f on S occur at the same point $\mathbf{u}$. Hence, the maximum and minimum values of f on S are equal. Thus, f must be constant on S.

Problems

6.34 A linear transformation $T : \mathbb{R}^2 \to \mathbb{R}^2$ is given as

$$T(x, y) = (Ax + By, Cx + Dy).$$

Find $\|T\|$ in terms of A, B, C, D. The norm on $\mathbb{R}^2$ is the standard Euclidean norm.

6.35 Find the maximum and minimum values of $x^2+y^2+z^2$ given that $x+y+z = 0$ and $(x-3)^2 + y^2 + z^2 = 9$.

6.36 Find the minimum value of $3x-y-3z$ where $x+y-z = 0$ and $x^2+2z^2 = 1$.

6.37 Find the maximum value of xyz where $x^2 + y^2 + z^2 = 3$.

6.38 Find the maximum and the minimum value of $x+y+z$ where $x^2+y^2+z^2 = 1$.

6.39 Find the minimum value of $5x - 2y + 7z$ where $x^2 + 2y + 4z^2 = 9$

6.40 Find the minimum value of xyz where $x^2 + 2y^2 + 3z^2 = 12$

6.41 Find the points on the curve $x^2 + xy + y^2 = 3$ closest to and farthest from the origin.

6.42 Find the maximum value of $yz + xy$ where $xy = 1$ and $y^2 + z^2 = 1$

CHAPTER 7

HIGHER-ORDER DERIVATIVES

Practically all computations are done in terms of polynomials. In fact, the scope of differential calculus is essentially limited to functions that can be approximated by polynomials. In these approximations the higher-order derivatives play a central role. We have already considered higher-order derivatives for functions of a real variable. Here we will discuss them in the general case.

7.1 DEFINITIONS

Let X and Y be two normed spaces. Let A be an open subset of X. We will define the higher-order derivatives of a function $f : A \to Y$.

Definition 7.1.1 Difference operators. New notations for directional derivatives. Let $\mathbf{a} \in A$ and $\mathbf{u} \in X$. The *difference operator* $\Delta_{\mathbf{u}}$ is defined as

$$\Delta_{\mathbf{u}} f(\mathbf{a}) = f(\mathbf{a} + \mathbf{u}) - f(\mathbf{a}).$$

Analysis in Vector Spaces.
By M. A. Akcoglu, P. F. A. Bartha and D. M. Ha

Here it is assumed that $(\mathbf{a} + \mathbf{u}) \in A$. We will also write $D_{\mathbf{u}} f(\mathbf{a})$ for the directional derivative of f at $\mathbf{a} \in A$ in the direction of $\mathbf{u} \in X$. Hence

$$D_{\mathbf{u}} f(\mathbf{a}) = f'(\mathbf{a};\, \mathbf{u}) = \lim_{t \to 0} \frac{f(\mathbf{a} + t\mathbf{u}) - f(\mathbf{a})}{t} = \lim_{t \to 0} \frac{\Delta_{t\mathbf{u}} f(\mathbf{a})}{t}.$$

Definition 7.1.2 Higher-order directional derivatives. Let $\mathbf{u}_i \in X$ be finitely many vectors. The higher-order directional derivatives

$$D_{\mathbf{u}_n} \cdots D_{\mathbf{u}_1} f(\mathbf{a}) \in Y$$

are defined inductively on $n \in \mathbb{N}$. The definition is known for $n = 1$. Assume that $F(\mathbf{a}) = D_{\mathbf{u}_n} \cdots D_{\mathbf{u}_1} f(\mathbf{a})$ is defined as a function $F : A \to Y$. Let $\mathbf{u} \in X$. Then we define

$$D_{\mathbf{u}} \cdot D_{\mathbf{u}_n} \cdots D_{\mathbf{u}_1} f(\mathbf{a}) = D_{\mathbf{u}} F(\mathbf{a}) = \lim_{t \to 0} \frac{\Delta_{t\mathbf{u}} F(\mathbf{a})}{t}.$$

The limit is taken in Y.

Definition 7.1.3 Higher-order derivatives. Let us now consider the main or 'full' derivative. Higher-order derivatives of a function $f : A \to Y$ are also defined inductively. The first derivative was already defined as a function $f' : A \to L(X, Y) = ML_1(X, Y)$. The second derivative of $f : A \to Y$ is the first derivative of $f' : A \to L(X, Y)$. Hence the second-order derivative is a function $f'' : A \to L(X, L(X, Y))$.

Recall that there is an isomorphism between $L(X, L(X, Y))$ and $ML_2(X^2, Y)$ as defined in Definition 3.3.7. Accordingly, we will consider the second derivative as a function

$$f'' : A \to ML_2(X^2, Y).$$

By induction we see that the nth order derivative will be defined as a function

$$f^{(n)} : A \to ML_n(X^n, Y).$$

Hence, the $(n + 1)$st order derivative will be a function

$$f^{(n+1)} : A \to L(X, ML_n(X^n, Y)) \cong ML_{n+1}(X^{n+1}, Y).$$

Notations 7.1.4 Values of higher-order derivatives. Notations for the values of higher-order derivatives depend on the way the spaces

$$L(X, ML_n(X^n, Y)) \text{ and } ML_{n+1}(X^{n+1}, Y)$$

are identified. This identification is made in such a way that the relation

$$f^{(n)}(\mathbf{a})(\mathbf{u}_n, \ldots, \mathbf{u}_1) = D_{\mathbf{u}_n} \cdots D_{\mathbf{u}_1} f(\mathbf{a}) \tag{7.1}$$

holds for all $n \in \mathbb{N}$.

Classes of Differentiable Functions

Definition 7.1.5 $\mathcal{C}^n$ functions. Let A be an open set in a vector space X. Then $\mathcal{C}^n = \mathcal{C}^n(A, Y)$, $n \in \mathbb{N}$, is the set of all functions $f : A \to Y$ for which

$$D^n f = f^{(n)} : A \to ML(X^n, Y)$$

exists and is continuous. Hence $\mathcal{C}^n(A, Y)$ is the class of all functions $f : A \to Y$ that are n-times continuously differentiable. Also, $\mathcal{C}^\infty = \mathcal{C}^\infty(A, Y)$ is the class of functions that have derivatives of all orders. A $\mathcal{C}^n$ diffeomorphism $f : A \to Y$ is a $\mathcal{C}^n$ function, and a diffeomorphism such that the reverse diffeomorphism $g : f(A) \to A$ is also a $\mathcal{C}^n$ function. Finally, the graph of a $\mathcal{C}^n$ function is a $\mathcal{C}^n$ graph, and $\mathcal{C}^n$ manifolds are locally $\mathcal{C}^n$ graphs. In the following statements about the properties of $\mathcal{C}^n$ functions, the superscript n will also stand for ∞ unless specified otherwise.

Theorem 7.1.6 Chain rule for $\mathcal{C}^n$ functions. *Let A and B be open sets in X and Y, respectively. If $f : A \to B$ and $g : B \to Z$ are $\mathcal{C}^n$ functions, then $h = g \cdot f : A \to Z$ is also a $\mathcal{C}^n$ function.*

Proof. The chain rule was given in Theorem 5.5.6. It shows that if f and g are differentiable, then h is also differentiable and

$$h'(\mathbf{x}) = g'(f(\mathbf{x})) \cdot f'(\mathbf{x}), \quad \mathbf{x} \in A.$$

We will express $h' : A \to L(X, Z)$ also as a composed function. Define

$$\begin{aligned} P &: A \to L(Y, Z) \times L(X, Y) \text{ and} && (7.2)\\ Q &: L(Y, Z) \times L(X, Y) \to L(X, Z) \text{ by} && (7.3)\\ P(\mathbf{a}) &= (g'(f(\mathbf{a})), f'(\mathbf{a})), \ \mathbf{a} \in A, \text{ and} && (7.4)\\ Q(S, T) &= ST, \ (S, T) \in L(Y, Z) \times L(X, Y). && (7.5) \end{aligned}$$

Then we see that $h'(\mathbf{a}) = Q(P(\mathbf{a}))$ for all $\mathbf{a} \in A$ or that

$$h' = Q \cdot P : A \to L(X, Z). \tag{7.6}$$

We proceed by induction on $n \in \mathbb{N}$. If $n = 1$, then the result is already obtained in Theorem 5.5.6. Now assume that the composition of two $\mathcal{C}^n$ functions is a $\mathcal{C}^n$ function. Let f and g be $\mathcal{C}^{n+1}$ functions. Then $g' : B \to L(Y, Z)$ is a $\mathcal{C}^n$ function. Therefore, $(g' \cdot f) : A \to L(Y, Z)$ is a $\mathcal{C}^n$ function by the induction hypothesis. In this case an easy check shows that

$$P = (g' \cdot f, f') : A \to L(Y, Z) \times L(X, Y)$$

is a $\mathcal{C}^n$ function. Also, Q is a bilinear function and, therefore, a $\mathcal{C}^\infty$ function. Hence the induction hypothesis shows that $h' = (Q \cdot P)$ is a $\mathcal{C}^n$ function. Therefore h is a $\mathcal{C}^{n+1}$ function. $\square$

Lemma 7.1.7 *Let $n \in \mathbb{N}$. Then $f : A \to Y$ belongs to $\mathcal{C}^n$ if and only if $D_{\mathbf{u}_n} \cdots D_{\mathbf{u}_1} f : A \to Y$ exists and is continuous for all $(\mathbf{u}_1, \ldots, \mathbf{u}_n) \in X^n$.*

Proof. If $f \in \mathcal{C}^n$, then clearly $D_{\mathbf{u}_n} \cdots D_{\mathbf{u}_1} f : A \to Y$ exists and is continuous for all $(\mathbf{u}_1, \ldots, \mathbf{u}_n) \in X^n$. To prove the converse, apply an induction on $n \in \mathbb{N}$. For $n = 1$ this result follows from the existence theorem for derivatives, Theorem 5.3.4. Now assume the result for $n \in \mathbb{N}$. Let

$$D_{\mathbf{v}} D_{\mathbf{u}_n} \cdots D_{\mathbf{u}_1} f : A \to Y$$

exist and be continuous for all $(\mathbf{u}_1, \cdots, \mathbf{u}_n, \mathbf{v}) \in X^{n+1}$. In particular,

$$D_{\mathbf{u}_n} \cdots D_{\mathbf{u}_1} f : A \to Y$$

also exists and is continuous. Hence $g = D_n f : A \to ML(X^n, Y)$ exists by the induction hypothesis. Then we see that $D_{\mathbf{v}} g : A \to ML(X^n, Y)$ exists and is continuous for all $\mathbf{v} \in X$. Therefore, the existence theorem, Theorem 5.3.4, shows that $Dg = D^{n+1} f : A \to ML(X^{n+1}, Y)$ also exists and is continuous. $\square$

7.2 CHANGE OF ORDER IN DIFFERENTIATION

We have defined $D_{\mathbf{u}_n} \cdots D_{\mathbf{u}_1} f : A \to Y$ as the result of n successive differentiations in the directions of $\mathbf{u}_1, \ldots, \mathbf{u}_n \in X$. It turns out that if f is a $\mathcal{C}^n$ function, then the result is independent of the order of these differentiations.

Lemma 7.2.1 Commutativity of the difference operators. *Let $\Delta_{\mathbf{u}_i}$ be n difference operators. Then for any permutation σ of $\{1, \ldots, n\}$*

$$\Delta_{\mathbf{u}_1} \cdots \Delta_{\mathbf{u}_n} f(\mathbf{x}) = \Delta_{\mathbf{u}_{\sigma(1)}} \cdots \Delta_{\mathbf{u}_{\sigma(n)}} f(\mathbf{x}),$$

whenever one of these expressions is defined.

Proof. It is enough to show that $\Delta_{\mathbf{v}} \Delta_{\mathbf{u}} f(\mathbf{x}) = \Delta_{\mathbf{u}} \Delta_{\mathbf{v}} f(\mathbf{x})$ whenever one side is defined. We have

$$\begin{aligned} \Delta_{\mathbf{v}} \Delta_{\mathbf{u}} f(\mathbf{x}) &= \Delta_{\mathbf{v}} (f(\mathbf{x} + \mathbf{u}) - f(\mathbf{x})) \\ &= f(\mathbf{x} + \mathbf{v} + \mathbf{u}) - f(\mathbf{x} + \mathbf{v}) - f(\mathbf{x} + \mathbf{u}) + f(\mathbf{x}) \\ &= \Delta_{\mathbf{u}} (f(\mathbf{x} + \mathbf{v}) - f(\mathbf{x})) = \Delta_{\mathbf{u}} \Delta_{\mathbf{v}} f(\mathbf{x}). \quad \square \end{aligned}$$

Remarks 7.2.2 Theorem 7.2.3 below establishes a relation between

$$\Delta_{\mathbf{u}_1} \cdots \Delta_{\mathbf{u}_n} f(\mathbf{x}) \text{ and } D_{\mathbf{u}_1} \cdots D_{\mathbf{u}_n} f(\mathbf{x}).$$

This result generalizes Lemma 5.3.2. Recall that if $g : A \to Y$ is a function and $E \subset A$ then

$$\Omega_E(g) = \sup \{ \|g(\mathbf{x}) - g(\mathbf{x}')\| \mid \mathbf{x},\, \mathbf{x}' \in E \}$$

is the oscillation of g over E if this supremum exists.

Theorem 7.2.3 *Assume that $f : A \to Y$ is a $\mathcal{C}^n$ function. Let $\mathbf{a} \in A$ and let $r > 0$ be such that $B_r(\mathbf{a}) \subset A$. Let $\mathbf{u}_1, \ldots, \mathbf{u}_n \in X$. Then*

$$\begin{aligned} \|\Delta_{\mathbf{u}_1} \cdots \Delta_{\mathbf{u}_n} f(\mathbf{a}+\mathbf{w}) - D_{\mathbf{u}_1} \cdots D_{\mathbf{u}_n} f(\mathbf{a})\| &\le \Omega_{B_r(\mathbf{a})}(D_{\mathbf{u}_1} \cdots D_{\mathbf{u}_n} f) \\ &\le \|\mathbf{u}_1\| \cdots \|\mathbf{u}_n\| \, \Omega_{B_r(\mathbf{a})}(D^n f) \end{aligned}$$

whenever $\|\mathbf{u}_1\| + \cdots + \|\mathbf{u}_n\| + \|\mathbf{w}\| < r$.

Proof. Define

$$\varphi(s) = \Delta_{\mathbf{u}_1} \cdots \Delta_{s\mathbf{u}_n} f(\mathbf{a}+\mathbf{w}) - D_{\mathbf{u}_1} \cdots D_{s\mathbf{u}_n} f(\mathbf{a}).$$

The first estimate of the theorem can be expressed as

$$\|\varphi(1)\| \le \Omega_{B_r(\mathbf{a})}(D_{\mathbf{u}_1} \cdots D_{\mathbf{u}_n} f).$$

We see easily that, as in the proof of Lemma 5.3.2,

$$\varphi'(s) = \Delta_{\mathbf{u}_1} \cdots \Delta_{\mathbf{u}_{n-1}} D_{\mathbf{u}_n} f(\mathbf{a}+\mathbf{w}+s\mathbf{u}_n) - D_{\mathbf{u}_1} \cdots D_{\mathbf{u}_{n-1}} D_{\mathbf{u}_n} f(\mathbf{a}).$$

Let $F(\mathbf{x}) = D_{\mathbf{u}_n} f(\mathbf{x})$ and assume that the conclusion of the theorem is true for $(n-1)$ as an induction hypothesis. This implies that, for all $s \in I = [\,0,\, 1\,]$,

$$\|\varphi'(s)\| \le \Omega_{B_r(\mathbf{a})}(D_{\mathbf{u}_1} \cdots D_{\mathbf{u}_{n-1}} F) = \Omega_{B_r(\mathbf{a})}(D_{\mathbf{u}_1} \cdots D_{\mathbf{u}_{n-1}} D_{\mathbf{u}_n} f).$$

Hence, by the mean value theorem, Theorem 5.1.13, we have

$$\|\varphi(1)\| = \|\varphi(1) - \varphi(0)\| \le (1-0)\Omega_{B_r(\mathbf{a})}(D_{\mathbf{u}_1} \cdots D_{\mathbf{u}_{n-1}} D_{\mathbf{u}_n} f).$$

This is the first estimate of the theorem. The second estimate follows from

$$\begin{aligned} \|D_{\mathbf{u}_1} \cdots D_{\mathbf{u}_{n-1}} D_{\mathbf{u}_n} f(\mathbf{x})\|_Y &= \|D^n f(\mathbf{x})(\mathbf{u}_1, \ldots, \mathbf{u}_n)\|_Y \\ &\le \|\mathbf{u}_1\|_X \cdots \|\mathbf{u}_n\|_X \, \|D^n f(\mathbf{x})\|_{ML(X^n, Y)}. \quad \square \end{aligned}$$

Theorem 7.2.4 *Assume that $f : A \to Y$ is a $\mathcal{C}^n$ function. Let $\mathbf{a} \in A$ and let $\mathbf{u}_i \in X$. Then for each $\varepsilon > 0$ there is a $\delta > 0$ such that*

$$\left\| \frac{\Delta_{t_n \mathbf{u}_n} \cdots \Delta_{t_1 \mathbf{u}_1} f(\mathbf{a})}{t_n \cdots t_1} - D_{\mathbf{u}_n} \cdots D_{\mathbf{u}_1} f(\mathbf{a}) \right\| < \varepsilon$$

whenever $0 < |t_i| < \delta$ for all $i = 1, \ldots, n$.

Proof. Choose $r > 0$ such that $B_r(\mathbf{a}) \subset A$ and such that

$$\|\mathbf{u}_1\| \cdots \|\mathbf{u}_n\| \, \Omega_{B_r(\mathbf{a})}(D^n f) < \varepsilon.$$

This can be done because of the continuity of $D^n f : A \to ML(X^n, Y)$. Then choose $\delta > 0$ such that

$$\|t_1 \mathbf{u}_1\| + \cdots + \|t_n \mathbf{u}_n\| < r$$

whenever $|t_i| < \delta$ for all $i = 1, \ldots, n$. In this case Theorem 7.2.3 shows that

$$\begin{aligned} & \|\Delta_{t_1 \mathbf{u}_1} \cdots \Delta_{t_n \mathbf{u}_n} f(\mathbf{a}) - (t_1 \cdots t_n) D_{\mathbf{u}_1} \cdots D_{\mathbf{u}_n} f(\mathbf{a})\| \\ = \; & \|\Delta_{t_1 \mathbf{u}_1} \cdots \Delta_{t_n \mathbf{u}_n} f(\mathbf{a}) - D_{t_1 \mathbf{u}_1} \cdots D_{t_n \mathbf{u}_n} f(\mathbf{a})\| \\ \le \; & \|t_1 \mathbf{u}_1\| \cdots \|t_n \mathbf{u}_n\| \, \Omega_{B_r(\mathbf{a})}(D^n f) \\ = \; & (t_1 \cdots t_n) \|\mathbf{u}_1\| \cdots \|\mathbf{u}_n\| \, \Omega_{B_r(\mathbf{a})}(D^n f) \\ \le \; & (t_1 \cdots t_n) \varepsilon. \end{aligned}$$

Then the proof follows easily. □

Theorem 7.2.5 *Assume that $f : A \to Y$ is a $\mathcal{C}^n$ function. Let $\mathbf{a} \in A$. Then*

$$D_{\mathbf{u}_1} \cdots D_{\mathbf{u}_n} f(\mathbf{a}) = f^{(n)}(\mathbf{a})(\mathbf{u}_n, \cdots, \mathbf{u}_1) \in Y$$

is independent of the ordering of the vectors $\mathbf{u}_1, \ldots, \mathbf{u}_n \in X$.

Proof. This follows from Theorem 7.2.4 above and from the commutativity of the difference operators, Lemma 7.2.1. □

Theorem 7.2.6 *Assume that $f : A \to Y$ is a $\mathcal{C}^n$ function. Then*

$$f^{(n)}(a) : X^n \to Y$$

is a symmetric multilinear function for each $\mathbf{a} \in A$.

Proof. This a reformulation of Theorem 7.2.5. □

7.3 SEQUENCES OF POLYNOMIALS

Recall that a homogeneous polynomial of degree $n \in N$ is a function $f : X \to Y$ of the form

$$f(\mathbf{x}) = Q(\mathbf{x}, \ldots, \mathbf{x}), \text{ where } Q \in ML_n(X^n, Y).$$

Hence, a homogeneous polynomial is defined in terms of an *associated multilinear function* Q. This multilinear function is defined on X^n with values in Y. The points in X are denoted as $\mathbf{x}$ and the points in X^n as

$$(\mathbf{x}_1, \ldots, \mathbf{x}_n), \ \mathbf{x}_i \in X, \ i = 1, \ldots, n.$$

The multilinear function associated with a polynomial is not unique. In particular, note that $f(\mathbf{x}) = Q(\mathbf{x}, \ldots, \mathbf{x}) = S(\mathbf{x}, \ldots, \mathbf{x})$ where

$$S(\mathbf{x}_1, \ldots, \mathbf{x}_n) = \frac{1}{n!} \sum_{\sigma \in \Sigma_n} Q(\mathbf{x}_{\sigma(1)}, \ldots, \mathbf{x}_{\sigma(n)})$$

is the symmetric part of Q, as in Definition 3.3.12.

Derivatives of Polynomials

Lemma 7.3.1 *Let $f(\mathbf{x}) = Q(\mathbf{x}, \ldots, \mathbf{x})$ be a homogeneous polynomial of degree n. Let $S \in ML_n(X^n, Y)$ be the symmetric part of Q. Then*

$$f'(\mathbf{x})(\mathbf{u}) = Q(\mathbf{u}, \mathbf{x}, \ldots, \mathbf{x}) + \ldots + Q(\mathbf{x}, \ldots, \mathbf{x}, \mathbf{u}) \quad (7.7)$$

$$= n\, S(\mathbf{x}, \ldots, \mathbf{x}, \mathbf{u}) \quad (7.8)$$

for all $\mathbf{x}, \mathbf{u} \in X$.

Proof. Define $P : X \to X^n$ by $P(\mathbf{x}) = (\mathbf{x}, \ldots, \mathbf{x}) \in X^n$, $\mathbf{x} \in X$. We see that $P : X \to X^n$ is a linear function. Hence

$$P'(\mathbf{x})(\mathbf{u}) = P(\mathbf{u}) = (\mathbf{u}, \ldots, \mathbf{u}) \text{ for all } \mathbf{x}, \mathbf{u} \in X.$$

Now $f : X \to Y$ is the composition $f = S \cdot P$. Now apply the chain rule, Theorem 5.5.6. The derivative of S is given by Theorem 5.6.2, together with Remarks 5.6.3. This gives (7.7). To obtain (7.8), note that $f(\mathbf{x}) = S(\mathbf{x}, \ldots, \mathbf{x})$ and use the symmetry of S in (7.7). □

Theorem 7.3.2 *Let $f(\mathbf{x}) = Q(\mathbf{x}, \ldots, \mathbf{x})$ be a homogeneous polynomial of degree n. Let $S \in ML_n(X^n, Y)$ be the symmetric part of Q. Then for each $k \in \mathbb{N}$ and for each $\mathbf{x} \in X$ the kth derivative $f^{(k)}(\mathbf{x}) \in ML_k(X^k, Y)$ exists. If $1 \leq k \leq n$, then*

$$f^{(k)}(\mathbf{x})(\mathbf{u}_1, \cdots, \mathbf{u}_k) = \frac{n!}{(n-k)!} S(\mathbf{x}, \ldots, \mathbf{x}, \mathbf{u}_1, \cdots, \mathbf{u}_k). \quad (7.9)$$

Here $\mathbf{x}$ *enters* $(n-k)$ *times as an argument of* $S : X^n \to Y$.

Proof. Lemma 7.3.1 gives (7.9) for $k = 1$. Now assume (7.9) for a k, $1 \leq k < n$. Let $\mathbf{u}_1, \ldots, \mathbf{u}_k$ be fixed vectors in X. Then, with $\ell = n - k$,

$$R(\mathbf{x}_1, \ldots, \mathbf{x}_\ell) = S(\mathbf{x}_1, \ldots, \mathbf{x}_\ell, \mathbf{u}_1, \ldots, \mathbf{u}_k)$$

is a symmetric multilinear function in $ML_\ell(X^\ell, Y)$. If g is the associated homogeneous polynomial of degree ℓ, then

$$\begin{aligned} f^{(k)}(\mathbf{x})(\mathbf{u}_1, \ldots, \mathbf{u}_k) &= (n!/\ell!)g(\mathbf{x}). \text{ Hence} \\ f^{(k+1)}(\mathbf{x})(\mathbf{u}_1, \ldots, \mathbf{u}_k, \mathbf{u}_{k+1}) &= (n!/\ell!)g'(\mathbf{x})(\mathbf{u}_{k+1}) \\ &= (n!/\ell!)\,\ell\, S(\mathbf{x}, \ldots, \mathbf{x}, \mathbf{u}_1, \cdots, \mathbf{u}_k, \mathbf{u}_{k+1}). \end{aligned}$$

This is (7.9) for $k+1$. Hence (7.9) is true for all k, $1 \leq k \leq n$. □

Corollary 7.3.3 *If* $f(\mathbf{x}) = S(\mathbf{x}, \ldots, \mathbf{x})$ *with symmetric* $S \in ML_n(X^n, Y)$, *then*

$$f^{(n)}(\mathbf{x})(\mathbf{u}_n, \cdots, \mathbf{u}_1) = n!\, S(\mathbf{u}_n, \cdots, \mathbf{u}_1) \tag{7.10}$$

is independent of $\mathbf{x}$. *Hence* $f^{(k)} : X \to ML_k(X^k, Y)$ *vanishes for all* $k > n$. *Also,* $f^{(k)}(\mathbf{0}) = \mathbf{0}$ *for all* $k \neq n$.

Proof. This follows directly from (7.9) of Theorem 7.3.2. If $k < n$, then

$$f^{(k)}(\mathbf{0})(\mathbf{u}_1, \cdots, \mathbf{u}_k) = \frac{n!}{(n-k)!} S(\mathbf{0}, \ldots, \mathbf{0}, \mathbf{u}_1, \cdots, \mathbf{u}_k) = \mathbf{0}.$$

If $k = n$, then S contains no $\mathbf{x}$ terms. Hence $f^{(n)} : X \to ML_n(X^n, Y)$ is a constant function and all higher derivatives vanish. □

Example 7.3.4 Define $F : L(X, X) \to L(X, X)$ by $F(T) = T^3$, $T \in L(X, X)$. This is a homogeneous polynomial of degree 3. In fact, $Q(A, B, C) = ABC$ defines a multilinear function and $F(T) = T^3 = Q(T, T, T)$. By three successive applications of Lemma 7.3.1 we obtain

$$\begin{aligned} F'(T)(A) &= AT^2 + TAT + T^2A \\ F''(T)(A, B) &= ABT + ATB + BAT + TAB + BTA + TBA \\ F'''(T)(A, B, C) &= ABC + ACB + BAC + CAB + BCA + CBA. \end{aligned}$$

The symmetric part of this polynomial is

$$S(A, B, C) = (ABC + BCA + CAB + ACB + CBA + BAC)/6.$$

Hence the relation $F'''(T)(A, B, C) = 6\, S(A, B, C)$ is verified.

Term-by-Term Differentiations

Consider a sequence of multilinear functions $Q_n \in ML_n(X^n, Y)$ and the associated homogeneous polynomials $f_n(x) = Q_n(\mathbf{x} \ldots, \mathbf{x})$. We define

$$F_n(\mathbf{x}) = \sum_{i=1}^{n} f_i(\mathbf{x}) \text{ and } G_n(\mathbf{x}) = F_n'(\mathbf{x}) = \sum_{i=1}^{n} f_i'(\mathbf{x}).$$

We assume that Q_ns are symmetric multilinear functions. This is not a loss of generality since $f_n(x) = Q_n(\mathbf{x} \ldots, \mathbf{x}) = S_n(\mathbf{x} \ldots, \mathbf{x})$, where S_n is the symmetric part of Q_n. We write $Q_n = S_n$. We also assume that there is an $R > 0$ for which the sequence $\|S_n\| R^n$ is bounded. As before,

$$B_r(\mathbf{0}) = \{ \mathbf{x} \mid \|\mathbf{x}\| < r \} \subset X$$

is the open ball of radius $r > 0$ about the origin.

Theorem 7.3.5 *If* $\mathbf{x} \in B_R(\mathbf{0})$, *then* $\lim_n F_n(\mathbf{x}) = F(\mathbf{x})$ *exists in* Y. *Also,* $F : B_R(\mathbf{0}) \to Y$ *is a continuous function.*

Proof. This a restatement of Theorem 4.4.32. □

Lemma 7.3.6 *If* $\mathbf{x} \in B_R(\mathbf{0})$ *and* $\mathbf{u} \in X$, *then*

$$G(\mathbf{x})(\mathbf{u}) = \lim_n G_n(\mathbf{x})(\mathbf{u}) = \sum_{n=1}^{\infty} f_n'(\mathbf{x})(\mathbf{u}) \tag{7.11}$$

exists in Y.

Proof. Theorem 7.3.2 shows that

$$\|f_n'(\mathbf{x})(\mathbf{u})\| = \|n\, S_n(\mathbf{u}, \mathbf{x}, \ldots, \mathbf{x})\| \le \|S_n\|\, n\, \|\mathbf{u}\|\, \|\mathbf{x}\|^{n-1}.$$

Let $\|\mathbf{x}\| < R$. Find an $r \in \mathbb{R}$ such that $\|\mathbf{x}\| < r < R$. Since $\|S_n\| R^n$ is bounded, we have $\sum_n \|S_n\| n r^{n-1} < \infty$ by Theorem 2.5.7. Hence

$$\sum_n \|S_n\|\, n\, \|\mathbf{u}\|\, \|\mathbf{x}\|^{n-1} < \infty.$$

Then, as in the proof of Theorem 4.4.32, we see that the limit in (7.11) exists. □

Notations 7.3.7 For fixed $\mathbf{x}, \mathbf{u} \in X$, $r > 0$, and $n \in \mathbb{N}$, define

$$h_n(s) = f_n(\mathbf{x} + s\mathbf{u}) - f_n(\mathbf{x}) - s\, f_n'(\mathbf{x})(\mathbf{u}) \in Y \text{ and} \tag{7.12}$$

$$\varphi_n(s) = (n(n-1)/2)\, \|S_n\|\, r^{n-2} \|\mathbf{u}\|^2\, s^2 \in \mathbb{R} \tag{7.13}$$

for all $s \in \mathbb{R}$.

Lemma 7.3.8 *If* $0 \leq r < R$ *then*

$$M = \sum_n (n(n-1)/2)\, \|S_n\|\, r^{n-2} \|\mathbf{u}\|^2 < \infty$$

and $\sum_n \varphi_n(s) \leq M s^2$ *for all* $s \in \mathbb{R}$.

Proof. Theorem 4.4.32 shows that $\sum_n (n(n-1)/2)\, \|S_n\|\, r^{n-2} \|\mathbf{u}\|^2 < \infty$ whenever $\|S_n\|\, R^n$ is a bounded sequence and $0 < r < R$. □

Lemma 7.3.9 *If* $(\|\mathbf{x}\| + \|\mathbf{u}\|\, |t|) < r$, *then* $\|h_n(t)\| \leq \varphi_n(t)$.

Proof. We see that $h_n : \mathbb{R} \to Y$ is a differentiable function and

$$\begin{aligned} h_n'(s) &= f_n'(\mathbf{x} + s\mathbf{u})(\mathbf{u}) - f_n'(\mathbf{x})(\mathbf{u}) \\ &= nQ_n(\mathbf{u},\, \mathbf{x} + s\mathbf{u},\, \ldots,\, \mathbf{x} + s\mathbf{u}) - nQ_n(\mathbf{u},\, \mathbf{x},\, \ldots,\, \mathbf{x}),\ s \in \mathbb{R}. \end{aligned}$$

For a fixed $\mathbf{u} \in X$, define $S_{n-1} \in ML_{n-1}(X^{n-1}, Y)$ by

$$S_{n-1}(\mathbf{x}_1, \ldots, \mathbf{x}_{n-1}) = S_n(\mathbf{u}, \mathbf{x}_1, \ldots, \mathbf{x}_{n-1}).$$

Then, clearly, $\|S_{n-1}\| \leq \|\mathbf{u}\|\, \|S_n\|$. Hence Theorem 4.3.14 on the increments of multilinear functions shows that, if $(\|\mathbf{x}\| + \|\mathbf{u}\|\, |s|) < R$, then

$$\begin{aligned} \|h_n'(s)\| &\leq n(n-1)\|S_{n-1}\|\, r^{n-2}\, \|s\mathbf{u}\| \\ &\leq n(n-1)\|S_n\|\, \|\mathbf{u}\|\, r^{n-2}\, \|s\mathbf{u}\| \\ &\leq n(n-1)\|S_n\|\, r^{n-2}\, \|\mathbf{u}\|^2\, |s| \\ &= \varphi_n'(|s|). \end{aligned}$$

Then Theorem 5.1.12, a version of the mean value theorem, shows that

$$\|h_n(t)\| = \|h_n(t) - h_n(0)\| \leq \varphi(|t|) - \varphi(0) = \varphi(t). \quad \square$$

Lemma 7.3.10 *Let* $f_n(x) = S_n(\mathbf{x}, \ldots, \mathbf{x})$, $\mathbf{x} \in X$ *be a sequence of polynomials* $X \to Y$. *For fixed* $\mathbf{x}, \mathbf{u} \in X$, *let*

$$F_n(\mathbf{x}) = \textstyle\sum_{k=1}^n f_k(\mathbf{x}) \text{ and } G_n(\mathbf{x})(\mathbf{u}) = \sum_{k=1}^n f_k'(\mathbf{x})(\mathbf{u}),\ n \in \mathbb{N}.$$

If there is an $R > 0$ *such that* $\|S_n\|\, R^n$ *is a bounded sequence, then*

$$\lim_n F_n(\mathbf{x}) = F(\mathbf{x}) \text{ and } \lim_n G_n(\mathbf{x})(\mathbf{u}) = G(\mathbf{x})(\mathbf{u})$$

both exist in Y for all $\mathbf{x} \in B_R(\mathbf{0})$ and $\mathbf{u} \in X$. Also, $F : B_R(\mathbf{0}) \to Y$ is a differentiable function and $F'(\mathbf{x})(\mathbf{u}) = G(\mathbf{x})(\mathbf{u})$ for all $\mathbf{u} \in X$.

Proof. The existence of $F(\mathbf{x})$ was obtained in Theorem 7.3.5 and the existence of $G(\mathbf{x})(\mathbf{u})$ in Lemma 7.3.6 above. Let $\mathbf{x} \in B_R(\mathbf{0})$, $\mathbf{u} \in X$. Let $c > 0$ be such that $(\|\mathbf{x}\| + c\|\mathbf{u}\|) < r < R$. If $|t| < c$, then

$$\begin{aligned} \|F_n(\mathbf{x} + t\mathbf{u}) - F_n(\mathbf{x}) - tG_n(\mathbf{x})(\mathbf{u})\| &\leq \sum\nolimits_{k=1}^{n} \|h_k(t)\| \\ &\leq \sum\nolimits_{n} \varphi_n(t) \leq M\,t^2. \end{aligned}$$

The last two inequalities follow from Lemmas 7.3.8 and 7.3.9 above. By taking the limit on n and using the the continuity of the norm function, we obtain

$$\|F(\mathbf{x} + t\mathbf{u}) - F(\mathbf{x}) - tG(\mathbf{x})(\mathbf{u})\| \leq M\,t^2$$

whenever $|t| < c$. This proves that $F'(\mathbf{x})(\mathbf{u}) = G(\mathbf{x})(\mathbf{u})$. □

Theorem 7.3.11 Term-by-term differentiation. *Let $f_n(x) = S_n(\mathbf{x}, \ldots, \mathbf{x})$ be a sequence of polynomials $X \to Y$. Assume that there is an $R > 0$ such that the sequence $\|S_n\|\, R^n$ is bounded. Then*

$$F(\mathbf{x}) = \sum\nolimits_{n} f_n(\mathbf{x})$$

defines a $\mathcal{C}^\infty$ function $F : B_R(\mathbf{0}) \to Y$. Also,

$$D_{\mathbf{u}_k} \cdots D_{\mathbf{u}_1} F(\mathbf{x}) = \sum\nolimits_{n} D_{\mathbf{u}_k} \cdots D_{\mathbf{u}_1} f_n(\mathbf{x}) \tag{7.14}$$

for all $\mathbf{u}_1, \ldots, \mathbf{u}_k \in X$.

Proof. This follows from Lemma 7.3.10 above by an induction on $k \in \mathbb{N}$. In fact, this lemma gives (7.14) for $k = 1$. Assume (7.14) for a fix set of vectors $\mathbf{u}_1, \ldots, \mathbf{u}_k \in X$. Let $\max_i \|\mathbf{u}_i\| = u$ Then we see easily that

$$\|D_{\mathbf{u}_k} \cdots D_{\mathbf{u}_1} f_n(\mathbf{x})\| \leq n^{k-1} \|S_n\|\, u^k\, \|\mathbf{x}\|^{n-k}.$$

But $\sum_n n^{k-1} \|S_n\|\, u^k\, r^{n-k} < \infty$ whenever $0 \leq r < R$. This is again by Theorem 4.4.32. Then Lemma 7.3.10 above shows that (7.14) is also true for $k + 1$. □

Example 7.3.12 The inversion operator. Let $L_{\mathrm{Inv}}(X, X)$ be the set of all invertible mappings in $L(X, X)$. Then the inversion operator

$$\vartheta = \mathrm{Inv} : L_{\mathrm{Inv}}(X, X) \to L_{\mathrm{Inv}}(X, X) \subset L(X, X)$$

takes $T \in L_{\text{Inv}}(X, X)$ to $\vartheta(T) = T^{-1}$. We claim that ϑ is a $\mathcal{C}^\infty$ function on $L_{\text{Inv}}(X, X)$. First, we show that all the derivatives $\vartheta^{(m)}(I)$ exist at the identity. This follows from the expansion

$$\vartheta(I + T) = I - T + T^2 + \cdots + (-1)^n T^n + \cdots$$

obtained in Theorem 4.4.38 and from Theorem 7.3.11 above. To see differentiability at a general point $A \in L_{\text{Inv}}(X, X)$, it is enough to let $A^{-1} = B$ and observe that

$$\vartheta(A + T) = \vartheta(I + BT) \cdot B$$

as in the proof of Theorem 4.4.38. Finally, the inversion operator

$$\text{Inv} : L_{\text{Inv}}(X, Y) \to L(Y, X)$$

between different spaces is also a $\mathcal{C}^\infty$ function. Again, to see this, it is enough to let let $A \in L_{\text{Inv}}(X, Y)$ with $B = A^{-1} \in L(Y, X)$ and observe that

$$\text{Inv}(A + T) = \vartheta(I + BT) \cdot B.$$

See Example 7.3.20 below on the computation of some derivatives of ϑ. △

Example 7.3.13 The exponentiation operator. The exponentiation operator exp : $L(X, X) \to L(X, X)$ takes $T \in L(X, X)$ to

$$\lim_n (I + T + (1/2!)T^2 + \cdots + (1/n!)T^n) = e^T \in L(X, X).$$

Theorem 7.3.11 shows that $\exp : L(X, X) \to L(X, X)$ is a $\mathcal{C}^\infty$ function. Note that the derivatives of e^T exist, but they may not have simple expressions. For example, if $A \in L(X, X)$ then

$$\begin{aligned} D_A e^T \quad &= \quad A + (AT + TA) + (1/2!)(AT^2 + TAT + TA^2) + (1/3!) \\ &\quad\quad +(1/3!)(AT^3 + TAT^2 + T^2AT + T^3A) + \cdots . \end{aligned}$$

This may not have a simple expression unless A and T commute. If $AT = TA$, then $D_A e^T = Ae^T = e^T A$. See Example 7.3.19 below for the derivatives of e^T at $T = \mathbf{0}$. Finally, let $X = \mathbb{R}$ and consider $t \in \mathbb{R}$ as the linear map that takes $x \in \mathbb{R}$ to $tx \in \mathbb{R}$. In this case e^t is the classical exponential function. We see that $D_a\, e^t = \lim_{r \to 0}(1/r)(e^{t+ra} - e^t) = ae^t$. △

Taylor Polynomials and Series

Taylor polynomials and Taylor series can be generalized from functions of a real variable to functions of a vector variable. No new arguments are needed here. These generalizations follow directly from the corresponding results for the functions of a real variable.

Definition 7.3.14 Let $f : A \to Y$ be a $\mathcal{C}^n$ function. Let $\mathbf{a} \in A$. Then

$$P_k(\mathbf{a}+\mathbf{x}) = \sum_{m=0}^{k} \frac{1}{m!} f^{(m)}(\mathbf{a})(\mathbf{x}, \ldots, \mathbf{x})$$

is the Taylor polynomial of f of degree k at $\mathbf{a} \in A$.

Theorem 7.3.15 Approximation by Taylor polynomials. *Let $f : A \to Y$ be a $\mathcal{C}^n$ function. Let $R > 0$ and $M > 0$ be such that $\|f^{(n)}(\mathbf{a}+\mathbf{x})\| \leq M$ whenever $\|\mathbf{x}\| < R$. Then*

$$\|f(\mathbf{a}+\mathbf{x}) - P_{n-1}(\mathbf{a}+\mathbf{x})\| \leq (1/n!)M \, \|\mathbf{x}\|^n \tag{7.15}$$

whenever $\|x\| < R$.

Proof. Let $\mathbf{x} \in X, 0 < \|\mathbf{x}\| < R$, be fixed. Define $\varphi(s) = f(\mathbf{a}+s\mathbf{x})$. Then Lemma 5.2.18 on the computation of directional derivatives shows that

$$\begin{aligned}
\varphi'(s) &= f'(\mathbf{a}+s\mathbf{x})(\mathbf{x}). \text{ By an induction on } k \text{ we obtain} \\
\varphi^{(k)}(s) &= f^{(k)}(\mathbf{a}+s\mathbf{x})(\mathbf{x}, \ldots, \mathbf{x}). \text{ Hence} \\
p_{n-1}(s) &= \sum_{k=0}^{n-1} (1/k!)\, \varphi^{(k)}(0)s^k \\
&= \sum_{k=0}^{n-1} (1/k!)\, f^{(k)}(\mathbf{a})(\mathbf{x}, \ldots, \mathbf{x})s^k.
\end{aligned}$$

Let $r = R/\|\mathbf{x}\|$. Then $r > 1$ and $\|f^{(n)}(\mathbf{a}+s\mathbf{x})\| < M$ whenever $0 \leq s < r$. Hence

$$|\varphi(n)(s)| = \|f^{(n)}(\mathbf{a}+s\mathbf{x})(\mathbf{x}, \ldots, \mathbf{x})\| \leq M \, \|\mathbf{x}\|^n$$

whenever $0 \leq s \leq 1 < r$. Then Taylor's theorem, Theorem 5.1.22, for functions of a real variable shows that

$$\|\varphi(1) - p_{n-1}(1)\| = \|f(\mathbf{a}+\mathbf{x}) - P_{n-1}(\mathbf{a}+\mathbf{x})\| \leq (1/n!)M \, \|\mathbf{x}\|^n.$$

Here we observed that $p_{n-1}(1) = P_{n-1}(\mathbf{a}+\mathbf{x})$. □

Notation 7.3.16 For each $n \in \mathbb{N}$ and $r > 0$, let

$$M_n(r) = \sup\left\{ \|f^{(n)}(\mathbf{a}+\mathbf{x})\| \mid \|\mathbf{x}\| \leq r \right\}$$

if it exists.

Theorem 7.3.17 Approximation by Taylor series. *Assume that there is an $R > 0$ such that $(1/n!)\, M_n(R)R^n$ is a bounded sequence. Then*

$$\begin{aligned} f(\mathbf{a}+\mathbf{x}) &= f(\mathbf{a}) + f'(\mathbf{a})(\mathbf{x}) + \cdots + \frac{1}{n!} f^{(n)}(\mathbf{a})(\mathbf{x}, \ldots, \mathbf{x}) + \cdots \\ &= \lim_n \sum_{k=0}^{n} \frac{1}{k!} f^{(k)}(\mathbf{a})(\mathbf{x}, \ldots, \mathbf{x}) = \lim_n P_n(\mathbf{a}+\mathbf{x}) \end{aligned}$$

whenever $\|\mathbf{x}\| < R$.

Proof. Let $(1/n!)\, M_n(R)R^n$ be a bounded sequence. We see that

$$\lim_n (1/n!)\, M_n(R) r^n = 0$$

whenever $|r| < R$. Then the result follows directly from Theorem 7.3.15. □

Derivatives in Terms of Taylor Series

Let $f : A \to Y$ be a function. If we can express $f(\mathbf{a}+\mathbf{x})$ as a convergent sequence of polynomials in $\mathbf{x}$, then we can easily find the derivatives of f at $\mathbf{a}$. We state this result as follows.

Theorem 7.3.18 *Let $f_n(\mathbf{x})$ be a sequence of homogeneous polynomials associated with multilinear functions $Q_n \in ML_n(X^n, Y)$. Assume that there is an $R > 0$ such that the sequence $\|Q_n\|\, R^n$ remains bounded. If*

$$F(\mathbf{x}) = f_1(\mathbf{x}) + f_2(\mathbf{x}) + \cdots$$

with $\|\mathbf{x}\| < R$, then $F^{(n)}(\mathbf{0})(\mathbf{u}_1, \cdots, \mathbf{u}_n) = n!\, S_n(\mathbf{u}_1, \cdots, \mathbf{u}_n)$. Here S_n is the symmetric part of Q_n, as in Definition 3.3.12.

Proof. By Theorem 7.3.11 we can differentiate $F(\mathbf{x})$ term by term. Hence

$$F^{(n)}(\mathbf{0})(\mathbf{u}_1, \cdots, \mathbf{u}_n) = f_1^{(n)}(\mathbf{0})(\mathbf{u}_1, \cdots, \mathbf{u}_n) + f_2^{(n)}(\mathbf{0})(\mathbf{u}_1, \cdots, \mathbf{u}_n) + \cdots.$$

But, by Corollary 7.3.3, $f_k^{(n)}(\mathbf{0})(\mathbf{u}_1, \cdots, \mathbf{u}_n) = \mathbf{0}$ unless $k = n$. Hence,

$$F^{(n)}(\mathbf{0})(\mathbf{u}_1, \cdots, \mathbf{u}_n) = f^{(n)}(\mathbf{0})(\mathbf{u}_1, \cdots, \mathbf{u}_n) = n!\, S_n(\mathbf{u}_1, \cdots, \mathbf{u}_n),$$

again by Corollary 7.3.3. □

Example 7.3.19 Derivatives of e^T at $T = \mathbf{0}$. From the definition

$$F(T) = e^T = I + T + (1/2!)T^2 + \cdots + (1/n!)T^n + \cdots$$

we see that $F^{(n)}(\mathbf{0})(A_1, \ldots, A_n) = n!\, S_n(A_1, \ldots, A_n)$. Here S_n is the symmetric part of $Q_n \in ML_n(X^n, Y)$ such that $(1/n!)\, T^n = Q_n(T, \ldots, T)$. Hence, for example,

$$F'''(\mathbf{0})(A, B, C) = (ABC + BCA + CAB + ACB + CBA + BAC). \quad \triangle$$

Example 7.3.20 The inversion operator. Let $L_{\text{Inv}}(X, X)$ be the set of all invertible mappings in $L(X, X)$. Then the inversion operator

$$\text{Inv} : L_{\text{Inv}}(X, X) \to L_{\text{Inv}}(X, X)$$

takes $T \in L_{\text{Inv}}(X, X)$ to $\text{Inv}\, T = T^{-1}$. We can find the derivatives of the function at $A \in L_{\text{Inv}}(X, X)$ as follows. Let $B = A^{-1}$. We have

$$(A + T)^{-1} = (A(I + BT))^{-1} = (I + BT)^{-1}A^{-1} = (I + BT)^{-1}B.$$

Theorem 4.4.38 shows that the series of polynomials

$$F(T) = B - (BT)B + (BT)^2B + \cdots + (-1)^n(BT)^nB + \cdots$$

converge to $F(T) = (A + T)^{-1}$ whenever $\|BT\| < 1$. In particular, we have convergence whenever $\|T\| < 1/\|B\|$. Hence we can find the derivatives of $F(T)$ at $T = \mathbf{0}$. We see easily that $F^{(n)}(\mathbf{0}) = (\text{Inv})^{(n)}(A)$. Hence

$$(\text{Inv})^{(n)}(A)(U_1, \ldots, U_n) = n!\, S_n(U_1, \ldots, U_n).$$

Here S_n is the symmetric part of a multilinear function that induces the polynomial $(BT)^nB$. The first three derivatives of the inversion operation are

$$\begin{aligned}
(\text{Inv})'(A)(U) &= -BUB \\
(\text{Inv})''(A)(U, V) &= (BUBVB + BVBUB) \\
(\text{Inv})'''(A)(U, V, W) &= -(BUBVBWB + BVBWBUB + BWBUBVB \\
&= +BUBWBVB + BWBVBUB + BVBUBWB).
\end{aligned}$$

In the classical case of $f(x) = (1/x)$, $x \neq 0$, these formulas are reduced to

$$f^{(n)}(x) = (-1)^n \frac{n!}{x^{n+1}}.$$

The verification of this is left as an exercise.

Inverse Function Theorem for Higher Derivatives

Theorem 7.3.21 Extended inverse function theorem. *Let A be an open set in X and let $f : A \to Y$ be a $\mathcal{C}^n$ function. Let $\mathbf{a} \in A$ and assume that $f'(\mathbf{a}) : X \to Y$ is an isomorphism. Then there is an open set G in X such that $\mathbf{a} \in G \subset A$ and such that the restriction of f to G is a $\mathcal{C}^n$-diffeomorphism.*

Proof. Let $f : A \to Y$ be a $\mathcal{C}^1$ function such that $f'(\mathbf{a}) : X \to Y$ is an isomorphism. Then the inverse function theorem, Theorem 6.1.4, shows that there is an open set G in X such that $\mathbf{a} \in G \subset A$ and such that the restriction of f to G is a diffeomorphism. Hence $f(G) = H$ is an open set in Y containing $\mathbf{b} = f(\mathbf{a})$, and the inverse function $g : H \to X$ is also a continuously differentiable function. Also, we know that

$$g'(\mathbf{y}) = (f'(g(y))^{-1} = (\vartheta \cdot f' \cdot g)(\mathbf{y}), \quad \mathbf{y} \in H, \tag{7.16}$$

since $g : H \to G$ is the reverse diffeomorphism of $f : G \to H$. Here

$$\vartheta : L_{\mathrm{Inv}}(X, Y) \to L(Y, X)$$

is the inversion operator that takes an invertible $T : X \to Y$ to its inverse $\vartheta(T) = T^{-1} : Y \to X$.

The additional part in the extended theorem is that if $f : G \to H$ is a $\mathcal{C}^n$ function, then $g : H \to G$ is also a $\mathcal{C}^n$ function. Proceed by induction on n. This result is true for $n = 1$ by the original theorem. Assume that it is true if f is a $\mathcal{C}^n$ function. Let f be a $\mathcal{C}^{n+1}$ function. Then $f' : G \to L(X, Y)$ and $g : H \to G$ are $\mathcal{C}^n$ functions by the induction hypothesis. Also, $\vartheta : L_{\mathrm{Inv}}(X, Y) \to L(Y, X)$ is a $\mathcal{C}^n$ function, since it is actually a $\mathcal{C}^\infty$ function by Example 7.3.12. Then (7.16) shows that $g' : H \to L(Y, X)$ is a composition of $\mathcal{C}^n$ functions. Hence, by Theorem 7.1.6, g' is also a $\mathcal{C}^n$ function. Therefore $g : H \to G$ is a $\mathcal{C}^{n+1}$ function. □

7.4 LOCAL EXTREMAL VALUES

Local extremal values of real-valued functions were defined in Definition 5.4.14. Taylor polynomials allow us to obtain a test to find these values. This is formulated in terms of positive or negative definite real-valued polynomials.

Definition 7.4.1 Positive definite polynomials. Let $p : X \to \mathbb{R}$ be a real-valued homogeneous polynomial of degree $n \in \mathbb{N}$. Then p is called a (strictly) *positive definite polynomial* if $p(\mathbf{x}) > 0$ whenever $\mathbf{x} \neq \mathbf{0}$.

Remarks 7.4.2 If X is an inner product space, then $p(\mathbf{x}) = \langle \mathbf{x}, \mathbf{x} \rangle$ is a positive definite homogeneous polynomial of degree 2. If p is a homogeneous polynomial of degree n, then $p(t\mathbf{x}) = t^n \, p(\mathbf{x})$. Hence, if n is odd, then p can never be positive definite. If n is even, then p may or may not be positive definite. If $p : \mathbb{R}^2 \to \mathbb{R}$ is a second-degree homogeneous polynomial, then there is a familiar necessary and sufficient condition for positive definiteness. This is given in Problem 7.1. In general, there may not be an easy way of finding out if a homogeneous polynomial is positive definite.

Lemma 7.4.3 *Let $p : X \to \mathbb{R}$ be a positive definite polynomial of degree n. Then there is a number $K > 0$ such that $K \, \|\mathbf{x}\|^n \le p(\mathbf{x})$ for all $\mathbf{x} \in X$.*

Proof. Let $S = \{\, \mathbf{u} \in X \mid \|\mathbf{u}\| = 1 \,\}$. Then S is a compact set in X. Also, $p : X \to \mathbb{R}$ is a continuous function. (One way of seeing this is to note that p is a differentiable function.) Hence p reaches a minimum value on the compact set S by Theorem 4.5.44. Therefore there is an $\mathbf{a} \in S$ such that $p(\mathbf{a}) \le p(\mathbf{u})$ for all $\mathbf{u} \in S$. Then $0 < p(\mathbf{a}) = K$ since p is positive definite and $\mathbf{a} \ne \mathbf{0}$. Now any $\mathbf{x} \in X$ can be expressed as $\mathbf{x} = \|\mathbf{x}\| \, \mathbf{u}$ with $\mathbf{u} \in S$. Hence $p(\mathbf{x}) = \|\mathbf{x}\|^n \, p(\mathbf{u})$. Then the proof follows. □

Theorem 7.4.4 *Let $f : A \to \mathbb{R}$ be a real-valued function defined on an open set A in X. Assume that f is differentiable as many times as needed. Let $\mathbf{a} \in A$ and let $m \in \mathbb{N}$ be the smallest integer such that $f^{(m)}(\mathbf{a}) \ne \mathbf{0}$. If m is odd, then f cannot have a local extremal value at $\mathbf{a}$. If m is even and if the polynomial*

$$p(\mathbf{x}) = f^{(m)}(\mathbf{a})(\mathbf{x}, \ldots, \mathbf{x})$$

is positive definite, then f has a local minimum at $\mathbf{a}$. If there are $\mathbf{x}_i$ such that $p(\mathbf{x}_1) > 0$ and $p(\mathbf{x}_2) < 0$, then f cannot have a local local extremal value at $\mathbf{a}$. If $p(\mathbf{x}) \ge 0$ or if $p(\mathbf{x}) \le 0$ for all $\mathbf{x} \in X$, then f may or may not have a local extremal value at $\mathbf{a}$.

Proof. Let $f^{(m)}(\mathbf{a})$ be the first nonzero derivative. Let

$$p(\mathbf{x}) = (1/m!) f^{(m)}(\mathbf{a})(\mathbf{x}, \ldots, \mathbf{x}), \quad \mathbf{x} \in X.$$

Then the mth order Taylor polynomial $P_m(\mathbf{a} + \mathbf{x})$ defined in Definition 7.3.14 becomes

$$P_m(\mathbf{a} + \mathbf{x}) = f(\mathbf{a}) + p(\mathbf{x}).$$

Let $R > 0$ and $M > 0$ be such that $\|f^{(m+1)}(\mathbf{a} + \mathbf{x})\| \le (m+1)! \, M$ whenever $\|\mathbf{x}\| < R$. Then Taylor's theorem, Theorem 7.3.15, shows that if $\|\mathbf{x}\| < R$ then

$$\begin{aligned} |f(\mathbf{a} + \mathbf{x}) - P_m(\mathbf{a} + \mathbf{x})| &= |f(\mathbf{a} + \mathbf{x}) - f(\mathbf{a}) - p(\mathbf{x})| & (7.17) \\ &\le M \|\mathbf{x}\|^{m+1}. & (7.18) \end{aligned}$$

Now assume that $m \in \mathbb{N}$ is odd. Let $\mathbf{r} \in X$ be such that $p(\mathbf{r}) \neq 0$. If $p(\mathbf{r}) < 0$, then $p(-\mathbf{r}) = (-1)^m\, p(\mathbf{r}) > 0$. Hence assume that $p(\mathbf{r}) > 0$ without loss of generality. Let $\mathbf{x} = t\mathbf{r}$ with $t \in \mathbb{R}$. Then $p(t\mathbf{r}) = t^m\, p(\mathbf{r})$ and $\|t\mathbf{r}\|^{m+1} = t^{m+1}\|\mathbf{r}\|$. Hence

$$t^m(p(\mathbf{r}) - tM\|\mathbf{r}\|) \leq f(\mathbf{a} + t\mathbf{r}) - f(\mathbf{a}) \leq t^m(p(\mathbf{r}) + tM\|\mathbf{r}\|)$$

whenever $|t|\,\|\mathbf{r}\| < R$. Take $t > 0$ such that $tM\|\mathbf{r}\| < (1/2)p(\mathbf{r})$. Then we see that

$$\begin{aligned} 0 < (1/2)t^m p(\mathbf{r}) < f(\mathbf{a} + t\mathbf{r}) - f(\mathbf{a}) \qquad &\text{and} \\ f(\mathbf{a} - t\mathbf{r}) - f(\mathbf{a}) < -(1/2)t^m p(\mathbf{r}) < 0. & \end{aligned}$$

Hence $f(\mathbf{a})$ cannot be a local extremal value for f.

Now assume that m is even and $p : X \to \mathbb{R}$ is positive definite. Lemma 7.4.3 shows that there is a $K > 0$ such that $K\|\mathbf{x}\|^m \leq p(\mathbf{x})$ for all $\mathbf{x} \in X$. In this case (7.18) shows that

$$\|\mathbf{x}\|^m(K - M\|\mathbf{x}\|) \leq f(\mathbf{a} + \mathbf{x}) - f(\mathbf{a})$$

whenever $\|\mathbf{x}\| < R$. But this implies $0 < f(\mathbf{a} + \mathbf{x}) - f(\mathbf{a})$ whenever

$$0 < \|\mathbf{x}\| < K/M \text{ and } \|\mathbf{x}\| < R.$$

Hence $f(\mathbf{a})$ is a local minimum value for f. Other parts of the theorem are left as Problem 7.2. □

Problems

7.1 Show that the polynomial $P(x, y) = Ax^2 + 2Bxy + Cy^2$ is positive definite if and only if $B^2 < AC$ and $0 < A$.

7.2 Give examples of the form

$$f(x, y) = Ax^2 + 2Bxy + Cy^2 + Dx^3 + Ex^2y + Fxy^2 + Gy^3$$

to cover all cases mentioned in Theorem 7.4.4.

PART III

INTEGRATION

CHAPTER 8

MULTIPLE INTEGRALS

Integration in $\mathbb{R}^n$ is a generalization of integration in $\mathbb{R} = \mathbb{R}^1$. In many ways, integration of functions of a single variable is a leading special case: it provides valuable guidance as we generalize to functions of several variables. Many of the basic definitions and theorems are essentially the same in $\mathbb{R}$ and in $\mathbb{R}^n$. In the general case, as in $\mathbb{R}$, integrals are defined as the limit of Riemann sums. Continuous functions are integrable. Integration is linear: if f, f' are integrable and α, $\alpha' \in \mathbb{R}$, then $\int(\alpha f + \alpha' f') = \alpha \int f + \alpha' \int f'$. Even many of the proofs that work in the one-variable case generalize to functions of several variables.

All the same, at least two significant complications emerge when we shift our focus from $\mathbb{R}$ to $\mathbb{R}^n$. The first of these has to do with computational techniques, and the second has to do with the contrast between length (in one dimension) and volume (in n dimensions). Let us take these two points in turn.

The first difference, as noted, pertains to computational techniques. The Fundamental Theorem of Calculus shows that differentiation and integration are inverse operations

Analysis in Vector Spaces.
By M. A. Akcoglu, P. F. A. Bartha and D. M. Ha

in $\mathbb{R}$. The theorem allows us to compute integrals in $\mathbb{R}$ by finding anti-derivatives. This simple method is not available when $n \geq 2$. Computations in $\mathbb{R}^n$ must usually be done by successive integrations, each involving the computation of an integral in $\mathbb{R}$ with familiar techniques. For this reason, integrals in $\mathbb{R}^n$ are usually referred to as *multiple integrals*. The main result that licenses this type of computation is *Fubini's Theorem* (Theorem 8.2.37), proven towards the end of Section 8.2.

The second difference relates to the fact that, while integration in $\mathbb{R}$ starts with the notion of the length of an interval, integration in $\mathbb{R}^n$ starts with higher-dimensional volumes. There is a basic difference between the definition of length and the definition of volume. If I is an interval with end points r and s, with $r < s$, then its length is given by $\ell(I) = (s - r)$. No such simple formula is available in higher dimensions, for we shall be interested in the volumes of many odd-shaped regions.

Volumes in higher dimensions — the main topic of Section 8.1 — are defined by a method that, in all essential respects, was developed by Archimedes (c. 287-212 B.C.). To estimate the area A of a two-dimensional region R, we can trace R on graph paper and count the number of squares of the graph paper that are completely in R. This gives a lower estimate for A. We can also count the number of squares that intersect R but may or may not be completely in R. This gives an upper estimate for A. To improve these estimates, we use finer graph paper with smaller squares. The region R has a well-defined area if and only if, as we let the squares on the paper shrink in size, the limit of the lower estimates agrees with the limit of the upper estimates.

The formal definition of area, or higher-dimensional volume, utilizes Archimedes' idea. We first define the volumes of certain basic sets, called *cubes*, and then the volumes of *unions* of cubes. In the one-dimensional case, these sets are intervals and their finite unions. In the the two-dimensional case described above, they are squares and finite unions of squares. And so on, for higher dimensions. Following the lead of Archimedes, we define the volume of other sets by forming upper and lower estimates with unions of cubes and then taking limits as the cubes become smaller and smaller. We shall denote volume in $\mathbb{R}^n$ by v^n or simply by v. One-dimensional volume, or length, will often be represented as ℓ instead of as v^1 or v.

It is quite important that the exact specifications of the squares in our sequence of grids, or the cubes employed in the limiting construction, turn out not to matter, so long as the edge lengths shrink to zero. Many different choices lead to the same limit and hence to the same notion of volume. We shall make use of *binary cubes*, whose edge lengths are always of the form 2^{-k} , as our basic type.

The approach we have just described for defining volume does not always work, since the upper and lower estimates may not converge to a common limit. In many texts, including this one, the term *Jordan set* is reserved for sets where the approach does lead to a well-defined volume. This is in honor of Camille Jordan (1838-1922), who formulated a general theory of volumes.

The ideas of Archimedes are enough to establish a complete theory of volumes. Actually, as we shall see, these methods also give a complete theory of integration. That is because an integral on $\mathbb{R}^n$ can be thought of as an $n+1$-dimensional volume, just as an integral on $\mathbb{R}$ is essentially the area of the two-dimensional region "under the curve" (for a positive function). The only part missing in this approach is an efficient general method for the computation of these volumes. Such a method was not obtained until almost two thousand years later through the development of calculus. Section 8.2 presents the basic theory of integration, together with the most important computational techniques.

Section 8.1, then, is devoted to providing a rigorous definition of volume, and Section 8.2 outlines the basic definitions and results for the theory of multiple integrals. The final two sections of the chapter provide a proof of the *Change of Variable Theorem.* This theorem, perhaps more than any other result, beautifully illustrates both the deep analogy and the contrast between integration on $\mathbb{R}$ and on $\mathbb{R}^n$. The one-variable version of the theorem is usually stated as follows:

$$\int_{g(a)}^{g(b)} f(y)\, dy = \int_a^b f(g(x))\, g'(x)\, dx. \tag{8.1}$$

The proof is an easy application of the Fundamental Theorem of Calculus. For functions on $\mathbb{R}^n$, the theorem takes the form

$$\int_B f(\mathbf{y})\, d\mathbf{y} = \int_A f(\varphi(\mathbf{x}))\ |\det(\varphi'(\mathbf{x}))|\ d\mathbf{x}, \tag{8.2}$$

where φ is a diffeomorphism, φ' is its derivative, and $B = \varphi(A)$. The formal analogy between the two equations is clear, with φ in the multidimensional case playing the role that g plays in the one-dimensional case. Unlike the one-dimensional case, the proof of this result is far from easy. That is a consequence of the complexity of the notion of volume as compared to length. In order to appreciate that a strong analogy between $\mathbb{R}$ and $\mathbb{R}^n$ still exists, one has to tackle this proof in three steps. The first step, in Section 8.3, is to prove the result where φ is a fixed linear transformation. The second and third steps, in Section 8.4, establish the result first where $f(\mathbf{y}) = 1$ and then for general f.

8.1 JORDAN SETS AND VOLUME

Recall that $\mathbb{Z} = \{0, \pm 1, \pm 2, \ldots\}$ and $\mathbb{Z}^+ = \{0, 1, 2, \ldots\}$ are the set of integers and the set of nonnegative integers. As explained in the introduction, we start by defining volume for a special class of cubes — a class that is big enough to construct the grids that we need to apply Archimedes' technique. We then define the volume of Jordan sets by employing that technique.

Binary Grids of Cubes

Definition 8.1.1 Binary grids in $\mathbb{R}^n$. Let $n \in \mathbb{N}$ and $k \in \mathbb{Z}^+$. The *kth-order binary grid* of $\mathbb{R}^n = \mathbb{R} \times \cdots \times \mathbb{R}$ is the collection $\mathcal{C}_k^n$ of cubes $C = I_1 \times \cdots \times I_n$ where each $I_i \subset \mathbb{R}$ is an interval of the form $I_i = [\, u_i,\, v_i\,)$ with $u_i = p_i\, 2^{-k}$, $v_i = (p_i + 1)\, 2^{-k}$, $p_i \in \mathbb{Z}$. Thus, $\mathcal{C}_k^n$ is a particular family of cubes in R^n with edge length 2^{-k}. The *center of C* is the point

$$\mathbf{c} = (1/2)(u_1 + v_1,\ \cdots\ ,\ u_n + v_n). \tag{8.3}$$

Let $\mathcal{Z}_k^n$ be the set of centers of the cubes in the grid $\mathcal{C}_k^n$. If the dimension n is understood from the context, we also write $\mathcal{C}_k$ and $\mathcal{Z}_k$ instead of $\mathcal{C}_k^n$ and $\mathcal{Z}_k^n$. If $n = 1$ or $n = 2$, then these cubes are certain intervals or squares. We will call the special cube $E_k = [\,0,\ 2^{-k})^n \in \mathcal{C}_k^n$ the *kth-order basic cube*. In particular, $E_0 = [\,0,\ 1\,)^n$ is called *the unit cube of* $\mathbb{R}^n$. Note that a binary grid divides $\mathbb{R}^n$ into mutually disjoint cubes $C \in \mathcal{C}_k^n$ by the hyper-planes $x_i = p_i\, 2^{-k}$, $p_i \in \mathbb{Z}$. (Taking half-open intervals is necessary for the cubes to be disjoint.)

Remarks 8.1.2 Translations of cubes. *Translation by* $\mathbf{s} \in \mathbb{R}^n$ is the transformation $\mathbb{R}^n \to \mathbb{R}^n$ that takes $\mathbf{x} \in \mathbb{R}^n$ to $(\mathbf{s} + \mathbf{x}) \in \mathbb{R}^n$. If $E \subset \mathbb{R}^n$, then $\mathbf{s} + E$ denotes the translation of E by $\mathbf{s}$. More explicitly,

$$\mathbf{s} + E = \{\, \mathbf{s} + \mathbf{x} \mid \mathbf{x} \in E\,\}.$$

Any two cubes in a grid $\mathcal{C}_k^n$ are translations of each other. In particular, each $C \in \mathcal{C}_k^n$ is the translation of the kth-order basic cube $E_k = [\,0,\ 2^{-k})^n \in \mathcal{C}_k^n$. These translations are by the *binary vectors* $\mathbf{s} = 2^{-k}(p_1,\ \ldots,\ p_n)$ with $p_i \in \mathbb{Z}$.

Cubes and Balls

Definition 8.1.3 The maximum norm. The Euclidean norm provides the standard notion of distance in $\mathbb{R}^n$. But in the theory of integration, the *maximum norm* is also a very convenient norm. It is defined as

$$\|(x_1,\ \ldots,\ x_n)\|_m = \max(|x_1|,\ \ldots,\ |x_n|).$$

With respect to this norm, balls are cubes. More precisely, the interior of a cube is an open ball and the closure of a cube is a closed ball. Restricting our attention to binary cubes, let $C \in \mathcal{C}_k$ be a kth-order cube with center $\mathbf{c}$, as in (8.3) of Definition 8.1.1. Then we see that

$$B_r(\mathbf{c}) = C^o \subset C \subset \overline{C} = \overline{B}_r(\mathbf{c})$$

where $r = (1/2)2^{-k} = 2^{-k-1}$ is the common radius of the cubes in $\mathcal{C}_k$. (Here both $B_r(\mathbf{c})$ and $\overline{B}_r(\mathbf{c})$ are balls with respect to the maximum norm, C^o is the interior of C, and $\overline{C}$ is the closure of C.) Hence

$$B_r(\mathbf{c}) = \{\, \mathbf{x} \mid \|\mathbf{x} - \mathbf{c}\|_m = \max_i |x_i - c_i| < r \,\}$$

and similarly for $\overline{B}_r(\mathbf{c})$, substituting $\leq$ for $<$. We shall often denote cubes using this notation.

Lemma 8.1.4 Density of the binary grids. *Any open set $G \subset \mathbb{R}^n$ contains cubes from a binary grid.*

Proof. As we proved in Chapter 4, the openness of a set is independent of the norm. Hence, given $\mathbf{u} \in G$, there is an $r > 0$ such that $\mathbf{x} \in G$ whenever $\|\mathbf{x} - \mathbf{u}\| < r$, where we are using the maximum norm. Hence,

$$(u_1 - r,\ u_1 + r) \times \cdots \times (u_n - r,\ u_n + r) \subset G.$$

Let $k \in \mathbb{N}$ be such that $2^{-k} < r$. In this case, each interval $(u_i - r,\ u_i + r)$ contains a binary interval $[\,p_i 2^{-k},\ (p_i + 1)2^{-k})$ with $p_i \in \mathbb{Z}$. Hence G contains a cube from the binary grid $\mathcal{C}_k$. $\square$

Cartesian Products of Cubes

In the following discussion, we use the integers n and m for the dimension of the space, while h and k are reserved for the order of the binary grid.

Notations 8.1.5 Sets in Cartesian products. Let $n, m \in \mathbb{N}$. If $A \subset \mathbb{R}^n$ and $B \subset \mathbb{R}^m$, then $A \times B \subset \mathbb{R}^n \times \mathbb{R}^m = \mathbb{R}^{n+m}$ is

$$A \times B = \{\, (\mathbf{a},\ \mathbf{b}) \mid \mathbf{a} \in A,\ \mathbf{b} \in B \,\}.$$

We can easily extend this idea to define Cartesian products of *collections* of sets. Let $\mathcal{A} \subset \mathcal{P}(\mathbb{R}^n)$ and $\mathcal{B} \subset \mathcal{P}(\mathbb{R}^m)$ be two collections of subsets of $\mathbb{R}^n$ and $\mathbb{R}^m$. Then $\mathcal{A} \times \mathcal{B}$ will denote the class of all subsets of $\mathbb{R}^n \times \mathbb{R}^m = \mathbb{R}^{n+m}$ that are of the form $A \times B$ with $A \in \mathcal{A}$ and $B \in \mathcal{B}$.

Lemma 8.1.6 Products of cubes. *Let n, $m \in \mathbb{N}$. Then $\mathcal{C}_k^{n+m} = \mathcal{C}_k^n \times \mathcal{C}_k^m$. Hence, $C \in \mathcal{C}_k^{n+m}$ if and only if there are $A \in \mathcal{C}_k^n$ and $B \in \mathcal{C}_k^m$ such that $C = A \times B$.*

Proof. This follows from Definition 8.1.1 above. □

The next result provides a partition of a binary cube into a disjoint union of smaller binary cubes.

Lemma 8.1.7 Cubes as the union of higher-order cubes. *Let h, $k \in \mathbb{Z}^+$ and $h \leq k$. Then each cube in $\mathcal{C}_h^n$ is the union of $2^{(k-h)n}$ cubes in $\mathcal{C}_k^n$.*

Proof. First, consider the result in $\mathbb{R}$, i.e., assume that $n = 1$. Let $q = (k - h)$. If $q = 1$, simply note that each interval $I = [\, r,\, s\,) \in \mathcal{C}_h^1$ is the union of two "cubes" $I_0 = [\, r,\, (r + s)/2\,)$ and $I_1 = [\,(r + s)/2,\, s\,)$ in $\mathcal{C}_k^1$, since $k = h + 1$. An induction proves the result for any $q \in \mathbb{N}$ and for $n = 1$.

Now assume the result holds for all $n \leq M$, where $M \in \mathbb{N}$, and for any $q \in \mathbb{N}$, where $q = k - h$ as before. If $C \in \mathcal{C}_h^{n+1}$, then $C = A \times B$ with $A \in \mathcal{C}_h^n$ and $B \in \mathcal{C}_h^1$, by Lemma 8.1.6. Hence, by the induction hypothesis, $A = \cup_i A_i$ and $B = \cup_j B_j$, where the first union contains 2^{qn} cubes $A_i \in \mathcal{C}_{h+q}^n = \mathcal{C}_k^n$ and the second union contains 2^q cubes $B_j \in \mathcal{C}_{h+q}^1 = \mathcal{C}_k^1$. Then

$$C = (A \times B) = \cup_i \cup_j (A_i \times B_j)$$

is the union of $2^{qn} \cdot 2^q = 2^{q(n+1)}$ cubes $(A_i \times B_j) \in \mathcal{C}_k^{n+1}$ So the result holds for $n + 1$ and thus for all n by induction. □

Volumes of the Unions of Cubes

The volume of an individual cube is the product of its edge lengths, the volume of a disjoint union of cubes is just the sum of the individual volumes, and the volume of a Cartesian product of cubes is the product of the individual volumes. First, we establish these basic facts for binary cubes.

Definition 8.1.8 Volumes of cubes in a grid. The *volume of* $C \in \mathcal{C}_k^n$ is defined as $v^n(C) = 2^{-kn}$. In particular, if $I = [\, p2^{-k},\, (p + 1)2^{-k}\,) \in \mathcal{C}_k^1$, then $v^1(I) = 2^{-k}$ is the length $\ell(I)$ of I. And if $C = (I_1 \times \cdots \times I_n) \in \mathcal{C}_k^n$ with $I_i \in \mathcal{C}_k^1$, then

$$v^n(C) = 2^{-kn} = (2^{-k})^n = \ell(I_1) \cdots \ell(I_n),$$

the product of n constant edge lengths. Note that $v^{n+m}(C) = v^{n+m}(A \times B) = v^n(A) \cdot v^m(B)$ whenever $C = (A \times B) \in \mathcal{C}_k^{n+m}$ with $A \in \mathcal{C}_k^n$ and $B \in \mathcal{C}_k^m$.

Lemma 8.1.9 Additivity of volume. *Let $h \leq k$. If $C \in \mathcal{C}_h$ and if $C = \cup_i A_i$ with $A_i \in \mathcal{C}_k$, then $v(C) = \sum_i v(A_i)$. (Note that we write v for v^n.)*

Proof. Lemma 8.1.7 shows that the number of cubes A_i in the union $C = \cup_i A_i$ is $2^{n(k-h)}$. Since $v(A_i) = 2^{-kn}$ for each i we have

$$\sum_i v(A_i) = 2^{n(k-h)} \cdot 2^{-kn} = 2^{-hn} = v(C). \quad \square$$

Definition 8.1.10 Finite unions of cubes in a grid. Let $\mathcal{U}_k$ denote the collection of sets consisting of the finite unions of cubes $C \in \mathcal{C}_k$. We assume that $\emptyset \in \mathcal{U}_k$ — it is the union of an empty collection of cubes. Lemma 8.1.7 shows that if $h \leq k$, then $\mathcal{C}_h \subset \mathcal{U}_k$. Hence we see that $\mathcal{U}_h \subset \mathcal{U}_k$ whenever $h \leq k$. We sometimes write $\mathcal{U}_k^n$ instead of $\mathcal{U}_k$ to identify the dimension of the underlying space $\mathbb{R}^n$.

Definition 8.1.11 Volumes of finite unions of cubes in a grid. Let $E \in \mathcal{U}_h$. Then $E \in \mathcal{U}_k$ for all $k \geq h$. Hence, for each $k \geq h$, there is a finite collection of cubes $\mathcal{E}_k \subset \mathcal{C}_k$ such that $E = \cup_{C \in \mathcal{E}_k} C$. Note that, by Lemma 8.1.7, $|\mathcal{E}_k| = 2^{(k-h)n}|\mathcal{E}_h|$, where $|\mathcal{E}_k|$ is the number of elements in $\mathcal{E}_k$. The *volume of E* is defined as

$$v(E) = v\left(\bigcup_{C \in \mathcal{E}_k} C\right) = \sum_{C \in \mathcal{E}_k} v(C) = 2^{-kn}|\mathcal{E}_k|. \tag{8.4}$$

The important point is that this number is well-defined: $v(E)$ is the same regardless of which $k \geq h$ we use in the definition. We sometimes write v^n instead of v to identify the dimension of the underlying space $\mathbb{R}^n$.

Lemma 8.1.12 *Let $E \in \mathcal{U}_h$ and $F \in \mathcal{U}_k$. Then $v(E \cup F) \leq v(E) + v(F)$. Also, if E and F are disjoint, then $v(E \cup F) = v(E) + v(F)$.*

Proof. Let $p = \max(h, k)$. Then $E, F \in \mathcal{U}_p$ by Definition 8.1.11. Hence, there are finite collections $\mathcal{E}, \mathcal{F} \subset \mathcal{C}_p$ such that $E = \cup_{C \in \mathcal{E}} C$ and $F = \cup_{C \in \mathcal{F}} C$. Then $G = E \cup F$ is the union of the collection $\mathcal{G} = \mathcal{E} \cup \mathcal{F}$. Hence the first part follows from the estimate that $|\mathcal{G}| \leq |\mathcal{E}| + |\mathcal{F}|$. If the sets E and F are disjoint, then $\mathcal{E}$ and $\mathcal{F}$ are also disjoint and $|\mathcal{G}| = |\mathcal{E}| + |\mathcal{F}|$. This proves the second part. $\square$

Lemma 8.1.13 Monotonicity of volume for unions of cubes. *Let $E \in \mathcal{U}_h$ and $F \in \mathcal{U}_k$ and suppose $E \subset F$. Then $v(E) \leq v(F)$.*

Proof. Once again, let $p = \max(h, k)$, so that $E, F \in \mathcal{U}_p$. Then there are finite collections $\mathcal{E}, \mathcal{F} \subset \mathcal{C}_p$ such that $E = \cup_{C \in \mathcal{E}} C$ and $F = \cup_{C \in \mathcal{F}} C$. Since $E \subset F$, we must have $\mathcal{E} \subset \mathcal{F}$. It follows that $|\mathcal{E}| \leq |\mathcal{F}|$, and hence $v(E) \leq v(F)$. $\square$

Lemma 8.1.14 Volumes in Cartesian products. *Let* $A \in \mathfrak{U}_k^n$ *and* $B \in \mathfrak{U}_k^m$. *Then* $(A \times B) \in \mathfrak{U}_k^{n+m}$ *and* $v^{n+m}(A \times B) = v^n(A) \cdot v^m(B)$.

Proof. Let $A = \cup_i A_i$ and $B = \cup_j B_j$ with finitely many $A_i \in \mathcal{C}_k^n$ and $B_j \in \mathcal{C}_k^m$. Then $A \times B = \cup_{ij}(A_i \times B_j)$ with finitely many $(A_i \times B_j) \in \mathcal{C}_k^{n+m}$. This follows from Definition 8.1.8. Then

$$\begin{aligned} v^{n+m}(A \times B) &= \sum_{ij} v^{n+m}(A_i \times B_j) = \sum_{ij} v^n(A_i) \cdot v^m(B_j) \\ &= \left(\sum_i v^n(A_i)\right) \cdot \left(\sum_j v^m(B_j)\right) = v^n(A) \cdot v^m(B), \end{aligned}$$

again by the same definition. □

Approximations by the Cubes of a Grid

Our immediate task is to show that the elementary results that we have just established for unions of cubes may be transferred to the class of sets that can be approximated by unions of cubes.

Definition 8.1.15 Approximating cubes of a bounded set. Let $\mathcal{C}_k^n = \mathcal{C}_k$ be the kth-order binary grid in $\mathbb{R}^n$. Let E be a bounded subset of $\mathbb{R}^n$.

Inner approximation. The collection of all cubes in $\mathcal{C}_k$ that are contained in E is denoted by $\mathcal{I}_k^n(E)$, or simply by $\mathcal{I}_k(E)$ if n is understood. The cubes in $\mathcal{I}_k(E)$ are called the *kth-order inner cubes of* E, and the union of all of these cubes is the *kth-order inner approximation of* E. It is denoted by $I_k(E)$.

Boundary approximation. The collection of all cubes in $\mathcal{C}_k$ that intersect both E and $E^c = \mathbb{R}^n \setminus E$ is denoted by $\mathcal{D}_k(E)$. The cubes in $\mathcal{D}_k(E)$ are called the *kth-order boundary cubes of* E, and the union of these cubes is the *kth-order boundary approximation of* E. It is denoted by $D_k(E)$.

Outer approximation. The set $O_k(E) = I_k(E) \cup D_k(E)$ is called the *the kth-order outer approximation of* E. Also, $\mathcal{O}_k(E) = \mathcal{I}_k(E) \cup \mathcal{D}_k(E)$ is the collection of all cubes in $\mathcal{C}_k$ that intersect E. The union of these *kth-order outer cubes* is $O_k(E)$.

Remarks 8.1.16 Note that

$$\mathcal{D}_k(E) = \mathcal{O}_k(E) \setminus \mathcal{I}_k(E) \text{ and } D_k(E) = O_k(E) \setminus I_k(E). \tag{8.5}$$

Also, $I_k(E)$ and $O_k(E)$ are the *best inner and outer approximations of a set* by the cubes of the grid $\mathcal{C}_k$ in the following sense. If $F, G \in \mathfrak{U}_k$ and if $F \subset E \subset G$, then also

$$F \subset I_k(E) \subset E \subset O_k(E) \subset G. \tag{8.6}$$

This is clear from the definitions. Finally, note that if $h \leq k$, then

$$I_h(E) \subset I_k(E) \subset E \subset O_k(E) \subset O_h(E). \tag{8.7}$$

No surprises here: the approximations improve as the grids become finer.

Lemma 8.1.17 *If E is a bounded set, then the limits*

$$\underline{v}(E) = \lim_k v(I_k(E)) \text{ and } \overline{v}(E) = \lim_k v(O_k(E)) \tag{8.8}$$

both exist and $\underline{v}(E) \leq \overline{v}(E)$.

Proof. The inclusions (8.7) above and the monotonicity of volume for unions of cubes, Lemma 8.1.13, show that

$$v(I_h(E)) \leq v(I_k(E)) \leq v(O_k(E)) \leq v(O_h(E))$$

whenever $h \leq k$. Hence $\underline{v}(E)$ and $\overline{v}(E)$ both exist and $\underline{v}(E) \leq \overline{v}(E)$. □

Definition 8.1.18 Inner and outer volumes of bounded sets. Let E be a bounded set in $\mathbb{R}^n$. Then the limits obtained above,

$$\underline{v}(E) = \lim_k v(I_k(E)) \text{ and } \overline{v}(E) = \lim_k v(O_k(E)), \tag{8.9}$$

are called, respectively, the *inner and outer volumes of E*.

Lemma 8.1.19 Monotonicity of the inner and outer volumes. *If A and B are bounded sets and if $A \subset B$, then $\underline{v}(A) \leq \underline{v}(B)$ and $\overline{v}(A) \leq \overline{v}(B)$.*

Proof. Definition 8.1.15 shows that if $A \subset B$, then

$$I_k(A) \subset I_k(B) \text{ and } O_k(A) \subset O_k(B). \tag{8.10}$$

Hence the proof follows. □

Jordan Sets

Definition 8.1.20 Jordan sets. A bounded set $E \subset \mathbb{R}^n$ is called a *Jordan set* if it has the same inner and outer volumes. In this case, $\underline{v}(E) = \overline{v}(E)$ is called *the volume of the Jordan set E* and is denoted by $v(E) = \underline{v}(E) = \overline{v}(E)$.

Theorem 8.1.21 Other definitions of Jordan sets. *Let $E \subset \mathbb{R}^n$ be a bounded set. Then the following are equivalent.*

(1) E is a Jordan set.

(2) For each $\varepsilon > 0$ there is a $k \in \mathbb{Z}^+$ such that $v(O_k(E)) - v(I_k(E)) < \varepsilon$.

(3) For each $\varepsilon > 0$ there is a $k \in \mathbb{Z}^+$ such that $v(D_k(E)) < \varepsilon$.

(4) For each $\varepsilon > 0$ there is a $k \in \mathbb{Z}^+$ and $F, G \in \mathfrak{U}_k$ such that $F \subset E \subset G$ and such that $v(G) - v(F) < \varepsilon$.

(5) For each $\varepsilon > 0$ there are Jordan sets F and G such that $F \subset E \subset G$ and such that $v(G) - v(F) < \varepsilon$.

Proof. The sets $I_k(E)$, $O_k(E)$, and $D_k(E)$ were defined in Definition 8.1.15. Note that $D_k(E) = O_k(E) \setminus I_k(E)$ by (8.6) and, therefore,

$$v(D_k(E)) = v(O_k(E)) - v(I_k(E))$$

by Lemma 8.1.12. This shows the equivalence of (2) and (3). The other parts of the theorem are left as exercises. □

Definition 8.1.22 Negligible sets. Jordan sets of zero volume are called *negligible sets*.

Lemma 8.1.23 *A set E is negligible if and only if for each $\varepsilon > 0$, there is a Jordan set G such that $E \subset G$ and $v(G) < \varepsilon$. In particular, the empty set is a negligible set.*

Proof. This is left as an exercise. □

Theorem 8.1.24 Additivity of volume. *If A and B are Jordan sets, then $A \cup B$, $A \cap B$, and $A \setminus B$ are also Jordan sets. Also, $v(A \cup B) \leq v(A) + v(B)$. If A and B are disjoint, then $v(A \cup B) = v(A) + v(B)$. Finally, if A_i are finitely many Jordan sets and $A = \cup_i A_i$, then A is a Jordan set, $v(A) \leq \sum_i v(A_i)$ in general, and $v(A) = \sum_i v(A_i)$ if the sets A_i are pairwise disjoint.*

Proof. We claim that $D_k(A \cup B) \subset D_k(A) \cup D_k(B)$. In fact, if E is a kth-order cube in $D_k(A \cup B)$, then E intersects both $A \cup B$ and its complement. Hence E contains a point from A or B and also a point not in A and not in B. Then we see that either $E \subset D_k(A)$ or $E \subset D_k(B)$. Then Lemma 8.1.12 shows that

$$v(D_k(A \cup B)) \leq v(D_k(A)) + v(D_k(B)).$$

This implies that $A \cup B$ is a Jordan set by Part (3) of Theorem 8.1.21. Similarly, $A \cap B$ and $A \setminus B$ are also Jordan sets. The other parts follows from Lemma 8.1.12. In fact, this lemma shows that

$$v(I_k(A \cup B)) \leq v(I_k(A)) + v(I_k(B)) \text{ in general and} \quad (8.11)$$
$$v(I_k(A \cup B)) = v(I_k(A)) + v(I_k(B)) \text{ if } A \text{ and } B \text{ are disjoint.} \quad (8.12)$$

The generalization to finitely many sets is by induction. □

Theorem 8.1.25 Cartesian products of Jordan sets. *Let $A \subset \mathbb{R}^n$ and $B \subset \mathbb{R}^m$ be Jordan sets. Then $A \times B$ is a Jordan set in $\mathbb{R}^n \times \mathbb{R}^m = \mathbb{R}^{n+m}$ and $v^{n+m}(A \times B) = v^n(A) \cdot v^m(B)$.*

Proof. The notation is as in Definition 8.1.15. By Lemma 8.1.14, $I_k(A) \times I_k(B)$ and $O_k(A) \times O_k(B)$ both belong to $\mathfrak{U}_k^{n+m}$ and

$$v^{n+m}(I_k(A) \times I_k(B)) = v^n(I_k(A)) \cdot v^m(I_k(B)) \quad (8.13)$$

and

$$v^{n+m}(O_k(A) \times O_k(B)) = v^n(O_k(A)) \cdot v^m(O_k(B)). \quad (8.14)$$

Hence both volumes in (8.13) and (8.14) approach the same limit $v^n(A) \cdot v^m(B)$ with increasing k. Since

$$I_k(A) \times I_k(B) \subset A \times B \subset O_k(A) \times O_k(B),$$

this shows that $A \times B$ is a Jordan set with $v^{n+m}(A \times B) = v^n(A) \cdot v^m(B)$. □

Blocks in $\mathbb{R}^n$

We can now show that not just binary cubes but also blocks in general are Jordan sets — our first important example.

Definition 8.1.26 Blocks in $\mathbb{R}^n$. A block $B \subset \mathbb{R}^n$ is a Cartesian product $B = J_1 \times \cdots \times J_n$ of intervals. Here each J_i is a bounded interval of arbitrary type: open, closed, or half-open. Empty intervals and empty blocks are allowed. Lemma 8.1.27 shows that each block is a Jordan set.

Lemma 8.1.27 Blocks are Jordan sets. *Any block $B = J_1 \times \cdots \times J_n$ in $\mathbb{R}^n$ is a Jordan set with $v(B) = \ell(J_1) \ldots \ell(J_n)$. Here $\ell(J_i) = 0$ if $J_i = \emptyset$. Otherwise, $\ell(J_i) = (s_i - r_i)$ where r_i is the left and s_i the right end point of J_i.*

Proof. If $n = 1$, then blocks are just intervals. We see easily that the conclusion is correct in this case because of the density of binary numbers, Lemma 8.1.4. The general case follows by induction on n, using Theorem 8.1.25. □

Lemma 8.1.28 *A Jordan set E with a nonempty interior is not negligible.*

Proof. Use the maximum norm on $\mathbb{R}^n$, as defined in Definition 8.1.3. Let $\mathbf{a} \in E^o$. Then there is a ball $B_r(\mathbf{a}) \subset E$. But in the maximum norm, $B_r(\mathbf{a})$ is a block with equal side lengths $2r$. Hence, $v(E) \geq v(B_r(\mathbf{a})) = (2r)^n > 0$. □

Jordan Sets and the Topology of $\mathbb{R}^n$

Topological considerations are important in many arguments about volume and integration. We will discuss the topology of Jordan sets in more detail in Section 8.3. Here we make some observations in order to deal with general partitions of $\mathbb{R}^n$.

Lemma 8.1.29 Blocks in $\mathbb{R}^n$ and the topology of $\mathbb{R}^n$. *Given a block E in $\mathbb{R}^n$ and $\varepsilon > 0$, there is a closed block F and an open block G such that $F \subset E^o \subset E \subset \overline{E} \subset G$ and such that $v(G) - v(F) < \varepsilon$.*

Proof. If $n = 1$, then closed (or open) blocks are closed (or open) intervals. The conclusion is correct in this case because of the density of binary numbers, Lemma 8.1.4. The general case follows by an easy induction on n on the basis of Theorem 8.1.25. □

Theorem 8.1.30 Jordan sets in $\mathbb{R}^n$ and the topology of $\mathbb{R}^n$. *Let E be a Jordan set and $\varepsilon > 0$. Then there is a closed Jordan set F and an open Jordan set G such that $F \subset E^o \subset E \subset \overline{E} \subset G$ and such that $v(G) - v(F) < \varepsilon$. Also, F can be taken as a finite union of closed blocks and G as a finite union of open blocks.*

Proof. Given $\varepsilon > 0$, find two unions of cubes $P = \cup_i B_i$ and $Q = \cup_j C_j$ such that $P \subset E \subset Q$ and $v(Q) - v(P) < \varepsilon/2$. Let $K \in \mathbb{N}$ be such that the number of B_i s and C_j s is less than K. By Lemma 8.1.29, we can find a closed block $F_i \subset B_i^o$ and an open block $G_j \supset \overline{C_j}$ such that

$$v(B_i) - v(F_i) < \varepsilon/(8K) \text{ and } v(G_j) - v(C_j) < \varepsilon/(8K) \tag{8.15}$$

for all i, j. Then $F = \cup_i F_i$ and $G = \cup_j G_j$ satisfy our requirements. □

Theorem 8.1.31 *Let E be a Jordan set and $\varepsilon > 0$. Then there are two Jordan sets F and G and a number $\delta > 0$ such that $F \subset E \subset G$, $v(G) - v(F) < \varepsilon$ and such* *that $B_\delta(\mathbf{u}) \subset E$ whenever $\mathbf{u} \in F$ and $B_\delta(\mathbf{v}) \subset G$ whenever $\mathbf{v} \in E$. Here $B_\delta(\mathbf{x})$ is the ball of radius δ around $\mathbf{x} \in \mathbb{R}^n$.*

Proof. Find F and G as in Theorem 8.1.30, so that F is a compact set contained in the open set E^o and $\overline{E}$ is a compact set contained in the open set G. Then the result follows from Theorem 4.5.48. □

General Partitions

As mentioned in the introduction, the method of binary cubes is not the only way to define volumes. Different collections of sets may be used to define Jordan sets and their volume (with the identical result). There are some natural conditions that any such collection must satisfy. These are formulated below.

Definition 8.1.32 Partitions. A collection $\mathcal{P}$ of Jordan sets in $\mathbb{R}^n$ is called a *(Jordan) partition (of $\mathbb{R}^n$)* if these sets are pairwise disjoint and if their union is $\mathbb{R}^n$. The pairwise disjointness of the sets means that any two distinct sets in $\mathcal{P}$ are disjoint.

Definition 8.1.33 Partitions of finite size. If E is a bounded set, then define its *size* as $(\operatorname{size} E) = \sup\{\, \|\mathbf{u}-\mathbf{v}\| \mid \mathbf{u},\, \mathbf{v} \in E \,\}$. A partition $\mathcal{P}$ is called *a partition of finite size* if the set $\{\, (\operatorname{size} P) \mid P \in \mathcal{P} \,\}$ is a bounded set in $\mathbb{R}$ and if any bounded set in $\mathbb{R}^n$ is contained in a finite union of sets in $\mathcal{P}$. If $\mathcal{P}$ is a partition of finite size, then

$$(\operatorname{size} \mathcal{P}) = \sup\{\, (\operatorname{size} P) \mid P \in \mathcal{P} \,\} \tag{8.16}$$

is called the *size of $\mathcal{P}$*.

Definition 8.1.34 Inner and outer approximations in a partition. Let $\mathcal{P}$ be a partition of finite size. Let E be a bounded set in $\mathbb{R}^n$. Then

$$I_{\mathcal{P}}(E) = \cup\{\, P \mid P \in \mathcal{P}, \text{ and } P \subset E \,\} \tag{8.17}$$

$$O_{\mathcal{P}}(E) = \cup\{\, P \mid P \in \mathcal{P}, \text{ and } P \cap E \neq \emptyset \,\} \tag{8.18}$$

will be called, respectively, the *inner and outer approximations of E in $\mathcal{P}$*. Definition 8.1.33 of partitions of finite size shows that both unions above are finite unions. Hence $I_{\mathcal{P}}(E)$ and $O_{\mathcal{P}}(E)$ are Jordan sets.

Theorem 8.1.35 Jordan sets in terms of partitions. *A bounded set E in $\mathbb{R}^n$ is a Jordan set if and only if it satisfies the following condition (A).*

(A) For each $\varepsilon > 0$ there is a $\delta > 0$ such that

$$v(O_{\mathcal{P}}(E)) - v(I_{\mathcal{P}}(E)) < \varepsilon \tag{8.19}$$

whenever $\mathcal{P}$ *is a partition of finite size and* $(\text{size}\,\mathcal{P}) < \delta$.

Proof. If E satisfies (A), then for each $\varepsilon > 0$ there are Jordan sets $I_{\mathcal{P}}(E)$ and $O_{\mathcal{P}}(E)$ such that $I_{\mathcal{P}}(E) \subset E \subset O_{\mathcal{P}}(E)$ and such that (8.19) is satisfied. In this case, Lemma 8.1.17 shows that E is a Jordan set.

To obtain the converse, let E be a Jordan set and $\varepsilon > 0$. Apply Theorem 8.1.31 to obtain the Jordan sets F and G and $\delta > 0$ such that $v(G) - v(F) < \varepsilon$ and such that $B_\delta(\mathbf{u}) \subset E$ whenever $\mathbf{u} \in F$ and $B_\delta(\mathbf{v}) \subset G$ whenever $\mathbf{v} \in E$. An easy check shows that if $\mathcal{P}$ is a partition of finite size with $(\text{size}\,\mathcal{P}) < \delta$, then

$$F \subset I_{\mathcal{P}}(E) \subset E \subset O_{\mathcal{P}}(E) \subset G.$$

In this case, $v(O_{\mathcal{P}}(E)) - v(I_{\mathcal{P}}(E)) \leq v(G) - v(F) < \varepsilon$. □

Problems

8.1 Find $I_k(E)$ and $O_k(E)$ for $k = 0, 1, 2, 3$, where

1. $E = \{ (x, y) \mid 0 \leq x,\ 0 \leq y,\ x^2 + y^2 \leq 1 \} \subset \mathbb{R}^2$,
2. $E = \{ (x, y) \mid 0 < x,\ 0 < y,\ x^2 + y^2 < 1 \} \subset \mathbb{R}^2$,
3. $E = \{ (x, y) \mid 0 \leq x,\ 0 \leq y,\ x + y \leq 1 \} \subset \mathbb{R}^2$,
4. $E = \{ (x, y) \mid 0 < x,\ 0 < y,\ x + y < 1 \} \subset \mathbb{R}^2$.

Here $I_k(E)$ and $O_k(E)$ are the inner and outer approximations of E, as in 8.1.15.

8.2 Find $I_k(E)$ and $O_k(E)$ for all $k \in \mathbb{Z}^+$, where E is the set of points in the unit square $[0, 1] \times [0, 1]$ with at least one rational coordinate. Find the inner and outer volumes (in this case areas) of E.

8.3 For each $n \in \mathbb{N}$ find a negligible set E_n such that $E = \cup_{n \in \mathbb{N}}$ is not negligible.

8.4 For each $n \in \mathbb{N}$ and for each $k = 0, 1, \ldots, n$ let

$$E_{nk} = \{ (x, y) \mid x = k/n,\ 0 \leq y \leq 1/n \} \subset \mathbb{R}^2,$$

$E_n = \cup_{k=0}^{n} E_{nk}$, and $E = \cup_{n \in \mathbb{N}} E_n$. Show that all these sets are negligible.

Problems on Cross-Sections

Definition 8.1.36 Cross-sections. Denote the points in $\mathbb{R}^{n+1} = \mathbb{R}^n \times \mathbb{R}$ as $(\mathbf{x}, y)$ with $\mathbf{x} \in \mathbb{R}^n$ and $y \in \mathbb{R}$. The *cross-section* of a set $E \subset \mathbb{R}^{n+1} = \mathbb{R}^n \times \mathbb{R}$ at $y \in \mathbb{R}$ is $E_y = \{ \mathbf{x} \mid \mathbf{x} \in \mathbb{R}^n,\ (\mathbf{x}, y) \in E \} \subset \mathbb{R}^n$.

Definition 8.1.37 Variations of cross-sections. Let E be a bounded set in $\mathbb{R}^{n+1}$ and J a bounded interval in $\mathbb{R}$. Then *the variation of the cross-sections of E over J* is $V_E(J) = \overline{v}^n((\cup_{y \in J} E_y) \setminus (\cap_{y \in J} E_y))$.

8.5 Let $E = \{ (x,\, y) \mid |x| + |y| \leq 1 \} \subset \mathbb{R} \times \mathbb{R}$. Find $V_E(J)$ for any interval J.

8.6 Let $E = \left\{ (x,\, y) \,\middle|\, x^2 + y^2 \leq 1 \right\} \subset \mathbb{R} \times \mathbb{R}$. Find $V_E(J)$ for any interval J.

8.7 Let E be the bounded region in the xy-plane between the parabolas $y = x^2$ and $y = 2x^2 - 1$. Find the variations of the cross-sections of E. Consider the cross-sections both with the vertical $x =$ constant lines and with the horizontal $y =$ constant lines.

8.8 Let $E \subset \mathbb{R} \times \mathbb{R}$ be the set of points in the unit square $[0,\, 1] \times [0,\, 1]$ with at least one rational coordinate. Find $V_E(J)$ for any interval J.

8.9 Let E be the bounded region in the xyz-space $\mathbb{R}^2 \times \mathbb{R}$ between the cylinders $x^2 + z^2 = 1$ and $y^2 + z^2 = 1$. Find the variation of the cross-sections of E with the horizontal $z =$ constant planes.

Definition 8.1.38 Continuously changing cross-sections. Let E be a bounded set in $\mathbb{R}^n \times \mathbb{R}$ and I an interval in $\mathbb{R}$. We will say that cross-sections of E change continuously on I if for each $p \in I$ and for each $\varepsilon > 0$ there is a $\delta > 0$ such that $V_E(I \cap (p - \delta,\, p + \delta)) < \varepsilon$.

8.10 Let $E \subset \mathbb{R}^{n+1} = \mathbb{R}^n \times \mathbb{R}$ be a bounded set. Let I be an interval such that $E_y = \emptyset$ for all $y \notin I$. Assume that each E_y is a Jordan set in $\mathbb{R}^n$ and that E has continuously changing cross-sections on $(I \cap J)$ for any interval J. Show that E is a Jordan set in $\mathbb{R}^{n+1}$. (Hint: First, assume that the continuity of the cross-sections is uniform on I in the following sense: for each $\varepsilon > 0$ there is a $\delta > 0$ such that if $V_E(I \cap J) < \varepsilon$ whenever $\ell(J) < \delta$.)

8.11 Show that both $|x| + |y| \leq 1$ and $x^2 + y^2 \leq 1$ are Jordan sets in $\mathbb{R}^2$.

8.12 Let I be an interval. Let $\mathbf{a} : I \to \mathbb{R}^n$ and $t : I \to \mathbb{R}$ be two continuous functions. Let A be a Jordan set in $\mathbb{R}^n$. Let $E \subset \mathbb{R}^n \times \mathbb{R}$ be defined in terms of its cross-sections E_y as $E_y = \emptyset$ if $y \notin I$ and $E_y = \mathbf{a}(y) + t(y)A$ if $y \in I$. Show that E is a Jordan set. (Hint: First, assume that A is a cube in $\mathcal{C}^n_k$.)

8.13 Cylinders. Let A be a Jordan set in $\mathbb{R}^n$ and $\mathbf{c} = (\mathbf{a},\, h) \in \mathbb{R}^n \times \mathbb{R}$ with $h \neq 0$. Then the set

$$C = C(A,\, \mathbf{c}) = \{ (\mathbf{x},\, 0) + t(\mathbf{a},\, h) \mid \mathbf{x} \in A,\, 0 \leq t \leq 1 \} \subset \mathbb{R}^{n+1}$$

is called a *cylinder*. Show that all cylinders are Jordan sets.

8.14 Cones. Let A be a Jordan set in $\mathbb{R}^n$ and $\mathbf{c} = (\mathbf{a},\, h) \in \mathbb{R}^n \times \mathbb{R}$ with $h \neq 0$. Then the set

$$K = K(A,\, \mathbf{c}) = \{\, (1-t)(\mathbf{x},\, 0) + t(\mathbf{a},\, h) \mid \mathbf{x} \in A,\; 0 \leq t \leq 1 \,\} \subset \mathbb{R}^{n+1}$$

is called a *cone*. Show that all cones are Jordan sets.

8.15 Show that all triangles in $\mathbb{R}^2$ are Jordan sets.

8.16 Show that the "general tetrahedra"

$$E = \left\{ \mathbf{x} = (x_1,\, \ldots,\, x_n) \in \mathbb{R}^n \;\middle|\; 0 \leq x_i,\; \sum_i x_i \leq 1 \right\}$$

are Jordan sets in $\mathbb{R}^n$.

8.17 Show that all Euclidean balls

$$E = \left\{ \mathbf{x} = (x_1,\, \ldots,\, x_n) \in \mathbb{R}^n \;\middle|\; \sum_i (x_i - c_i)^2 \leq R^2 \right\}$$

are Jordan sets in $\mathbb{R}^n$. Here $\mathbf{c} = (c_1,\, \ldots,\, c_n)$ is the center of the ball. (Hint: First, assume that $\mathbf{c} = \mathbf{0}$ and proceed by induction on $n \in \mathbb{N}$.)

8.18 Show that all ellipsoids

$$E = \left\{ \mathbf{x} = (x_1,\, \ldots,\, x_n) \in \mathbb{R}^n \;\middle|\; \sum_i r_i^{-2}(x_i - c_i)^2 \leq R^2 \right\}$$

are Jordan sets in $\mathbb{R}^n$. Here $\mathbf{c} = (c_1,\, \ldots,\, c_n)$ is the center of the ellipsoid and $(r_1,\, \ldots,\, r_n)$ is a fixed vector in $\mathbb{R}^n$ with $r_i \neq 0$.

8.19 Show that the region $E = \{\, (x,\, y,\, z) \mid x^2 + z^2 \leq 1,\; y^2 + z^2 \leq 1 \,\}$ is a Jordan set in $\mathbb{R}^3$. This is the region between two (ordinary circular) cylinders.

8.20 Let $E_n = \{\, (x,\, y) \mid x = p/n,\; y = q/n,\; p,\, q = 0, 1,\, \ldots,\, n \,\} \subset \mathbb{R}^2$, with $n \in \mathbb{N}$. Show that $E = \cup_n E_n$ is not a Jordan set in $\mathbb{R}^2$, but all of its cross-sections are Jordan sets in $\mathbb{R}$.

8.21 Give an example of a bounded set E in R^2 such that E is not a Jordan set in $\mathbb{R}^2$, but all of its cross-sections $E_y \subset \mathbb{R}$ are Jordan sets in $\mathbb{R}$ with $v(E_y) = \ell(E_y) = 1/2$ for all y in the interval $[\,0,\, 1\,]$.

8.2 INTEGRALS

A nonnegative function $f : \mathbb{R}^n \to \mathbb{R}^+$ is said to be an integrable function if the region $E_f \subset \mathbb{R}^{n+1} = \mathbb{R}^n \times \mathbb{R}$ under the graph of this function is a Jordan set in $\mathbb{R}^{n+1}$. In this case, the $(n+1)$-dimensional volume of E_f is called the *integral of* f. Hence, the main problems in integration are to find out whether this special type of set is a Jordan set and, if it is, to compute its volume.

Definition 8.2.1 Regions under graphs. Given $f : \mathbb{R}^n \to \mathbb{R}$, define

$$E_f = \{ (\mathbf{x}, y) \in \mathbb{R}^n \times \mathbb{R} \mid 0 \le y < f(\mathbf{x}) \} \subset \mathbb{R}^{n+1}$$

as *the region under the graph of* f. Note that only the positive values of f are important in this definition.

Definition 8.2.2 Integrals of nonnegative functions. A nonnegative function $f : \mathbb{R}^n \to \mathbb{R}^+$ is said to be *integrable* if E_f is a Jordan set in $\mathbb{R}^{n+1}$. In this case, the *integral of* f is defined as $\int f = v^{n+1}(E_f)$.

Definition 8.2.3 Positive and negative parts of a function. Given a real-valued function $f : X \to \mathbb{R}$ on a set X, define

$$f^+(x) = \max(f(x), 0) \quad \text{and} \quad f^-(x) = \max(-f(x), 0) = -\min(f, 0),$$

the positive part and *the negative part of* f. Note that both the positive and the negative parts are nonnegative functions and $f = f^+ - f^-$.

Definition 8.2.4 Integrals of general functions. A function $f : \mathbb{R}^n \to \mathbb{R}$ is said to be integrable if its positive and negative parts f^+ and f^- are both integrable. In this case, the *integral of* f is defined as $\int f = \int f^+ - \int f^-$.

Definition 8.2.5 Bounded functions of bounded support. A function $f : \mathbb{R}^n \to \mathbb{R}$ may be integrable only if both E_{f^+} and E_{f^-} are bounded sets in $\mathbb{R}^{n+1}$. This happens if and only if there is a number $M \in \mathbb{R}$ and a bounded set $S \subset \mathbb{R}^n$ such that $-M \le f(\mathbf{x}) \le M$ for all $\mathbf{x} \in \mathbb{R}^n$ and $f(\mathbf{x}) = 0$ if $\mathbf{x} \notin S$. Such a function is called *a bounded function of bounded support*. The set S is called *a support for* f. Any bounded set is contained in a finite union of cubes. Hence we may always assume that a function of bounded support has its support in a finite union of cubes.

Step Functions

Step functions constitute an especially important class of integrable functions. They are simple to integrate, and it is easy to establish their basic properties. Yet it turns out that any integrable function can be approximated by step functions. As a result, most of the fundamental properties of integrable functions can be obtained from the corresponding properties of step functions.

Definition 8.2.6 Characteristic functions. Let $E \subset \mathbb{R}^n$. The *characteristic (or indicator) function* of E is the function $\chi_E : \mathbb{R}^n \to \mathbb{R}$ defined by $\chi_E(\mathbf{x}) = 1$ if $\mathbf{x} \in E$ and $\chi_E(\mathbf{x}) = 0$ if $\mathbf{x} \notin E$.

Definition 8.2.7 Step functions. A function $\varphi : \mathbb{R}^n \to \mathbb{R}$ is called an *kth-order step function* if it can be expressed as a linear combination $\varphi = \sum_i \alpha_i \chi_{C_i}$ of the characteristic functions of a finite number of cubes $C_i \in \mathcal{C}_k^n$.

A step function has the constant value α_i on the set C_i and the value 0 everywhere else. For instance, a step function on $\mathbb{R}$ takes constant non-zero values on finitely many binary intervals and is zero elsewhere.

Lemma 8.2.8 Step functions are integrable. *If $\varphi = \sum_i \alpha_i \chi_{C_i}$ is a step function, then φ is integrable and $\int \varphi = \sum_i \alpha_i v^n(C_i)$.*

Proof. First, assume that $\alpha_i \geq 0$ for all i. In this case, we have

$$E_\varphi = \cup_i \left(C_i \times [0, \alpha_i)\right).$$

This is a Jordan set in $\mathbb{R}^{n+1}$, since it is a finite union of the blocks $(C_i \times [0, \alpha_i))$. These blocks are pairwise disjoint because the cubes C_i in the grid $\mathcal{C}_k^n$ are pairwise disjoint. Hence, by the additivity of volume, Theorem 8.1.24,

$$\int \varphi = v^{n+1}(E_\varphi) = \sum_i v^{n+1}(C_i \times [0, \alpha_i)) = \sum_i \alpha_i \, v^n(C_i).$$

In general, let $\beta_i = \max(0, \alpha_i)$ and $\gamma_i = \max(0, -\alpha_i)$. Then $\lambda = \sum_i \beta_i \chi_{C_i}$ is the positive part of φ and $\mu = \sum_i \gamma_i \chi_{C_i}$ is the negative part of φ. Both of these parts are nonnegative step functions, and therefore integrable. We verify easily that the integral of φ is still $\int \varphi = \sum_i \alpha_i v^n(C_i)$. $\square$

Lemma 8.2.9 Integration of step functions as a positive linear operator. *The class $\mathcal{S}$ of step functions $\mathbb{R}^n \to \mathbb{R}$ is a vector space, and integration $\int : \mathcal{S} \to \mathbb{R}$ is a positive linear operator. More explicitly, if $\varphi, \varphi' \in \mathcal{S}$ and $\alpha, \alpha' \in \mathbb{R}$, then*

$(\alpha\varphi + \alpha'\varphi') \in \mathcal{S}$ and $\int(\alpha\varphi + \alpha'\varphi') = \alpha\int\varphi + \alpha'\int\varphi'$. Also, if $\varphi \leq \psi$, then $\int\varphi \leq \int\psi$.

Proof. If φ and φ' are of the same order, then the lemma is clear. If the orders of φ and φ' are h and h', then we see that φ and φ' are both of order $k = \max(h, h')$. Hence the proof follows. □

Riemann Condition for Integrability

In general, there are no efficient algorithms to decide if a given set is a Jordan set (though Problem 8.51 helps in many cases) and no simple way to compute the volume of a Jordan set. In the case of a region under the graph of a function, however, the situation is different. A bounded function of bounded support can be approximated by step functions. A condition for the integrability of a function can be formulated, and its integral computed, in terms of these approximating step functions.

Definition 8.2.10 Approximations by step functions. Let $f : \mathbb{R}^n \to \mathbb{R}$ be a bounded function of a bounded support S. For any nonempty $F \subset \mathbb{R}^n$ let

$$\inf(f, F) = \inf\{\, f(\mathbf{x}) \mid \mathbf{x} \in F \,\} \text{ and } \sup(f, F) = \sup\{\, f(\mathbf{x}) \mid \mathbf{x} \in F \,\}.$$

For each $k \in \mathbb{Z}^+$ define

$$\varphi_k = \sum\nolimits_{C \in \mathcal{C}_k^n} \inf(f, C)\chi_C \text{ and } \psi_k = \sum\nolimits_{C \in \mathcal{C}_k^n} \sup(f, C)\chi_C, \tag{8.20}$$

respectively the *kth-order lower and upper approximations of f by step functions.* These functions really are step functions, since the sums in their definitions are finite. In fact, if C does not intersect S, then $\inf(f, C) = \sup(f, C) = 0$. Since the support S of f is a bounded set, the number of $C \in \mathcal{C}_k^n$ that intersect S is finite.

Lemma 8.2.11 Monotonicity of the lower and upper approximations. *If $h \leq k$, then $\varphi_h \leq \varphi_k \leq f \leq \psi_k \leq \psi_h$.*

Proof. This follows from the definitions. The approximations get better as the grid on which the step functions are defined becomes finer. □

Definition 8.2.12 Lower and upper sums. The integrals

$$\int \varphi_k = \sum\nolimits_{C \in \mathcal{C}_k} \inf(f, C)\, v^n(C) = \sum\nolimits_{C \in \mathcal{C}_k} \inf(f, C)\, 2^{-kn} \tag{8.21}$$

$$\int \psi_k = \sum\nolimits_{C \in \mathcal{C}_k} \sup(f, C)\, v^n(C) = \sum\nolimits_{C \in \mathcal{C}_k} \sup(f, C)\, 2^{-kn} \tag{8.22}$$

are called the *kth-order lower and upper (Riemann) sums for* f. We also write

$$L_k(f) = \int \varphi_k \quad \text{and} \quad U_k(f) = \int \psi_k.$$

Theorem 8.2.13 Limits of lower and upper sums. *If* $f : \mathbb{R}^n \to \mathbb{R}$ *is a bounded function of bounded support, then* $\lim_k L_k(f)$ *and* $\lim_k U_k(f)$ *exist.*

Proof. Lemma 8.2.9 shows that integration is a positive linear operator on the vector space of step functions. Combining this fact with Lemma 8.2.11, we see that $L_h(f) \leq L_k(f) \leq U_k(f) \leq U_h(f)$ whenever $0 \leq h \leq k$. Hence both sequences are monotone and bounded, and therefore convergent. □

Definition 8.2.14 Lower and upper integrals. Let $f : \mathbb{R}^n \to \mathbb{R}$ be a bounded function of bounded support. Then $\lim_k L_k(f)$ and $\lim_k U_k(f)$ obtained in Theorem 8.2.13 are called *the lower integral of* f and *the upper integral of* f. They are denoted as

$$\lim_k L_k(f) = \lim_k \int \varphi_k = \underline{\int} f \text{ and } \lim_k U_k(f) = \lim_k \int \psi_k = \overline{\int} f.$$

Theorem 8.2.15 Riemann condition for integrability. *Let* $f : \mathbb{R}^n \to \mathbb{R}$ *be a bounded function of bounded support. Then* f *is integrable if and only if* $\underline{\int} f = \overline{\int} f$. *In this case,* $\int f$ *is the common value of these limits.*

Proof. Recall that E_f is the region under the graph of f as defined in Definition 8.2.1. First, consider a nonnegative function f. Assume that $\underline{\int} f = \overline{\int} f$. Given $\varepsilon > 0$, find a $k \in \mathbb{N}$ such that

$$\int \psi_k - \int \varphi_k = v^{n+1}(E_{\psi_k}) - v^{n+1}(E_{\varphi_k}) < \varepsilon. \tag{8.23}$$

Hence, E_{ψ_k} and E_{φ_k} are two Jordan sets in $\mathbb{R}^{n+1}$ such that $E_{\varphi_k} \subset E_f \subset E_{\psi_k}$ and such that $v^{n+1}(E_{\psi_k}) - v^{n+1}(E_{\varphi_k}) < \varepsilon$. Then Theorem 8.1.21 shows that E_f is a Jordan set in $\mathbb{R}^{n+1}$. Hence f is integrable. Also,

$$\int \psi_k = v^{n+1}(E_{\psi_k}) \leq v^{n+1}(E_f) \leq v^{n+1}(E_{\varphi_k}) = \int \varphi_k$$

shows that $v^{n+1}(E_f) = \int f = \underline{\int} f = \overline{\int} f$.

Conversely, assume that f is (nonnegative and) integrable. This means that E_f is a Jordan set in $\mathbb{R}^{n+1}$. Given $\varepsilon > 0$, find a $k \in \mathbb{N}$ so that the inner and the outer approximations of E_f by the kth-order cubes in $\mathbb{R}^{n+1}$ satisfy

$$v^{n+1}(O_k(E_f)) - v^{n+1}(I_k(E_f)) < \varepsilon.$$

Now any kth-order cube $H \in \mathcal{C}_k^{n+1}$ is of the form $H = C \times [r, s)$, where $C \in \mathcal{C}_k^n$ and $r = 2^{-k}p$, $s = 2^{-k}(p+1)$, with $p \in \mathbb{Z}$. If $H \subset I_k(E_f)$, then we see that $0 \leq r < s \leq \inf(f, C)$. Otherwise H would contain points outside of E_f. Hence, we see that $H \subset E_{\varphi_k}$. Therefore, $I_k(E_f) \subset E_{\varphi_k}$. Similarly, we see that $E_{\psi_k} \subset O_k(E_f)$. Hence

$$\int \psi_k - \int \varphi_k \leq v^{n+1}(O_k(E_f)) - v^{n+1}(I_k(E_f)) < \varepsilon.$$

Therefore, $\underline{\int} f = \overline{\int} f$. If f is not necessarily nonnegative, then one considers the positive and negative parts separately. This is left as an exercise. □

Supports of Lower and Upper Approximations

Let f be a bounded function with a bounded support S. Lemma 8.2.16 relates S and the supports of its lower and upper approximations, as defined in Definition 8.2.10. In this lemma $O_k(S)$ is the outer approximation of S defined in Definition 8.1.15. Hence $O_k(S)$ is the union of all kth-order cubes that intersect S.

Lemma 8.2.16 *Let S be a support for a bounded function f and let $h, k \in \mathbb{N}$, $h \leq k$. Then $O_h(S)$ is a support for the kth-order lower and upper approximations φ_k and ψ_k of f.*

Proof. Each $\mathbf{x} \in X$ belongs to a unique cube $C \in \mathcal{C}_k$. If $\mathbf{x} \notin O_k(S)$, then C is disjoint from $O_k(S)$. Hence, C is also disjoint from S. Therefore,

$$\sup(f, C) = \inf(f, C) = 0 \text{ and } \varphi_k(\mathbf{x}) = \psi_k(\mathbf{x}) = 0.$$

This means that $O_k(S)$ is a support for φ_k and ψ_k. Then $O_h(S)$ is also a support for φ_k and ψ_k, since $O_k(S) \subset O_h(S)$. □

Theorem 8.2.17 *Let K be a compact support for a bounded function f. If K is contained in an open set G, then there is an $h \in \mathbb{N}$ and a compact set K' such that $K \subset K' \subset G$ and K' is a support of φ_k and ψ_k for all $k \geq h$.*

Proof. Theorem 4.5.48 shows that if K is compact, G is open, and if $K \subset G$ then there is a compact K' and an $h \in \mathbb{N}$ such that $O_k(S) \subset K' \subset G$ for all $k \geq h$. Then the proof follows from Lemma 8.2.16 above. □

Integration as a Positive Linear Operator

Lemma 8.2.18 *Let f and f' be two bounded functions of bounded support. If $f \leq f'$, then $\underline{\int} f \leq \underline{\int} f'$ and $\overline{\int} f \leq \overline{\int} f'$.*

Proof. If $f \leq f'$, then also $\varphi_k \leq \varphi'_k$ and $\psi_k \leq \psi'_k$ for the respective approximating step functions. Then the proof follows. □

Lemma 8.2.19 *Let $f : \mathbb{R}^n \to \mathbb{R}$ be a function. If for each $\varepsilon > 0$ there are integrable functions f', f'' such that $f' \leq f \leq f''$ and such that $\int f'' - \int f' < \varepsilon$, then f is integrable.*

Proof. From Lemma 8.2.18 we obtain $\int f' \leq \underline{\int} f \leq \overline{\int} f \leq \int f''$. Hence

$$0 \leq \overline{\int} f - \underline{\int} f \leq \int f'' - \int f' < \varepsilon$$

for all $\varepsilon > 0$. Therefore, $\underline{\int} f = \overline{\int} f$ and f is integrable. □

Lemma 8.2.20 *A function $f : \mathbb{R}^n \to \mathbb{R}$ is integrable if and only if for each $\varepsilon > 0$ there are step functions φ and ψ such that $\varphi \leq f \leq \psi$ and $\int \psi - \int \varphi < \varepsilon$.*

Proof. If f is integrable, then $\int \varphi_k - \int \psi_k < \varepsilon$ for a sufficiently large k, where φ_k and ψ_k are approximating step functions. In the other direction, by Lemma 8.2.18, $\varphi \leq f \leq \psi$ implies $\int \varphi \leq \underline{\int} f \leq \overline{\int} f \leq \int \psi$. Hence, if $\int \psi - \int \varphi < \varepsilon$, then $\overline{\int} f - \underline{\int} f < \varepsilon$. □

Theorem 8.2.21 Integral as a limit of sums. *Let $f : \mathbb{R}^n \to \mathbb{R}$ be an integrable function. Then*

$$\int f = \lim_k \sum\nolimits_{\mathbf{c} \in \mathcal{Z}_k} f(\mathbf{c})\, 2^{-kn}. \tag{8.24}$$

Here $\mathcal{Z}_k$ is the set of centers of the cubes in the grid $\mathcal{C}_k$, defined in Definition 8.1.1. It is understood that the above sum is taken over the centers with $f(\mathbf{c}) \neq 0$.

Proof. Let $\vartheta_k = \sum_{C \in \mathcal{C}_k} f(\mathbf{c})\chi_C$. We see that $\varphi_k \leq \vartheta_k \leq \psi_k$. Therefore, by Lemma 8.2.18 above,

$$L_k(f) \leq \int \vartheta_k = \sum\nolimits_{\mathbf{c} \in \mathcal{Z}_k} f(\mathbf{c})\, 2^{-kn} \leq U_k(f).$$

Then the result follows by Theorem 8.2.15. □

Remarks 8.2.22 If K is a support for the function f in (8.24), then the collection $\mathcal{Z}_k$ of all kth-order centers can be replaced by $\mathcal{K}_k \subset \mathcal{Z}_k$, the collection of all kth-order centers that are contained in K. In fact, f vanishes on the remaining centers in $(\mathcal{Z}_k \setminus \mathcal{K}_k)$.

Remarks 8.2.23 The role of the centers. In Theorem 8.2.21 above, the centers of the cubes do not have any particular significance. If $\mathbf{u}_C$ is an arbitrary point from each cube $C \in \mathcal{C}_k$, then $\int f = \lim_k \sum_{C \in \mathcal{C}_k} f(\mathbf{u}_C)\, 2^{-kn}$. In fact, if $\lambda_k = \sum_{C \in \mathcal{C}_k} f(\mathbf{u}_C)\chi_C$, one still has $\varphi_k \le \lambda_k \le \psi_k$. Then the proof follows by the same arguments as before.

Theorem 8.2.24 Positivity and the linearity of integration. *The class $\mathfrak{I}$ of integrable functions $\mathbb{R}^n \to \mathbb{R}$ is a vector space and integration $\int : \mathfrak{I} \to \mathbb{R}$ is a positive linear operator. More explicitly, if $f,\, f' \in \mathfrak{I}$ and $\alpha,\, \alpha' \in \mathbb{R}$, then*

$$(\alpha f + \alpha' f') \in \mathfrak{I} \text{ and } \int(\alpha f + \alpha' f') = \alpha \int f + \alpha' \int f'.$$

Also, if $f \le f'$, then $\int f \le \int f'$.

Proof. Given $\varepsilon > 0$, use Corollary 8.2.20 to find the step functions $\varphi,\, \psi,\, \varphi',\, \psi'$ such that $\varphi \le f \le \psi$, $\varphi' \le f' \le \psi'$, $\int \psi - \int \varphi < \varepsilon/2$, and $\int \psi' - \int \varphi' < \varepsilon/2$. Then $(\varphi + \varphi') \le (f + f') \le (\psi + \psi')$ and $\int(\psi + \psi') - \int(\varphi + \varphi') < \varepsilon$. The last inequality follows from the linearity of integration on step functions, Lemma 8.2.9. Hence $(f + f')$ is integrable. The integrability of αf follows easily from $\alpha\varphi \le \alpha f \le \alpha\psi$ if $\alpha \ge 0$, or from $\alpha\psi \le \alpha f \le \alpha\varphi$ if $\alpha < 0$. This shows that $\mathfrak{I}$ is a vector space. The linearity and positivity of integration $\int : \mathcal{S} \to \mathbb{R}$ on step functions are known from Lemma 8.2.9. Then we verify easily, using Theorem 8.2.21, that integration $\int : \mathfrak{I} \to \mathbb{R}$ on integrable functions is also a positive and linear operator. □

Theorem 8.2.25 *The class $\mathfrak{I}$ of integrable functions $\mathbb{R}^n \to \mathbb{R}$ is closed under multiplication and under taking absolute values, minima, and maxima.*

Proof. To show that $f \cdot g$ is integrable, first assume that f and g are nonnegative integrable functions. Since both are bounded functions, there is an M such that $0 \le f(\mathbf{x}) \le M$ and $0 \le g(\mathbf{x}) \le M$ for all $\mathbf{x} \in \mathbb{R}^n$. Given $\eta > 0$, find the step functions $\varphi,\, \psi,\, \lambda,\, \mu$ such that $0 \le \varphi \le f \le \psi$ and $0 \le \lambda \le g \le \mu$ and such that $\int \psi - \int \varphi < \eta$ and $\int \mu - \int \lambda < \eta$. Then $0 \le \varphi \cdot \lambda \le f \cdot g \le \psi \cdot \mu$. Also,

$$\psi \cdot \mu - \varphi \cdot \lambda = (\psi - \varphi)\mu + \varphi(\mu - \lambda) \le M(\psi - \varphi) + M(\mu - \lambda)$$

shows that $\int(\psi \cdot \mu) - \int(\varphi \cdot \lambda) \le M(\int \psi - \int \varphi) + M(\int \mu - \int \lambda) \le 2M\eta$. This can be made less than given any $\varepsilon > 0$. Hence $f \cdot g$ is integrable. In the general case, one has $f \cdot g = (f' - f'')(g' - g'')$, with the respective positive and negative parts. These parts are all integrable by Definition 8.2.4. Hence, $f \cdot g$ is integrable. The integrability of $|f|$ follows from $|f| = f' + f''$. Then

$$\begin{aligned} \max(f,\, g) &= (1/2)((f + g) + |f - g|) \text{ and} \\ \min(f,\, g) &= (1/2)((f + g) - |f - g|) \end{aligned}$$

are also integrable. □

Integrals of Characteristic Functions

Theorem 8.2.26 Integrability of characteristic functions. *A characteristic function $\chi_A : \mathbb{R}^n \to \mathbb{R}^+$ is integrable if and only if A is a Jordan set in $\mathbb{R}^n$. In this case, $\int \chi_A = v^n(A)$.*

Proof. Let φ_k and ψ_k be the kth-order approximating step functions for χ_A, as in Definition 8.2.10. An easy check shows that $\varphi_k = \chi_{I_k(A)}$ and $\psi_k = \chi_{O_k(A)}$. Hence, $\int \psi_k - \int \varphi_k = v^n(O_k(A)) - v^n(I_k(A))$ converges to zero if and only if A is a Jordan set. $\square$

Definition 8.2.27 Integration over a set. Let $f : \mathbb{R}^n \to \mathbb{R}$ be a function and E a set in $\mathbb{R}^n$. If $f \cdot \chi_E$ is integrable, then $\int (f\chi_E)$ is called *the integral of f over E*. It is also denoted as $\int_E f$ or as $\int_E f(\mathbf{x})d\mathbf{x}$ if one wants to indicate that $\mathbf{x}$ denotes a general point in $\mathbb{R}^n$.

Integrals of Continuous Functions

Remarks 8.2.28 Continuous functions on compact sets. Recall from Definition 4.4.2 that a function $f : \mathbb{R}^n \to \mathbb{R}$ is said to be continuous on $E \subset \mathbb{R}^n$ if for each $\mathbf{a} \in E$ and for each $\varepsilon > 0$, there is a $\delta > 0$ such that $|f(\mathbf{x}) - f(\mathbf{a})| < \varepsilon$ whenever $\|\mathbf{x} - \mathbf{a}\| < \delta$ and $\mathbf{x} \in E$. Theorem 4.5.46 shows that if $E = K$ is a compact set, then the continuity of f on K is uniform. That is, for each $\varepsilon > 0$, there is a $\delta > 0$ such that $|f(\mathbf{x}) - f(\mathbf{x}')| < \varepsilon$ whenever $\|\mathbf{x} - \mathbf{x}'\| < \delta$ and $\mathbf{x}$, $\mathbf{x}' \in K$. Also, a continuous function on a compact set is bounded, as shown in Theorem 4.5.44.

Theorem 8.2.29 Integrability of continuous functions. *Any continuous function on a compact Jordan set is integrable on that set.*

Proof. The idea of the proof is to exploit the uniform continuity of f. If we take small enough cubes, then the lower and upper approximations for f will be almost the same, except on boundary cubes. But the boundary cubes don't matter in computing the integral, since their volume can be made as small as desired.

In detail: let K be a compact Jordan set and $g : K \to \mathbb{R}$ a continuous function. The theorem states that $f = g\chi_K$ is integrable. Here $f(\mathbf{x}) = g(\mathbf{x})$ if $\mathbf{x} \in K$ and $f(\mathbf{x}) = 0$ otherwise. Now g is bounded on K. If $-M \leq g(\mathbf{x}) \leq M$ for all $\mathbf{x} \in K$, then $-M \leq f(\mathbf{x}) \leq M$ for all $\mathbf{x} \in \mathbb{R}^n$. Hence f is a bounded function and vanishes outside of the bounded set K.

Let $\eta > 0$ be given. Use the uniform continuity of g on K to find a $\delta > 0$ such that $|f(\mathbf{x}) - f(\mathbf{x}')| < \eta$ whenever $\|\mathbf{x} - \mathbf{x}'\| < \delta$ and $\mathbf{x}$, $\mathbf{x}' \in K$. Find a $k_0 \in \mathbb{Z}^+$ such that if $k \geq k_0$ and if $C \in \mathfrak{C}_k$, then $\|\mathbf{x} - \mathbf{x}'\| < \delta$ for any $\mathbf{x}, \mathbf{x}' \in C$. For a fixed $k \geq k_0$, let C_i s be the kth-order inner cubes of K and B_j s the kth-order boundary cubes of K. Also, let $\alpha_i = \inf(f, C_i)$ and $\beta_i = \sup(f, C_i)$. Then by our choice of k_0, $0 \leq \beta_i - \alpha_i \leq \eta$. If φ_k and ψ_k are the kth-order lower and upper approximations of f, as in Definition 8.2.10, then we see that

$$\sum_i \alpha_i \chi_{C_i} \leq \varphi_k \leq \psi_k \leq \sum_i \beta_i \chi_{C_i} + \sum_j M \chi_{B_j}.$$

All these functions are step functions. Hence

$$\begin{aligned}\int \psi_k - \int \varphi_k &\leq \sum_i (\beta_i - \alpha_i) v(C_i) + \sum_j M v(B_j) \\ &\leq \eta\, v(I_k(K)) + M\, v(D_k(K)) \\ &\leq \eta\, v(K) + M\, v(D_k(K)).\end{aligned}$$

Given $\varepsilon > 0$, choose $\eta > 0$ so that $\eta\, v(K) < \varepsilon/2$. Then choose $k \geq k_0$ so that $2M\, v(D_k(K)) < \varepsilon/2$. This is possible since K is a Jordan set and therefore $\lim_k v(D_k(K)) = 0$ by Lemma 8.1.17. It follows that $\int \psi_k - \int \varphi_k < \varepsilon$. Hence g is integrable by Theorem 8.2.15 or by Lemma 8.2.20. □

Fundamental Theorem of Calculus

Theorem 8.2.30 *Let $f : \mathbb{R} \to \mathbb{R}$ be continuous on a closed interval $I = [a, b]$. Define $F : I \to \mathbb{R}$ by $F(x) = \int f \cdot \chi_{[a, x]}$ for all $x \in I^o = (a, b)$. Then F is continuous on I, differentiable on I^o, and $F'(x) = f(x)$ for all $x \in I^o$.*

Proof. Continuity of f on the compact set I implies that f is bounded on I. Let $0 \leq f(x) \leq M$ for all $x \in I$. If $a \leq r \leq s \leq b$, then we see that $F(s) - F(r) = \int f \cdot \chi_{(r, s]}$. Hence $|F(s) - F(r)| \leq M|s - r|$. Therefore F is continuous on I.

Next, given $x \in I^o$ and $\varepsilon > 0$, use the continuity of f at x to find r_0 and s_0 such that $a \leq r_0 < x < s_0 \leq b$ and such that $|f(x) - f(y)| < \varepsilon$ whenever $r_0 \leq y \leq s_0$. Let $p = (f(x) - \varepsilon)$ and $q = (f(x) + \varepsilon)$. Then $p \leq f(y) \leq q$ for all $y \in [r_0, s_0]$. Therefore, if $r_0 \leq r < s \leq s_0$, then

$$p = (f(x) - \varepsilon) \leq (F(s) - F(r))/(s - r) \leq (f(x) + \varepsilon) = q.$$

Hence $F'(x)$ exists and $F'(x) = f(x)$. □

Corollary 8.2.31 *Let $f : I \to \mathbb{R}$ be as in Theorem 8.2.30. Let $G : I \to \mathbb{R}$ be continuous on I and differentiable in I^o with $G'(x) = f(x)$ for $x \in I^o$. Then $G(s) - G(r) = \int f \cdot \chi_J$ whenever $a \leq r \leq s \leq b$ and J is any type of interval with the end points r and s.*

Proof. Let $\varphi(x) = F(x) - G(x) + G(a)$ for $x \in I$. Then φ is continuous on I, vanishes at a, and is differentiable in I^o with $\varphi'(x) = 0$ for all $x \in I^o$. Then the mean value theorem, Theorem 5.1.13, shows that $\varphi(x) = 0$ for all $x \in I$. Hence $G(s) - G(r) = F(s) - F(r) = \int f \cdot \chi_{(r,\, s]}$ by Theorem 8.2.30. □

Integrals in Product Spaces

Notations 8.2.32 Functions of two variables. Let $n, m \in \mathbb{N}$. As usual, we may consider $\mathbb{R}^{n+m}$ as $\mathbb{R}^n \times \mathbb{R}^m$. Points in $W = \mathbb{R}^{n+m}$ are denoted by $\mathbf{w} = (\mathbf{u}, \mathbf{v})$ with $\mathbf{u} \in U = \mathbb{R}^n$ and $\mathbf{v} \in V = \mathbb{R}^m$. The coordinate projections $P : W \to U$ and $Q : W \to V$ are defined as $P(\mathbf{u}, \mathbf{v}) = \mathbf{u}$ and $Q(\mathbf{u}, \mathbf{v}) = \mathbf{v}$. Functions $f : W \to \mathbb{R}$ on $W = \mathbb{R}^{n+m}$ are considered as functions of two variables $f : U \times V \to \mathbb{R}$ with values $f(\mathbf{u}, \mathbf{v}) \in \mathbb{R}$. For each fixed $\mathbf{u} \in U$, we have a function $f(\mathbf{u}, \cdot\,) : V \to \mathbb{R}$ and for each fixed $\mathbf{v} \in V$, we have a function $f(\,\cdot, \mathbf{v}) : U \to \mathbb{R}$.

Notations 8.2.33 Iterated integrals. Let $f : W \to \mathbb{R}$ be a bounded function of a bounded support $S \subset W$. Let $A = PS$ and $B = QS$ be the projections of S on U and V. These sets are also bounded. Hence $f(\mathbf{u}, \cdot\,) : V \to \mathbb{R}$ is a bounded function of bounded support for each $\mathbf{u} \in U$ and $f(\cdot, \mathbf{v}\,) : U \to \mathbb{R}$ is a bounded function of bounded support for each $\mathbf{v} \in V$. Therefore these two functions have lower and upper integrals. The lower integral of $f(\cdot, \mathbf{v}\,) : U \to \mathbb{R}$ is denoted as

$$\underline{\int} f(\mathbf{u}, \mathbf{v})\, d\mathbf{u}$$

and similarly for the other integrals. Here the symbol $d\mathbf{u}$ is only used to indicate that this lower integral is taken with respect to $\mathbf{u}$, keeping $\mathbf{v}$ fixed. The result depends on $\mathbf{v}$.

This integration defines a new function $g : V \to \mathbb{R}$ by $g(\mathbf{v}) = \underline{\int} f(\mathbf{u}, \mathbf{v})\, d\mathbf{u}$, $\mathbf{v} \in V$. This function is also a bounded function of bounded support. In fact, if $|f(\mathbf{w})| \leq M$ for all $\mathbf{w} \in W$, then $|g(\mathbf{v})| \leq M v^n(A)$. Also, if $\mathbf{v} \notin B = QS$, then $f(\mathbf{u}, \mathbf{v}) = 0$. Therefore the integrals of $g : V \to \mathbb{R}$ also exist. The lower integral of $g : V \to \mathbb{R}$ is denoted as

$$\underline{\int} g(\mathbf{v})\, d\mathbf{v} = \underline{\int}\, \underline{\int} f(\mathbf{u}, \mathbf{v})\, d\mathbf{u}\, d\mathbf{v}.$$

In this last expression, the order of $d\mathbf{u}\, d\mathbf{v}$ is important. It indicates that the first integration is with respect to $\mathbf{u}$ and the second integration is with respect to $\mathbf{v}$.

The same observations apply to upper integrals. In fact, we see that eight different iterated integrals can be defined, taking into account the four possible combinations of upper and lower integrals and the two possible orders of integration. If any one of the integrals in these expressions exists, then lower and upper integral signs may be replaced by the integral sign.

These observations give us our first special case of Fubini's theorem.

Lemma 8.2.34 *Let $C \in \mathcal{C}_k^{n+m}$ be a cube in $\mathbb{R}^{n+m}$ and $f = \chi_C$. Then*

$$\int f(\mathbf{w})\, d\mathbf{w} = \int\int f(\mathbf{u}, \mathbf{v})\, d\mathbf{u}\, d\mathbf{v}.$$

Proof. We have, by Lemma 8.1.6, that $\mathcal{C}_k^{n+m} = \mathcal{C}_k^n \times \mathcal{C}_k^m$. Hence $C = A \times B$ with $A \in \mathcal{C}_k^n$ and $B \in \mathcal{C}_k^m$. Hence $f(\mathbf{u}, \mathbf{v}) = \chi_C(\mathbf{u}, \mathbf{v}) = \chi_A(\mathbf{u}) \cdot \chi_B(\mathbf{v})$, and

$$\begin{aligned} \int\int \chi_{A\times B}(\mathbf{u}, \mathbf{v})\, d\mathbf{u}\, d\mathbf{v} &= \int\int \chi_A(\mathbf{u})\, \chi_B(\mathbf{v})\, d\mathbf{u}\, d\mathbf{v} \\ &= \upsilon^n(A) \int \chi_B(\mathbf{v})\, d\mathbf{v} \\ &= \upsilon^n(A) \cdot \upsilon^m(B) = \upsilon^{n+m}(A \times B) \\ &= \upsilon^{n+m}(C) = \int \chi_C(\mathbf{w})\, d\mathbf{w}. \end{aligned}$$

Here we have used Theorem 8.2.26 on the integration of characteristic functions. The relation $\upsilon^{n+m}(C) = \upsilon^n(A) \cdot \upsilon^m(B)$ is given in Definition 8.1.8 of the volumes of cubes. □

By combining this special case with the technique of approximation by step functions, we prove the general version of Fubini's theorem, Theorem 8.2.37 below.

Lemma 8.2.35 *Let $f_i : \mathbb{R}^{n+m} \to \mathbb{R}$ be integrable functions such that*

$$\int f_i(\mathbf{w})\, d\mathbf{w} = \int\int f_i(\mathbf{u}, \mathbf{v})\, d\mathbf{u}\, d\mathbf{v}. \tag{8.25}$$

Then (8.25) also holds for a (finite) linear combination $f = \sum_i \alpha_i f_i$, $\alpha_i \in \mathbb{R}$.

Proof. This follows directly from the linearity of integration, Theorem 8.2.24. Note that this theorem is applied three times here for the three separate integrals that appear in (8.25). □

Corollary 8.2.36 **Fubini's theorem for step functions.** *Let $\varphi : \mathbb{R}^{n+m} \to \mathbb{R}$ be a step function. Then*

$$\int \varphi(\mathbf{w})\, d\mathbf{w} = \int\int \varphi(\mathbf{u}, \mathbf{v})\, d\mathbf{u}\, d\mathbf{v}.$$

Proof. Follows directly from Lemmas 8.2.35 and 8.2.34 above. □

Theorem 8.2.37 Fubini's theorem. *If $f : W \to \mathbb{R}$ is integrable, then*

$$\int f(\mathbf{w})\, d\mathbf{w} = \int \underline{\int} f(\mathbf{u}, \mathbf{v})\, d\mathbf{u}\, d\mathbf{v} = \int \overline{\int} f(\mathbf{u}, \mathbf{v})\, d\mathbf{u}\, d\mathbf{v}, \tag{8.26}$$

with the corresponding results in the other order of integration.

Proof. Lemma 8.2.20 shows that given $\varepsilon > 0$, there are two step functions φ and ψ such that $\varphi \le f \le \psi$ and such that $\int (\psi - \varphi) < \varepsilon$. Then

$$\int \varphi(\mathbf{u}, \mathbf{v})\, d\mathbf{u} \le \underline{\int} f(\mathbf{u}, \mathbf{v})\, d\mathbf{u} \le \int \psi(\mathbf{u}, \mathbf{v})\, d\mathbf{u}. \tag{8.27}$$

Here we have used the positivity properties of lower integrals, Lemma 8.2.18, and the fact that the first and last integrals exist. We obtain, again by using the positivity properties of lower and upper integrals and the fact that the first and last functions in (8.27) are integrable functions $V \to \mathbb{R}$,

$$\begin{aligned} \int \varphi(\mathbf{w})\, d\mathbf{w} &= \int \int \varphi(\mathbf{u}, \mathbf{v})\, d\mathbf{u}\, d\mathbf{v} \le \underline{\int} \underline{\int} f(\mathbf{u}, \mathbf{v})\, d\mathbf{u}\, d\mathbf{v} \\ &\le \overline{\int} \underline{\int} f(\mathbf{u}, \mathbf{v})\, d\mathbf{u}\, d\mathbf{v} \le \int \int \psi(\mathbf{u}, \mathbf{v})\, d\mathbf{u}\, d\mathbf{v} = \int \psi(\mathbf{w})\, d\mathbf{w}. \end{aligned}$$

The equalities follow from Corollary 8.2.36. Now we also have that

$$\int \varphi(\mathbf{w})\, d\mathbf{w} \le \int f(\mathbf{w})\, d\mathbf{w} \le \int \psi(\mathbf{w})\, d\mathbf{w}$$

and that $(\int \psi(\mathbf{w})\, d\mathbf{w} - \int \varphi(\mathbf{w})\, d\mathbf{w}) < \varepsilon$. Hence we see that $\int \underline{\int} f(\mathbf{u}, \mathbf{v})\, d\mathbf{u}\, d\mathbf{v}$ exists and is equal to $\int f(\mathbf{w})\, d\mathbf{w}$. The proof of the second equality in (8.26) uses the same arguments. □

Notations 8.2.38 Iterated integrals. Fubini's theorem has an obvious generalization to decompositions into more than two components. In particular, $\mathbb{R}^n$ may be decomposed into n one-dimensional subspaces spanned by the vectors in the standard basis of $\mathbb{R}^n$. In this case, Fubini's theorem shows that

$$\int f(\mathbf{x})\, d\mathbf{x} = \int \cdots \int f(x_1, \ldots, x_n)\, dx_1 \cdots dx_n \tag{8.28}$$

if all these integrals exist. On the right-hand side of (8.28), there are n integrals of functions of one variable. The convention is that they have to be performed in the following order. For the first integral consider f as a function of x_1 only, keeping the other variables fixed. The integral of this function results in a function of $(n-1)$ variables

$$f_1(x_2, \ldots, x_n) = \int f(x_1, \ldots, x_n)\, dx_1.$$

For the second integral consider f_1 as a function of x_2 only, keeping the other variables fixed. The integral of this function results in a function of $(n-2)$ variables

$$f_2(x_3, \ldots, x_n) = \int f_1(x_2, \ldots, x_n)\, dx_2.$$

After $(n-1)$ integrals, the last integral to be performed is $\int f_{n-1}(x_n)\, dx_n$. Fubini's theorem states that if $\int f(\mathbf{x})\, d\mathbf{x}$ exists, then it is equal to the result of this last integral. Note that the existence of these n iterated integrals does not imply the existence of $\int f(\mathbf{x})\, d\mathbf{x}$; this has to be assumed separately. Problem 8.29 asks for a counter-example.

Notations 8.2.39 Integrals on $\mathbb{R}^2$. If f is an integrable function on $\mathbb{R}^2$, then

$$\int f = \int_{\mathbb{R}^2} f(x, y)\, dx\, dy = \int f(x, y)\, dx\, dy = \int \int f(x, y)\, dx\, dy.$$

Here the first three integrals show some of the common notations used for the integral of f on $\mathbb{R}^2$. The last integral is an iterated integral. It shows two integrals to be computed in the order explained in Notations 8.2.38 above.

If E is a Jordan set in the xy-plane, then

$$\int_E f = \int_E f(x, y)\, dx\, dy = \int_I \int_{E_x} f(x, y)\, dy\, dx$$

are some of the notations for the integral of $f\chi_E$. In the iterated integral, the first integration is with respect to y. It is the integral of $f(x, \cdot) : \mathbb{R} \to \mathbb{R}$ over the cross-section E_x of E with the $x =$ constant line. The second integration can be performed over any interval I in the x-axis such that $E_x = \emptyset$ whenever $x \notin I$. Hence I is any interval that contains the projection of E on the x-axis.

In most examples E_x is an interval in the y-axis, and E_x is not empty if and only if $a \leq x \leq b$. The initial and final points of the interval E_x depend on x. If these points are denoted, respectively, as $g_1(x)$ and $g_2(x)$, then

$$\int_E f(x, y)\, dx\, dy = \int_a^b \int_{g_1(x)}^{g_2(x)} f(x, y)\, dy\, dx. \tag{8.29}$$

This form is applicable in many examples.

Notations 8.2.40 Integrals on $\mathbb{R}^3$. Let $f : \mathbb{R}^3 \to \mathbb{R}$ be a function and let G be a Jordan set in $\mathbb{R}^3$. Assume that f is integrable over G. Then in many applications, $\int_G f$ can be expressed as

$$\int_G f(x,\, y,\, z)\, dx\, dy\, dz = \int_a^b \int_{g_1(x)}^{g_2(x)} \int_{h_1(x,\, y)}^{h_1(x,\, y)} f(x,\, y,\, z)\, dz\, dy\, dx. \qquad (8.30)$$

This form is similar to (8.29) for integration in $\mathbb{R}^2$. Here the first integral is with respect to z. Integration is performed on the portion of the $x,\, y =$ constants line that lies in the region G. In (8.30) it is assumed that if this portion is not empty, then it is an interval with the initial point $h_1(x,\, y)$ and the final point $h_2(x,\, y)$. After the first integration one obtains an integral in $\mathbb{R}^2$. It is over a set E in $\mathbb{R}^2$. This set consists of all points $(x,\, y)$ for which the line $x,\, y =$ constants intersects G. Hence E is the projection of G on the xy-plane. The remaining two integrals are integrals on $\mathbb{R}^2$. Then one uses the notations in 8.2.39 above.

Notations 8.2.41 Another decomposition of $\mathbb{R}^3$. There is another common form for integrals on $\mathbb{R}^3$. It is obtained by applying Fubini's theorem to the decomposition of $W = \mathbb{R}^3$ as $U \times V = \mathbb{R}^2 \times \mathbb{R}$. One obtains

$$\int_G f(x,\, y,\, z)\, dx\, dy\, dz = \int_p^q \left(\int_{G_z} f(x,\, y,\, z)\, dx\, dy \right) dz. \qquad (8.31)$$

In the first integral, z is a constant, and this integral is performed on the set

$$G_z = \{\, (x,\, y) \mid (x,\, y,\, z) \in G \,\} \subset \mathbb{R}^2.$$

This is the cross-section of G with the $z =$ constant plane. The integral on G_z is computed as an integral in $\mathbb{R}^2$. The result is a function of z only. One obtains (8.31) by taking $[\, p,\, q\,]$ an interval on the z axis such that $G_z = \emptyset$ for $z \notin [\, p,\, q\,]$.

Example 8.2.42 Volume between two cylinders. We compute the volume of the region E bounded by the two cylinders $x^2 + z^2 = 1$ and $y^2 + z^2 = 1$.

The region E is the set of all $(x,\, y,\, z) \in \mathbb{R}^3$ such that $x^2 + z^2 \leq 1$ and $y^2 + z^2 \leq 1$. We know that this is a Jordan set by Problem 8.19. We would like to compute $\int \chi_E$. We apply Fubini's theorem as expressed in (8.31) above. Hence

$$v^3(E) = \int \chi_E(x,\, y,\, z)\, ddx\, dy\, dz = \int_p^q \left(\int_{G_z} f(x,\, y,\, z)\, dx\, dy \right) dz.$$

The first integral is on the xy-plane. In the first integral the z coordinate is fixed and the integration is over the cross-section of E with a $z =$ constant plane. Hence $G_z = \{\, (x,\, y) \mid x^2 \leq (1 - z^2),\ \text{and } y^2 \leq (1 - z^2) \,\}$. We see that $G_z = \emptyset$ if $z^2 > 1$, and

G_z is a square with side length $2(1-z^2)^{1/2}$ if $z^2 \le 1$. Therefore the first integral is the area of this square. Hence we obtain

$$v^3(E) = 4\int_{-1}^{1}(1-z^2)dz = 8-(8/3) = 16/3.$$

As an exercise, let us also decompose the xyz-space in a different way. Let U be the yz-plane and V the x-axis. In this case, the cross-section of E with the $x=$ constant plane is

$$F(x) = \left\{ (y,\, z) \mid z^2 \le (1-x^2) \text{ and } y^2+z^2 \le 1 \right\}.$$

This is the part of the unit disc $y^2+z^2 \le 1$ between the lines $z = \pm(1-x^2)^{1/2}$. This area can be computed in different ways. We apply Fubini's theorem again and obtain

$$\begin{aligned} v^2(F(x)) &= \int_{-(1-x^2)^{1/2}}^{(1-x^2)^{1/2}} \int_{-(1-z^2)^{1/2}}^{(1-z^2)^{1/2}} dy\, dz \\ &= 2\int_{-(1-x^2)^{1/2}}^{(1-x^2)^{1/2}} (1-z^2)^{1/2}\, dz \\ &= 2(\cos^{-1} x + x(1-x^2)^{1/2}), \text{ and,} \\ v^3(E) &= \int_{-1}^{1} v^2(F(x))\, dx \\ &= 4\int_0^1 (\cos^{-1} x + x(1-x^2)^{1/2})\, dx = 16/3. \end{aligned}$$

This example illustrates how the complexity of the computations may depend upon the order of integration. △

Example 8.2.43 All Euclidean balls are Jordan sets. This fact is proved in one way in Problem 8.17. Here is another argument. Proceed by induction on the number of dimensions. The unit ball in $\mathbb{R}$ is $(-1,\, 1)$, which is a Jordan set. Now assume that the unit ball $A \subset \mathbb{R}^n$ is a Jordan set in $\mathbb{R}^n$. The function $f : A \to \mathbb{R}^+$ defined by $f(\mathbf{x}) = (1-\|\mathbf{x}\|^2)^{1/2}$ is continuous on A. Hence the region under its graph is a Jordan set E_f in $\mathbb{R}^n$. This is the upper half of the unit ball. Similarly, the lower half is also a Jordan set. Hence the unit ball is a Jordan set, since the union of two Jordan sets is still a Jordan set. Finally, any ball is obtained from the unit ball by scaling and a translation, and therefore the same arguments apply to any ball. △

Example 8.2.44 Volume of the unit ball in $\mathbb{R}^n$. The volume of the unit ball in a Euclidean space can be computed by an induction on the number of dimensions. Let ϑ_k be the volume of the unit ball in $\mathbb{R}^n$. Note that the volume of any ball of radius r in $\mathbb{R}^n$ is $r^n\vartheta_k$. Also, $\vartheta_1 = \ell(-1,\, 1) = 2$. Assume that ϑ_k is known. Let

$W = \mathbb{R}^{n+1} = R^n \times \mathbb{R} = U \times V$ and write $\mathbf{w} = (\mathbf{u},\, v) \in W$ with $\mathbf{u} \in U = \mathbb{R}^n$ and $v \in V = \mathbb{R}$. Let

$$E = \{\, (\mathbf{u},\, v) \in W \mid \|\mathbf{u}\|^2 + v^2 < 1 \,\}$$

be the unit ball in $\mathbb{R}^{n+1}$. The cross-sections of E with $v =$ constant subspaces are $E_v = \{\, \mathbf{u} \in U \mid \|\mathbf{u}\|^2 < 1 - v^2 \,\}$. Hence E_v is a ball of radius $(1 - v^2)^{1/2}$ in $\mathbb{R}^n$. Therefore $v^n(E_v) = (1 - v^2)^{n/2}\vartheta_k$. This gives

$$\begin{aligned} \vartheta_{n+1} &= \int \chi_E(\mathbf{w})\, d\mathbf{w} = \int\!\!\int \chi_E(\mathbf{u},\, v)\, d\mathbf{u}\, dv \\ &= \int_{-1}^{1} v^n(E_v)\, dv = 2\vartheta_k \int_0^1 (1 - v^2)^{n/2}\, dv. \end{aligned}$$

The computation of the last integral is left as an exercise. We obtain

$$\vartheta_{2n} = \pi^n/n! \quad \text{and} \quad \vartheta_{2n-1} = 2^n \pi^{n-1}/(1 \cdot 3 \cdot 5 \cdots (2n-1))$$

for all $n \in \mathbb{N}$. △

Problems

8.22 Volumes of cylinders. Let A be a Jordan set in $\mathbb{R}^n$ and $\mathbf{c} = (\mathbf{a},\, h) \in \mathbb{R}^n \times \mathbb{R}$ with $h \neq 0$. Show that the volume of cylinder

$$C = C(A,\, \mathbf{c}) = \{\, (\mathbf{x},\, 0) + t(\mathbf{a},\, h) \mid \mathbf{x} \in A,\ 0 \le t \le 1 \,\} \subset \mathbb{R}^{n+1}$$

is $v^{n+1}(C) = |h|\, v^n(A)$. (Cylinders are Jordan sets by Problem 8.13.)

8.23 Volumes of cones. Let A be a Jordan set in $\mathbb{R}^n$ and $\mathbf{c} = (\mathbf{a},\, h) \in \mathbb{R}^n \times \mathbb{R}$ with $h \neq 0$. Show that the volume of cone

$$K = K(A,\, \mathbf{c}) = \{\, (1 - t)(\mathbf{x},\, 0) + t(\mathbf{a},\, h) \mid \mathbf{x} \in A,\ 0 \le t \le 1 \,\} \subset \mathbb{R}^{n+1}$$

is $v^{n+1}(K) = |h|\, v^n(A)/(n + 1)$. (Cones are Jordan sets by Problem 8.14.)

8.24 Show that the volume of the tetrahedron

$$E = \left\{ \mathbf{x} = (x_1,\, \ldots,\, x_n) \in \mathbb{R}^n \;\middle|\; 0 \le x_i,\ \textstyle\sum_i x_i \le 1 \right\}$$

is $v^n(E) = 1/n!$. (Tetrahedra are Jordan sets by Problem 8.16.)

8.25 Find the volume of the ellipsoid

$$E = \left\{ \mathbf{x} = (x_1,\, \ldots,\, x_n) \in \mathbb{R}^n \;\middle|\; \textstyle\sum_i r_i^{-2}(x_i - c_i)^2 \le R^2 \right\}.$$

Here $\mathbf{c} = (c_1, \ldots, c_n)$ is the center of the ellipsoid and $(r_1, \ldots, r_n)$ is a fixed vector in $\mathbb{R}^n$ with $r_i \neq 0$. (Ellipsoids are Jordan sets by Problem 8.18.)

8.26 Integrate the function $f(x, y, z) = x^2 + y^2 + z^2$ over the tetrahedron $E = \{ (x, y, z) \mid |x| + |y| + |z| \leq 1 \}$.

8.27 Let $p \in \mathbb{R}$. Find the volume V of

$$E = \{ (x, y, z) \mid x^2 + y^2 + z^2 \leq 1, \ p \leq z \}.$$

How do we know that E is a Jordan set?

8.28 Integrate $f(x, y, z) = |z|$ over E defined in Problem 8.27 above.

8.29 Give an example of a nonintegrable function $f : \mathbb{R}^2 \to \mathbb{R}$ for which the integrals $\int \int f(x, y)\, dx\, dy$ exist. (This means that $g(y) = \int f(x, y)\, dx$ exists for each y and defines an integrable function $g : \mathbb{R} \to \mathbb{R}$. Such an example shows that the converse of Fubini's theorem is false.)

8.30 Let $f : \mathbb{R}^n \to \mathbb{R}$ be an integrable function. Points in $\mathbb{R}^{n+1} = \mathbb{R}^n \times \mathbb{R}$ are denoted as $\mathbf{z} = (\mathbf{x}, y)$. Let

$$\begin{aligned} A &= \{ (\mathbf{x}, y) \in \mathbb{R}^n \times \mathbb{R} \mid \mathbf{x} \in \mathbb{R}^n, \ f(\mathbf{x}) < y \leq 0 \} \text{ and} \\ B &= \{ (\mathbf{x}, y) \in \mathbb{R}^n \times \mathbb{R} \mid \mathbf{x} \in \mathbb{R}^n, \ 0 \leq y < f(\mathbf{x}) \}. \end{aligned}$$

Let $D \subset \mathbb{R}^{n+1}$ be a compact set containing $A \cup B$. Let $G : D \to \mathbb{R}$ be an integrable function. Show that

$$\int_B G(\mathbf{z})\, d\mathbf{z} - \int_A G(\mathbf{z})\, d\mathbf{z} = \int_X \int_0^{f(\mathbf{x})} G(\mathbf{x}, y)\, dy\, d\mathbf{x}.$$

Here the integral with respect to $y \in \mathbb{R}$ is expressed in familiar notations. If $f(\mathbf{x}) > 0$ then this is an integral over the interval $[0, f(\mathbf{x})]$ as defined in this course. If $f(\mathbf{x}) < 0$ then this is the negative of the integral over the interval $[f(\mathbf{x}), 0]$.

8.31 If $f : \mathbb{R}^2 \to \mathbb{R}$ is integrable, show that for any $\varepsilon > 0$,

$$E_\varepsilon = \left\{ y \in \mathbb{R} \;\middle|\; \overline{\int} f(x, y)\, dx - \underline{\int} f(x, y)\, dx > \varepsilon \right\}$$

is a Jordan subset of $\mathbb{R}$ with $v(E_\varepsilon) = 0$.

8.32 Let $f : \mathbb{R}^n \to \mathbb{R}$ be a bounded function of compact support. For each $\varepsilon > 0$, let $E_\varepsilon = \{ \mathbf{z} \mid \omega(f, \mathbf{z}) \geq \varepsilon \}$, where $\omega(f, \mathbf{z})$ is the oscillation of f at $\mathbf{z}$ (Problem 4.96). Show that f is integrable if and only if E_ε is a negligible set for all $\varepsilon > 0$.

8.33 Show that if a bounded function of compact support is integrable, then the set E of its discontinuities is a countable union of negligible sets E_i. That is, there is a sequence of negligible sets E_i, $i \in \mathbb{N}$, such that $E = \cup_{i \in \mathbb{N}} E_i$. Give an example to show that the set E of discontinuities need not itself be negligible.

8.34 Let $f : \mathbb{R}^n \to \mathbb{R}$ be a bounded function of compact support. If the set E of discontinuities of f is a countable union of negligible sets E_i (that is, $E = \cup_{i \in \mathbb{N}} E_i$), then f is integrable. (Problems 8.33 and 8.34 give a necessary and sufficient condition for the integrability of a bounded function of compact support. This is known as *Lebesgue's theorem.*)

The following set of three problems provides a proof of the *bounded convergence theorem.*

8.35 Let $0 < \beta$. Let K and B_k be compact Jordan sets such that $B_k \subset K$ and $\beta \leq v(B_k)$ for all $k \in \mathbb{N}$. Then show that there is an $\mathbf{x} \in K$ that belongs to infinitely many B_k s.

8.36 Let f be a step function with compact support in a Jordan set K. Assume that $0 \leq f(\mathbf{x}) \leq M$. Given $\alpha > 0$, let D be the set of $\mathbf{x} \in K$ such that $\alpha \leq f(\mathbf{x})$. Then show that $\int f \leq Mv(D) + \alpha v(K)$.

8.37 Let f_n be a sequence of integrable functions. Assume that all f_n s have support in a compact Jordan set K and that $|f_n(\mathbf{x})| \leq M$ for all $n \in \mathbb{N}$ and for all $\mathbf{x} \in K$. If $\lim_n f_n(\mathbf{x}) = f(\mathbf{x})$ for all $\mathbf{x} \in K$ and if f is also integrable, then show that $\lim_n \int f_n = \int f$.

8.38 Give a counterexample to show that the conclusion in Problem 8.37 is false if there is no compact set that contains the support of all f_ns.

8.39 Give a counterexample to show that the conclusion in Problem 8.37 is false if there is no number M which is an upper bound for all $|f_n|$s.

8.40 Give an example to show that the limit function in Problem 8.37 need not be integrable.

8.41 Give examples of f_n and f to show that several of the hypotheses in Problem 8.37 are not necessary for the conclusion of the problem to be true.

8.3 IMAGES OF JORDAN SETS

The image of a Jordan set under a diffeomorphismis a Jordan set and its volume can be computed by the change of variables formula. In this section, we establish the first point and we prove the second point for the special case of an invertible linear transformation.

We begin with Theorem 8.3.2, which provides a nice topological characterization of Jordan sets. This establishes an important connection between Jordan sets and the topology of $\mathbb{R}^n$. All further results on volume and integration depend upon this connection.

It is convenient to repeat here a few definitions that will be useful in what follows. A set in $\mathbb{R}^n$ is called negligible if it is a Jordan set of zero volume, as stated in Definition 8.1.22. The kth-order boundary cubes of a set E, as defined in Definition 8.1.15, are the kth-order cubes that intersect both E and its complement $E^c = \mathbb{R}^n \setminus E$. The union of these cubes is $D_k(E) = O_k(E) \setminus I_k(E)$, the kth-order boundary approximation of E.

Topological Definition of Jordan Sets

Lemma 8.3.1 Boundary cubes. *Any boundary cube of E intersects ∂E. Hence $D_k(E) \subset O_k(\partial E)$. The converse is not true. A cube that intersects ∂E is not necessarily a boundary cube of E.*

Proof. Any boundary cube contains points $\mathbf{a} \in E$ and $\mathbf{b} \notin E$. The line segment joining these two points contains a point from ∂E, and this point also belongs to the boundary cube because cubes are convex. To disprove the converse: any cube contains points from its boundary but, trivially, does not intersect its complement. □

Theorem 8.3.2 Boundaries of Jordan sets. *A bounded set is a Jordan set if and only if its boundary is a negligible set.*

Proof. Let E be a Jordan set. Given $\varepsilon > 0$, use Theorem 8.1.30 to find a closed Jordan set F and and an open Jordan set G such that

$$F \subset E^o \subset E \subset \overline{E} \subset G \text{ and } v(G) - v(F) < \varepsilon.$$

Hence $\partial E = \overline{E} \setminus E^o \subset G \setminus F$. Also, $G \setminus F$ is a Jordan set and

$$v(G \setminus F) = v(G) - v(F) < \varepsilon$$

by Theorem 8.1.24. Hence ∂E is between the Jordan sets $\emptyset$ and $G \setminus F$ with $v(G \setminus F) - v(\emptyset) = v(G \setminus F)$. This implies that ∂E is a Jordan set of zero volume.

Conversely, assume that E is a bounded set of negligible boundary. Given $\varepsilon > 0$, find an integer k such that $v(O_k(\partial E)) < \varepsilon$. Hence, by Lemma 8.3.1,

$$D_k(E) \subset O_k(\partial E) \text{ and, therefore, } v(D_k(E)) < \varepsilon.$$

Then Theorem 8.1.21 shows that E is a Jordan set. □

Corollary 8.3.3 Interiors and closures of Jordan sets. *A bounded set is a Jordan set if and only if both its interior and closure are Jordan sets of the same volume.*

Proof. This is left as an exercise. □

Jordan Sets Under Diffeomorphisms

We begin by showing that diffeomorphisms transform Jordan sets into Jordan sets. It is convenient to recall a few facts about diffeomorphisms. First, if $f : A \to Y$ is a $\mathcal{C}^1$ diffeomorphism, then A and $B = f(A)$ are both open sets, $f : A \to B$ has an inverse $g : B \to A$, and both f and g have continuous derivatives $f' : A \to L(X, Y)$ and $g' : B \to L(Y, X)$. Second, $f : A \to B$ preserves the topological properties of sets in A. Open sets, compact sets and boundaries are transformed into corresponding open sets, compact sets and boundaries. In particular, $\partial f(E) = f(\partial E)$ for any E such that $\overline{E} \subset A$. This last condition is necessary to ensure that not only E but also ∂E are contained in A.

In the context of Jordan sets we will mostly use the maximum norm

$$\|\mathbf{x}\| = \|(x_1, \ldots, x_n)\|_m = \max(|x_1|, \ldots, |x_n|).$$

This norm was introduced in Definition 8.1.3, where we noted that all balls in this norm are open blocks. Hence, all such balls are Jordan sets. The interior of a cube C is a ball B such that $B \subset C \subset \overline{B}$. We will call the radius of B the radius of C. In particular, the cubes in the kth-order grid are all cubes of radius 2^{-k-1}.

Lemma 8.3.4 *Let $A \subset \mathbb{R}^n$ be an open set and $f : A \to \mathbb{R}^n$ a $\mathcal{C}^1$ mapping. Let $K \subset A$ be a compact set. Then there is a number M with the following property: if a cube C of radius r is contained in K, then $f(C)$ is contained in a cube of radius Mr.*

Proof. The continuous function $f' : A \to L(\mathbb{R}^n, \mathbb{R}^n)$ is bounded on the compact set K. Hence there is an M such that $\|f'(\mathbf{x})\|_{L(\mathbb{R}^n, \mathbb{R}^n)} < M$ for all $\mathbf{x} \in K$. Let C

be a cube contained in K. If $\mathbf{u}, \mathbf{v} \in C$, then the line segment joining these points is in $C \subset K$. Therefore $\|f'(\mathbf{w})\| < M$ for every $\mathbf{w}$ on this segment. Then the mean value theorem 5.1.13 implies that $\|f(\mathbf{u}) - f(\mathbf{v})\| < M\|\mathbf{u} - \mathbf{v}\|$. We see that this implies the conclusion of the lemma. □

Theorem 8.3.5 *Let A be an open set in $\mathbb{R}^n$ and $f : A \to \mathbb{R}^n$ a diffeomorphism. Let E be a Jordan set in A with $\overline{E} \subset A$. Then $f(E)$ is also a Jordan set.*

Proof. The basic idea is to combine the fact that ∂E is negligible with the preceding lemma. A cover for ∂E of small volume will be mapped by f to a cover for $\partial f(E)$ of small volume. So $\partial f(E)$ is also negligible, which shows that $f(E)$ is a Jordan set.

In detail: the set $\overline{E}$ is compact (as it is closed and bounded), and contained in the open set A. Theorem 4.5.48 shows that there is a compact set K and a number $\delta > 0$ such that $\overline{E} \subset K^o \subset K \subset A$. Note that in this case there is a number $\delta > 0$ such that $\mathbf{x} \in K$ whenever $\|\mathbf{x} - \mathbf{a}\| < \delta$ and $\mathbf{a} \in E$. This follows again from Theorem 4.5.48. Hence we see that if a cube of radius less than $\delta/2$ intersects E, then it is contained in K.

Now E is a Jordan set. Hence it has a negligible boundary ∂E. Given $\varepsilon > 0$, find an outer approximation $O_k(\partial E)$ of this boundary with volume $v(O_k(\partial E)) < \varepsilon$. Then $O_k(\partial E) = \cup_i C_i$ is a finite union of kth-order cubes C_i of radius $r = 2^{-k-1}$. Without loss of generality, assume that $2r = 2^{-k} < \delta$. In this case we see that all C_i s are contained in K. Therefore all $f(C_i)$ s are contained in cubes of radius Mr, where M is the number obtained in Lemma 8.3.4. Now $f(\partial E) \subset f(O_k(\partial E)) = f(\cup_i C_i) = \cup_i f(C_i)$ shows that $f(\partial E)$ is contained in a set of volume

$$v(\cup_i f(C_i)) \leq \sum_i v(f(C_i)) \leq \sum_i M^n v(C_i) \tag{8.32}$$

$$= M^n v(O_k(\partial E)) < M^n \varepsilon. \tag{8.33}$$

Hence $f(\partial E) = \partial f(E)$ is a negligible set and $f(E)$ is a Jordan set. □

Jordan Sets Under Isomorphisms

An isomorphism $T : \mathbb{R}^n \to \mathbb{R}^n$ is an invertible linear transformation. Therefore it is also a diffeomorphism. It follows that TE is a Jordan set whenever E is a Jordan set, by Theorem 8.3.5. In this case, more is true. We show that there is a fixed number $\rho(T) > 0$ such that $v(TE) = \rho(T)\, v(E)$ for all Jordan sets E. This number will be called the *volume multiplier* of T.

Definition 8.3.6 Translations. Let $\mathbf{a} \in \mathbb{R}^n$. Then *translation by* $\mathbf{a}$ is the transformation $T_\mathbf{a} : \mathbb{R}^n \to \mathbb{R}^n$ given by $T_\mathbf{a}(\mathbf{x}) = \mathbf{a} + \mathbf{x}$. If $E \subset \mathbb{R}^n$, then $\mathbf{a} + E$ denotes the image of E under translation by $\mathbf{a}$. That is, $\mathbf{a}+E$ consists of all vectors $(\mathbf{a}+\mathbf{x})$ with $\mathbf{x} \in E$.

Theorem 8.3.7 Volumes under translations. *If $E \subset \mathbb{R}^n$ is a Jordan set and $\mathbf{a} \in \mathbb{R}^n$, then $\mathbf{a}+E$ is also a Jordan set and $v(\mathbf{a}+E) = v(E)$.*

Proof. Since a translation is a diffeomorphism, we know that $\mathbf{a}+E$ is a Jordan set whenever E is a Jordan set. Translation of a block is also a block with the same side lengths. Hence the volumes of blocks are preserved under translations. We see that this is also true for finite disjoint unions of blocks, by the additivity of volume. Hence the volumes of inner and outer approximations of sets are preserved under translations, since they consist of finite disjoint unions of cubes. It follows easily that $v(\mathbf{a}+E) = v(E)$ for any Jordan set E. □

Lemma 8.3.8 *Let $T : \mathbb{R}^n \to \mathbb{R}^n$ be an isomorphism. Let $C_0 = [0, 1)^n$ be the unit cube in $\mathbb{R}^n$. Let $\rho(T) = v(TC_0)$. Then $\rho(T) > 0$ and $v(TC) = \rho(T)v(C)$ for any cube C in any kth-order grid $\mathcal{C}_k$, $k \in \mathbb{N}$.*

Proof. All cubes $C \in \mathcal{C}_k$ are translations of each other. The same is true for their images TC. Hence $v(TC)$ is the same for all $C \in \mathcal{C}_k$. The unit cube C_0 is the disjoint union of 2^{kn} cubes in $\mathcal{C}_k$. Therefore TC_0 is the disjoint union of 2^{kn} translates of TC, for any $C \in \mathcal{C}_k$. Hence $v(TC) = 2^{-kn}v(C_0) = v(C)\rho(T)$. Also, TC_0 has a nonempty interior since the interior of C_0 is nonempty and $T : \mathbb{R}^n \to \mathbb{R}^n$ is an isomorphism. Hence $\rho(T) = v(TC_0) > 0$, by Lemma 8.1.28. □

Lemma 8.3.9 *Let $T : \mathbb{R}^n \to \mathbb{R}^n$ be an isomorphism. Let $C_0 = [0, 1)^n$ be the unit cube in $\mathbb{R}^n$ and let $k \in \mathbb{N}$. Let $\rho(T) = v(TC_0)$. If $F = \cup_i F_i$ is a finite union of cubes $F_i \in \mathcal{C}_k$, then $v(TF) = \rho(T)v(F)$.*

Proof. The cubes in a grid are pairwise disjoint. So if $F_i \in \mathcal{C}_k$, then the F_i are disjoint and therefore the images TF_i are also disjoint, since T is an isomorphism. Hence $v(F) = \sum_i v(F_i)$ and $v(TF) = \sum_i v(TF_i)$. Then the result follows from Lemma 8.3.8. □

Theorem 8.3.10 *Let $T : \mathbb{R}^n \to \mathbb{R}^n$ be an isomorphism. Then there is a number $\rho(T) > 0$ such that $v(TE) = \rho(E)\, v(E)$ for all Jordan sets $E \subset \mathbb{R}^n$.*

Proof. Let $\rho(T) = v(TC_0)$, where C_0 is the unit cube of $\mathbb{R}^n$. Let E be a Jordan set and $\varepsilon > 0$. Find inner and outer approximations of E, F and G such that

$F \subset E \subset G$ and $v(G) - v(F) < \varepsilon$. Here F and G are both finite disjoint unions of cubes F_i, $G_j \in \mathcal{C}_k$. Then, by Lemma 8.3.9,

$$\rho(T)v(F) = v(TF) \leq v(TE) \leq v(TG) = \rho(T)v(G).$$

We see that both numbers $v(TE)$ and $\rho(T)v(E)$ are between $\rho(T)v(F)$ and $\rho(T)v(G)$. Hence

$$|v(TE) - \rho(T)v(E)| \leq (\rho(T)v(G) - \rho(T)v(F)) \leq \varepsilon\rho(T).$$

Since $\varepsilon > 0$ is arbitrary, this implies that $v(TE) = \rho(T)v(E)$. □

Definition 8.3.11 Volume multipliers of isomorphisms. Let T be an isomorphism $\mathbb{R}^n \to \mathbb{R}^n$. Then the *volume multiplier* of T is defined as the number $\rho(T) > 0$ obtained in Theorem 8.3.10. Hence $v(TE) = \rho(T)v(E)$ for all Jordan sets E in $\mathbb{R}^n$.

Theorem 8.3.12 Volume multipliers of isometries. *Let $T : \mathbb{R}^n \to \mathbb{R}^n$ be an isometry of the Euclidean space $\mathbb{R}^n$. Then $\rho(T) = 1$.*

Proof. The unit ball $B = \{ \mathbf{x} \mid \|\mathbf{x}\| \leq 1 \}$, with the usual Euclidean norm (rather than the maximum norm), is a Jordan set and is invariant under any isometry T. Also, it has nonzero volume since its interior is not empty. Therefore $v(TB) = v(B) = \rho(T)v(B)$, which shows that $\rho(T) = 1$. □

Corollary 8.3.13 *Let $(\mathbf{u}_1, \ldots, \mathbf{u}_n)$ be an orthonormal basis for $\mathbb{R}^n$ and $\lambda_i \geq 0$. Then the "rectangular box"*

$$B = \{ \mathbf{x} = s_1\mathbf{u}_1 + \cdots + s_n\mathbf{u}_n \mid 0 \leq s_i \leq \lambda_i, \;\; i = 1, \ldots, n, \}$$

is a Jordan set in $\mathbb{R}^n$ and $v(B) = \lambda_1 \cdots \lambda_n$.

Proof. Let $(\mathbf{e}_1, \ldots, \mathbf{e}_n)$ be the standard basis of $\mathbb{R}^n$. Then the transformation defined by $T\mathbf{u}_i = \mathbf{e}_i$ is an isometry and takes B into a rectangular block with side lengths λ_i. Hence $v(B) = v(TB) = \lambda_1 \cdots \lambda_n$. □

Corollary 8.3.14 Volume multipliers of scaling transformations. *Let $t > 0$. If $T : \mathbb{R}^n \to \mathbb{R}^n$ is defined by $T\mathbf{x} = t\,\mathbf{x}$, then $\rho(T) = t^n$.*

Proof. Apply Corollary 8.3.13 with $\lambda_i = t$, $i = 1, \ldots, n$. □

Determinants and Volume Multipliers

Remarks 8.3.15 A review of determinants. If X is an n-dimensional vector space, then a determinant on X is any nonzero alternating multilinear function $\psi : X^n \to \mathbb{R}$. Each ordered basis $\mathbb{E} = (\mathbf{e}_1, \ldots, \mathbf{e}_n)$ defines a unique determinant $\psi_{\mathbb{E}}$ on X such that $\psi_{\mathbb{E}}(\mathbf{e}_1, \ldots, \mathbf{e}_n) = 1$. Theorem C.7.2 shows that if $T : X \to X$ is a linear transformation and if $\mathbb{E} = (\mathbf{e}_1, \ldots, \mathbf{e}_n)$ is a basis for X, then the number $\psi(T\mathbf{e}_1, \ldots, T\mathbf{e}_n)/\psi(\mathbf{e}_1, \ldots, \mathbf{e}_n)$ is independent of the choice of the determinant ψ and of the basis $\mathbb{E}$. It is called the *determinant* of T and denoted by $\det T$. Note that $\det T = \psi_{\mathbb{E}}(T\mathbf{e}_1, \ldots, T\mathbf{e}_n)$ for any basis $\mathbb{E}$.

Remarks 8.3.16 Euclidean determinants. There is a special situation in Euclidean spaces. If $\mathbb{E}$ and $\mathbb{U}$ are two orthonormal bases for a Euclidean space, then they define the same determinant up to a factor of ± 1. Such a determinant is called a *Euclidean determinant*. Hence there are exactly two Euclidean determinants in a Euclidean space.

Theorem 8.3.17 Determinants as volume multipliers. *Let T be an isomorphism of the Euclidean space $\mathbb{R}^n$. Then $\rho(T) = |\det T|$.*

Proof. By the spectral theorem, Theorem 3.6.4, T has an *eigenbasis* $\mathbb{E} = (\mathbf{e}_1, \ldots, \mathbf{e}_n)$. Recall from Definition 3.6.3 what this means: $\mathbb{E}$ is an orthonormal basis for $\mathbb{R}^n$ such that $T\mathbf{e}_i \perp T\mathbf{e}_j$ whenever $i \neq j$. It follows that if we put $\mathbf{u}_i = T\mathbf{e}_i/\|T\mathbf{e}_i\|$, then $\mathbb{U} = (\mathbf{u}_1, \ldots, \mathbf{u}_n)$ is another orthonormal basis for $\mathbb{R}^n$ such that $T\mathbf{e}_i = \lambda_i \mathbf{u}_i$ for all i, with scalars $\lambda_i \geq 0$. In fact, each $\lambda_i > 0$ since T is an isomorphism. Therefore

$$\begin{aligned}
\det T &= \psi_{\mathbb{E}}(T\mathbf{e}_1, \ldots, T\mathbf{e}_n) && (8.34)\\
&= \pm\psi_{\mathbb{U}}(T\mathbf{e}_1, \ldots, T\mathbf{e}_n) && (8.35)\\
&= \pm\psi_{\mathbb{U}}(\lambda_1\mathbf{u}_1, \ldots, \lambda_k\mathbf{u}_k) && (8.36)\\
&= \pm(\lambda_1 \cdots \lambda_k)\psi_{\mathbb{U}}(\mathbf{u}_1, \ldots, \mathbf{u}_n) && (8.37)\\
&= \pm(\lambda_1 \cdots \lambda_k). && (8.38)
\end{aligned}$$

Here (8.35) follows from the fact that determinants with respect to any two orthonormal bases differ by a factor of ± 1 only. To obtain (8.37), we use the multilinearity of the determinant. The last step is by the definition of $\psi_{\mathbb{U}}$.

Now T takes the rectangular box

$$B = \{\, \mathbf{x} = s_1\mathbf{e}_1 + \cdots + s_n\mathbf{e}_n \mid 0 \leq s_i \leq 1, \;\; i = 1, \ldots, n \,\}$$

with volume $v(B) = 1$ to the rectangular box

$$TB = \{\, \mathbf{x} = s_1\mathbf{u}_1 + \cdots + s_n\mathbf{u}_k \mid 0 \leq s_i \leq \lambda_i, \;\; i = 1, \ldots, n \,\}$$

with volume $v(TB) = (\lambda_1 \cdots \lambda_k)$. Hence

$$\rho(T) = v(TB) = (\lambda_1 \cdots \lambda_k) = |\det T|.$$

□

Problems

8.42 A plane $Ax + By + Cz = D$ divides an ellipsoid

$$(x/a)^2 + (y/b)^2 + (z/c)^2 = 1$$

into two parts. Find the volumes of these parts. (One part may be empty.)

8.43 Find the volume of $E = \{ (x, y) \mid (2x + y)^2 + (x - y)^2 \leq 1 \} \subset \mathbb{R}^2$.

8.44 Find the volume of $E = \{ (x, y) \mid |2x + y| + |x - y| \leq 1 \} \subset \mathbb{R}^2$.

8.45 Find the volume of

$$E = \{ (x, y, z) \mid |x| + |x + y| + |x + y + z| \leq 1 \} \subset \mathbb{R}^3.$$

8.46 Let $H \subset B \subset A \subset \mathbb{R}^n$. Assume that A is compact, H is open, and $tA \subset H$ whenever $0 < t < 1$. Show that A, B, and H are all Jordan sets of the same volume.

8.47 Show that any open ball or any closed ball with respect to any norm on $\mathbb{R}^n$ is a nonnegligible Jordan set.

8.48 Let $T : \mathbb{R}^n \to \mathbb{R}^n$ be an isometry with respect to an arbitrary norm on $\mathbb{R}^n$. Then show that $\rho(T) = 1$.

8.49 Denote the vectors in $\mathbb{R}^{n+1} = \mathbb{R}^n \times \mathbb{R}$ as $(\mathbf{x}, y)$ with $\mathbf{x} \in \mathbb{R}^n$ and $y \in \mathbb{R}$. Let $\mathbf{a} \in \mathbb{R}^n$ be a fixed vector. Define $R : \mathbb{R}^{n+1} \to \mathbb{R}^{n+1}$ by $R(\mathbf{x}, y) = (\mathbf{x} + y\mathbf{a}, y)$ for all $(\mathbf{x}, y) \in \mathbb{R}^{n+1}$. Then show that $\rho(R) = 1$.

8.50 Let E be a bounded set in $\mathbb{R}^n$ such that if $0 < t < 1$, then $tE \subset E$. Prove or disprove that E is a Jordan set.

8.51 Show that any bounded convex set is a Jordan set.

8.52 Let $F \subset \mathbb{R}^n$ be a bounded set and $r > 0$. Show that $\cup_{\mathbf{x} \in F} B_r(\mathbf{x})$ is a Jordan set.

8.53 Let E be an open set in $X = \mathbb{R}^m$ and F an open set in $Y = \mathbb{R}^n$. Let $\mathbf{u} : E \to U = \mathbb{R}^m$ and $\mathbf{v} : F \to V = \mathbb{R}^n$ be two diffeomorphisms. Show that $\mathbf{w}(\mathbf{x}, \mathbf{y}) = (\mathbf{u}(\mathbf{x}), \mathbf{v}(\mathbf{y}))$ defines a diffeomorphism

$$\mathbf{w} : (E \times F) \to W = \mathbb{R}^{m+n} = \mathbb{R}^m \times \mathbb{R}^n.$$

Also show that $\rho(\mathbf{w}'(\mathbf{x}, \mathbf{y})) = \rho(\mathbf{u}'(\mathbf{x}))\, \rho(\mathbf{v}'(\mathbf{y}))$ for $\mathbf{x} \in E$, $\mathbf{y} \in F$.

8.4 CHANGE OF VARIABLES

Introduction 8.4.1 Changing the integration variable is a familiar technique from basic calculus courses. Let $y : A \to \mathbb{R}$ be a continuously differentiable function defined on an open interval A. Assume that $y'(x) > 0$ for all $x \in A$. If $[\,a,\, b\,] \subset A$ and if g is integrable over $[\,y(a),\, y(b)\,]$, then

$$\int_{y(a)}^{y(b)} g(y)\, dy = \int_a^b g(y(x))\, y'(x)\, dx \tag{8.39}$$

The generalization of this result to multiple integrals is one of the major theorems of this course. This result is stated as Theorem 8.4.16 below. The key step in the proof of this theorem is an approximation theorem. It compares the volumes of the images of small sets under a diffeomorphism and under the derivative of that diffeomorphism. Such an approximation is fairly easy to obtain locally for small balls about a fixed point. We need a uniform version of this result. The proof of this uniform approximation theorem depends on a uniform mean value theorem. Hence we first review some mean value theorems.

Notations 8.4.2 Diffeomorphisms and compact sets. For this section, we review a few basic facts and some standard notation. Let A be an open set in $X = \mathbb{R}^n$. Let $\varphi : A \to Y = \mathbb{R}^n$ be a diffeomorphism. This means that $\varphi : A \to Y$ is a continuously differentiable function, the image of A under φ is an open set $B = \varphi(A) \subset Y$, and there is a continuously differentiable inverse function $\psi = \varphi^{-1} : B \to X$. If K is a compact set and if $K \subset A$, then Theorem 4.5.43 shows that K is mapped to a compact set $H = \varphi(K) \subset B$ by the continuous function φ. Also, the derivatives $\varphi' : A \to L(X, Y)$ and $\psi' : B \to L(Y, X)$ are continuous functions. Hence, by Theorem 4.5.44, they are bounded on the compact sets K and H respectively. Finally, by Theorem 4.5.48, if K is a compact set, A is an open set, and $K \subset A$, then there is a $\delta > 0$ such that $B_\delta(\mathbf{a}) \subset A$ for all $\mathbf{a} \in K$.

A Review of Mean Value Theorems

We review several mean value theorems. Throughout the discussion, φ is a diffeomorphism.

Theorem 8.4.3 Basic mean value theorem. *Assume that*

$$\mathbf{w}(t) = (t\mathbf{u} + (1-t)\mathbf{v}) \in A \text{ and that } \|\varphi'(\mathbf{w}(t))\| \leq M$$

for all $t \in [0, 1]$. *Then* $\|\varphi(\mathbf{v}) - \varphi(\mathbf{u})\| \leq M \|\mathbf{v} - \mathbf{u}\|$.

Proof. This is a restatement of the basic mean value theorem, Theorem 5.1.13. □

Theorem 8.4.4 Mean value theorem on convex sets. *Let* C *be a convex set and* $C \subset A$. *If* $\|\varphi'(\mathbf{x})\| \leq M$ *for all* $\mathbf{x} \in C$ *then*

$$\|\varphi(\mathbf{v}) - \varphi(\mathbf{u})\| \leq M \|\mathbf{v} - \mathbf{u}\|$$

for all $\mathbf{u}, \mathbf{v} \in C$.

Proof. This follows from Theorem 8.4.3. □

Theorem 8.4.5 Local mean value theorem. *For each* $\mathbf{a} \in A$ *and for each* $\varepsilon > 0$, *there is a* $\delta > 0$ *such that* $B_\delta(\mathbf{a}) \subset A$ *and*

$$\|\varphi(\mathbf{v}) - \varphi(\mathbf{u}) - \varphi'(\mathbf{a})(\mathbf{v} - \mathbf{u})\| \leq \varepsilon \|\mathbf{v} - \mathbf{u}\|$$

for all $\mathbf{u}, \mathbf{v} \in B_\delta(\mathbf{a})$.

Proof. Let $\mathbf{a} \in A$. Since A is open, there is a $\delta_0 > 0$ such that $B_{\delta_0}(\mathbf{a}) \subset A$. Also, there is a δ such that $0 < \delta \leq \delta_0$ and such that

$$\|\varphi'(\mathbf{u}) - \varphi'(\mathbf{a})\| < \varepsilon \text{ whenever } \mathbf{u} \in B_\delta(\mathbf{a}).$$

This follows from the continuity of $\varphi' : A \to L(X, Y)$. Define $\lambda : A \to Y$ as

$$\lambda(\mathbf{x}) = \varphi(\mathbf{x}) - \varphi(\mathbf{a}) - \varphi'(\mathbf{a})(\mathbf{x} - \mathbf{a}), \quad \mathbf{x} \in A. \tag{8.40}$$

We see that $\|\lambda'(\mathbf{x})\| = \|\varphi'(\mathbf{x}) - \varphi'(\mathbf{a})\| < \varepsilon$ for all $\mathbf{x} \in B_\delta(\mathbf{a})$. Now $B_\delta(\mathbf{a})$ is a convex set. Then Theorem 8.4.4, the mean value theorem on convex sets, shows that

$$\|\lambda(\mathbf{v}) - \lambda(\mathbf{u})\| = \|\varphi(\mathbf{v}) - \varphi(\mathbf{u}) - \varphi'(\mathbf{a})(\mathbf{v} - \mathbf{u})\| \leq \varepsilon \|\mathbf{v} - \mathbf{u}\| \tag{8.41}$$

for all $\mathbf{u}, \mathbf{v} \in B_\delta(\mathbf{a})$. □

Theorem 8.4.6 Uniform mean value theorem. *For each $\varepsilon > 0$ and for each compact set $K \subset A$ there is a $\delta > 0$ such that*

$$B_\delta(\mathbf{a}) \subset A \text{ and } \|\varphi(\mathbf{v}) - \varphi(\mathbf{u}) - \varphi'(\mathbf{a})(\mathbf{v} - \mathbf{u})\| \leq \varepsilon \|\mathbf{v} - \mathbf{u}\| \tag{8.42}$$

for all $\mathbf{a} \in K$ *and for all* $\mathbf{u}, \mathbf{v} \in B_\delta(\mathbf{a})$.

Proof. Let K be a compact set, $K \subset A$, and $\varepsilon > 0$. Use Theorem 4.5.48 to find another compact set $K_0 \subset A$ and a number $\delta_0 > 0$ such that $B_{\delta_0}(\mathbf{a}) \subset K_0$ for all $\mathbf{a} \in K$. The continuous function $\varphi' : A \to L(X, Y)$ is uniformly continuous on the compact set $K_0 \subset A$. That means we can find a δ such that $0 < \delta \leq \delta_0$ and such that $\|\varphi'(\mathbf{v}) - \varphi'(\mathbf{u})\| < \varepsilon$ whenever $\mathbf{u}, \mathbf{v} \in K_0$ and $\|\mathbf{v} - \mathbf{u}\| < \delta$. Now if $\mathbf{a} \in K$, then $B_\delta(\mathbf{a}) \subset B_{\delta_0}(\mathbf{a}) \subset K_0 \subset A$. Define $\lambda : A \to Y$ as in (8.40). As before, we see that $\|\lambda'(\mathbf{x})\| < \varepsilon$ for all $\mathbf{x} \in B_\delta(\mathbf{a})$. Then (8.42) follows as in (8.41) above. □

Definition 8.4.7 Affine approximations. The *affine approximation* of φ at $\mathbf{a} \in A$ is the affine mapping $\vartheta_\mathbf{a} : X \to Y$ given as

$$\vartheta_\mathbf{a}(\mathbf{x}) = \varphi(\mathbf{a}) + \varphi'(\mathbf{a})(\mathbf{x} - \mathbf{a}), \ \mathbf{x} \in X. \tag{8.43}$$

Note that $\vartheta_\mathbf{a}(\mathbf{a}) = \varphi(\mathbf{a})$ and $\vartheta_\mathbf{a}(\mathbf{v}) - \vartheta_\mathbf{a}(\mathbf{u}) = \varphi'(\mathbf{a})(\mathbf{v} - \mathbf{u})$, $\mathbf{a} \in A$, $\mathbf{u}, \mathbf{v} \in X$.

Lemma 8.4.8 Inverses of affine approximations. *Let* $\mathbf{a} \in A$ *and* $\mathbf{b} = \varphi(\mathbf{a})$. *Let* $\vartheta_\mathbf{a}$ *be the affine approximation of* φ *at* $\mathbf{a}$. *Then* $\vartheta_\mathbf{a}$ *is invertible and*

$$\vartheta_\mathbf{a}^{-1}(\mathbf{y}) = \mathbf{a} + (\varphi'(\mathbf{a}))^{-1}(\mathbf{y} - \mathbf{b}) = \psi(\mathbf{b}) + \psi'(\mathbf{b})(\mathbf{y} - \mathbf{b}), \ \mathbf{y} \in Y. \tag{8.44}$$

Also, $\vartheta_\mathbf{a}^{-1} : B \to X$ *is the affine approximation of* $\psi : B \to A$ *at* $\mathbf{b} = \varphi(\mathbf{a})$.

Proof. Note that $(\varphi'(\mathbf{a}))^{-1} = \psi'(\mathbf{b})$, since $(\psi \cdot \varphi) : A \to A$ is the identity. Then the proof follows directly from the definitions. □

The affine approximation $\vartheta_\mathbf{a}$ allows us to approximate f in a neighborhood of $\mathbf{a}$. The next theorem shows that the 'local' multiplying effect of f on Jordan sets is also well approximated by the multiplying effect of $\vartheta_\mathbf{a}$, which (as we know) is expressed by the value $|\det \varphi'(\mathbf{a})|$.

Uniform Approximations Theorem

The uniform approximations theorem states that the image under φ of a suitably small ball around $\mathbf{a}$ can be approximated both from inside and outside by images under $\vartheta_\mathbf{a}$ of slightly reduced and slightly enlarged balls around $\mathbf{a}$. The radius of the original ball is r; the radii of the reduced and enlarged balls are $(1-t)r$ and $(1+t)r$ for a specified parameter t with $0 < t < 1$. (Nothing in the argument will depend upon the particular choice of norm.)

Theorem 8.4.9 Uniform approximations. *Let $\varphi : A \to Y$ be a diffeomorphism and $K \subset A$ a compact set. Let $0 < t < 1$. Then there is a $\delta > 0$ such that if $\mathbf{a} \in K$ and if $0 < r < \delta$, then $B_r(\mathbf{a}) \subset A$ and*

$$\vartheta_{\mathbf{a}}(B_{(1-t)r}(\mathbf{a})) \subset \varphi(B_r(\mathbf{a})) \subset \vartheta_{\mathbf{a}}(B_{(1+t)r}(\mathbf{a})). \tag{8.45}$$

Proof. *Choosing α and M.* Apply Theorem 4.5.48 to find an $\alpha > 0$ and a compact set K_0 such that $B_\alpha(\mathbf{a}) \subset K_0 \subset A$ for all $\mathbf{a} \in K$. Also, find a number $M > 0$ such that $\|\varphi'(\mathbf{x})\| \leq M$ for all $\mathbf{x} \in K_0$. Such a number exists, since the continuous function $\varphi' : A \to L(X, Y)$ is bounded on the compact set $K_0 \subset A$. Note that

$$\|\varphi(\mathbf{v}) - \varphi(\mathbf{u})\| \leq M \|\mathbf{v} - \mathbf{u}\| \tag{8.46}$$

whenever $\mathbf{a} \in K$ and $\mathbf{u}, \mathbf{v} \in B_\alpha(\mathbf{a})$. This follows from Theorem 8.4.4, the mean value theorem on convex sets. Also,

$$\|\vartheta_{\mathbf{a}}(\mathbf{u}) - \vartheta_{\mathbf{a}}(\mathbf{v})\| = \|\varphi'(\mathbf{a})(\mathbf{v} - \mathbf{u})\| \leq M \|\mathbf{v} - \mathbf{u}\| \tag{8.47}$$

for all $\mathbf{u}, \mathbf{v} \in K$.

Choosing β. Let $H = \varphi(K)$. Then H is a compact set contained in $B = \varphi(A)$. Use Theorem 4.5.48 again to find a number $\beta_0 > 0$ and a compact set H_0 such that $B_{\beta_0}(\mathbf{b}) \subset H_0 \subset B$ for all $\mathbf{b} \in H$. Now apply Theorem 8.4.6, the uniform mean value theorem, to find a β such that $0 < \beta \leq \beta_0$ and such that

$$\|\psi(\mathbf{y}) - \psi(\mathbf{b}) - \psi'(\mathbf{b})(\mathbf{y} - \mathbf{b})\| \leq (t/M) \|\mathbf{y} - \mathbf{b}\| \tag{8.48}$$

whenever $\mathbf{b} \in H$ and $\mathbf{y} \in B_\beta(\mathbf{b})$.

A basic estimate. Let $\mathbf{b} \in H$, $\mathbf{y} \in B_\beta(\mathbf{b})$, and $\mathbf{a} = \psi(\mathbf{b}) = \vartheta_{\mathbf{a}}^{-1}(\mathbf{b})$. Then

$$\|\mathbf{v} - \mathbf{u}\| \leq (t/M) \|\mathbf{y} - \mathbf{b}\| \tag{8.49}$$

with $\mathbf{v} = \varphi^{-1}(\mathbf{y}) = \psi(\mathbf{y})$ and $\mathbf{u} = \vartheta_{\mathbf{a}}^{-1}(\mathbf{y})$. This is just a reformulation of (8.48). In fact, using the expression of $\vartheta_{\mathbf{a}}^{-1} : Y \to X$ in (8.44), we see that

$$\psi(\mathbf{y}) - \psi(\mathbf{b}) - \psi'(\mathbf{b})(\mathbf{y} - \mathbf{b}) = \mathbf{v} - \mathbf{a} - \vartheta_{\mathbf{a}}^{-1}(\mathbf{y}) + \vartheta_{\mathbf{a}}^{-1}(\mathbf{b}) \tag{8.50}$$

$$= \mathbf{v} - \mathbf{a} - \mathbf{u} + \mathbf{a} = \mathbf{v} - \mathbf{u}. \tag{8.51}$$

Completion of the proof. We will show that the inclusions (8.45) in Theorem 8.4.9 are satisfied with $\delta = \min(\alpha, \beta/M)$. To obtain the first inclusion, let $\mathbf{b} = \varphi(\mathbf{a})$ and $\mathbf{y} = \vartheta_{\mathbf{a}}^{-1}(\mathbf{u}) \in \vartheta_{\mathbf{a}}(B_{(1-t)r}(\mathbf{a}))$ with $\mathbf{u} \in B_{(1-t)r}(\mathbf{a})$. Note that

$$\|\mathbf{y} - \mathbf{b}\| \leq M\|\mathbf{u} - \mathbf{a}\| < M(1-t)r < M(1-t)\delta \tag{8.52}$$

$$\leq M(1-t)(\beta/M) < \beta. \tag{8.53}$$

Hence (8.49) is applicable. Let $\mathbf{v} = \psi(\mathbf{y})$. Then

$$\begin{aligned} \|\mathbf{v} - \mathbf{a}\| &\leq \|\mathbf{u} - \mathbf{a}\| + \|\mathbf{v} - \mathbf{u}\| \leq (1-t)r + (t/M)\|\mathbf{y} - \mathbf{b}\| \quad &(8.54) \\ &< (1-t)r + t(1-t)r < r. &(8.55) \end{aligned}$$

This shows that if $\mathbf{y} = \vartheta_{\mathbf{a}}^{-1}(\mathbf{u}) \in \vartheta_{\mathbf{a}}(B_{(1-t)r}(\mathbf{a}))$, then also $\mathbf{y} = \varphi(\mathbf{v}) \in \varphi(B_r(\mathbf{a}))$. Hence the first inclusion follows. To obtain the second inclusion in (8.45), let $\mathbf{y} = \varphi(\mathbf{v}) \in \varphi(B_r(\mathbf{a}))$. We see that, as in (8.52), $\mathbf{y} \in B_\beta(\mathbf{b})$. Hence (8.49) is again applicable. Let $\mathbf{u} = \vartheta_{\mathbf{a}}^{-1}(\mathbf{y})$. Then

$$\begin{aligned} \|\mathbf{u} - \mathbf{a}\| &\leq \|\mathbf{v} - \mathbf{a}\| + \|\mathbf{v} - \mathbf{u}\| \leq r + (t/M)\|\mathbf{y} - \mathbf{b}\| \quad &(8.56) \\ &< r + tr = (1+t)r. &(8.57) \end{aligned}$$

This shows that if $\mathbf{y} = \varphi(\mathbf{v}) \in \varphi(B_r(\mathbf{a}))$, then also $\mathbf{y} = \vartheta_{\mathbf{a}}(\mathbf{u}) \in \vartheta_{\mathbf{a}}(B_{(1+t)r}(\mathbf{a}))$. Hence the second inclusion follows. □

Change of Volumes

Remarks 8.4.10 Cubes and balls. We will use the maximum norm

$$\|(x_1, \ldots, x_n)\|_m = \max_i |x_i| \text{ on } \mathbb{R}^n.$$

This is a natural norm here since the balls in the maximum norm are the interiors of cubic blocks, as was observed in Definition 8.1.3.

Notations 8.4.11 Cubes as unions of smaller cubes. Let C be a cube in a grid $\mathcal{C}_{k_0}$. If $k \geq k_0$, then C is a union of cubes in $\mathcal{C}_k$ as $C = \cup_i C_i$, $C_i \in \mathcal{C}_k$. The center of C_i is $\mathbf{a}_i$. Note that the interior of C_i is the open ball $C_i^o = B_r(\mathbf{a}_i)$ with $r = 2^{-k-1}$ and the closure of C_i is the closed ball $\overline{C}_i = B_r(\mathbf{a}_i)$.

Lemma 8.4.12 *Let $\varphi : A \to Y = \mathbb{R}^n$ be a diffeomorphism. Let $C \in \mathcal{C}_{k_0}$, $\overline{C} \subset A$, and $0 < t < 1$. Then there is a $k_1 \geq k_0$ such that*

$$\begin{aligned} (1-t)^n \sum_i \rho(\varphi'(\mathbf{a}_i)) v(C_i) &\leq \underline{v}(\varphi(C)) \leq \overline{v}(\varphi(C)) \quad &(8.58) \\ &\leq (1+t)^n \sum_i \rho(\varphi'(\mathbf{a}_i)) v(C_i) &(8.59) \end{aligned}$$

whenever $k \geq k_1$, $C_i \in \mathcal{C}_k$, and $C_i \subset C$, with the center at $\mathbf{a}_i$ and radius $r = 2^{-k-1}$.

Proof. Find $\delta > 0$ from the uniform approximations theorem, Theorem 8.4.9, so that (8.45) is satisfied whenever $\mathbf{a} \in C$ and $0 < r < \delta$. Find $k_1 \geq k_0$ so that $r = 2^{-k_1-1} < \delta$. Since $B_r(\mathbf{a}_i) \subset C_i \subset \overline{B}_r(\mathbf{a}_i)$ and $C = \cup_i C_i$, we obtain

$$\cup_i \vartheta_{\mathbf{a}_i}(B_{(1-t)r}(\mathbf{a}_i)) \subset \cup_i \varphi(C_i) \subset \cup_i \vartheta_{\mathbf{a}_i}(B_{(1+t)r}(\mathbf{a}_i)) \quad (8.60)$$

from (8.45) of Theorem 8.4.9. But $C = \cup_i C_i$ implies that $\varphi(C) = \cup_i \varphi(C_i)$ since φ is a diffeomorphism and, therefore, one-to-one. Also, the sets $\varphi(C_i)$ are pairwise disjoint. Therefore

$$\sum_i v(\vartheta_{\mathbf{a}_i}(B_{(1-t)r}(\mathbf{a}_i))) \leq \underline{v}(\varphi(C)) \leq \overline{v}(\varphi(C)) \tag{8.61}$$

$$\leq \sum_i v(\vartheta_{\mathbf{a}_i}(B_{(1+t)r}(\mathbf{a}_i))). \tag{8.62}$$

Also, by Definition 8.3.11 and Corollary 8.3.14,

$$v(\vartheta_{\mathbf{a}_i}(B_{(1-t)r}(\mathbf{a}_i))) = \rho(\varphi'(\mathbf{a}_i))\, v(B_{(1-t)r}(\mathbf{a}_i)) \tag{8.63}$$

$$= (1-t)^n \rho(\varphi'(\mathbf{a}_i))\, v(B_r(\mathbf{a}_i)), \tag{8.64}$$

and similarly for $v(\vartheta_{\mathbf{a}_i}(B_{(1+t)r}(\mathbf{a}_i)))$. Then (8.58) and (8.59) follow. □

Lemma 8.4.13 Images of cubes. *Let $\varphi : A \to Y = \mathbb{R}^n$ be a diffeomorphism. Let C be a cube in a grid $\mathcal{C}_{k_0}$ and $\overline{C} \subset A$. Then $\varphi(C)$ is a Jordan set and*

$$v(\varphi(C)) = \int_C \rho(\varphi'(\mathbf{x}))\, d\mathbf{x}.$$

Proof. First, $\rho(\varphi'(\mathbf{x}))$ is continuous and therefore integrable on a cube C. Therefore, $\lim_k \sum_i \rho(\varphi'(\mathbf{a}_i)) v(C_i) = \int_C \rho(\varphi'(\mathbf{x}))\, d\mathbf{x}$ by Theorem 8.2.21. Here k is the order of the grid $\mathcal{C}_k$ that contains the cubes C_i, and the sum is extended over all $C_i \in \mathcal{C}_k$ contained in C. Then Lemma 8.4.12 shows that

$$(1-t)^n \int_C \rho(\varphi'(\mathbf{x}))\, d\mathbf{x} \leq \underline{v}(\varphi(C)) \leq \overline{v}(\varphi(C)) \tag{8.65}$$

$$\leq (1+t)^n \int_C \rho(\varphi'(\mathbf{x}))\, d\mathbf{x}. \tag{8.66}$$

The conclusion follows, since these inequalities are true for all t, $0 < t < 1$. □

Theorem 8.4.14 Change of volumes. *Let $A \subset X = \mathbb{R}^n$ be an open set and $\varphi : A \to Y = \mathbb{R}^n$ a diffeomorphism. If E is a Jordan set with closure $\overline{E}$ contained in A, then $F = \varphi(E)$ is a Jordan set contained in $\varphi(A) = B$ and*

$$v(F) = v(\varphi(E)) = \int_E \rho(\varphi'(\mathbf{x}))\, d\mathbf{x}. \tag{8.67}$$

Proof. Lemma 8.4.13 above shows that the conclusion of this theorem is true for cubes with closures contained in A. Then the linearity of integrals and the additivity of volumes show that this conclusion is also true for Jordan sets E that are the finite disjoint unions of such cubes. Let E be any Jordan set with closure $\overline{E} \subset A$. Let

$F_k = I_k(E)$ and $G_k = O_k(E)$ be the kth-order inner and outer approximations of E, as defined in Definition 8.1.15. Use Theorem 4.5.48 to obtain a compact set $K \subset A$ and a $k_0 \in \mathbb{N}$ such that $G_k \subset K$ for all $k \geq k_0$. The conclusion of the change of volumes theorem is valid for all F_k and G_k, $k \geq k_0$. Since $F_k \subset E \subset G_k \subset K$, we obtain

$$v(\varphi(F_k)) = \int_{F_k} \rho(\varphi'(\mathbf{x}))\, d\mathbf{x} \leq \underline{v}(\varphi(E)) \tag{8.68}$$

$$\leq \overline{v}(\varphi(E)) \leq \int_{G_k} \rho(\varphi'(\mathbf{x}))\, d\mathbf{x} = v(\varphi(G_k)). \tag{8.69}$$

The difference between these outer and inner estimates is

$$v(\varphi(G_k)) - v(\varphi(F_k)) = \int_{(G_k \setminus F_k)} \rho(\varphi'(\mathbf{x}))\, d\mathbf{x} \leq M\, v(G_k \setminus F_k),$$

where M is an upper bound for the positive function $\rho(\varphi'(\mathbf{x}))$ on K. Such an upper bound exists, since $(\rho \cdot \varphi') : A \to \mathbb{R}$ is a continuous function and, therefore, bounded on the compact set $K \subset A$. But $\lim_k v(G_k \setminus F_k) = 0$ since E is a Jordan set. Hence $\varphi(E)$ is also a Jordan set and $\int_E \rho(\varphi'(\mathbf{x}))\, d\mathbf{x}$ exists. Then this integral is $v(\varphi(E))$ since both of these numbers are between $v(\varphi(F_k))$ and $v(\varphi(G_k))$ for all $k \geq k_0$. □

Change of Integrals

Lemma 8.4.15 *Let A be an open set in $X = \mathbb{R}^n$. Let $\varphi : A \to \mathbb{R}^n$ be a diffeomorphism. Then $\Phi(\mathbf{x}, t) = (\varphi(\mathbf{x}), t)$, $(\mathbf{x}, t) \in A \times \mathbb{R}$, defines a diffeomorphism $\Phi : (A \times \mathbb{R}) \to (\mathbb{R}^n \times \mathbb{R})$. Also, $\rho(\Phi'(\mathbf{x}, t)) = \rho(\varphi'(\mathbf{x}))$ for all $(\mathbf{x}, t) \in A \times \mathbb{R}$.*

Proof. We see that Φ is a continuously differentiable function. Its range

$$\Phi(A \times \mathbb{R}) = B \times \mathbb{R}$$

is open in $R^n \times \mathbb{R}$, since $B = \varphi(A)$ is open in $\mathbb{R}^n$. Also, $\Psi(\mathbf{y}, t) = (\psi(\mathbf{y}), t)$ for $(\mathbf{y}, t) \in (B \times \mathbb{R})$ defines the inverse of Φ, where ψ is the inverse of φ. Clearly, Ψ is a continuously differentiable function. Finally, if E is a Jordan set in $\mathbb{R}^n$ and J is an interval, then $\Phi'(\mathbf{x}, t)(E \times J) = (\varphi'(\mathbf{x})(E)) \times J$. Hence we see that the volume multipliers are related as stated in the lemma. □

Theorem 8.4.16 Change of integrals. *Let $A \subset X = \mathbb{R}^n$ be an open set and $\varphi : A \to \mathbb{R}^n$ a diffeomorphism. If $f : Y \to \mathbb{R}$ is integrable and has a compact support $S \subset B = \varphi(A)$, then $(f \cdot \varphi) : A \to \mathbb{R}$ is integrable on A and*

$$\int_B f(\mathbf{y})\, d\mathbf{y} = \int_A f(\varphi(\mathbf{x}))\, \rho(\varphi'(\mathbf{x}))\, d\mathbf{x}. \tag{8.70}$$

Proof. First, assume that $f : Y \to \mathbb{R}$ is a nonnegative function. Let

$$E_f = \{ (\mathbf{y}, t) \mid 0 \le t < f(\mathbf{y}), \mathbf{y} \in B \} \subset (B \times \mathbb{R})$$

be the region under the graph of f. Since f is an integrable function, the set E_f is a Jordan set in $\mathbb{R}^n \times \mathbb{R}$. Also, its closure is contained in $(S \times \mathbb{R}) \subset (B \times \mathbb{R})$. If $G = \{ (\mathbf{x}, t) \mid 0 \le t < f(\varphi(\mathbf{x})), \mathbf{x} \in A \}$, then we see that $E_f = \Phi(G)$, with the notations in Lemma 8.4.15 above. Hence, by the change of volumes theorem, Theorem 8.4.14,

$$\begin{aligned}\int_B f(\mathbf{y})\, d\mathbf{y} &= v^{n+1}(E_f) = v^{n+1}(\Phi(G)) \\ &= \int_G \rho(\Phi'(\mathbf{x}, t))\, dt d\mathbf{x} = \int_G \rho(\varphi'(\mathbf{x}))\, dt d\mathbf{x} \\ &= \int_A f(\mathbf{x})\rho(\varphi'(\mathbf{x})) d\mathbf{x}.\end{aligned}$$

Here the last step follows from Fubini's theorem, Theorem 8.2.37. □

Some Useful Notation

In practice, changing the variable of integration in multiple integrals requires a careful distinction between the old and new spaces and careful attention to the *direction* of the transformation between these spaces. The following notation is helpful. Suppose we want to compute an integral $\int_E f(\mathbf{x})\, d\mathbf{x}$ in the X-space. We want to make a change of variables and express $\mathbf{x}$ as $\mathbf{x} = \varphi(\mathbf{y})$. Note that this is a transformation from Y to X, using $\varphi : Y \to X$. The new integration region is the set of all $\mathbf{y}$ such that $\varphi(\mathbf{y}) = \mathbf{x} \in E$. In other words, the new integration region is $\varphi^{-1}(E)$. We start by writing

$$\int_E f(\mathbf{x})\, d\mathbf{x} = \int_{\varphi^{-1}(E)} f(\varphi(\mathbf{y}))U(\mathbf{y})\, d\mathbf{y}.$$

Here, we know that U must be the volume multiplier for either $\varphi : Y \to X$ or $\varphi^{-1} : X \to Y$. But which one? To get the correct answer, write $U(\mathbf{y})$ as

$$U(\mathbf{y}) = \left| \frac{d\mathbf{x}}{d\mathbf{y}} \right|$$

so that the $d\mathbf{y}$ s cancel out and we are left with $d\mathbf{x}$, the original variable of integration. It follows, of course, that the correct multiplier is

$$U(\mathbf{y}) = \left| \frac{\partial(x_1, \ldots, x_n)}{\partial(y_1, \ldots, y_n)} \right|,$$

the usual notation for the determinant of the Jacobian matrix.

Problems

8.54 Integrate $f(x, y) = 2x^2 + y^2$ over the region bounded by the lines $3x + 2y = 5$, $3x + 2y = 8$, $6x - 4y = 2$, and $6x - 4y = 7$.

8.55 Integrate $f(x, y) = 2x^2 + y^2$ over the region in the first quadrant (that is, $x \geq 0$ and $y \geq 0$) bounded by the curves $xy = 2$, $xy = 4$, $y = 3x$, and $y = 5x$.

8.56 The unit disc $x^2 + y^2 \leq 1$ is divided into two parts by the circle $x^2 + (y-1)^2 = 1$. Use polar coordinates $x = r\cos\theta$, $y = r\sin\theta$ to find the areas of these parts.

8.57 The elliptical region $(x/a)^2 + (y/b)^2 \leq 1$ is divided into two parts by the ellipse $(x/a)^2 + ((y-b)/b)^2 = 1$. Find the areas of these parts.

8.58 Let $f : \mathbb{R} \to \mathbb{R}$ be a continuous function, $0 < p < q$, $0 < r < s$, and

$$E = \{\, (x, y) \mid px \leq x^2 + y^2 \leq qx, \;\; ry \leq x^2 + y^2 \leq sy \,\} \subset \mathbb{R}^2.$$

Show that one can define a function $F : \mathbb{R}^2 \to \mathbb{R}$ so that

$$\int_E f(y/x)\, dx\, dy = \int_p^q \int_r^s F(u, v)\, du\, dv$$

for all choices of p, q, r, and s. Find F explicitly in terms of f. Contrive some examples for $f \neq 0$ to compute this integral without too much work.

8.59 Let $f(x, y) = \exp(-x^2 - y^2)$. Compute the integral of f over a disc $x^2 + y^2 \leq R^2$.

8.60 Let $F(R)$ be the integral of $f(x, y) = \exp(-x^2 - y^2)$ over $x^2 + y^2 \leq R^2$.

1. Show that $\lim_{R\to\infty} F(R) = K \in \mathbb{R}$ exists. What is K?
2. Let $\alpha > 0$. Show that $\lim_{R\to\infty} F(\alpha R)$ exists and is equal to K.
3. Let $\alpha \neq 0$ and $\beta \neq 0$. Show that $\lim_{R\to\infty} \int_{-R}^{R} \int_{-R}^{R} \exp(-\alpha^2 x^2 - \beta^2 y^2)\, dx\, dy$ exists and is equal to $K/|\alpha\beta|$.
4. Show that $\lim_{R\to\infty} \int_{-R}^{R} \exp(-x^2)\, dx$ exists and is equal to $\sqrt{K}$.

8.61 Let $T : \mathbb{R}^2 \to \mathbb{R}^2$ be an isomorphism. For each R, let E_R be the elliptical region $(x/a)^2 + (y/b)^2 \leq R^2$. Here $a \neq 0$ and $b \neq 0$ are fixed. Show that $\lim_{R\to\infty} \int_{E_R} \exp(-\langle T\mathbf{z}, T\mathbf{z}\rangle)\, d\mathbf{z}$ exists and compute its value.

8.62 Let $n \in \mathbb{N}$ and let $T : \mathbb{R}^n \to \mathbb{R}^n$ be an isomorphism. For each R, let E_R be the elliptical region $\sum_i (x_i/a_i)^2 \leq R^2$. Here $a_i \neq 0$ are fixed. Show that $\lim_{R\to\infty} \int_{E_R} \exp(-\langle T\mathbf{z}, T\mathbf{z}\rangle)\, d\mathbf{z}$ exists and compute its value.

8.63 **Centroid.** Let E be a Jordan set in $\mathbb{R}^n$. Show that there is a unique vector $\mathbf{c} \in \mathbb{R}^n$, called the *centroid* of E, such that

$$\langle \mathbf{a}, \mathbf{c}\rangle = \frac{1}{v(E)} \int_E \langle \mathbf{a}, \mathbf{z}\rangle\, d\mathbf{z}$$

for all $\mathbf{a} \in \mathbb{R}^n$.

8.64 **Pappus' Theorem.** Let E be a Jordan set in the xz-plane. Let $\mathbf{c} = (a, b)$ be the centroid of E. Assume that if $(x, z) \in E$, then $x > 0$. Rotate E about the z-axis to get a solid R in $\mathbb{R}^3$. Show that $v^3(R) = 2\pi\, a\, v^2(E)$.

8.65 Find the volume of the torus obtained by rotating the disc $(x-2)^2 + z^2 \leq 1$ around the z-axis.

CHAPTER 9

INTEGRATION ON MANIFOLDS

The main objective of this chapter is to present and investigate definitions of content (volume) and integration on manifolds: curves, surfaces and the like. We work up to these definitions, beginning with the simpler problem of defining volume on vector spaces and subspaces other than $\mathbb{R}^n$.

Any n-dimensional vector space X is isomorphic to $\mathbb{R}^n$. One natural way to define volume on X is to use an isomorphism $T : X \to \mathbb{R}^n$. $E \subset X$ is a Jordan set in X if TE is a Jordan set in $\mathbb{R}^n$, and its volume is then defined as $v(TE)$. This definition of volume on X depends on T, albeit in a minor way: any two volume functions defined in this manner are constant multiples of each other. That is a consequence of Theorem 8.3.10.

We shall restrict our attention to the problem of defining volume on a Euclidean space X. In this case, there is a particularly natural choice for T which defines a volume on X called the *Euclidean volume*. But there is also a way to define this volume as an intrinsic feature of X, without making use of the mapping to $\mathbb{R}^n$.

Analysis in Vector Spaces.
By M. A. Akcoglu, P. F. A. Bartha and D. M. Ha

The two approaches, extrinsic and intrinsic, also give us (equivalent) ways to define k-dimensional volume on k-dimensional subspaces of X.

We move to the definitions of content and integration of real-valued functions on manifolds in section 9.2. These definitions are extrinsic; they utilize a parametric characterization to reduce integration over a manifold to integration over an underlying (Euclidean) parameter space. This theory of content and integration is then extended, in sections 9.3 and 9.4, to the integration of vector functions and tensor fields. In both cases, the vector or tensor function is used to generate a real-valued function which is then integrated over the manifold.

The final section raises the question of whether we can provide an intrinsic characterization of content and integration on a manifold. We show that the answer is "yes" for the special case of manifolds that can be represented as graphs in some coordinate system. We prove that the intrinsically defined 'geometric' content agrees with the standard, extrinsically defined notion of earlier sections.

9.1 EUCLIDEAN VOLUMES

Definition 9.1.1 Euclidean volumes. Let Z be an n-dimensional Euclidean space. If $(\mathbf{e}_1, \ldots, \mathbf{e}_n)$ is an orthonormal basis, then

$$T\mathbf{z} = T(z_1\mathbf{e}_1 + \cdots + z_n\mathbf{e}_n) = (z_1, \ldots, z_n) \tag{9.1}$$

defines an isomorphism $T : Z \to \mathbb{R}^n$. The *Euclidean volume* of a set $E \subset Z$ is defined as $v_Z(E) = v^n(TE)$ (or simply $v(TE)$), where v^n (or v) is volume in $\mathbb{R}^n$. Hence $v_Z(E)$ is defined whenever $TE \subset \mathbb{R}^n$ is a Jordan set.

This definition has to be justified by showing that $v_Z(E)$ is independent of the choice of orthonormal basis. This is left as an exercise. It also follows from an intrinsic characterization of Euclidean volume which does not rely upon any mapping from Z to $\mathbb{R}^n$. We develop this idea by offering a second definition of Euclidean volume in terms of Euclidean determinants on Z (Theorem 9.1.7 and Remarks 9.1.8 below).

A Review of Euclidean Determinants

Determinants were introduced in section 3.6. Here, we review and expand upon that earlier discussion.

For any nonnegative integer k, let $\Lambda_k(Z) = AML_k(Z^k, \mathbb{R})$ be the linear space of all alternating multilinear functions $Z^k \to \mathbb{R}$, as defined in Definition 3.6.21. Let $n = \dim Z$. A basic fact about $\Lambda_n(Z)$ is that it is a one-dimensional space. Any nonzero element of $\Lambda_n(Z)$ is called a determinant (or a determinant function) on Z. Hence any determinant is a basis for $\Lambda_n(Z)$. In particular any determinant is a multiple of any other determinant.

Definition 9.1.2 Determinant of a basis. Determinant functions are related to the familiar determinant $\det\{a_{ij}\}$ of an $n \times n$ matrix $\{a_{ij}\}$. In fact, let $\mathbb{U} = (\mathbf{u}_1, \ldots, \mathbf{u}_n)$ be an (ordered) basis for Z and let $\mathbb{A} = (\mathbf{a}_1, \ldots, \mathbf{a}_n)$ be any (ordered) set of n vectors in Z^n. Then the coordinates $\{a_{ij}\}$ of $\mathbf{a}_i$ s with respect to $\mathbf{u}_j$ s form an $n \times n$ matrix. The determinant of this matrix defines a determinant function $\psi_{\mathbb{U}} : Z^n \to \mathbb{R}$. We say that $\psi_{\mathbb{U}}$ is the *determinant of the ordered basis* $\mathbb{U}$. Note that $\psi_{\mathbb{U}}$ is uniquely defined by the condition that $\psi_{\mathbb{U}}(\mathbb{U}) = 1$. Also, note that any determinant function is a multiple of $\psi_{\mathbb{U}}$.

Definition 9.1.3 Euclidean determinants. If Z is a Euclidean space, then a determinant function ϑ is called a *Euclidean determinant* if $\vartheta(\mathbb{E}) = \pm 1$ for any orthonormal basis $\mathbb{E} = (\mathbf{e}_1, \ldots, \mathbf{e}_n)$. If we fix an orthonormal basis $\mathbb{E}$, then there are exactly two Euclidean determinants: $\psi_{\mathbb{E}}$ and $-\psi_{\mathbb{E}}$. Any (ordered) basis $\mathbb{U}$, however, may be used to characterize these two Euclidean determinants, since

$$\psi_{\mathbb{E}} = \psi_{\mathbb{U}} / \psi_{\mathbb{U}}(\mathbb{E}),$$

where $\psi_{\mathbb{U}}$ is the determinant of $\mathbb{U}$ (in the sense of Definition 9.1.2).

Definition 9.1.4 Oriented Euclidean spaces. A Euclidean space Z together with one of its Euclidean determinants ϑ is called an *oriented Euclidean space*. An oriented Euclidean space may be also denoted as (Z, ϑ). Each Euclidean space has exactly two orientations.

Remarks 9.1.5 Computation of Euclidean determinants. Let $\vartheta \in \Lambda_n(Z)$ be a Euclidean determinant and $\mathbb{A} = (\mathbf{a}_1, \ldots, \mathbf{a}_n) \in Z^n$. Then $\vartheta(\mathbb{A}) \in \mathbb{R}$ can be computed in two different ways:

(1) $\vartheta(\mathbb{A}) = \pm \det\langle \mathbf{a}_i, \mathbf{e}_j \rangle$, in terms of coordinates over an orthonormal basis $(\mathbf{e}_1, \ldots, \mathbf{e}_n)$;

(2) $\vartheta(\mathbb{A}) = \pm(\det\langle \mathbf{a}_i, \mathbf{a}_j \rangle)^{1/2}$, intrinsically.

The equivalence of (1) and (2) derives from the following facts. First, if A is the matrix with entries $\langle \mathbf{a}_i, \mathbf{e}_j \rangle$, then the conjugate or transpose matrix A^* has the same determinant (Appendix C, Theorem C.2.4). Second, $\langle \mathbf{a}_i, \mathbf{a}_j \rangle$ are the entries of the product AA^*. Third, $\det AA^* = \det A \det A^* = (\det A)^2$.

Euclidean determinants and volume

Remarks 9.1.6 Volume of a box in $\mathbb{R}^n$. Let $X = \mathbb{R}^n$ with its standard inner product. If $\mathbb{A} = (\mathbf{a}_1, \ldots, \mathbf{a}_n) \in X^n$, let

$$B(\mathbb{A}) = \{ \alpha_1\mathbf{a}_1 + \cdots + \alpha_n\mathbf{a}_n \mid 0 \leq \alpha_i \leq 1, \;\; i = 1, \ldots, n \} \subset X \qquad (9.2)$$

be the *box spanned by* $\mathbb{A}$. Then $B(\mathbb{A})$ is a Jordan set and $v(B(\mathbb{A})) = |\vartheta(\mathbb{A})|$. This follows from our work in chapter 8.

Theorem 9.1.7 *Let Z be an n-dimensional Euclidean space. Let*

$$B(\mathbb{A}) = \{ \alpha_1\mathbf{a}_1 + \cdots + \alpha_n\mathbf{a}_n \mid 0 \leq \alpha_i \leq 1, \;\; i = 1, \ldots, n \} \qquad (9.3)$$

be the box spanned by $\mathbb{A} = (\mathbf{a}_1, \ldots, \mathbf{a}_n) \in Z^n$. Then $B(\mathbb{A})$ is a Jordan set. Its Euclidean volume is $v_Z(B(\mathbb{A})) = (\det\langle \mathbf{a}_i, \mathbf{a}_j\rangle_Z)^{1/2} = |\vartheta(\mathbb{A})|$, where $\vartheta \in \Lambda_n(Z)$ is a Euclidean determinant on Z.

Proof. Let $T : Z \to \mathbb{R}^n = X$ be an isomorphism as defined in 9.1.1. Then $TB(\mathbb{A}) \subset X$ is the box spanned by $(T\mathbf{a}_1, \ldots, T\mathbf{a}_n)$. Since $TB(\mathbb{A})$ is a Jordan set, so is $B(\mathbb{A})$, and

$$v_Z(B(\mathbb{A})) = v(TB(\mathbb{A})) = v(B(T\mathbb{A})) = (\det\langle T\mathbf{a}_i, T\mathbf{a}_j\rangle_X)^{1/2} \qquad (9.4)$$

$$= (\det\langle \mathbf{a}_i, \mathbf{a}_j\rangle_Z)^{1/2} = |\vartheta(\mathbb{A})|. \qquad (9.5)$$

The reason for the first equality in (9.5) is that $T : Z \to X = \mathbb{R}^n$ preserves inner products. □

Remarks 9.1.8 Intrinsic definition of Euclidean volumes. The boxes in a Euclidean space are "geometrical" objects, independent of the choice of coordinates. As we have just seen, their volumes can be defined in terms of Euclidean determinants, which are also independent of coordinates. Hence the volume of a box can be defined purely in terms of the inner product. Since the inner product defines the "geometry" of the space, volume is a geometrically determined quantity.

Note also that in defining the Euclidean volume of a box in Z, we have also defined the volumes of unions of boxes and hence the volumes of all Jordan sets. In fact, Jordan sets are those that can be approximated by inner and outer unions of boxes that are close in volume.

Integrals on Euclidean Spaces

Let $f : Z \to \mathbb{R}$ be a real-valued function defined on a Euclidean space Z. For convenience we will assume that all functions are defined on the whole space. This is not a restriction of generality: any function can be extended to the whole space by making it vanish outside its original domain of definition. This does not affect the integration.

The integral of such a function is denoted by

$$\int f \text{ or } \int_Z f \text{ or } \int_Z f(\mathbf{z})\, d\mathbf{z} \tag{9.6}$$

or by a similar notation. The definition is obvious. Let $T : X = \mathbb{R}^n \to Z$ be an isomorphism that takes the standard basis of $X = \mathbb{R}^n$ to an orthonormal basis $\mathbb{E}$ of Z. Then define $\int_Z f(\mathbf{z})\, d\mathbf{z}$ as $\int_X f(T\mathbf{x})\, d\mathbf{x}$, if this latter integral exists. We see that the result is independent of the choice of the orthonormal basis $\mathbb{E}$.

Volumes on Subspaces

The preceding approach gives us an intrinsic notion of volume for subspaces of a Euclidean space.

Definition 9.1.9 Euclidean volumes on subspaces. A subspace X of a Euclidean space Z is also a Euclidean space with the same inner product (restricted to X). It follows that X has its own two Euclidean determinants $\pm\vartheta_X$, as well as its own volume v_X defined on its own family of Jordan sets. We sometimes write v_k for this k-dimensional volume. Note that for any k-dimensional box $B(\mathbb{A})$ in X, where $\mathbb{A} = (\mathbf{a}_1, \ldots, \mathbf{a}_k)$, Theorem 9.1.7 tells us that

$$v_k(B(\mathbb{A})) = (\det\langle \mathbf{a}_i, \mathbf{a}_j\rangle)^{1/2}.$$

Definition 9.1.10 Lower-dimensional volume of a box. Let $\mathbb{A} = (\mathbf{a}_1, \ldots, \mathbf{a}_k) \in Z^k$ be a k-tuple of vectors from Z. It is convenient to define

$$v_k(B(\mathbb{A})) = (\det\langle \mathbf{a}_i, \mathbf{a}_j\rangle)^{1/2} \tag{9.7}$$

as the *k-dimensional volume* of the box spanned by these vectors. Of course, this agrees with the volume in Definition 9.1.9, when $B(\mathbb{A})$ is considered as a Jordan subset of a k-dimensional subspace. Sometimes, however, it is convenient to talk about the k-dimensional volume of a box without specifying any subspace. Note that $v_k(B(\mathbb{A}))$ is nonzero if and only if the vectors in $\mathbb{A}$ form a linearly independent set.

Theorem 9.1.11 Euclidean determinants on coordinate systems. *Let (X, Y) be a coordinate system in Z with the respective Euclidean determinants ϑ_X, ϑ_Y, and ϑ_Z. Then there is a γ such that*

$$\vartheta_Z(\mathbb{A}, \mathbb{B}) = \gamma\, \vartheta_X(\mathbb{A})\, \vartheta_Y(\mathbb{B}) \tag{9.8}$$

for all $\mathbb{A} = (\mathbf{a}_1, \ldots, \mathbf{a}_k) \in X^k$ *and* $\mathbb{B} = (\mathbf{b}_1, \ldots, \mathbf{b}_\ell) \in Y^\ell$, *where*

$$(\mathbb{A}, \mathbb{B}) = (\mathbf{a}_1, \ldots, \mathbf{a}_k; \mathbf{b}_1, \ldots, \mathbf{b}_\ell) \in Z^n. \tag{9.9}$$

Here $\dim X = k$, $\dim Y = \ell$, *and* $\dim Z = n = k + \ell$. *If* $X \perp Y$, *then* $\gamma = \pm 1$.

Proof. This is a restatement of Theorems C.6.5 and C.5.3. □

Theorem 9.1.12 Euclidean volumes on coordinate systems. *Let* (X, Y) *be a coordinate system in* Z. *Then there is a* $\beta = \beta(X, Y) > 0$ *such that*

$$v_Z(A \times B) = \beta\, v_X(A)\, v_Y(B) \tag{9.10}$$

for all Jordan sets $A \subset X$ *and* $B \subset Y$. *Also, if* $X \perp Y$ *then* $\beta(X, Y) = 1$.

Proof. If A and B are boxes, then the first part is a restatement of Theorem C.6.5. The result for general Jordan sets follows by approximations. If $X \perp Y$, then take the boxes A and B as the unit boxes spanned by orthonormal bases in X and Y. Then $A \times B$ is also a unit box spanned by an orthonormal basis in Z. Hence in this case all three determinants will be equal to 1. Then $\beta(X, Y) = 1$ follows. □

Corollary 9.1.13 *Let* X *be a proper subspace of* Z. *Then any bounded set* E *in* X *is a Jordan set in* Z *and* $v_Z(E) = 0$.

Proof. Let A be a Jordan set in X and $E \subset A$. Let $Y = X^\perp$. Then Y is a Euclidean space. Let B_r be the ball of radius r in Y. Then

$$E = E \times \{\mathbf{0}\} \subset A \times B_r, \text{ for all } r > 0. \tag{9.11}$$

But $v_Z(A \times B_r) = v_X(A)\, v_Y(B_r) = r^\ell v_X(A)\, v_Y(B_1) \to 0$ as $r \to 0$. (Here $\ell = \dim Y$.) This shows that E is a negligible set in Z. □

Volume Multipliers of Linear Maps

Lemma 9.1.14 *Let* W *and* Z *be two Euclidean spaces of the same dimension. Let* $T : W \to Z$ *be a linear map. Then there is a number* $\rho(T) \geq 0$ *such that*

$$v_Z(TE) = \rho(T)\, v_W(E) \tag{9.12}$$

for all Jordan sets $E \subset W$. *Also* $\rho(T) = |\vartheta_Z(T\mathbb{E})|$, *where* $\mathbb{E} = (\mathbf{e}_1, \ldots, \mathbf{e}_n)$ *is an orthonormal basis for* W *and* ϑ_Z *is a Euclidean determinant for* Z.

Proof. Theorem 8.3.17 shows that if $W = Z = \mathbb{R}^n$ and if $T : W \to Z$ is an isomorphism, then this result holds with $\rho(T) = v_Z(TU)$, where $U = [0, 1)^n$ is the unit cube in W. The arguments for the general case are the same.

If T is not an isomorphism, then the range $X = TW$ of T is a proper subspace of Z. Therefore $TE \subset X$ is a negligible set in Z for all bounded $E \subset W$, by Corollary 9.1.13. In this case, let $\rho(T) = 0$. □

Definition 9.1.15 General volume multipliers. Let $T : W \to Z$ be a linear transformation between any two Euclidean spaces. Let $\mathbb{A} = (\mathbf{a}_1, \ldots, \mathbf{a}_k)$ be an orthonormal basis for W. Then the number

$$\rho(T) = v(B(T\mathbb{A})) = (\det\langle T\mathbf{a}_i, T\mathbf{a}_j\rangle)^{1/2} \tag{9.13}$$

is called the *general volume multiplier* of T. If W and Z have the same dimension, then this agrees with the earlier definition. Note that $\rho(T) > 0$ if and only if $\dim W = \dim(TW)$, that is, if and only if T is one-to-one.

Problems

9.1 Let (X, Y) and (U, V) be two orthogonal coordinate systems (Definition 3.1.42) in Z. Assume that (U, Y) is also a coordinate system for Z. Let $P : U \to X$ be the orthogonal projection of U on X and let $Q : Y \to V$ be the orthogonal projection of Y on V. Show that $\rho(P) = \rho(Q)$.

9.2 Let U be a two-dimensional Euclidean space. Let $T : U \to \mathbb{R}^3$ be a linear mapping. Show that $\rho(T) = \|T\mathbf{u}_1 \times T\mathbf{u}_2\|$, where $(\mathbf{u}_1, \mathbf{u}_2)$ is an orthonormal basis for U and $\times$ denotes the cross product in $\mathbb{R}^3$.

9.2 INTEGRATION ON MANIFOLDS

A manifold M consists of the "local images" of subspaces under diffeomorphisms. For our discussion of integration on manifolds, it is also convenient to work with the reverse diffeomorphisms. They will be called charts on M.

Charts for Manifolds

The following notation is used in this chapter and Chapter 10.

Definition 9.2.1 Charts. Let W and Z be two spaces, U a subspace of W, and M a subset of Z. Let $\Psi : G \to W$ be a diffeomorphism defined on an open set $G \subset Z$. Then Ψ is called a *chart* for M if

$$\Psi(M \cap G) = U \cap H \text{ with } H = \Psi(G). \tag{9.14}$$

Hence M is a manifold if each $\mathbf{m} \in M$ is contained in the domain of a chart for M.

If $\Psi : G \to H = \Psi(G)$ is a chart for M, then $\Psi^{-1} = \Phi : H \to G$ is the *reverse chart.* Note that the restriction of $\Phi : H \to Z$ to $U \cap H$ is a parametric representation $\Phi|_{U\cap H} = \varphi : U \cap H \to Z$ for $M \cap G$.

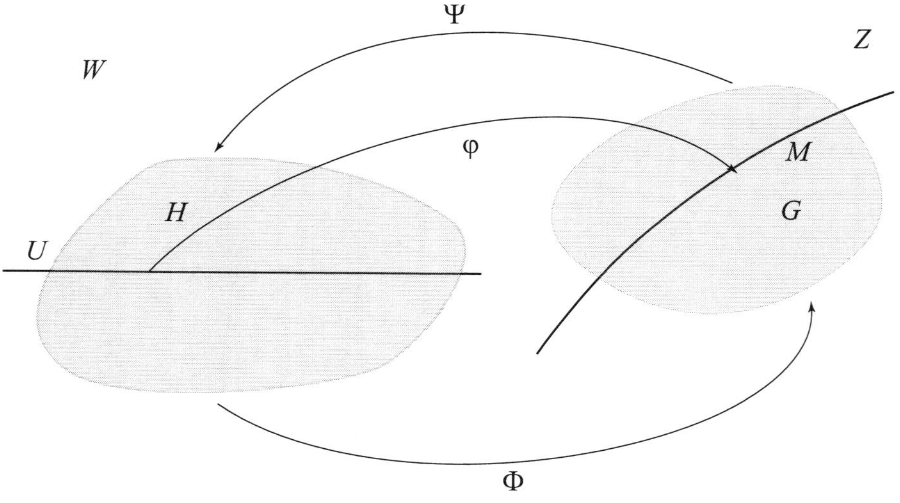

Figure 9.1. Charts as in Definition 9.2.1.

Definition 9.2.2 Atlases for a manifold. A collection $\mathcal{A}$ of charts

$$\Psi_\alpha : G_\alpha \to H_\alpha (= \Psi(G_\alpha)), \quad \alpha \in \mathcal{A}, \tag{9.15}$$

is called an *atlas* for M if the collection of their domains $\{G_\alpha\}$ forms an open cover for M. Hence $M \subset \cup_\alpha G_\alpha$ and $\Psi_\alpha(M \cap G_\alpha) = U_\alpha \cap H_\alpha$. We shall usually assume that $U_\alpha = U$ is the same subspace for all α, which (as we shall see) involves no loss of generality. Note that every manifold has an atlas; this follows from the definition of a manifold.

Definition 9.2.3 Derivatives and tangent spaces. Let $\Phi : H \to G$ be a reverse chart. If $\mathbf{u} \in U \cap H$ and $\Phi(\mathbf{u}) = \mathbf{m}$, then $\mathbf{m} \in M$ and

$$T_{\mathbf{m}} = \Phi'(\mathbf{u})U$$

is called the *tangent space* of M at $\mathbf{m}$. If we let $\varphi = \Phi|_{U \cap H}$, then $\varphi'(\mathbf{u}) : U \to T_{\mathbf{m}}$ is the restriction of $\Phi'(\mathbf{u}) : W \to Z$ to the subspace U.

Note that $\varphi'(\mathbf{u})$ is an isomorphism between U and $T_{\mathbf{m}}$. The volume multiplier $\rho(\Phi'(\mathbf{u})|_U) = \rho(\varphi'(\mathbf{u}))$ of this isomorphism plays a central role in the theory (and practical applications) of integration on manifolds. Also note that

$$\Psi'(\mathbf{m})|_{T_{\mathbf{m}}} : T_{\mathbf{m}} \to U \text{ is the inverse of } \Phi'(\mathbf{u})|_U : U \to T_{\mathbf{m}}. \tag{9.16}$$

Hence $\rho(\Psi'(\mathbf{m})|_{T_{\mathbf{m}}}) = 1/\rho(\Phi'(\mathbf{u})|_U)$.

Local Integrals

Remarks 9.2.4 Local integrals. Integrals on manifolds are usually computed by means of parametric representations (see Definition 6.3.11). When one computes an integral using a parametric representation, it is understood that this is only a partial integration over the part of the manifold covered by this representation. We will call such an integral a *local integral*. This generates a problem: how do we integrate a function whose support does not lie entirely within a region covered by a single parametric representation?

Here is an efficient way to handle this situation. Let $\Psi : G \to H$ be a chart for M, with the associated parametric representation $\varphi : U \cap H \to Z$. (It helps to keep Figure 9.1 handy.) The part of M covered by this representation is $M \cap G$. When working with this particular parametric representation, assume that any function to be integrated has compact support contained in G. In this case, we can use φ to obtain the complete integral of the function on M.

The integration of a general function is defined by using a technique called *partitions of unity*. In essence, the idea is to re-write an arbitrary function f as a sum of functions f_i, each of which has compact support covered by single parametric representation, and to define the integral of f as the sum of the (local) integrals of the functions f_i. This technique is discussed later in this section, starting with Definition 9.2.10. We shall begin with local integrals, working with a fixed chart $\Psi : G \to H$ and the associated parametric representation $\varphi : U \cap H \to Z$.

Definition 9.2.5 Local integration. Let $\Phi : H \to G$ be a reverse chart for a manifold M. If $f : Z \to \mathbb{R}$ is a function with compact support $K \subset G$, then its integral over M is defined as

$$\int_M f = \int_U f(\Phi(\mathbf{u}))\, \rho(\Phi'(\mathbf{u})|_U)\, d\mathbf{u}, \tag{9.17}$$

whenever the integral on the right exists. This integral involves ordinary integration over a Euclidean space, and is computed by the usual methods of multiple integrals.

We can re-state (9.17) in terms of the parametric representation $\varphi = \Phi|_U : U \to Z$:

$$\int_M f = \int_U f(\varphi(\mathbf{u}))\, \rho(\varphi'(\mathbf{u}))\, d\mathbf{u}. \tag{9.18}$$

To justify this definition, we must show that if there are two reverse charts $\Phi_i : H_i \to G_i$ such that the support of $f : Z \to \mathbb{R}$ is contained in $G_1 \cap G_2$, then the definitions of the integral of f using these two charts give the same result. That is the content of our next theorem.

Theorem 9.2.6 *Let $\Phi_i : H_i \to G_i$ be two reverse charts for M. Let $f : Z \to \mathbb{R}$ be a function with support in $G_1 \cap G_2$. Then*

$$\int_U f(\Phi_1(\mathbf{u}))\, \rho(\Phi_1'(\mathbf{u})|_U)\, d\mathbf{u} = \int_U f(\Phi_2(\mathbf{u}))\, \rho(\Phi_2'(\mathbf{u})|_U)\, d\mathbf{u} \tag{9.19}$$

whenever either one of the integrals exists.

Proof. Let Ψ_1 and Ψ_2 be the charts associated with Φ_1 and Φ_2. Let $G = G_1 \cap G_2$. The restriction of each Ψ_i to G is another chart for M. We see that the integrals in (9.19) for the original charts are the same integrals for the restricted charts. Hence, we may assume that $G_1 = G_2 = G$, without loss of generality. Then $\Theta = \Psi_2 \cdot \Phi_1 : H_1 \to H_2$ is a diffeomorphism. We see that $\theta = \Theta|_{U \cap H_1} : U \cap H_1 \to U \cap H_2$ is also a diffeomorphism, and $\varphi_1 = \varphi_2 \cdot \theta$. (Here we have put $\varphi_i = \Phi_i|_U$.) Hence $\varphi_1'(\mathbf{u}) = \varphi_2'(\theta(\mathbf{u})) \cdot \theta'(\mathbf{u})$, and therefore,

$$\rho(\varphi_1'(\mathbf{u})) = \rho(\varphi_2'(\theta(\mathbf{u}))) \cdot \rho(\theta'(\mathbf{u})). \tag{9.20}$$

The change of variables theorem, Theorem 8.4.16, shows that

$$\int_{U \cap H_2} g(\mathbf{u}_2)\, d\mathbf{u}_2 = \int_{U \cap H_1} g(\theta(\mathbf{u}_1))\rho(\theta'(\mathbf{u}_1))\, d\mathbf{u}_1 \tag{9.21}$$

whenever the first integral exists. If $g(\mathbf{u}_2) = f(\varphi_2(\mathbf{u}_2))\, \rho(\varphi_2'(\mathbf{u}_2))$, then

$$g(\theta(\mathbf{u}_1))\rho(\theta'(\mathbf{u}_1)) = f(\varphi_2(\theta(\mathbf{u}_1)))\, \rho(\varphi_2'(\theta(\mathbf{u}_1)))\rho(\theta'(\mathbf{u}_1)) \tag{9.22}$$

$$= f(\varphi_1(\mathbf{u}_1))\rho(\varphi_1'(\mathbf{u}_1)). \tag{9.23}$$

To obtain (9.23) we used (9.20). The conclusion now follows. □

Contents on Manifolds

Definition 9.2.7 Jordan sets in manifolds. A set E in a manifold M is called a *Jordan set in M* if its characteristic function $\chi_E : M \to \mathbb{R}$ is integrable on M. In this case, $\sigma(E) = \int_M \chi_E$ is called the *content* of E. The term 'content' stands for k-dimensional volume on a k-dimensional manifold.

Remarks 9.2.8 Notations for integrals. The integral of $f : M \to \mathbb{R}$ on a manifold is denoted by expressions such as

$$\int_M f = \int_M f(\mathbf{m})\, d\mathbf{m} = \int_M f(\mathbf{m})\, \sigma(d\mathbf{m}) = \int_M f\, d\sigma \tag{9.24}$$

or by similar expressions. In general, none of these expressions is suitable for computations. To compute an integral on a manifold, we have to use a chart and the corresponding parametric representations $\varphi : U \cap H \to Z$ to reduce a problem of integration on a manifold to a problem of integration on a Euclidean space. Then we can use the usual techniques of multiple integration to compute the integral. With this practical focus on computation in mind, a useful notation for the integral is

$$\int_M f\, d\sigma = \int_{U\cap H} f(\varphi(\mathbf{u}))\, \rho(\varphi'(\mathbf{u}))\, d\mathbf{u} = \int_{U\cap H} f(\varphi(\mathbf{u}))\, d\sigma(\mathbf{u}), \tag{9.25}$$

where $\rho(\varphi'(\mathbf{u}))\, d\mathbf{u} = d\sigma(\mathbf{u})$ is considered as the content of a small part of the manifold. This small part is the image of a small cube in the Euclidean space U.

Remarks 9.2.9 Content and integration. Integration on M can be defined in terms of the content on M. In fact, we see that if $f : M \to \mathbb{R}$ is a function of compact support $K \subset M$, then $\int_M f$ exists if and only if for each $\varepsilon > 0$ there are finitely many pairwise disjoint Jordan sets $E_i \subset M$ such that $K \subset \cup_i E_i$ and such that

$$\sum_i (\sup\{\, f(\mathbf{z}) \mid \mathbf{z} \in E_i \,\} - \inf\{\, f(\mathbf{z}) \mid \mathbf{z} \in E_i \,\})\, \sigma(E_i) < \varepsilon. \tag{9.26}$$

The details are left as an exercise.

General Integrals

The above definition of integrals is only for functions $M \to \mathbb{R}$ with supports contained in the domain of a single chart. Integrals of more general functions are defined by a technique called the partitions of unity. The basis of this technique is the following theorem.

Definition 9.2.10 Partitions of unity. Let A be a set in a Euclidean space Z. Let G_i s be a finite collection of bounded open sets such that $A \subset \cup_i G_i$. Then a (finite) set of $\mathcal{C}^\infty$ functions $\lambda_i : Z \to [\,0,\, 1\,]$ is called *a partition of unity for A subordinate to* *the covering* G_i if each λ_i has a compact support contained in G_i and if $\sum_i \lambda_i(\mathbf{z}) = 1$ for all $\mathbf{z} \in A$.

Theorem 9.2.11 Existence of partitions of unity. *Given a finite covering G_i of a set A by bounded open sets, there is a partition of unity for A subordinate to the covering G_i.*

The proof of this theorem is fairly elementary and self-contained. The requirement that λ_i s are $\mathcal{C}^\infty$ functions does not cause any additional complications in the proof. It is made for convenience so that the result can be used in other applications that require a high degree of differentiability. Theorem 9.2.11 is re-stated and proved in Appendix D as Theorem D.1.8.

Definition 9.2.12 Integrals on manifolds. Let M be a manifold in Z. Let $f : M \to \mathbb{R}$ be a function of compact support $S \subset M$. By the definition of a manifold, each point $\mathbf{m} \in S$ is in the domain G of some chart, and each such G is an open set. Use the compactness of S to find finitely many of these domains G_i such that $S \subset \cup_i G_i$. Then find a partition of unity $\lambda_i : Z \to [0, 1]$ for S subordinate to G_i and define the *surface integral* of $f : M \to \mathbb{R}$ as

$$\int_M f = \sum_i \int_M f\,\lambda_i \tag{9.27}$$

if each $f\lambda_i : M \to \mathbb{R}$ is integrable. Note that the integrals on the right-hand side are obtained by local integration (as in Definition 9.2.5), since each $f\lambda_i : M \to \mathbb{R}$ has a compact support contained in $M \cap G_i$.

Nevertheless, our definition has to be justified by showing that different partitions of unity lead to the same result.

Lemma 9.2.13 *Let $f : M \to \mathbb{R}$ be a function of compact support $S \subset M$. Let G_i and H_j be two finite coverings of S by bounded open sets. Let α_i and β_j be two partitions of unity for S subordinate to G_i and to H_j respectively. If*

$$\sum_i \int_M f\alpha_i \text{ exists, then } \sum_j \int_M f\beta_j \text{ also exists,} \tag{9.28}$$

and these two sums are equal.

Proof. Note that $\alpha_i(\mathbf{z}) = \sum_j \alpha_i(\mathbf{z})\beta_j(\mathbf{z})$ for all $\mathbf{z} \in S$. If $\int_M f\alpha_i$ exists, then $\int_M f\alpha_i\beta_j$ also exists and

$$\int_M f\alpha_i = \sum_j \int_M f\alpha_i\beta_j. \text{ Hence} \tag{9.29}$$

$$\sum_i \int_M f\alpha_i = \sum_i \sum_j \int_M f\alpha_i\beta_j. \tag{9.30}$$

The double summation above is also equal to $\sum_j \int_M f\beta_j$, by symmetry. $\square$

Problems

9.3 Find the surface area of a sphere of radius R.

9.4 A plane $z = h$, $0 \le h \le 1$, divides the unit sphere $x^2 + y^2 + z^2 = 1$ into two parts. Compute the surface areas of these parts.

9.5 Let E be a Jordan set in the rectangle $(-\pi, \pi) \times (0, \pi)$. Map this region by the spherical coordinates to get a region $\Phi(E)$ in the unit sphere. Hence $\Phi(E)$ consists of all points with the Cartesian coordinates

$$x = \cos\theta \,\sin\varphi, \quad y = \sin\theta \,\sin\varphi, \quad z = \cos\varphi$$

with $(\theta, \varphi) \in E$. Express the surface area of $\Phi(E)$ as an integral over E. Find the areas corresponding to rectangles $-\pi < p \le \theta \le q < \pi$, $0 < r \le \varphi \le s < \pi$. Also apply this result to give another solution of Problem 9.3.

9.6 Find the surface area of the part of the sphere $x^2 + y^2 + z^2 = 4$ that lies inside the cylinder $(x-1)^2 + y^2 \le 1$.

9.7 Compute the surface area of the helicoidal surface

$$x = r\cos\theta, \quad y = r\sin\theta, \quad z = \theta,$$

where $1 \le r \le 2$, $0 \le \theta \le 2\pi$.

9.8 Find the surface area of the part of the cylinder $x^2 + y^2 = 1$ that lies between the planes $z = 0$ and $z = 2x + 3y + 10$.

9.9 Find the surface area of the part of the cylinder $x^2 + z^2 = a^2$ that lies above the xy-plane and inside the cylinder $x^2 + y^2 = a^2$.

9.10 Integrate $f(x, y, z) = |z|$ over the surface of the sphere

$$x^2 + y^2 + z^2 = 1.$$

9.11 Compute $\int_G (x^2 z + y^2 z)\,d\sigma$ where G is the upper half of the sphere

$$x^2 + y^2 + z^2 = 4.$$

9.12 Compute $\int_G (y^2 + z^2)\, d\sigma$ where G is the part of the surface

$$x = 4 - y^2 - z^2$$

that lies in the region $x \geq 0$.

9.13 Compute $\int_G (y^2 + z^2)\, d\sigma$ where G is the part of the surface

$$x^2 = 4 - y^2 - z^2$$

that lies in the region $x \geq 0$.

9.14 Compute $\int_G yz\, d\sigma$ where G is the part of the plane $z = y + 3$ that lies in the cylinder $x^2 + y^2 = 1$.

9.15 Compute $\int_C xy^4\, d\sigma$ where C is the right half of the circle

$$x^2 + y^2 = 16.$$

9.16 Compute $\int_C xy d\sigma$ where C is the line segment joining $(-1,\, 1)$ to $(2,\, 3)$.

9.17 Compute $\int_C xyz\, d\sigma$ where C is the curve

$$x = \sin 2t,\ y = 3t,\ z = \cos 2t,\ 0 \leq t \leq \pi/4.$$

9.18 **Centroid.** Let E be a Jordan set on a manifold M in a Euclidean space$Z$. Show that there is a unique vector $\mathbf{c}_E$ such that

$$\langle \mathbf{a},\, \mathbf{c}_E \rangle \sigma(E) = \int_E \langle \mathbf{a},\, \mathbf{z} \rangle d\sigma$$

for all $\mathbf{a} \in Z$. This vector is called the *centroid* of E.

9.19 Find the centroid (Problem 9.18) of the upper-half of the sphere

$$x^2 + y^2 + z^2 = 1.$$

9.20 Find the centroid (Problem 9.18) of the helix

$$x = \cos t,\ y = \sin t,\ z = t,\ 0 \leq t \leq a.$$

9.21 Find the centroid (Problem 9.18) of the helicoidal surface

$$x = r\cos t,\ y = r\sin t,\ z = t,\ 0 \leq t \leq a,\ 1 \leq r \leq 2.$$

9.22 Pappus' theorem. Let E be a curve in the half of rz-plane corresponding to $r > 0$. Let $\ell(E)$ be the length of E and let $\mathbf{c}_E = (a, b)$ be the centroid of E. Rotate C around the z-axis to obtain a surface S in $\mathbb{R}^3$. Show that the surface area of S is $2\pi a \ell(E)$. (See also Problem 8.64.)

9.23 Find the surface area of the torus obtained by rotating the circle

$$(r-2)^2 + z^2 = 1$$

around the z-axis. (See also Problem 8.65.)

9.3 ORIENTED MANIFOLDS

We will formulate definitions for the integral of a vector field and the integral of a tensor field over a manifold. Both definitions require the concept of an *oriented manifold*. In this section, we define the orientation of a manifold. We will distinguish between *local orientation* and *global orientation*. Almost all computations of integrals on manifolds are done locally, in terms of charts. For these computations, we only need the local orientations induced by charts. Global orientations, however, are needed for Stokes' theorem, which is discussed in Chapter 10.

As a preliminary to defining the orientation of a manifold, it is useful to begin with some facts about the orientation of a vector space, as defined in Definition 9.1.4. In particular, the idea of charts that preserve orientation becomes important. We will define orientation-preserving diffeomorphisms and then show that an atlas of orientation-preserving charts always exists.

Notations 9.3.1 Review of charts. Charts were defined in Definition 9.2.1, but it helps to repeat the main definitions and the standard notation. Recall that Z and W are two Euclidean spaces and U is a subspace of W. A chart for a manifold M is a diffeomorphism $\Psi : G \to H = \Psi(G)$ such that $\Psi(G \cap M) = H \cap U$. The reverse chart is $\Phi : H \to G$ and the corresponding parametric representation is $\varphi = \Phi|_{H \cap U} : H \cap U \to G \cap M$. If $\mathbf{u} \in H \cap U$ and $\mathbf{m} = \Phi(\mathbf{u}) = \varphi(\mathbf{u})$, then $T_{\mathbf{m}} = \Phi'(\mathbf{u})U = \varphi'(\mathbf{u})U$ is the tangent space of M at $\mathbf{m} \in M$. We will assume that $\dim W = \dim Z = n \geq 2$ and $1 \leq k = \dim U \leq n$. An atlas for a manifold M is a collection of charts whose domains cover M.

Notations 9.3.2 Orientations of U, W, and Z. We will assume that W and Z are oriented, respectively, by the Euclidean determinants ϱ and ϑ. Assume that U is also oriented by a positive orthonormal basis $\mathbb{E}$. These are all arbitrarily chosen but fixed orientations. If $U = W$, however, then we assume that U and W have the same orientations. This is a trivial case that is only important if we wish to consider an open subset of Z as an n-dimensional manifold.

Orientation-Preserving Charts

Definition 9.3.3 Orientation-preserving isomorphisms. An isomorphism T between the two oriented spaces (W, ϱ) and (Z, ϑ) is called *orientation-preserving* if there is a positive basis of W that is mapped by T to a positive basis of Z. In this case, T maps all positive bases of W to positive bases of Z (Problem 9.25). If T is not orientation-preserving, then it is called *orientation-reversing*.

Definition 9.3.4 Orientation-preserving diffeomorphisms. Let X and Y be two oriented spaces. Let A be an open set in X and let $\theta : A \to Y$ be a diffeomorphism. Let $\mathbf{a} \in A$. If $\theta'(\mathbf{a}) : X \to Y$ is an orientation-preserving isomorphism, then $\theta : A \to Y$ is called an *orientation-preserving diffeomorphism at* $\mathbf{a}$. If $\theta : A \to Y$ is an orientation-preserving diffeomorphism at every $\mathbf{x} \in A$, then it is called an *orientation-preserving diffeomorphism*.

Lemma 9.3.5 *Let $\theta : A \to Y$ be an orientation-preserving diffeomorphism at $\mathbf{a} \in A$. Then there is an open set G such that $\mathbf{a} \in G \subset A$ and such that the restriction of θ to G is an orientation-preserving diffeomorphism.*

Proof. Let $\mathbb{E}$ be a positive basis for X. Let Y be oriented by a Euclidean determinant ξ. We see that $f(\mathbf{x}) = \xi(\theta'(\mathbf{x})\mathbb{E})$, $\mathbf{x} \in A$, defines a continuous function $f : A \to \mathbb{R}$. If θ is orientation-preserving at $\mathbf{a} \in A$, then $f(\mathbf{a}) > 0$. So there is an open neighborhood G of $\mathbf{a} \in A$ such that $\mathbf{a} \in G \subset A$ and such that $f(\mathbf{x}) > 0$ for all $\mathbf{x} \in G$. This means that the restriction of θ to G is an orientation-preserving diffeomorphism. □

Notations 9.3.6 Bases for U and W. Let $V = U^{\perp}$ be the orthogonal complement of U. Let $\mathbb{U} = (\mathbf{u}_1, \ldots, \mathbf{u}_k)$ be a basis for U and let $\mathbb{V} = (\mathbf{v}_1, \ldots, \mathbf{v}_\ell)$ be a basis for V. Then we see that $(\mathbb{V}, \mathbb{U}) = (\mathbf{v}_1, \ldots, \mathbf{v}_\ell; \mathbf{u}_1, \ldots, \mathbf{u}_k)$ is a basis for W.

Definition 9.3.7 Orientation-preserving charts. A chart $\Psi : G \to H$ for a manifold M is orientation-preserving if Ψ is an orientation-preserving diffeomorphism on G.

Theorem 9.3.8 shows that there is no loss of generality in assuming that all charts are orientation-preserving.

Theorem 9.3.8 *Any manifold has an atlas of orientation-preserving charts.*

Proof. Let M be a manifold and let $\mathbf{m} \in M$. Let $\Psi_0 : G_0 \to H_0$ be a chart for M such that $\mathbf{m} \in G_0$. Let $\Phi_0 : H_0 \to G_0$ be the reverse chart. Let $\mathbb{E}$ be a positive basis for U. Let $\mathbb{V}$ be a basis for the orthogonal complement of U such that $(\mathbb{V}, \mathbb{U})$ is a positive basis for W. Let $\Psi_0(\mathbf{m}) = \mathbf{u}$. Then $(\Phi_0'(\mathbf{u})\mathbb{V}, \Phi_0'(\mathbf{u})\mathbb{U})$ is a basis for Z. If this is a positive basis for Z, then $\Phi_0 : H_0 \to G_0$ is orientation-preserving at $\mathbf{u}$. In this case, we see that $\Psi_0 : G_0 \to H_0$ is orientation-preserving at $\mathbf{m}$. Hence. Lemma 9.3.5 shows that there is an open set G such that $\mathbf{m} \in G \subset G_0$ and such that the restriction of Ψ_0 to G is orientation-preserving. This restriction is still a chart for M. Hence $\mathbf{m}$ is contained in the domain of an orientation-preserving chart.

If $(\Phi_0'(\mathbf{u})\mathbb{V}, \Phi_0'(\mathbf{u})\mathbb{U})$ is not a positive basis for Z, then replace $\Psi_0 : G_0 \to H_0$ by $\Psi_1 = R \cdot \Psi_0 : G_0 \to RH_0$, where $R : W \to W$ is defined as follows. With the notations in Notations 9.3.6, let

$$R\mathbf{u}_1 = -\mathbf{u}_1,\ R\mathbf{u}_i = \mathbf{u}_i,\ R\mathbf{v}_j = \mathbf{v}_j,\ 1 < i \le k,\ 1 \le j \le \ell. \tag{9.31}$$

We see that $R : W \to W$ is an isomorphism and $RU = U$. An easy verification shows that $\Psi_1 : G_0 \to H_1 = RH_0$ is another chart for M and that Ψ_1 is orientation-preserving at $\mathbf{m}$. Then a restriction of Ψ_1 to a neighborhood of $\mathbf{m}$ is an orientation-preserving chart. □

Local Orientations

Definition 9.3.9 Local orientation of a manifold. A *local orientation* of a manifold M is defined by a chart $\Psi : G \to H$. It is an orientation of each of the tangent spaces $T_{\mathbf{m}}$ at the points $\mathbf{m} \in G \cap M$. If $\Psi(\mathbf{m}) = \mathbf{u}$, then the orientation of $T_{\mathbf{m}}$ is determined by taking a positive basis $\mathbb{E}$ for U and declaring $\mathbb{B}_{\mathbf{m}} = \Phi'(\mathbf{u})\mathbb{E} = \varphi'(\mathbf{u})\mathbb{E}$ a positive basis for $T_{\mathbf{m}}$. We see that these orientations are independent of the choice of the positive basis $\mathbb{E}$ in U. This point is formulated as Problem 9.24.

Example 9.3.10 If M is a curve, then its tangent spaces are one-dimensional. In this case, the orientations of these tangent lines define a positive direction on the curve. The curve $M = \{ (x, y) \mid x^2 + y^2 = 1,\ y \ne 0 \}$ in this example consists of two halves of a circle. Here $Z = \mathbb{R}^2$ is represented by the xy-plane. Represent $W = \mathbb{R}^2$ by the θr-plane. The following are four different charts $\Psi_i : G \to W$ for M with the same domain $G \subset Z$ and with the same range $H \subset W$. They are expressed in terms of the reverse charts $\Phi_i : H \to G$. Let

$$\begin{aligned} H_1 &= \{ (\theta, r) \mid -\pi < \theta < 0,\ -1/2 < r < 1/2 \}, \\ H_2 &= \{ (\theta, r) \mid 0 < \theta < \pi,\ -1/2 < r < 1/2 \}, \end{aligned}$$

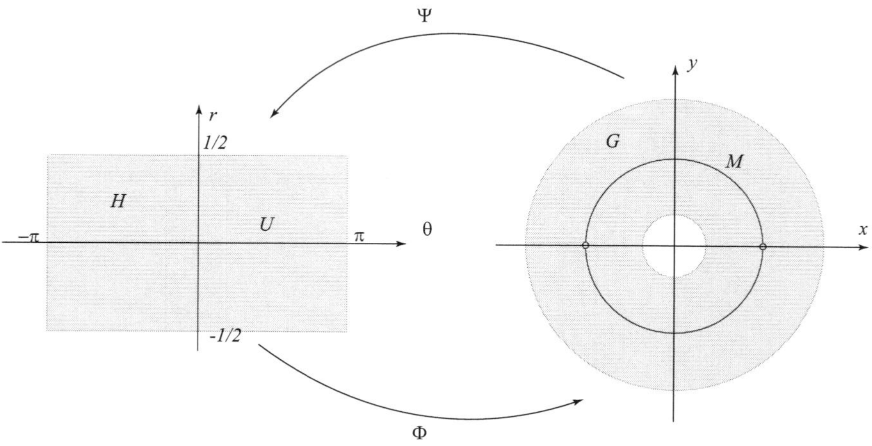

Figure 9.2. For Example 9.3.10.

and $H = H_1 \cup H_2$. See Figure 9.2. Define

$$\begin{aligned}
\Phi_1(\theta,\, r) &= ((1+r)\cos\theta,\ (1+r)\sin\theta), \quad (\theta,\, r) \in H,\\
\Phi_2(\theta,\, r) &= ((1+r)\cos\theta,\ -(1+r)\sin\theta), \quad (\theta,\, r) \in H,\\
\Phi_3(\theta,\, r) &= \begin{cases} (-(1+r)\cos\theta,\ -(1+r)\sin\theta), & \text{if } (\theta,\, r) \in H_1,\\ ((1+r)\cos\theta,\ -(1+r)\sin\theta), & \text{if } (\theta,\, r) \in H_2 \end{cases}\\
\Phi_4(\theta,\, r) &= \begin{cases} ((1+r)\cos\theta,\ -(1+r)\sin\theta), & \text{if } (\theta,\, r) \in H_1,\\ (-(1+r)\cos\theta,\ -(1+r)\sin\theta), & \text{if } (\theta,\, r) \in H_2 \end{cases}
\end{aligned}$$

We see that $\Phi_i : H \to G$ are reverse charts for M. Here U is the $r = 0$ line, which is the θ-axis in the θr-plane. Let U be oriented in the standard way, with the basis consisting of the unit vector $(1,\, 0)$. These four charts induce four different orientations on M. In intuitive terms they can be described as the four possible choices of the clockwise or the counterclockwise orientations for the upper and the lower halves of M.

Incidentally, in this example all Ψ_is are charts with domains covering all of M. Hence, they orient M completely; they also give examples of global orientations of a manifold.

It may seem pedantic to use the diffeomorphisms $\Phi_i : H \to G$ rather than the simpler and more natural parametric representations $\varphi_i : H \cap U \to G \cap M$. One reason for using the full diffeomorphism approach is to verify that φ_i is indeed the restriction of a diffeomorphism to a subspace. Another reason is that the charts and reverse charts play an important role in some applications, most notably Stokes' theorem, to be discussed in Chapter 10. $\triangle$

Remarks 9.3.11 Charts with connected domains. Connected sets were defined in Definition 4.5.32. We see that in Example 9.3.10, the domains of Φ_i and Ψ_i are not connected sets. This is the reason for having so many possible local orientations. Problem 9.31 states that a collection of charts with the same connected domain can induce at most two different local orientations.

Global Orientations

A global orientation of a manifold is a collection of local orientations that cover the whole manifold and agree on their common domains. The precise definition is formulated in terms of atlases of compatible local orientations. Recall that atlases for manifolds were defined in Definition 9.2.2.

Definition 9.3.12 Compatible charts. Let $\Psi_i : G_i \to H_i$, $i = 1, 2$, be two diffeomorphisms for a manifold M. They are called *compatible charts* if for every $\mathbf{m} \in G_1 \cap G_2 \cap M$, they induce the same orientation on the tangent space $T_{\mathbf{m}}$.

Definition 9.3.13 Orientable manifolds. Global orientations. A manifold is called an *orientable manifold* if it has an atlas of compatible charts. An atlas of compatible charts for a manifold is called a *global orientation* of this manifold.

Remarks 9.3.14 Not all manifolds are orientable. There are surfaces for which the unit normal vectors cannot be chosen in a continuous way. These surfaces are not orientable. A standard example is the *Möbius strip.* (See Problem 9.30.)

Remarks 9.3.15 Outer boundary-surfaces. In this course we will consider only one example of a global orientation. This will be the *outer boundary-surface* of a set. It is introduced in Section 10.5, in connection with Stokes' theorem.

Orientations of Surfaces and the Right-Hand Rule

Surfaces are $(n-1)$-dimensional manifolds in an n-dimensional space. In particular, planes are the $(n-1)$-dimensional subspaces of an n-dimensional space. Tangent spaces of surfaces are planes. An orientation of a surface consists of the orientations of its tangent planes.

There is a useful way of visualizing the orientation of a plane. The normal space of a plane is one-dimensional. Hence, a plane has two unit normal vectors. We can establish a one-to-one correspondence between these two unit normal vectors and the two orientations of a plane. There is no intrinsic correspondence between these elements; such a correspondence is established by a convention. In physics this convention is usually referred to as the *right-hand rule*.

Definition 9.3.16 The right-hand rule. Assume that T is an oriented plane in an oriented space Z. Let $\mathbb{B}$ be a positive basis for T. Then the corresponding unit normal vector $\mathbf{n}$ is specified by the requirement that $(\mathbf{n},\, \mathbb{B})$ should be a positive basis for Z. Here, if $\mathbb{B} = (\mathbf{b}_2,\, \ldots,\, \mathbf{b}_n)$, then $(\mathbf{n},\, \mathbb{B}) = (\mathbf{n},\, \mathbf{b}_2,\, \ldots,\, \mathbf{b}_n)$.

Remarks 9.3.17 Orientation of the tangent planes. Let $\Psi : G \to H$ be a chart for a surface M in Z. Then $G \cap M$ is oriented by this chart. Therefore, the tangent planes $T_{\mathbf{m}}$ at $\mathbf{m} \in G \cap M$ are oriented planes. It is assumed that the underlying space Z is also an oriented space. Hence, the right-hand rule applied to $T_{\mathbf{m}}$ associates a unit normal vector $\mathbf{n}_{\mathbf{m}}$ to each point $\mathbf{m} \in G \cap M$. This defines a function $\mathbf{n} : G \cap M \to Z$. The value of this function at $\mathbf{m} \in G \cap M$ is a unit normal vector of $T_{\mathbf{m}}$. Theorem 9.3.20 shows that this is a continuous function.

Definition 9.3.18 Outer normals of an oriented surface. The normal vectors obtained in Remarks 9.3.17 are called *the outer normals* of the oriented surface. This is in reference to the canonical case of orienting the outer boundary-surface of a set. This case is discussed in Section 10.5. Otherwise, the name of outer normal is used only as a convenient way of referring to these *orienting normal vectors*.

Notations 9.3.19 Outer normal vector of U. The orientations of U, W, and Z were introduced in Notations 9.3.2. They are arbitrary but fixed orientations. When working with surfaces and with the right-hand rule, the outer unit normal vector of U is also important. We denote this (fixed) vector by $\mathbf{e} = \mathbf{e}_1$. Also, from now on, $\mathbb{E} = (\mathbf{e}_2,\, \ldots,\, \mathbf{e}_n)$ is a positive orthonormal basis for U. Hence, $(\mathbf{e},\, \mathbb{E}) = (\mathbf{e}_1,\, \mathbf{e}_2,\, \ldots,\, \mathbf{e}_n)$ is a positive orthonormal basis for W. Recall that ϑ and ϱ are, respectively, the positive Euclidean determinants of Z and W.

Theorem 9.3.20 An expression for outer normals. *Let $\Psi : G \to H$ be an orientation-preserving chart for M. Let $F(\mathbf{z}) = \mathbf{e} \cdot \Psi(\mathbf{z})$, $\mathbf{z} \in G$. (Here we write $\mathbf{x} \cdot \mathbf{y}$ for the inner product $\langle \mathbf{x},\, \mathbf{y} \rangle$.) Then*

$$\mathbf{n}_{\mathbf{m}} = \nabla F(\mathbf{m}) / \|\nabla F(\mathbf{m})\|,\ \mathbf{m} \in G \cap M, \tag{9.32}$$

is the outer unit normal vector of M at $\mathbf{m} \in G \cap M$, with respect to the orientation induced by the chart $\Psi : G \to H$.

Proof. Let $\mathbf{a} \in G$. We claim first that $\nabla F(\mathbf{a}) \neq \mathbf{0}$. The linear map $\Psi'(\mathbf{a}) : Z \to W$ is an isomorphism. Let $\mathbf{c} = \Psi'(\mathbf{a})^{-1}\mathbf{e}$. Then $\nabla F(\mathbf{a}) \cdot \mathbf{c} = \mathbf{e} \cdot \Psi'(\mathbf{a})\mathbf{c} = \mathbf{e} \cdot \mathbf{e} = 1$, by Remarks 5.4.11 and Example 5.6.5. Hence $\mathbf{n_m}$ as defined in the statement of the theorem is well-defined. Next, observe that $F(\mathbf{z}) = 0$ is an implicit equation for $G \cap M$. Then Theorem 6.4.5 shows that $F'(\mathbf{m})\mathbf{z} = 0$ is an implicit equation for the tangent space of $G \cap M$ at $\mathbf{m}$. But $F'(\mathbf{m})\mathbf{z} = 0$ means that $\nabla F(\mathbf{m}) \cdot \mathbf{z} = 0$, again by Remarks 5.4.11. This shows that $\mathbf{n_m}$ is a unit normal vector of $G \cap M$ at every $\mathbf{m} \in G \cap M$.

The next step is to show that $\mathbf{n_m}$ is the outer unit normal, i.e., that $(\mathbf{n_m}, \mathbb{B}_\mathbf{m})$ is a positive basis for Z where $\mathbb{B}_\mathbf{m}$ is a positive basis for the tangent space $T_\mathbf{m}$. Now we know that $(\Phi'(\mathbf{b})\mathbf{e},\ \Phi'(\mathbf{b})\mathbb{E}) = (\mathbf{c},\ \mathbb{B}_\mathbf{m})$ is a positive basis for Z, since Ψ is orientation-preserving. Hence $\vartheta(\mathbf{c},\ \mathbb{B}_\mathbf{m}) > 0$. If $\mathbf{c} = (\mathbf{c} \cdot \mathbf{n_m})\mathbf{n_m} + \mathbf{c}_1$, then $\mathbf{c}_1 \in T_\mathbf{m}$. Since $\mathbb{B}_\mathbf{m}$ is a basis for $T_\mathbf{m}$ we see that $\vartheta(\mathbf{c}_1,\ \mathbb{B}_\mathbf{m}) = 0$. By multilinearity, this implies

$$\vartheta(\mathbf{c},\ \mathbb{B}_m) = (\mathbf{c} \cdot \mathbf{n_m})\vartheta(\mathbf{n_m},\ \mathbb{B}_\mathbf{m}).$$

Since $\mathbf{c} \cdot \mathbf{n_m} > 0$ and $\vartheta(\mathbf{c},\ \mathbb{B}_\mathbf{m}) > 0$, we see that $\vartheta(\mathbf{n_m},\ \mathbb{B}_\mathbf{m}) > 0$. Hence $\mathbf{n_m}$ is indeed the outer unit normal vector. □

Remarks 9.3.21 Continuity of the outer normals. Theorem 9.3.20 shows that the outer normal function $\mathbf{n} : G \cap M \to Z$ is continuous. The converse is also true. This is stated as Problem 9.32.

The following theorem is useful in computations. Here ϑ is a positive Euclidean determinant for Z and $\mathbb{E}$ is a positive orthonormal basis for U.

Theorem 9.3.22 Volume multipliers and outer normals. *If $\Psi : G \to H$ is a chart, then $\rho(\varphi'(\mathbf{u})) = \vartheta(\mathbf{n_m},\ \mathbb{B}_\mathbf{m})$. Here $\mathbf{m} = \varphi(\mathbf{u})$ and $\mathbb{B}_\mathbf{m} = \varphi'(\mathbf{u})\mathbb{E}$.*

Proof. By Definition 9.3.9 of the orientation on M, the basis

$$\varphi'(\mathbf{u})\mathbb{E} = \Phi'(\mathbf{u})\mathbb{E} = \mathbb{B}_\mathbf{m}$$

is a positive basis for the tangent space $T_\mathbf{m}$. Then by Definition 9.3.18 of the outer normals, $(\mathbf{n_m},\ \mathbb{B}_\mathbf{m})$ is a positive basis for Z. Therefore $\vartheta(\mathbf{n_m},\ \mathbb{B}_\mathbf{m}) > 0$. Also, since $\mathbf{n_m} \perp T_\mathbf{m}$, Theorem 9.1.12 shows that

$$\vartheta(\mathbf{n_m},\ \mathbb{B}_\mathbf{m}) = \|\mathbf{n_m}\| \cdot |\vartheta_{T_\mathbf{m}}(\mathbb{B}_\mathbf{m})|\,.$$

Here $\vartheta_{T_\mathbf{m}}$ is a Euclidean determinant for $T_\mathbf{m}$. But $\mathbb{B}_\mathbf{m}$ is the image of an orthonormal basis for U. In this case, Lemma 9.1.14 shows that

$$\rho(\varphi'(\mathbf{u})) = |\vartheta_{T_\mathbf{m}}(\mathbb{B}_\mathbf{m})|.$$

Then the conclusion follows. □

Problems

9.24 Let $S : X \to Y$ be an isomorphism between two oriented Euclidean spaces. If the bases $\mathbb{E}$ and $\mathbb{E}'$ have the same orientation in X, then show that $S\mathbb{E}$ and $S\mathbb{E}'$ have the same orientation in Y.

9.25 Let W and Z be oriented spaces. Show that an isomorphism $T : W \to Z$ is orientation-preserving (Definition 9.3.3) if and only if it maps any positive bases of W to a positive basis of Z.

9.26 Let α be a determinant on W and let β be a determinant on Z. Let $T : W \to Z$ be an isomorphism. Show that $\alpha(\mathbb{E}) \cdot \beta(T\mathbb{E})$ is nonzero and has the same sign for all bases $\mathbb{E}$ of W.

9.27 Show that an isomorphism is orientation-preserving if and only if its inverse is orientation-preserving.

9.28 **Compatible isomorphisms.** Let $P_i : W \to Z$ be two isomorphisms between two oriented spaces. If they are both orientation-preserving or both orientation-reversing, then they are called *compatible isomorphisms*. Show that two isomorphisms are compatible if and only if their inverses are compatible. Also show that compatibility is independent of the orientations of the spaces.

9.29 Let $T : W \to Z$ be an orientation-preserving isomorphism. Is $-T$ also orientation-preserving? Why?

9.30 **Möbius strip.** Define a surface Σ in cylindrical coordinates (r, ϑ, z) as follows. For each $\alpha \in \mathbb{R}$ let

$$H_\alpha = \{\, (r, \vartheta,\, z) \mid r \geq 0,\ \vartheta = \alpha,\ z \in \mathbb{R} \,\}$$

be the $\vartheta = \alpha$ half-plane. Let the intersection of Σ with H_α be the line segment

$$L_\alpha = \{\, (r,\, z) \mid r = 2 + t\cos(\alpha/2),\ z = t\sin(\alpha/2),\ t \in (-1,\, 1) \,\}.$$

Show that Σ is not an orientable surface.

9.31 Let G be a connected set and let $\Psi_i : G \to H_i$ be two charts for a manifold M. Let $T_{\mathbf{m}}$ be the tangent space of M at $\mathbf{m} \in M$. Show that the local orientations $(\Psi_i,\, \mathbb{E}_i)$ induce either the same orientation on $T_{\mathbf{m}}$ for all $\mathbf{m} \in G \cap M$, or the opposite orientations on $T_{\mathbf{m}}$ for all $\mathbf{m} \in G \cap M$.

9.32 Let M be a surface. Let $\mathbf{n} : M \to Z$ be a continuous function such that $\mathbf{n}_{\mathbf{m}}$ is a unit normal vector of M at every $\mathbf{m} \in M$. Show that the vectors $\mathbf{n}_{\mathbf{m}}$ are the outer normals of M with respect to some orientation.

9.4 INTEGRALS OF VECTOR FIELDS

Vector-valued functions are usually called *vector fields*. Vector fields can be integrated over surfaces or along curves. Such integrals are called, respectively, *surface integrals* and *line integrals* of vector fields. They depend upon the orientation of the manifold over which integration takes place.

In the context of our discussion of the integration of vector fields, we will denote the inner product in Z as the dot product to conform to common usage. Hence we write $\mathbf{a} \cdot \mathbf{b}$ for $\langle \mathbf{a}, \mathbf{b} \rangle$, where $\mathbf{a}, \mathbf{b} \in Z$.

Definition 9.4.1 Vector fields. Let D be a set in a Euclidean space Z. A function $\mathbf{h} : D \to Z$ is referred to as a *vector field* on D. Hence a vector field $\mathbf{h}$ attaches a vector $\mathbf{h}(\mathbf{z}) \in Z$ to every point $\mathbf{z} \in D$. We shall be mainly concerned with vector fields $\mathbf{f} : C \to Z$ defined on a curve C, and with vector fields $\mathbf{f} : S \to Z$ defined on a surface S.

Line Integrals of Vector Fields

Curves are one-dimensional manifolds. Integrals of vector fields along curves are called *line integrals*. These integrals are usually computed by employing parametric representations, as discussed in 9.2.4.

Remarks 9.4.2 Parametric representations of curves. Parametric representations of manifolds were defined in 9.2.1. For one-dimensional manifolds, the subspace U is one-dimensional. It is oriented by a unit vector $\mathbf{u} \in U$. We identify $t\mathbf{u} \in U$ with $t \in \mathbb{R}$. Then a parametric representation of C becomes a function of a real variable, $\varphi : L \to Z$, defined on an open set $L \subset \mathbb{R}$. To conform to common notation, we denote this function as $\mathbf{r} : L \to Z$. The derivative $\mathbf{r}'(t) = \Phi'(t\mathbf{u})\,\mathbf{u} \in Z$ is the directional derivative of the reverse chart $\Phi : H \to Z$ at the point $t\mathbf{u} \in U \subset W$ in the direction of $\mathbf{u}$. Hence $\mathbf{r}'(t) \neq 0$ as it is the directional derivative of a diffeomorphism.

Remarks 9.4.3 Integrals on curves. Integrals on manifolds were defined in 9.2.5. Let $\mathbf{r} : L \to Z$, $L \subset \mathbb{R}$, be a parametric representation for a curve C. Let $f : C \to \mathbb{R}$ be a function. In this case we see that

$$\int_C f = \int_L f(\mathbf{r}(t))\rho(\mathbf{r}'(t))\, du \tag{9.33}$$

is an integral on $\mathbb{R}$. Note that $\mathbf{r}'(t) \in Z$ is the image of the unit vector $\mathbf{u} \in U$ under the linear map $\Phi'(t\mathbf{u})|_U : U \to Z$. Then, by Definition 9.1.15, $\rho(\mathbf{r}'(t)) = \|\mathbf{r}'(t)\|_Z$.

The content on C is called the *arc-length* and usually denoted as s. Hence, by the notations in 9.2.8,

$$\int_C f\,ds = \int_L f(\mathbf{r}(t))\rho(\mathbf{r}'(t))\,dt \tag{9.34}$$

$$= \int_L f(\mathbf{r}(t))\|\mathbf{r}'(t)\|dt = \int_L f(\mathbf{r}(t))ds(t). \tag{9.35}$$

Definition 9.4.4 Orientation induced by $\mathbf{r} : L \to Z$. The line U in W is oriented by the unit vector $\mathbf{u}$. Hence $\mathbf{r} : L \to C \subset Z$ orients the tangent line at $\mathbf{r}(t) = \mathbf{c} \in C$ by the derivative vector $\mathbf{r}'(t)$. Then

$$\mathbf{t_c} = \mathbf{r}'(t)/\|\mathbf{r}'(t)\|_Z \tag{9.36}$$

is the positive unit tangent vector of C at $\mathbf{c} \in C$.

Definition 9.4.5 Line integrals of vector fields. Let C be a curve oriented by $\mathbf{r} : L \to C$ as in Definition 9.4.4. The line integral of a vector field $\mathbf{f}$ on C is defined as the integral of the real valued function $f(\mathbf{c}) = \mathbf{f}(\mathbf{c}) \cdot \mathbf{t_c}$, on C. Hence, by the remarks in Remarks 9.4.3 above,

$$\int_C \mathbf{f} = \int_L \mathbf{f}(\mathbf{r}(t)) \cdot \mathbf{t}_{\mathbf{r}(t)}\, \|\mathbf{r}'(t)\|\,dt \tag{9.37}$$

$$= \int_L \mathbf{f}(\mathbf{r}(t)) \cdot \mathbf{r}'(t)\,dt. \tag{9.38}$$

Other notations for this integral are

$$\int_C \mathbf{f} \cdot \mathbf{t} = \int_C \mathbf{f} \cdot \mathbf{t}\,ds = \int_C \mathbf{f} \cdot\, d\mathbf{s}. \tag{9.39}$$

They suggest that $\|\mathbf{r}'(t)\|\,dt = ds$ may be considered as the length of a small segment of the curve and

$$\mathbf{t}_{\mathbf{r}(t)}\, \|\mathbf{r}'(t)\|\,dt = \mathbf{t}_{\mathbf{r}(t)}\,ds = \mathbf{r}'(t)\,du = d\mathbf{s}(t) \tag{9.40}$$

as a small displacement in the positive direction of the curve. One well-known physical interpretation of $\int_C \mathbf{f} \cdot\, d\mathbf{s}$ is that it is the work done by the force $\mathbf{f}$ when a particle travels along C in the positive direction.

Surface Integrals of Vector Fields

Let $\mathbf{f} : M \to Z$ be a vector field defined on a surface M. We will assume that $\mathbf{f}$ has a compact support K contained in the domain G of a chart $\Psi : G \to H$ for M. This means we need only consider local integration on M. Full generality is obtained by applying the partitions of unity theorem, Theorem 9.2.11, as discussed in Definition 9.2.12.

Notations 9.4.6 Outer vectors of the surface. The surface M is locally oriented by the chart $\Psi : G \to H$, as in Definition 9.3.9. Let $\mathbf{n} : G \cap M \to Z$ be the outer normal function of this orientation, as defined in Definition 9.3.18.

Definition 9.4.7 Surface integrals of vector fields. Let M be a surface oriented by a chart $\Psi : G \to H$. Let $\mathbf{n} : M \to Z$ be the outer normal function for this orientation. Let $\mathbf{f} : M \to Z$ be a vector field with a compact support $K \subset G$. Then the surface integral of $\mathbf{f}$ on M is defined as the integral of the real valued function $f(\mathbf{m}) = \mathbf{f}(\mathbf{m}) \cdot \mathbf{n}_{\mathbf{m}}$, $\mathbf{m} \in M$, on M. Hence, by Definition 9.2.5 of local integrals,

$$\int_M \mathbf{f} = \int_U \mathbf{n}_{\varphi(\mathbf{u})} \cdot \mathbf{f}(\varphi(\mathbf{u}))\, \rho(\varphi'(\mathbf{u}))\, d\mathbf{u}. \tag{9.41}$$

The second integral is the integral of a real-valued function on a Euclidean space.

Theorem 9.4.8 Computations of surface integrals. *Let $\mathbb{E}$ be a positive orthonormal basis for U. Let ϑ be the positive Euclidean determinant in Z. Then*

$$\int_M \mathbf{f} = \int_U \vartheta(\mathbf{f}(\mathbf{m}), \mathbb{B}_{\mathbf{m}})\, d\mathbf{u}.$$

Here $\mathbf{m} = \varphi(\mathbf{u})$ and $\mathbb{B}_{\mathbf{m}} = \varphi'(\mathbf{u})\mathbb{E}$, where $\mathbb{E}$ is a positive orthonormal basis for U.

Proof. We have

$$\int_M \mathbf{f} = \int_U \mathbf{n}_{\varphi(\mathbf{u})} \cdot \mathbf{f}(\varphi(\mathbf{u}))\, \rho(\varphi'(\mathbf{u}))\, d\mathbf{u} \tag{9.42}$$

$$= \int_U \mathbf{n}_{\varphi(\mathbf{u})} \cdot \mathbf{f}(\varphi(\mathbf{u}))\, \vartheta(\mathbf{n}_{\varphi(\mathbf{u})}, \varphi'(\mathbf{u})\mathbb{E})\, d\mathbf{u} \tag{9.43}$$

$$= \int_U \mathbf{n}_{\mathbf{m}} \cdot \mathbf{f}(\mathbf{m})\, \vartheta(\mathbf{n}_{\mathbf{m}}, \mathbb{B}_{\mathbf{m}})\, d\mathbf{u} \tag{9.44}$$

$$= \int_U \vartheta(\mathbf{f}(\mathbf{m}), \mathbb{B}_{\mathbf{m}})\, d\mathbf{u}. \tag{9.45}$$

Here (9.42) is by Definition 9.4.7, and (9.43) follows from Theorem 9.3.22. To obtain (9.44) let $\mathbf{m} = \varphi(\mathbf{u})$ and $\mathbb{B}_{\mathbf{m}} = \varphi'(\mathbf{u})\mathbb{E}$. Finally, for (9.45), let

$$\mathbf{f}(\mathbf{m}) = (\mathbf{n}_{\mathbf{m}} \cdot \mathbf{f}(\mathbf{m}))\mathbf{n}_{\mathbf{m}} + \mathbf{f}_1(\mathbf{m}), \tag{9.46}$$

where $\mathbf{f}_1(\mathbf{m}) \in T_{\mathbf{m}}$ is the tangential component of $\mathbf{f}(\mathbf{m})$. Substitute (9.46) into $\vartheta(\mathbf{f}(\mathbf{m}), \mathbb{B}_{\mathbf{m}})$ and note that $\vartheta(\mathbf{f}_1(\mathbf{m}), \mathbb{B}_{\mathbf{m}}) = 0$. We obtain (9.45). □

Remarks 9.4.9 Special expressions in $\mathbb{R}^3$. Let $Z = \mathbb{R}^3$ be the xyz-space and $U = \mathbb{R}^2$ the uv-plane. Euclidean determinants can be expressed in a familiar way in $\mathbb{R}^3$. One Euclidean determinant in $\mathbb{R}^3$ is

$$\vartheta(\mathbf{a}, \mathbf{b}, \mathbf{c}) = \mathbf{a} \cdot (\mathbf{b} \times \mathbf{c}). \tag{9.47}$$

Here $\mathbf{b} \times \mathbf{c}$ is the usual cross product. Assume that (9.47) is the positive Euclidean determinant of Z. The orthonormal basis $\mathbb{E}$ in U determines the coordinates (u, v). Let us denote the parametric representation $\varphi : H \cap U \to Z$ of $G \cap M$ as $\mathbf{r}(u, v)$ to use more conventional notation. Here (u, v) are the coordinates with respect to the orthonormal basis $\mathbb{E} = (\mathbf{e}_2, \mathbf{e}_3)$ of U. Hence

$$\varphi'(\mathbf{u})\mathbf{e}_1 = \mathbf{r}_u(u, v) \text{ and } \varphi'(\mathbf{u})\mathbf{e}_2 = \mathbf{r}_v(u, v)$$

are the partial derivatives with respect to u and v. Then

$$\vartheta(\mathbf{n}_{\varphi(\mathbf{u})}, \varphi'(\mathbf{u})\mathbb{E}) = \mathbf{n}_{\mathbf{r}(u, v)} \cdot (\mathbf{r}_u(u, v) \times \mathbf{r}_v(u, v)) > 0. \tag{9.48}$$

Let $\mathbf{f} : \mathbb{R}^3 \to \mathbb{R}^3$ be a $\mathcal{C}^1$ vector field of compact support $K \subset G$. Then

$$\int_M \mathbf{f} = \int_M \mathbf{f} \cdot \mathbf{n} \tag{9.49}$$

$$= \int_U (\mathbf{f} \cdot \mathbf{n}) \, \|\mathbf{r}_u \times \mathbf{r}_v\| \, du \, dv \tag{9.50}$$

$$= \int_U \mathbf{f}(\mathbf{r}(u, v)) \cdot \mathbf{r}_u(u, v) \times \mathbf{r}_v(u, v) \, du \, dv. \tag{9.51}$$

Other familiar expressions for this integral are

$$\int_M \mathbf{f} \cdot \mathbf{n} \, dS = \int_M \mathbf{f} \cdot d\mathbf{S}. \tag{9.52}$$

Here $dS = \|\mathbf{r}_u \times \mathbf{r}_v\| \, du \, dv$ denotes the surface area element on S. To find the area of a part of S this is the expression we have to integrate over that part. The expression $d\mathbf{S} = \mathbf{n} dS$ is sometimes called the vectorial area element. To obtain one physical interpretation of $\int_S \mathbf{f} \cdot d\mathbf{S}$, consider $\mathbf{f}$ as the stationary velocity of a body of fluid in motion. Then this integral gives the volume of fluid that passes through S per unit time. Material in the next chapter on Stokes' theorem might make this interpretation more plausible.

Problems

9.33 Define $\mathbf{f} : \mathbb{R}^2 \to \mathbb{R}^2$ by $\mathbf{f}(x, y) = (xy, xy^2)$. Compute $\int_C \mathbf{f}$ from the point $(0, 0)$ to the point $(1, 1)$, where C is

1. the parabola $y = x^2$;
2. the circle $x^2 + (y - 1)^2 = 1$, oriented counterclockwise;
3. the circle $x^2 + (y - 1)^2 = 1$, oriented clockwise;
4. the line $y = x$.

9.34 Define $\mathbf{f} : \mathbb{R}^3 \to \mathbb{R}^3$ by $\mathbf{f}(x, y, z) = (xyz, xy^2z, y)$. Compute $\int_C \mathbf{f}$ from the point $(0, 0, 0)$ to the point $(1, 1, 1)$, where C is

1. the intersection of the parabolic cylinder $z = x^2$ with the plane $y = z$;
2. the shorter arc of the intersection of the circular cylinder $x^2 + (y - 1)^2 = 1$ with the plane $y = z$;
3. the longer arc of the intersection of the circular cylinder $x^2 + (y - 1)^2 = 1$ with the plane $y = z$;
4. the line $x = y = z$.

9.35 Let A be a connected open set in a Euclidean space Z. Let $\mathbf{f} : A \to Z$ be a continuous vector field. Show that the following are equivalent.

1. There is a $\mathcal{C}^1$ function $F : A \to \mathbb{R}$ such that $\mathbf{f} = \nabla F$.
2. If C is a curve in A, then the line integral $\int_C \mathbf{f}$ depends only on the initial and the final points of C. More explicitly, if C_i are two curves in A with the same initial point P and the same final point Q, then $\int_{C_1} \mathbf{f} = \int_{C_2} \mathbf{f}$.

9.36 Let $B = \{ (x, y) \mid x \leq 0, \ y = 0 \}$ and let $A = \mathbb{R}^2 \setminus B$. Define $\mathbf{f} : A \to \mathbb{R}^2$ as $\mathbf{f}(x, y) = (-y, x)/(x^2 + y^2)$. Show that there is an $F : A \to \mathbb{R}$ such that $\mathbf{f} = \nabla F$. Also, extend $\mathbf{f} : A \to \mathbb{R}^2$ to a vector field on $\mathbb{R}^2 \setminus \{ \mathbf{0} \}$ in an obvious way and compute the line integral of this extended vector field over the circle $x^2 + y^2 = 1$, oriented counterclockwise.

9.37 Define $\mathbf{f} : R^3 \to \mathbb{R}^3$ as $\mathbf{f}(x, y, z) = (xyz, xy^2z, z)$. Compute $\int_S \mathbf{f}$ over the following surfaces S, oriented by taking normals with positive z-coordinates.

1. Upper half of the sphere $x^2 + y^2 + z^2 = 1$.

2. The part of the paraboloid $z = 1 - (x^2 + y^2)$.

3. The part of the cone $z = (x^2 + y^2)^{1/2}$ between the planes $z = 1$ and $z = 2$.

4. The helicoidal surface $x = r\cos\vartheta$, $y = r\sin\vartheta$, $z = \vartheta$, where $1 \leq r \leq 2$ and $0 \leq \vartheta \leq 2\pi$.

9.38 A curve in $\mathbb{R}^2$ is also a surface. Let C be the unit circle $x^2 + y^2 = 1$ in $\mathbb{R}^2$ oriented by the outer normals as a surface and oriented counterclockwise as a curve. Let $F(x, y) = P(x, y)\mathbf{i} + Q(x, y)\mathbf{j}$ be a $\mathcal{C}^1$ vector field. Express the line integral $\int_C \mathbf{f} \cdot d\mathbf{s}$ and the surface integral $\int_C \mathbf{f} \cdot d\mathbf{S}$ as ordinary integrals on $\mathbb{R}$.

9.5 INTEGRALS OF TENSOR FIELDS

We have considered the integrals of scalar-valued functions on manifolds and the integrals of vector valued functions on lines and surfaces. These special cases are unified (and simplified) by introducing the integration of *tensor-valued* functions. Indeed, the change of variables formula becomes more transparent when stated in terms of tensor functions. Also, tensor-valued functions are essential in Stokes' theorem.

We will not discuss the algebraic theory of tensors; instead, we use this term in a restricted sense as a name for alternating multilinear functions. (Commonly, any multilinear function may be referred to as a tensor.)

Definition 9.5.1 Tensors. Let Z be a vector space. For any $k \in \mathbb{N}$ let

$$\Lambda_k(Z) = AML_k(Z^k, \mathbb{R}) \tag{9.53}$$

be the vector space of all alternating multilinear functions $\lambda : Z^k \to \mathbb{R}$.

Functions in $\Lambda_k(Z)$ are called *alternating tensors*, or *alternating k-tensors*, on Z. As we shall consider only alternating tensors, we will simply refer to them as *tensors* or *k-tensors*.

Definition 9.5.2 Tensor fields. Let $D \subset Z$. A function $\omega : D \to \Lambda_k(Z)$ is called a *tensor field* or *a tensor field of order k* on D. In this case

$$\omega(\mathbf{z}) \in \Lambda_k(Z) \text{ for } \mathbf{z} \in D \text{ and} \tag{9.54}$$

$$\omega(\mathbf{z})(\mathbb{Z}) = \omega(\mathbf{z})(\mathbf{z}_1, \ldots, \mathbf{z}_k) \in \mathbb{R} \text{ for } \mathbf{z} \in D \text{ and } \mathbb{Z} \in Z^k. \tag{9.55}$$

A tensor field $\omega : Z \to \Lambda_k(Z)$ takes values in a (finite-dimensional) normed vector space. Hence the continuity and the differentiability of these functions are well defined.

Remarks 9.5.3 Integrations of tensor fields. Tensor fields of order k are integrated over k-dimensional oriented manifolds. The process is the same as for vector fields. We first have to produce a real-valued function, defined in terms of the tensor field and the given oriented manifold, and then we can integrate this real-valued function over the manifold.

For convenience, we shall assume that all integrals are local integrals, computed in terms of a single chart. This amounts to assuming that the tensor field has a compact support contained in the domain of one chart. The general case is obtained by applying the partitions of unity theorem, Theorem 9.2.11, as discussed in Definition 9.2.12.

We will also assume that $\Psi : G \to H$ is an orientation-preserving chart, as defined in Definition 9.3.4. Theorem 9.3.8 shows that this assumption implies no loss of generality.

Definition 9.5.4 Integrals of tensor fields. Let M be a k-dimensional manifold. Let $\Psi : G \to H$ be a chart for M that orients $G \cap M$. For each $\mathbf{m} \in G \cap M$, let $\mathbb{B}_{\mathbf{m}}$ be a positive orthonormal basis for the tangent space $T_{\mathbf{m}}$. Let $\omega : M \to \Lambda_k(Z)$ be a $\mathcal{C}^1$ tensor field of compact support $K \subset G$. The integral of ω on $M \cap G$ is defined as the integral of the real-valued function $f(\mathbf{m}) = \omega(\mathbf{m})(\mathbb{B}_{\mathbf{m}})$ over $G \cap M$. Hence

$$\int_M \omega = \int_U \omega(\varphi(\mathbf{u}))(\mathbb{B}_{\varphi(\mathbf{u})})\, \rho(\varphi'(\mathbf{u}))\, d\mathbf{u}.$$

Theorem 9.5.5 Computation of integrals. *Let $\mathbb{E}$ be a positive orthonormal basis for U. Let $\mathbb{B}_{\mathbf{m}} = \mathbb{B}_{\varphi(\mathbf{u})} = \varphi'(\mathbf{u})\mathbb{E}$, where $\mathbf{m} = \varphi(\mathbf{u})$. Then*

$$\int_M \omega = \int_U \omega(\mathbf{m})(\mathbb{B}_{\mathbf{m}})\, d\mathbf{u} \tag{9.56}$$

$$= \int_U \omega(\varphi(\mathbf{u}))(\varphi'(\mathbf{u})\mathbb{E})\, d\mathbf{u}. \tag{9.57}$$

Proof. Let $\mathbf{m} \in G \cap M$. The function $\omega(\mathbf{m}) : Z^k \to \mathbb{R}$ restricted to ${T_{\mathbf{m}}}^k$ becomes a member of $\Lambda_k(T_{\mathbf{m}})$. This space is one-dimensional since $\dim T_{\mathbf{m}} = k$. Hence this restricted function is a multiple of the positive Euclidean determinant $\vartheta_{\mathbf{m}} \in \Lambda_k(T_{\mathbf{m}})$. Let $\omega(\mathbf{m})(\mathbb{T}) = g(\mathbf{m})\vartheta_{\mathbf{m}}(\mathbb{T})$ for all k-tuples $\mathbb{T} \in {T_{\mathbf{m}}}^k$. By letting $\mathbb{T} = \mathbb{B}_{\mathbf{m}}$, we see that $\omega(\mathbf{m})(\mathbb{B}_{\mathbf{m}}) = g(\mathbf{m}) = f(\mathbf{m})$. Hence

$$\omega(\mathbf{m})(\mathbb{T}) = f(\mathbf{m})\vartheta_{\mathbf{m}}(\mathbb{T}) \tag{9.58}$$

for all $\mathbb{T} \in T_{\mathbf{m}}{}^{k}$. Therefore

$$\begin{aligned}\int_{M} \omega &= \int_{U} f(\mathbf{m})\, \rho(\varphi'(\mathbf{u}))\, d\mathbf{u} & (9.59)\\ &= \int_{U} f(\mathbf{m})\, \vartheta_{\mathbf{m}}(\mathbb{B}_{\mathbf{m}})\, d\mathbf{u} & (9.60)\\ &= \int_{U} \omega(\mathbf{m})(\mathbb{B}_{\mathbf{m}})\, d\mathbf{u}. & (9.61)\end{aligned}$$

To obtain (9.60), note that $\rho(\varphi'(\mathbf{u})) = |\,\vartheta_{\mathbf{m}}(\mathbb{B}_{\mathbf{m}})\,| = \vartheta_{\mathbf{m}}(\mathbb{B}_{\mathbf{m}})$. Here, the first equality follows as in the proof of Theorem 9.3.22 and the second equality holds because $\vartheta_{\mathbf{m}}(\mathbb{B}_{\mathbf{m}}) > 0$, by the definition of $\vartheta_{\mathbf{m}}$. The inference to (9.61) then follows from (9.58). □

Connection with Vector Fields

Each vector field $\mathbf{f} : D \to Z$, $D \subset Z$, defines two tensor fields, a tensor field ξ of order one and a tensor field η of order $(n-1)$, where $n = \dim Z$. The ξ field is defined in terms of inner products only. For the definition of the η field, Z must be a space oriented by a Euclidean determinant $\vartheta \in \Lambda_n(Z)$. Then the integrals of ξ on one-dimensional manifolds are the line integrals of $\mathbf{f}$ and the integrals of η on $(n-1)$-dimensional manifolds are the surface integrals of $\mathbf{f}$.

Definition 9.5.6 Tensor fields associated with a vector field. Let (Z, ϑ) be an n-dimensional oriented Euclidean space. Given a vector field $\mathbf{f} : D \to Z$, where $D \subset Z$, define $\xi : D \to \Lambda_1(Z)$ and $\eta : D \to \Lambda_{n-1}(Z)$ as follows. For all $\mathbf{a} \in D$:

$$\begin{aligned}\xi(\mathbf{a})(\mathbb{Z}) &= \mathbf{f}(\mathbf{a}) \cdot \mathbf{z}_1, & \mathbb{Z} = (\mathbf{z}_1) \in Z^1, \text{ and}\\ \eta(\mathbf{a})(\mathbb{Z}) &= \vartheta(\mathbf{f}(\mathbf{a}), \mathbb{Z}), & \mathbb{Z} = (\mathbf{z}_2, \ldots, \mathbf{z}_n) \in Z^{n-1}.\end{aligned} \tag{9.62}$$

Theorem 9.5.7 *Let $\mathbf{f}$ be a vector field with the associated tensor fields ξ and η. If C is a curve oriented by a parametric representation $\varphi : U \cap H \to Z$, then*

$$\int_{C \cap G} \xi = \int_{C \cap G} \mathbf{f}. \tag{9.63}$$

If M is a surface oriented by a parametric representation $\varphi : U \cap H \to Z$ then

$$\int_{M \cap G} \eta = \int_{M \cap G} \mathbf{f}. \tag{9.64}$$

In each case, $G = \Phi(H)$ where $\Phi : H \to G$ is the inverse of the chart $\Psi : G \to H$ associated with the parametric representation $\varphi = \Phi|_{U \cap H}$.

Proof. This is left as an exercise. □

Remarks 9.5.8 Integrals of real-valued functions. Let (Z, ϑ) be an oriented n-dimensional Euclidean space. Associate a real-valued function$f : D \to \mathbb{R}$ and a tensor field $\zeta : D \to \Lambda_n(Z)$ through the relation

$$\zeta(\mathbf{a})(\mathbb{Z}) = f(\mathbf{a})\vartheta(\mathbb{Z}), \mathbf{a} \in D, \mathbb{Z} = (\mathbf{z}_1, \ldots, \mathbf{z}_n) \in Z^n. \tag{9.65}$$

An open set $G \subset Z$ is an n-dimensional manifold in Z. An obvious chart for G is the identity map. Let $\mathbb{E} \in Z^n$ be a positive orthonormal basis. All tangent spaces of G are Z, and all are oriented by ϑ. Then we see that

$$\int_G f(\mathbf{z})\, d\mathbf{z} = \int_G \zeta(\mathbf{z})(\mathbb{E})\, d\mathbf{z} = \int_G \zeta. \tag{9.66}$$

This is rather trivial but still very useful.

Definition 9.5.9 Vector fields associated with tensor field. The relations between the vectors and tensors are symmetrical. Let $\lambda \in \Lambda_1(Z)$ be a tensor of order one. Then $\lambda : Z \to \mathbb{R}$ is a linear function. Hence, by the representation theorem for linear functions on Euclidean spaces, there is a vector $\mathbf{f} \in Z$ such that $\lambda(\mathbf{z}) = \mathbf{f} \cdot \mathbf{z}$ for all $\mathbf{z} \in Z$. If $\xi : D \to \Lambda_1(Z)$ is a tensor field, then there is a vector field $\mathbf{f} : D \to Z$ such that

$$\xi(\mathbf{a})(\mathbb{Z}) = \mathbf{f}(\mathbf{a}) \cdot \mathbf{z}_1, \ \mathbb{Z} = (\mathbf{z}_1) \in Z^1 \tag{9.67}$$

for all $\mathbf{a} \in D$. Lemma 9.5.10 gives a similar representation for $\lambda \in \Lambda_{n-1}(Z)$.

Lemma 9.5.10 *Let $\lambda \in \Lambda_{n-1}(Z)$ with $n = \dim Z$. Let $\vartheta \in \Lambda_n(Z)$ be a determinant. Then there is a unique vector $\mathbf{p} \in Z$ such that*

$$\lambda(\mathbb{Z}) = \vartheta(\mathbf{p}, \mathbb{Z}), \ \textit{for all } \mathbb{Z} \in Z^{n-1}. \tag{9.68}$$

Proof. The case of $n = 2$ is left as an exercise. Note that the representation is different from the one in (9.67), even though $n - 1 = 1$ in this case.

Now let $n \geq 3$ and $\lambda \in \Lambda_{n-1}(Z)$. Let $\mathbb{E} = (\mathbf{e}_1, \ldots, \mathbf{e}_n)$ be a basis for Z such that $\vartheta(\mathbb{E}) = 1$. Then for each $\mathbf{e}_i \in \mathbb{E}$, there is an $\mathbb{E}_i \in Z^{n-1}$ such that $\mathbb{E}_i$ is a permutation of $\mathbf{e}_j$ with $j \neq i$ and such that $\vartheta(\mathbf{e}_i, \mathbb{E}_i) = 1$. Let $p_i = \lambda(\mathbb{E}_i)$ and $\mathbf{p} = \sum_i p_i \mathbf{e}_i$. Then an easy verification shows that (9.68) is satisfied. □

Change of Variables

Let M be a manifold in Z. To simplify things, assume as usual that M is covered by a single chart $\Psi : G \to H$ with $M \subset G$. (Otherwise, replace M by $M \cap G$ for this section, and use Partitions of Unity for the general case.)

Let X be another space and $\Omega : G \to X$ a diffeomorphism with $F = \Omega(G)$ and with the reverse diffeomorphism $\Gamma : F \to G$. Then $L = \Omega(M)$ is manifold in X. In fact, $\Psi \cdot \Gamma : F \to H$ is a chart for L.

Let M be oriented by the chart $\Psi : G \to H$ and by a fixed orthonormal basis $\mathbb{E}$ in U, as in Definition 9.3.9. Let L be oriented by the chart $\Psi \cdot \Gamma : F \to H$.

Definition 9.5.11 Pullbacks of tensor fields. With the notation above, the *pullback* of the tensor field $\omega : G \to \Lambda_k(Z)$ by the diffeomorphism $\Gamma : F \to G$ is defined as

$$\Gamma^*(\omega)(\mathbf{x})(\mathbb{X}) = \omega(\Gamma(\mathbf{x}))(\Gamma'(\mathbf{x})\mathbb{X}) \tag{9.69}$$

for $\mathbf{x} \in F$ and $\mathbb{X} \in X^k$. This is another tensor field $\Gamma^*(\omega) : F \to \Lambda_k(X)$.

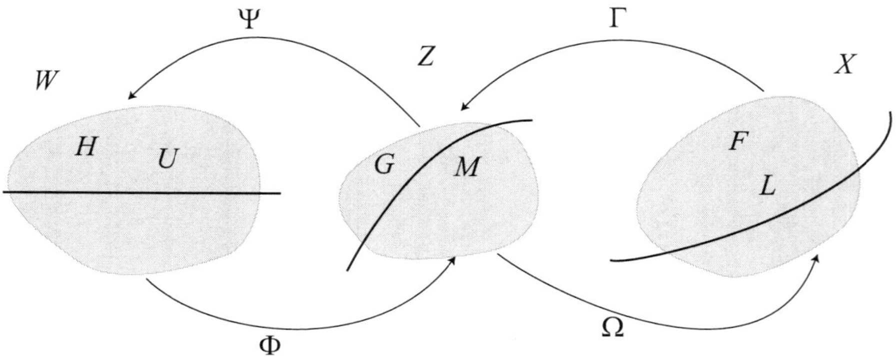

Figure 9.3. Diffeomorphisms in Theorem 9.5.12.

Theorem 9.5.12 Change of variables for tensor fields. *We have*

$$\int_M \omega = \int_L \Gamma^*(\omega), \tag{9.70}$$

with the notations and assumptions above.

Proof. Let $\tau = \Gamma^*(\omega)$. Let $\Phi : H \to G$ be the reverse chart of $\Psi : G \to H$ and $\Theta = \Omega \cdot \Phi : H \to F$ the reverse chart of $\Psi \cdot \Gamma : F \to H$. Let $\mathbf{z} = \Phi(\mathbf{u})$ and $\mathbf{x} = \Omega(\mathbf{z})$. Then $\mathbf{x} = \Theta(\mathbf{u})$. Also $\Theta'(\mathbf{u}) = \Omega'(\mathbf{z})\Phi'(\mathbf{u})$. Then

$$\begin{aligned} \tau(\Theta(\mathbf{u}))(\Theta'(\mathbf{u})\mathbb{E}) &= \omega(\Gamma(\Theta(\mathbf{u}))(\Gamma'(\mathbf{x})\Theta'(\mathbf{u})\mathbb{E}) && (9.71)\\ &= \omega(\Phi(\mathbf{u}))(\Phi'(\mathbf{u})\mathbb{E}). && (9.72)\end{aligned}$$

The integral of ω on M is the integral of $\omega(\varphi(\mathbf{u}))(\varphi'(\mathbf{u})\mathbb{E}) = \omega(\Phi(\mathbf{u}))(\Phi'(\mathbf{u})\mathbb{E})$ on $U \cap H$, and the integral of $\tau = \Gamma^*(\omega)$ is the integral of $\tau(\Theta(\mathbf{u}))(\Theta'(\mathbf{u})\mathbb{E})$ on the same set $U \cap H$. Then the proof follows. □

Lemma 9.5.13 Composition of pullbacks. *Let A, B, and C be open sets in the Euclidean spaces X, Y, and Z, respectively. Let $\Psi : A \to Y$ and $\Phi : B \to Z$ be diffeomorphisms with $\Psi(A) = B$. Then $(\Phi \cdot \Psi)^*(\omega) = \Psi^*(\Phi^*(\omega))$ for any tensor field $\omega : Z \to \Lambda_k(Z)$.*

Proof. This follows from the definitions. The details are left as an exercise. Note the change of order in $(\Phi \cdot \Psi)^*(\omega) = \Psi^*(\Phi^*(\omega))$. □

Problems

9.39 Let Z be an n-dimensional oriented Euclidean space. Definition 9.5.6 shows that a vector defines a tensor of order one and another tensor of order $(n-1)$. If $n = 2$, then these orders are the same; show, however, that the associated tensors are different. What are the tensors on $\mathbb{R}^2$ associated with the vector $(1, 0)$? Conversely, each tensor of order one on $\mathbb{R}^2$ is associated with two vectors. What are the vectors in $\mathbb{R}^2$ associated with the tensor $\tau : \mathbb{R}^2 \to \mathbb{R}$ defined by $\tau(x, y) = y$? (You may assume that $\mathbb{R}^2$ has its standard orientation.)

9.40 Define $\mathbf{f} : \mathbb{R}^3 \to \mathbb{R}^3$ by $\mathbf{f}(x, y, z) = (x^2 z, y^2 z, x^2 + y^2)$. Let M be the part of the cone $z = (x^2 + y^2)^{1/2}$ between the planes $z = 1$ and $z = 2$. Let M be oriented by the normals with positive z-coordinates. Let $B = \{ (x, y, z) \mid x = y = 0 \}$ be the z-axis and $G = \mathbb{R}^3 \setminus B$. Define $\Omega : G \to G$ by

$$\Omega(x, y, z) = (x, y, z - (x^2 + y^2)^{1/2} + (x^2 + y^2)).$$

1. Show that Ω is a diffeomorphism and find its inverse $\Gamma : G \to G$.
2. Find the tensor field $\omega : \mathbb{R}^3 \to \Lambda_2(\mathbb{R}^3)$ so that $\int_M \mathbf{f} = \int_M \omega$.
3. Let $L = \Omega(M)$. Compute $\int_M \omega$ and $\int_L \Gamma^*(\omega)$ directly and verify the change of variables theorem, Theorem 9.5.12.

9.6 INTEGRATION ON GRAPHS

Graphs form a simple class of manifolds. A graph is defined by a single special type of chart. Furthermore, content and integration on graphs can be interpreted geometrically.

Notations 9.6.1 Graphs and coordinate systems. For this section, let $(X,\, Y)$ be an orthogonal coordinate system for Z. Hence

$$Z = X \oplus Y \simeq X \times Y, \tag{9.73}$$

where $X \perp Y$. Let $P : Z \to X$ and $Q : Z \to Y$ be the corresponding coordinate projections. Let $A \subset X$ be an open set and $h : A \to Y$ a $\mathcal{C}^1$ map. The graph of h is

$$M = \{\, (\mathbf{x},\, h(\mathbf{x})) \mid \mathbf{x} \in A \,\}. \tag{9.74}$$

Remarks 9.6.2 Graphs are manifolds. The graph of h is a manifold in Z. To see this, let $G = H = A \times Y$. Then

$$\Psi(\mathbf{x},\, \mathbf{y}) = (\mathbf{x},\, \mathbf{y} - h(\mathbf{x})) \;\text{ and }\; \Phi(\mathbf{x},\, \mathbf{y}) = (\mathbf{x},\, \mathbf{y} + h(\mathbf{x})) \tag{9.75}$$

define a chart $\Psi : G \to H$ for M and the reverse chart $\Phi : H \to G$. We have

$$\Psi(M \cap G) = X \cap H = A. \tag{9.76}$$

Hence $\varphi(\mathbf{x},\, \mathbf{0}) = (\mathbf{x},\, h(\mathbf{x}))$, $\mathbf{x} \in A$, is a parametric representation for M.

Remarks 9.6.3 Tangent spaces are graphs. Let $\mathbf{a} \in A$ and let $T_{\mathbf{a}}$ be the (linear) tangent space of M at $(\mathbf{a},\, h(\mathbf{a})) \in M$. Then $T_{\mathbf{a}}$ is the graph of the linear function $h'(\mathbf{a}) : X \to Y$.

Notations 9.6.4 Orthogonal projections of tangent spaces onto X. For each $\mathbf{a} \in A$ let

$$P|_{T_{\mathbf{a}}} = P_{\mathbf{a}} : T_{\mathbf{a}} \to X \tag{9.77}$$

be the restriction of the orthogonal projection $P : Z \to X$ to $T_{\mathbf{a}}$. Then

$$T_{\mathbf{a}} = \{\, (\mathbf{x},\, h'(\mathbf{a})\mathbf{x}) \mid \mathbf{x} \in X \,\} \;\text{ and } \tag{9.78}$$

$$P_{\mathbf{a}}(\mathbf{x},\, h'(\mathbf{a})\mathbf{x}) = (\mathbf{x},\, \mathbf{0}) \;\text{ for all } \mathbf{x} \in X. \tag{9.79}$$

We see that $P_{\mathbf{a}} : T_{\mathbf{a}} \to X$ is an isomorphism. Its inverse $P_{\mathbf{a}}^{-1} : X \to T_{\mathbf{a}}$ is given by $P_{\mathbf{a}}^{-1}(\mathbf{x},\, \mathbf{0}) = (\mathbf{x},\, h'(\mathbf{a})\mathbf{x})$, $\mathbf{x} \in X$.

Lemma 9.6.5 *Let $\varphi(\mathbf{x},\, \mathbf{0}) = (\mathbf{x},\, h(\mathbf{x}))$, $\mathbf{x} \in A$, be the parametric representation of M obtained above. Then $\varphi'(\mathbf{a},\, \mathbf{0}) = P_{\mathbf{a}}^{-1} : X \to T_{\mathbf{a}}$ for all $\mathbf{a} \in A$.*

Proof. We verify that the application of $\varphi'(\mathbf{a},\, \mathbf{0}) : X \to Z$ to $(\mathbf{x},\, \mathbf{0}) \in X$ is

$$\varphi'(\mathbf{a},\, \mathbf{0})(\mathbf{x},\, \mathbf{0}) = (\mathbf{x},\, h'(\mathbf{a})\mathbf{x}) = P_{\mathbf{a}}^{-1}(\mathbf{x},\, \mathbf{0}). \;\square \tag{9.80}$$

Lemma 9.6.6 *Let $f : M \to \mathbb{R}$. Then*

$$\int_M f = \int_A f(\mathbf{x}, h(\mathbf{x}))/\rho(P_{\mathbf{x}})\, d\mathbf{x} \tag{9.81}$$

if the second integral exists.

Proof. We have $\int_M f = \int_A f(\varphi(\mathbf{x}, \mathbf{0}))\, \rho(\varphi'(\mathbf{x}, \mathbf{0}))\, d\mathbf{x}$ by the definitions in 9.2.5. Also, $\varphi'(\mathbf{x}, \mathbf{0}) = P_{\mathbf{x}}^{-1}$ implies $\rho(\varphi'(\mathbf{x}, \mathbf{0})) = 1/\rho(P_{\mathbf{x}})$. □

Remarks 9.6.7 A geometric interpretation. Assume that the integral

$$\int_A f(\mathbf{x}, h(\mathbf{x}))/\rho(P_{\mathbf{x}})\, d\mathbf{x}$$

exists. Then it can be approximated by a finite sum

$$\sum\nolimits_i f(\mathbf{a}_i, h(\mathbf{a}_i))\, v_X(E_i)/\rho(P_{\mathbf{a}_i}), \tag{9.82}$$

where E_i s are pairwise disjoint sets in X and $\mathbf{a}_i \in A \cap E_i$. Let $B_i = P_{\mathbf{a}_i}^{-1} E_i$, which is a subset of $T_{\mathbf{a}_i}$. Then $v_X(E_i)/\rho(P_{\mathbf{a}_i}) = v_i(B_i)$, where v_i is the volume on the tangent space $T_{\mathbf{a}_i}$. Hence $\int_M f$ can be approximated by

$$\sum\nolimits_i f(\mathbf{a}_i, h(\mathbf{a}_i))\, v_i(B_i). \tag{9.83}$$

Here $(\mathbf{a}_i, h(\mathbf{a}_i)) \in M$ and $B_i \subset T_{\mathbf{a}_i}$.

Now the sets B_i are pairwise disjoint, since their projections $E_i = P_{\mathbf{a}_i} B_i$ on X are pairwise disjoint. Therefore we can imagine the following procedure to compute the integral $\int_M f$. The manifold M is partitioned into finitely many small sets $M_i = \varphi(E_i)$, each of these sets is replaced by the 'flat' set $B_i = \varphi'(\mathbf{a}_i, \mathbf{0})E_i$ (a piece of the tangent space at $\mathbf{a}_i$), and then the sum in (9.83) is formed using the Euclidean volumes of the B_i s.

Remarks 9.6.8 Formulations in terms of normals. The normal space $N_{\mathbf{a}}$ of M at $(\mathbf{a}, h(\mathbf{a}))$ is defined as $N_{\mathbf{a}} = T_{\mathbf{a}}{}^{\perp}$, the orthogonal complement of the tangent space. Let $Q_{\mathbf{a}} = Q|_{N_{\mathbf{a}}} : N_{\mathbf{a}} \to Y$ be the restriction of the orthogonal projection $Q : Z \to Y$ to the normal space $N_{\mathbf{a}}$. It is a useful fact that $P_{\mathbf{a}}$ and $Q_{\mathbf{a}}$ have the same volume multipliers. This result follows easily from Theorem 3.6.20. It is also stated as Problem 3.92.

Remarks 9.6.9 Computations of volume multipliers. Let $S : X \to Y$ be a linear transformation. Let $T = \{\, (\mathbf{x}, S\mathbf{x}) \mid \mathbf{x} \in X \,\}$ be the graph of S. Let $L : X \to T$ be the mapping $L(\mathbf{x}, \mathbf{0}) = (\mathbf{x}, S\mathbf{x})$. The arguments above show that the volume multiplier $\rho(L)$ plays an important role in integration on graph-manifolds. With the notations we have been employing, $\rho(L)$ corresponds to $1/\rho(P_{\mathbf{a}})$ at a particular point $\mathbf{a} \in A$ with $S = h'(\mathbf{a}) : X \to Y$. How do we compute $\rho(L)$?

Let $\mathbb{E} = (\mathbf{e}_1, \ldots, \mathbf{e}_k)$ be an orthonormal basis for X. We know that

$$\rho(L) = (\det\langle L\mathbf{e}_i, L\mathbf{e}_j\rangle)^{1/2}. \tag{9.84}$$

But $L\mathbf{e}_i = (\mathbf{e}_i, S\mathbf{e}_i)$ and $\langle L\mathbf{e}_i, L\mathbf{e}_j\rangle_Z = \delta_{ij} + \langle S\mathbf{e}_i, S\mathbf{e}_j\rangle_Y$. Computations are simplified if we can exhibit an eigenbasis for $S : X \to Y$: an orthonormal basis $\mathbb{E}$ for X such that $S\mathbb{E}$ is an orthogonal set in Y. In this case, $\langle S\mathbf{e}_i, S\mathbf{e}_j\rangle = 0$ if $i \neq j$ and $\langle S\mathbf{e}_i, S\mathbf{e}_i\rangle = \lambda_i \geq 0$ for some scalars λ_i. Then we obtain

$$\rho(L)^2 = (1 + \lambda_1) \cdots (1 + \lambda_k). \tag{9.85}$$

Here the values λ_i are also the eigenvalues of the nonnegative definite self-adjoint transformation $S^*S : X \to X$.

The case of $\dim Y = 1$ is especially important. In this case, the computations are simple. Assume that $Y = \mathbb{R}$ and $S : X \to \mathbb{R}$ is a real valued linear map. Hence there is a unit vector $\mathbf{a} \in X$ and a scalar α such that $S\mathbf{x} = \alpha\langle \mathbf{a}, \mathbf{x}\rangle$ for all $\mathbf{x} \in X$. Complete $\mathbf{a}$ to an orthonormal basis $\mathbb{U}$ for X. This is an eigenbasis for S. All vectors in $\mathbb{U}$ are mapped to 0, except $\mathbf{a}$. Hence $\rho(L)^2 = (1 + \alpha^2)$ in this case.

Geometric Content

Let M be a manifold in a Euclidean space Z. As we noted in Remarks 9.2.8, our definitions of integration and content on M depend upon a parametric representation that allows us to reduce integration on M to integration on a Euclidean space. To be sure, we showed that such integrals do not vary if we change the parametric representation. Nevertheless, there is no escaping the fact that, on these definitions, we rely upon extrinsic elements (charts and subsets of a Euclidean space) to define the existence and value of an integral on M.

In the remainder of this chapter, we provide an intrinsic definition of content (and hence integration) on M in terms of the volume on the space Z in which the manifold is embedded. We shall call the content defined in this way *geometric content*. Then we prove the non-obvious fact that geometric content agrees with the standard notion of content defined earlier.

The definition of geometric content depends upon the concept of *enlargements* of sets.

Definition 9.6.10 Enlargements of a set. For any set K in Z and for any $r > 0$, let $K_r = \cup_{\mathbf{k}\in K} B_r(\mathbf{k})$ be the *enlargement* of K in Z by $r > 0$. Hence

$$K_r = \{\, \mathbf{z} \in Z \mid \exists \mathbf{k} \in K,\ \|\mathbf{z} - \mathbf{k}\| < r \,\}. \tag{9.86}$$

Problem 8.52 shows that if K is a bounded set in Z, then K_r is always a Jordan set in Z. We will assume this fact for convenience. Otherwise, our definitions have to be formulated in terms of inner and outer volumes.

Definition 9.6.11 Geometric content. Let E be a bounded set in an n-dimensional Euclidean space Z. Let k be an integer, $0 \le k \le n$, and $\ell = (n - k)$. Let S_ℓ be the volume of the unit ball in $\mathbb{R}^\ell$ if $\ell \ge 1$ and $S_0 = 1$. Then the *k-dimensional geometric content* of E is defined as

$$\tau_k(E) = \lim_{r \to 0+} \frac{1}{r^\ell\, S_\ell}\, v_Z(E_r) \tag{9.87}$$

if this limit exists.

For example, if E is a curve in $\mathbb{R}^3$, we find the 1-dimensional geometric content of E by considering sausage-like sets in $\mathbb{R}^3$ consisting of all points within r of E. We calculate the volume of the sausage, divide by πr^2 (the area of the circle or radius r), and note whether this ratio converges to some value.

Our main purpose in this section is to prove the following theorem.

Theorem 9.6.12 Agreement of standard content with geometric content. *Let (X, Y) be an orthogonal coordinate system for $Z = X \times Y$ with $k = \dim X$ and $\ell = \dim Y$. Let A be an open set in X, $f : A \to Y$ a $\mathcal{C}^1$ function, and M the graph of f. Let $\varphi(\mathbf{x}) = (\mathbf{x}, f(\mathbf{x}))$, $\mathbf{x} \in A$, be a parametric equation for M. If E is a Jordan set in X such that $\overline{E} \subset A$ and if $L = \varphi(E)$, then $\tau_k(L) = \sigma(L)$.*

The proof of this theorem is given after a few lemmas. The following comments might be helpful in following the arguments in those lemmas.

Remarks 9.6.13 Comments on the proof of Theorem 9.6.12. Fubini's theorem allows us to compute $v_Z(L_r)$ (the volume of the enlargement of L) by integration over E (essentially), as $\int_E \lambda(\mathbf{x})\, d\mathbf{x}$. Here $\lambda(\mathbf{a}) = v_Y(C(L_r, \mathbf{a}))$ is the volume in Y of the cross-section $C(L_r, \mathbf{a})$ of L_r with the "vertical space" $Y(\mathbf{a})$. This vertical space consists of vectors with constant X components $\mathbf{x} = \mathbf{a}$.

The cross-section $C(L_r, \mathbf{a})$ consists of all vectors in $Y(\mathbf{a})$ that have a distance less than r to L. If L is replaced by the affine tangent space at $(\mathbf{a}, f(\mathbf{a}))$, then the orthogonal projection of $C(L_r, \mathbf{a})$ on the normal plane $N(\mathbf{a})$ is a ball of radius r. Therefore

$$v_Y(C(L_r, \mathbf{a})) \simeq v_Y(B_r(\mathbf{0}))/\rho(S(\mathbf{a})) = r^\ell\, S_\ell/\rho(S(\mathbf{a})). \tag{9.88}$$

Here $\rho(S(\mathbf{a}))$ is the volume multiplier of the orthogonal projection $S(\mathbf{a}) : Y \to N(\mathbf{a})$. This is the same as for the orthogonal projection $Y(\mathbf{a}) \to N(\mathbf{a})$.

As observed in Remarks 9.6.8, we have $\rho(S(\mathbf{a})) = \rho(R(\mathbf{a}))$. Hence

$$\begin{aligned} v_Z(L_r) &\simeq \int_E r^\ell\, S_\ell/\rho(S(\mathbf{x}))\, d\mathbf{x} = \int_E r^\ell\, S_\ell/\rho(R(\mathbf{x}))\, d\mathbf{x} \qquad (9.89)\\ &= r^\ell\, S_\ell \int_E (\rho(R(\mathbf{x})))^{-1}\, d\mathbf{x} = r^\ell\, S_\ell \sigma(L). \qquad (9.90)\end{aligned}$$

The formal proof of Theorem 9.6.12 consists of justifications for the various approximations we have made in these comments.

Definition 9.6.14 Cross-sections. Let H be a set in $Z = X \times Y$. If $\mathbf{a} \in X$, then the *cross-section* of H at $\mathbf{a} \in X$ is

$$C(H,\, \mathbf{a}) = \{\mathbf{y} \in Y \mid (\mathbf{a},\, \mathbf{y}) \in H\} \subset Y. \qquad (9.91)$$

The following two lemmas are not stated in terms of cross-sections, but they will be used to obtain estimates on various cross-sections.

Lemma 9.6.15 *Let A be an open set in X and let $f,\, g : A \to Y$ be two functions. Let $B_{r_0}(\mathbf{a}) \subset A$. For each r, $0 < r < r_0$, define*

$$\begin{aligned} A(\mathbf{a},\, r) &= \{\mathbf{y} \in Y \mid \exists\, \mathbf{x} \in X,\ \|(\mathbf{x},\, f(\mathbf{x})) - (\mathbf{a},\, \mathbf{y})\|_Z < r\}, \qquad (9.92)\\ B(\mathbf{a},\, r) &= \{\mathbf{y} \in Y \mid \exists\, \mathbf{x} \in X,\ \|(\mathbf{x},\, g(\mathbf{x})) - (\mathbf{a},\, \mathbf{y})\|_Z < r\}. \qquad (9.93)\end{aligned}$$

Assume that there are $\varepsilon > 0$ and δ, $0 < \delta < r_0$, such that

$$\|f(\mathbf{x}) - g(\mathbf{x})\|_Y \le \varepsilon \|\mathbf{x} - \mathbf{a}\|_X \text{ whenever } \|\mathbf{x} - \mathbf{a}\|_X < \delta.$$

Then $B(\mathbf{a},\, (1-\varepsilon)r) \subset A(\mathbf{a},\, r) \subset B(\mathbf{a},\, (1+\varepsilon)r)$ whenever $0 < r < \delta$.

Proof. Let $0 < r < \delta$ and $\mathbf{y} \in A(\mathbf{a},\, r)$. Then there is an $\mathbf{x} \in X$ such that

$$\|(\mathbf{x},\, f(\mathbf{x})) - (\mathbf{a},\, \mathbf{y})\|_Z^2 = \|\mathbf{x} - \mathbf{a}\|_X^2 + \|f(\mathbf{x}) - \mathbf{y}\|_Y^2 < r^2 < \delta^2.$$

Hence $\|\mathbf{x} - \mathbf{a}\|_X < \delta$ and $\|f(\mathbf{x}) - g(\mathbf{x})\|_Y \le \varepsilon \|\mathbf{x} - \mathbf{a}\|_X < \varepsilon\, r$. Therefore

$$\begin{aligned} \|(\mathbf{x},\, g(\mathbf{x})) - (\mathbf{a},\, \mathbf{y})\|^2 &= \|\mathbf{x} - \mathbf{a}\|^2 + \|g(\mathbf{x}) - \mathbf{y}\|^2\\ &\le \|\mathbf{x} - \mathbf{a}\|^2 + (\|f(\mathbf{x}) - \mathbf{y}\| + \|g(\mathbf{x}) - f(\mathbf{x})\|)^2\\ &\le \|\mathbf{x} - \mathbf{a}\|^2 + (\|f(\mathbf{x}) - \mathbf{y}\| + \varepsilon\, r)^2\\ &= \|\mathbf{x} - \mathbf{a}\|^2 + \|f(\mathbf{x}) - \mathbf{y}\|^2 + 2\varepsilon\, r \|f(\mathbf{x}) - \mathbf{y}\| + \varepsilon^2 r^2\\ &\le r^2 + 2\varepsilon\, r^2 + \varepsilon^2 r^2 = (1+\varepsilon)^2 r^2.\end{aligned}$$

This shows that $\mathbf{y} \in B(\mathbf{a},\, (1+\varepsilon)r)$. Hence $A(\mathbf{a},\, r) \subset B(\mathbf{a},\, (1+\varepsilon)r)$ whenever $0 < r < \delta$. By the symmetry between these sets, we also have

$$B(\mathbf{a},\, (1-\varepsilon)r) \subset A(\mathbf{a},\, (1+\varepsilon)(1-\varepsilon)r) \subset A(\mathbf{a},\, r)$$

whenever $0 < r < \delta$. Then the conclusion follows. □

Lemma 9.6.16 *Let $T \subset Z = X \times Y$ be the graph of a linear function $F : X \to Y$, where Y is an ℓ-dimensional space. For each $\mathbf{a} \in X$ and $r > 0$, let*

$$B(\mathbf{a},\, r) = \{\, \mathbf{y} \in Y \mid \exists\, \mathbf{x} \in X,\ \|(\mathbf{x},\, F\mathbf{x}) - (\mathbf{a},\, \mathbf{y})\|_Z < r \,\}.$$

Then $v_Y(B(\mathbf{a},\, r)) = r^\ell\, S_\ell\, \rho(R)^{-1}$. Here S_ℓ is the volume of the unit ball in Y, $R = P(T,\, X) : T \to X$ is the orthogonal projection of T to X, and $\rho(R)$ is the volume multiplier of R.

Proof. We see that $\mathbf{y} \in B(\mathbf{a},\, r)$ if and only if there is a $\mathbf{t} \in T$ such that $\|(\mathbf{a},\, \mathbf{y}) - \mathbf{t}\| < r$. Let $N = T^\perp$ and let $S : Y \to N$ be the orthogonal projection of Y to N. Then, by the properties of orthogonal projections,

$$\|S(\mathbf{a},\, \mathbf{y})\| = \min_{\mathbf{t} \in T} \|(\mathbf{a},\, \mathbf{y}) - \mathbf{t}\|.$$

Therefore $(\mathbf{a},\, \mathbf{y}) \in B(\mathbf{a},\, r)$ if and only if $S(\mathbf{a},\, \mathbf{y}) \in B_r(\mathbf{0})$. Here $B_r(\mathbf{0}) \subset N$ is the ball of radius r in N about the origin of N. Hence

$$\rho(S) v_Y(B(\mathbf{a},\, r)) = v_N(B_r(\mathbf{0})) = r^l S_l.$$

Then the proof is concluded by $\rho(S) = \rho(R)$, as observed in Remarks 9.6.8. □

Remarks 9.6.17 Review of assumptions and notation. Let X and Y be two Euclidean spaces with $\dim X = k$ and $\dim Y = \ell$. Let A be an open set in X and $f : A \to Y$ a $\mathcal{C}^1$ function. Then $\mathbf{y} = f(\mathbf{x})$ is an explicit equation for the manifold

$$M = \{\, (\mathbf{x},\, \mathbf{y}) \in Z = X \times Y \mid \mathbf{x} \in A,\ \mathbf{y} = f(\mathbf{x}) \,\}. \tag{9.94}$$

A parametric equation for M is $\varphi : A \to Z$, with $\varphi(\mathbf{x}) = (\mathbf{x},\, f(\mathbf{x})) \in Z$, $\mathbf{a} \in A$. At the point $(\mathbf{x},\, f(\mathbf{x})) \in M$, the linear tangent space of M is $T(\mathbf{x})$. Also, $\rho(R(\mathbf{x}))$ is the volume multiplier of the orthogonal projection $R(\mathbf{x}) : T(\mathbf{x}) \to X$. Finally, $L = \varphi(E) \subset M$, where E is a Jordan set such that $\overline{E} \subset A$.

Lemma 9.6.18 *Let P, Q, R, and E be Jordan sets in X such that*

$$\overline{P} \subset E^o \subset \overline{E} \subset Q^o \subset \overline{Q} \subset R^o \subset \overline{R} \subset A. \tag{9.95}$$

Let $L = \varphi(E)$. *Then for each* $\varepsilon > 0$ *there is a* $\delta > 0$ *such that if* $0 < r < \delta$, *then*

$$(1-\varepsilon)^{\ell} \int_P \rho(R(\mathbf{x}))^{-1}\, d\mathbf{x} \le r^{-\ell} S_\ell^{-1} v_Z(L_r) \le (1+\varepsilon)^{\ell} \int_Q \rho(R(\mathbf{x}))^{-1}\, d\mathbf{x}. \quad (9.96)$$

Proof. First, find an $r_0 > 0$ such that if $0 < r < r_0$, then

$$P_r \subset E \subset E_r \subset Q \subset Q_r \subset \overline{R}. \quad (9.97)$$

This can be done because of the assumptions in (9.95). Here, as implicit in (9.97), all enlargements are in X. Now use the uniform continuity of $f' : A \to L(X, Y)$ on the compact set $\overline{R}$ to find a $\delta > 0$ such that $0 < \delta < r_0$ and such that

$$\|f'(\mathbf{u}) - f'(\mathbf{v})\|_{L(X,Y)} < \varepsilon \text{ whenever } \mathbf{u},\, \mathbf{v} \in \overline{R} \text{ and } \|\mathbf{u} - \mathbf{v}\|_X < \delta. \quad (9.98)$$

We claim that if $\mathbf{a} \in Q$ and if $0 < r < \delta$, then

$$\|f(\mathbf{x}) - f(\mathbf{a}) - f'(\mathbf{a})(\mathbf{x} - \mathbf{a})\| \le \varepsilon \|\mathbf{x} - \mathbf{a}\| \quad (9.99)$$

whenever $\|\mathbf{x} - \mathbf{a}\| < \delta$. This follows from the mean value theorem applied to $\vartheta(\mathbf{x}) = f(\mathbf{x}) - f'(\mathbf{a})\mathbf{x}$. Since $\vartheta'(\mathbf{x}) = f'(\mathbf{x}) - f'(\mathbf{a})$, we see that $\|\vartheta'(\mathbf{x})\| < \varepsilon$ whenever $\|\mathbf{x} - \mathbf{a}\| < \delta$. Therefore

$$\|f(\mathbf{x}) - f(\mathbf{a}) - f'(\mathbf{a})(\mathbf{x} - \mathbf{a})\| = \|\vartheta(\mathbf{x}) - \vartheta(\mathbf{a})\| \le \varepsilon \|\mathbf{x} - \mathbf{a}\| \quad (9.100)$$

whenever $\|\mathbf{x} - \mathbf{a}\| < \delta$. Now at every point $\mathbf{a} \in H$, apply Lemma 9.6.15 with $f(\mathbf{x})$ and $g(\mathbf{x}) = f(\mathbf{a}) + f'(\mathbf{a})(\mathbf{x} - \mathbf{a})$. We obtain

$$B(\mathbf{a},\, (1-\varepsilon)r) \subset A(\mathbf{a},\, r) \subset B(\mathbf{a},\, (1+\varepsilon)r) \quad (9.101)$$

whenever $\mathbf{a} \in H$ and $0 < r < \delta$. Also, $v_Y(B(\mathbf{a},\, r)) = r^{\ell} S_\ell \rho(R(\mathbf{a}))^{-1}$ by Lemma 9.6.16. Hence

$$(1-\varepsilon)^{\ell} r^{\ell} S_\ell \rho(R(\mathbf{a}))^{-1} \le v_Y(A(\mathbf{a},\, r)) \le (1+\varepsilon)^{\ell} r^{\ell} S_\ell \rho(R(\mathbf{a}))^{-1} \quad (9.102)$$

whenever $\mathbf{a} \in Q$ and $0 < r < \delta$. Now if $0 < r < \delta$, then we see that

$$C(L_r,\, \mathbf{a}) = A(\mathbf{a},\, r) \text{ if } \mathbf{a} \in P, \quad (9.103)$$

$$C(L_r,\, \mathbf{a}) \subset A(\mathbf{a},\, r) \text{ if } \mathbf{a} \in Q, \text{ and} \quad (9.104)$$

$$C(L_r,\, \mathbf{a}) = \emptyset \text{ if } \mathbf{a} \notin Q. \quad (9.105)$$

We have $v_Z(L_r) = \int_A v_Y(C(L_r,\, \mathbf{x}))\, d\mathbf{x}$ by Fubini's theorem. Hence

$$\int_P v_Y(A(\mathbf{x},\, r))\, d\mathbf{x} \le v_Z(L_r) \le \int_Q v_Y(A(\mathbf{x},\, r))\, d\mathbf{x}. \quad (9.106)$$

Then the proof follows by the estimates in (9.102). $\square$

Proof of Theorem 9.6.12. Let P, Q, and R be as in Lemma 9.6.18 above. We have

$$\sigma(L) = \int_E \rho(R(\mathbf{x}))^{-1}\, d\mathbf{x},$$

as observed in Remarks 9.6.8. Hence

$$\int_P \rho(R(\mathbf{x}))^{-1}\, d\mathbf{x} \leq \sigma(L) \leq \int_Q \rho(R(\mathbf{x}))^{-1}\, d\mathbf{x}. \tag{9.107}$$

The difference between these two integrals can be made arbitrarily small. In fact, the continuous function $\rho(R(\,\cdot\,))^{-1} : A \to \mathbb{R}$ is bounded on the compact set $\overline{R} \subset A$. If M is an upper bound for $\rho(R(\,\cdot\,))^{-1}$, then

$$0 \leq \int_Q \rho(R(\mathbf{x}))^{-1}\, d\mathbf{x} - \int_P \rho(R(\mathbf{x}))^{-1}\, d\mathbf{x} \leq M v_X(Q \setminus P). \tag{9.108}$$

Given any $\xi > 0$, we can choose the sets P and Q in Lemma 9.6.18 to make $v_X(Q \setminus P) < \xi$. Then the proof follows by comparing the estimates in (9.96) and (9.107) above. $\square$

Example 9.6.19 Surface area of a sphere. Let $\Sigma_n(R)$ be the surface of the sphere $B_R(\mathbf{0}) \subset \mathbb{R}^n$. We will show that it has a positive $(n-1)$-dimensional geometric content. The enlargements of $\Sigma_n(R)$ are

$$\begin{aligned} \Sigma_n(R)_r &= B_{R+r}(\mathbf{0}) \setminus B_{R-r}(\mathbf{0}). \text{ Hence} && (9.109)\\ v^n(\Sigma_n(R)) &= v^n(B_{R+r}(\mathbf{0})) - v^n(B_{R-r}(\mathbf{0})) && (9.110)\\ &= ((R+r)^n - (R-r)^n)S_n. && (9.111) \end{aligned}$$

Since $\ell = n - (n-1) = 1$ in this case and since $S_1 = 2$, we have

$$\tau_{(n-1)}(\Sigma_n(R)) = \lim_{r \to 0+} \frac{(R+r)^n - (R-r)^n}{2r} S_n = nR^{n-1}S_n. \tag{9.112}$$

Hence $\tau_1(\Sigma_2(R)) = 2\pi R$ and $\tau_2(\Sigma_3(R)) = 4\pi R^2$, since $S_2 = \pi$ and $S_3 = (4/3)\pi$.

CHAPTER 10

STOKES' THEOREM

Stokes' theorem generalizes the fundamental theorem of calculus to functions of several variables. In this chapter, we prove this major result in two different ways. The first way is via direct generalization of the fundamental theorem of calculus for functions of one variable. This approach leads to a proof that is concise but difficult to motivate. Our second approach employs the concept of flows. This second method arguably comes closer than the first to replicating the intuitions and ideas behind the original proofs of Stokes' theorem. Although the proof is easier to motivate, it is somewhat lengthy because it requires background preparation pertaining to flows.

The first step in both proofs is to establish Stokes' theorem for a special case. Indeed, the two proofs differ only at this first step. We shall refer to this special case as the *basic Stokes' theorem*. The general case is obtained by the application of two different tools. The first is to pass from special regions to more general regions by using diffeomorphisms. The second is to move from local results to global results by employing the partitions of unity. Both of these steps are fairly routine, and they operate in the same manner in all applications.

Analysis in Vector Spaces.
By M. A. Akcoglu, P. F. A. Bartha and D. M. Ha

Stokes' theorem is stated in terms of a vector field and the "divergence" of that vector field. These are, respectively, vector-valued and real-valued functions. In determining how these functions change under diffeomorphisms, the most fruitful strategy is to represent them as tensor fields and use the pullbacks of these fields. The basic idea of this strategy was developed in chapter 9.

Throughout this chapter, inner products are denoted as dot products, as is customary in discussions of Stokes' theorem.

10.1 BASIC STOKES' THEOREM

Let Z be a Euclidean space. The basic Stokes' theorem is stated for a $\mathcal{C}^1$ vector field $\mathbf{f} : G \to Z$ that is defined on an open set $G \subset Z$ and has a compact support $K \subset G$. Recall that K contains the closure of the set $\{ \mathbf{z} \mid \mathbf{f}(\mathbf{z}) \neq \mathbf{0} \}$. We may assume that K is a Jordan set. Also, if one defines $\mathbf{f}(\mathbf{z}) = \mathbf{0}$ for $\mathbf{z} \notin G$, then the extended function is still a $\mathcal{C}^1$ function. Hence we may assume, without loss of generality, that the vector field is defined on the whole space Z.

Divergence of a Vector Field

Definition 10.1.1 Divergence of a vector field. Let $\mathbf{f} : Z \to Z$ be a $\mathcal{C}^1$ vector field. Its *divergence* $\operatorname{div} \mathbf{f} : Z \to \mathbb{R}$ is defined as

$$(\operatorname{div} \mathbf{f})(\mathbf{z}) = \operatorname{Tr} \mathbf{f}'(\mathbf{z}), \quad \mathbf{z} \in Z. \tag{10.1}$$

Here $\mathbf{f}'(\mathbf{z}) : Z \to Z$ is a linear transformation and $\operatorname{Tr} \mathbf{f}'(\mathbf{z})$ is its *trace*.

Remarks 10.1.2 Trace of a linear transformation. Suppose that $(\mathbf{e}_1, \ldots, \mathbf{e}_n)$ is an orthonormal basis of Z and φ is a Euclidean determinant with $\varphi(\mathbf{e}_1, \ldots, \mathbf{e}_n) = 1$. For any linear transformation $T \in L(Z, Z)$,

$$\operatorname{Tr} T = \varphi(T\mathbf{e}_1, \ldots, \mathbf{e}_n) + \cdots + \varphi(\mathbf{e}_1, \ldots, T\mathbf{e}_n) \tag{10.2}$$

(See Definition C.7.3 in Appendix C.)

Note that $\operatorname{Tr} : L(Z, Z) \to \mathbb{R}$ is a linear function. It follows that divergence is a linear operation on vector fields:

$$\operatorname{div} (\alpha \mathbf{f} + \beta \mathbf{g}) = \alpha \operatorname{div} \mathbf{f} + \beta \operatorname{div} \mathbf{g}. \tag{10.3}$$

Another basic property of the trace is that

$$\operatorname{Tr} T = \lim_{t \to 0} \frac{\det(I + tT) - 1}{t}, \tag{10.4}$$

where $I \in L(Z, Z)$ is the identity. This result is obtained in Theorem C.7.4.

Lemma 10.1.3 *Suppose* $T\mathbf{z} = S(\mathbf{z})\,\mathbf{p}$, *where* $S : Z \to \mathbb{R}$ *is a linear transformation and* $\mathbf{p} \in Z$ *is a fixed point. Then* $T : Z \to Z$ *is linear and* $\operatorname{Tr} T = S(\mathbf{p})$.

Proof. The linearity of T is obvious. Since $S : Z \to \mathbb{R}$ is linear, there is a fixed $\mathbf{a} \in Z$ such that $S(\mathbf{z}) = \mathbf{a} \cdot \mathbf{z}$ for all $\mathbf{z} \in Z$. Let $(\mathbf{e}_1, \ldots, \mathbf{e}_n)$ be an orthonormal basis such that $\mathbf{a} = \alpha\mathbf{e}_1$. Then $T\mathbf{e}_1 = \alpha\mathbf{p}$ and $T\mathbf{e}_i = \mathbf{0}$ if $i \neq 1$. Hence

$$
\begin{aligned}
\operatorname{Tr} T &= \varphi(\alpha\mathbf{p}, \mathbf{e}_2, \ldots, \mathbf{e}_n) && (10.5)\\
&= \varphi(\alpha(\mathbf{e}_1 \cdot \mathbf{p})\mathbf{e}_1, \mathbf{e}_2, \ldots, \mathbf{e}_n) && (10.6)\\
&= \alpha\mathbf{e}_1 \cdot \mathbf{p} = \mathbf{a} \cdot \mathbf{p} = S(\mathbf{p}). && (10.7)
\end{aligned}
$$

Here (10.5) follows from the definition of trace in (10.3) and the fact that $T\mathbf{e}_i = \mathbf{0}$ if $i \neq 1$. To obtain (10.6), write

$$\mathbf{p} = (\mathbf{p} \cdot \mathbf{e}_1)\mathbf{e}_1 + \mathbf{p}_2,$$

where $\mathbf{p}_2$ lies in the space spanned by $(\mathbf{e}_2, \ldots, \mathbf{e}_n)$. Then use the multilinearity of φ and the fact that $\varphi(\mathbf{p}_2, \mathbf{e}_2, \ldots, \mathbf{e}_n) = 0$. For (10.7), simply note that $\varphi(\mathbf{e}_1, \ldots, \mathbf{e}_n) = 1$. □

Lemma 10.1.4 *Suppose* $\mathbf{f} = f\,\mathbf{p}$, *where* $f : Z \to \mathbb{R}$ *is a* $\mathcal{C}^1$ *function and* $\mathbf{p} \in Z$ *is fixed. Then for any* $\mathbf{u} \in Z$, $\operatorname{div} \mathbf{f}(\mathbf{u}) = f'(\mathbf{u})\,\mathbf{p} = f'(\mathbf{u}; \mathbf{p})$, *the directional derivative of* f *at* $\mathbf{u}$ *along* $\mathbf{p}$.

Proof. Let $\mathbf{u} \in Z$. Then $S = f'(\mathbf{u}) : Z \to \mathbb{R}$ is a linear transformation and $\mathbf{f}'(\mathbf{u})\,\mathbf{z} = (f'(\mathbf{u})\,\mathbf{z})\mathbf{p} = S(\mathbf{z})\mathbf{p}$. Hence the result follows from Lemma 10.1.3. □

Corollary 10.1.5 *Let* $(\mathbf{e}_1, \ldots, \mathbf{e}_n)$ *be an orthonormal basis for* Z *with the corresponding coordinate functions* $(z_1, \ldots, z_n)$. *If* $\mathbf{f}$ *is expressed as*

$$\mathbf{f} = \sum\nolimits_i f_i \mathbf{e}_i, \tag{10.8}$$

then

$$\operatorname{div} \mathbf{f}(\mathbf{z}) = \sum\nolimits_i \partial f_i / \partial z_i. \tag{10.9}$$

Equation (10.9) is the most common way to express $\operatorname{div} \mathbf{f}$.

Proof. If $\mathbf{f} = f_i\,\mathbf{e}_i$, then the result follows from Lemma 10.1.4. The general case follows from the linearity of the divergence operator, as stated in (10.3). □

Basic Stokes' Theorem: direct version

Definition 10.1.6 Two sides of a plane. Let $\mathbf{n} \in Z$ be a unit vector. Let U be the plane $\mathbf{z} \cdot \mathbf{n} = 0$. The sets

$$A = \{\, \mathbf{z} \mid \mathbf{z} \cdot \mathbf{n} < 0 \,\} \text{ and } B = \{\, \mathbf{z} \mid 0 < \mathbf{z} \cdot \mathbf{n} \,\} \tag{10.10}$$

are called, respectively, *the lower and the upper sides of* U. The vector $\mathbf{n}$ is called the *outer unit normal* vector of U.

Notations 10.1.7 Two types of integrals. The following arguments involve integrals over two Euclidean spaces. There are integrals in Z, and integrals on planes like U. (If $\dim Z = n$, then the two types are essentially integrals on $\mathbb{R}^n$ and integrals on $\mathbb{R}^{n-1}$.) These integrals will be denoted by expressions of the form $\int_Z f(\mathbf{z})\, d\mathbf{z}$ and $\int_U f(\mathbf{u})\, d\mathbf{u}$. The second integral involves only values $f(\mathbf{z})$ for $\mathbf{z} = \mathbf{u} \in U$.

Theorem 10.1.8 Basic Stokes' theorem. *Let $\mathbf{f} : Z \to Z$ be a $\mathcal{C}^1$ vector field of compact support. If A and U are as in Definition 10.1.7, then*

$$\int_A \operatorname{div} \mathbf{f}(\mathbf{z})\, d\mathbf{z} = \int_U \mathbf{n} \cdot \mathbf{f}(\mathbf{u})\, d\mathbf{u}. \tag{10.11}$$

Proof. Complete $\mathbf{n} = \mathbf{e}_1$ to an orthonormal basis $(\mathbf{e}_1, \ldots, \mathbf{e}_n)$. Let the corresponding coordinate functions be $(z_1, \ldots, z_n)$. Since $\mathbf{f}$ has compact support, we can find an $M > 0$ such that if $\mathbf{f}(\mathbf{z}) \neq \mathbf{0}$, then $-M < z_i < M$ for all i. Both sides of (10.11) are linear in $\mathbf{f}$. Hence it is enough to prove the result for $\mathbf{f} = f_i \mathbf{e}_i$. But we have to treat the cases $i = 1$ and $i \neq 1$ separately. We see that (10.11) states

$$\int_A \frac{\partial f_i}{\partial z_i}(\mathbf{z})\, d\mathbf{z} = 0 \quad \text{if } i \neq 1 \text{ and} \tag{10.12}$$

$$\int_A \frac{\partial f_1}{\partial z_1}(\mathbf{z})\, d\mathbf{z} = \int_U f_1(\mathbf{u})\, d\mathbf{u}. \tag{10.13}$$

We evaluate the integrals on the left-hand side by Fubini's Theorem. First, let $i \neq 1$ be fixed. Decompose Z as $Z = V \times X$, where X is the one-dimensional space spanned by $\mathbf{e}_i$ and $V = X^{\perp}$. Write each point in Z as $\mathbf{z} = \mathbf{v} + x\mathbf{e}_i = (\mathbf{v}, x)$, with $\mathbf{v} \in V$ and $x = z_i$. Then we see that

$$\int_A \frac{\partial f_i}{\partial z_i}(\mathbf{z})\, d\mathbf{z} = \int_{V \cap A} \int_{\mathbb{R}} \frac{\partial f_i}{\partial x}(\mathbf{v}, x)\, dx\, d\mathbf{v} \tag{10.14}$$

$$= \int_{V \cap A} \int_{-M}^{M} \frac{\partial f_i}{\partial x}(\mathbf{v}, x)\, dx\, d\mathbf{v} \tag{10.15}$$

$$= \int_{V \cap A} (f_i(\mathbf{v}, M) - f_i(\mathbf{v}, -M))\, d\mathbf{v} \tag{10.16}$$

$$= \int_{V \cap A} (0 - 0)\, d\mathbf{v} = 0. \tag{10.17}$$

This proves the result for $i \neq 1$.

For $i = 1$, decompose Z as $U \times Y$ where Y is the one-dimensional space spanned by $\mathbf{e}_1 = \mathbf{n}$. Denote the points in Z as $\mathbf{z} = \mathbf{u} + y\mathbf{n} = (\mathbf{u},\, y)$ with $\mathbf{u} \in U$ and $y = z_1$. Then we see that

$$\int_A \frac{\partial f_1}{\partial z_1}(\mathbf{z})\, d\mathbf{z} = \int_U \int_{Y \cap A} \frac{\partial f_1}{\partial y}(\mathbf{u},\, y)\, dy\, d\mathbf{u} \tag{10.18}$$

$$= \int_U \int_{-M}^{0} \frac{\partial f_1}{\partial y}(\mathbf{u},\, y)\, dy\, d\mathbf{u} \tag{10.19}$$

$$= \int_U (f_1(\mathbf{u},\, 0) - f_1(\mathbf{u},\, -M))\, d\mathbf{u} \tag{10.20}$$

$$= \int_U (f_1(\mathbf{u},\, 0) - 0)\, d\mathbf{u} \tag{10.21}$$

$$= \int_U f_1(\mathbf{u})\, d\mathbf{u}. \tag{10.22}$$

For the last step, the point $\mathbf{z} = (\mathbf{u},\, 0)$ is expressed as $\mathbf{z} = \mathbf{u} + 0\mathbf{n} = \mathbf{u}$. This completes the proof of the theorem in the general case. □

Problems

10.1 On the basis of Corollary 10.1.5, we use $\nabla \cdot \mathbf{f}$ as another notation for

$$\operatorname{div} \mathbf{f} = \operatorname{div} \sum_i f_i \mathbf{e}_i = \sum_i \frac{\partial f_i}{\partial z_i}.$$

Here $(\mathbf{e}_1,\, \ldots,\, \mathbf{e}_n)$ is an orthonormal basis for Z and z_is are the corresponding coordinate functions. If $\mathbf{f} = \nabla F$, then show that $\operatorname{div} \mathbf{f} = \sum_i (\partial^2 F / \partial z_i^2)$. This is also expressed as

$$\operatorname{div} \mathbf{f} = \nabla \cdot \nabla F = \nabla^2 F = \Delta F.$$

The expression $\nabla^2 F = \Delta F$ is called the *Laplacian* of F. A function F is called a *harmonic function* if $\Delta F = 0$.

10.2 Show that $\nabla \cdot (F\, \nabla G) = (\nabla F) \cdot (\nabla G) + F \nabla^2 G$.

10.3 Let Z be an n-dimensional Euclidean space. Let $r = \|\mathbf{z}\|$. Show that $F(\mathbf{z}) = r^{2-n}$ is a harmonic function (Problem 10.1) in $Z \setminus \{\, \mathbf{0} \,\}$. Also, if $n = 2$, then show that $F(\mathbf{z}) = \log r$ is a harmonic function in $Z \setminus \{\, \mathbf{0} \,\}$.

10.4 Let $B = \{\, (x,\, y) \mid x \leq 0,\ y = 0 \,\}$ and let $A = \mathbb{R}^2 \setminus B$. Define

$$\vartheta : A \to (-\pi,\, \pi) \text{ by } \cos \vartheta = x(x^2 + y^2)^{-1/2} \text{ and } \sin \vartheta = y(x^2 + y^2)^{-1/2}.$$

Show that ϑ is a harmonic function. See also Problem 9.36.

10.2 FLOWS

Definition 10.2.1 Flows of compact support. Let Z be a Euclidean space. Let $J \subset \mathbb{R}$ be an open interval containing $0 \in \mathbb{R}$. A *flow (of compact support)* is a mapping $F : (Z \times J) \to Z$ that satisfies the following conditions.

(1) There is a compact set $K \subset Z$, called *a support* of F, such that

$$F(\mathbf{z},\, t) = \mathbf{z} \text{ for all } (\mathbf{z},\, t) \in K^c \times J = (Z \setminus K) \times J.$$

(2) $F(\mathbf{z},\, 0) = \mathbf{z}$ for all $\mathbf{z} \in Z$.

(3) $F : (Z \times J) \to Z$ is a $\mathcal{C}^1$ function.

(4) For each fixed $t \in J$, the mapping

$$F(\,\cdot\,,\, t) : Z \to Z \text{ is a diffeomorphism of } Z \textit{ onto } Z.$$

A flow can be considered as specifying the motion of a set of particles, such as the particles in a fluid (think of a swirling pond). The particle that was at $\mathbf{z}$ at initial time $t = 0$ moves to the position $F(\mathbf{z},\, t)$ at time $t \in J$. The particles outside the compact set K do not move at all. This may seem like an important restriction, but it is assumed only for technical convenience. See 10.2.7 for further remarks.

Notations 10.2.2 Derivatives of flows. Velocity fields. The derivative of a flow $F : (Z \times J) \to Z$ at a point $(\mathbf{a},\, \alpha) \in (Z \times J)$ is a linear transformation $F'(\mathbf{a},\, \alpha) : (Z \times \mathbb{R}) \to Z$. We express the application of this linear transformation to a vector $(\mathbf{z},\, t) \in (Z \times \mathbb{R})$ as

$$F'(\mathbf{a},\, \alpha)(\mathbf{z},\, t) = (D_Z F)(\mathbf{a},\, \alpha)\mathbf{z} + t\,(\partial F/\partial t)(\mathbf{a},\, \alpha). \tag{10.23}$$

Here $D_Z F : (Z \times J) \to L(Z,\, Z)$ is the derivative of F restricted to Z and $(\partial F/\partial t) : (Z \times J) \to Z$ is the partial derivative of F with respect to $t \in J$. Hence $D_Z F$ can be considered as the space derivative of F and $(\partial F/\partial t)$ the time derivative of F. More explicitly, $D_Z F(\mathbf{a},\, \alpha) : Z \to Z$ is the unique linear transformation that satisfies

$$\lim_{\mathbf{z} \to \mathbf{0}} \frac{\|F(\mathbf{a} + \mathbf{z},\, \alpha) - F(\mathbf{a},\, \alpha) - D_Z F(\mathbf{a},\, \alpha)\mathbf{z}\|}{\|\mathbf{z}\|} = 0, \tag{10.24}$$

and $(\partial F/\partial t)(\mathbf{a},\, \alpha) = (\partial F)/(\partial t)(\mathbf{a},\, \alpha) \in Z$ is the vector defined by

$$\frac{\partial F}{\partial t}(\mathbf{a},\, \alpha) = \lim_{t \to 0} \frac{F(\mathbf{a},\, \alpha + t) - F(\mathbf{a},\, \alpha)}{t}. \tag{10.25}$$

We call $(\partial F/\partial t)(\,\cdot\,, t) : Z \to Z$ the *velocity field* of the flow at the time $t \in J$. If a particle is at the point $\mathbf{z} \in Z$ at initial time $t = 0$, then at a general time $t \in J$ it is at the point $F(\mathbf{z}, t) \in Z$ with the velocity of $(\partial F/\partial t)(\mathbf{z}, t) \in Z$.

Definition 10.2.3 The initial velocity field. Velocities at the initial time $t = 0$ will be important. We let $\mathbf{f}(\mathbf{z}) = (\partial F/\partial t)(\mathbf{z}, 0)$ and define $\mathbf{f} : Z \to Z$ as the *initial velocity field*.

Smooth Flows

One usually calls a function a smooth function if it has derivatives of all orders. For our arguments below, we need much less.

Definition 10.2.4 Smooth flows. A function $F : (Z \times J) \to Z$ is said to satisfy the *smoothness condition* if the second-order mixed partial derivatives

$$\partial(D_Z F)/\partial t \text{ and } D_Z(\partial F/\partial t) \tag{10.26}$$

exist, are equal to each other, and are continuous functions $Z \times J \to L(Z, Z)$. A function $F : (Z \times J) \to Z$ is said to be a *smooth flow* if it is a flow and it satisfies the smoothness condition. From now on, all flows are assumed to be smooth flows even if this is not explicitly stated.

Smooth Flows with Given Initial Velocities

If $F : Z \times J \to Z$ is a smooth flow, then its initial velocity field

$$\mathbf{f}(\,\cdot\,) = (\partial F/\partial t)(\,\cdot\,, 0) : Z \to Z \tag{10.27}$$

is a $\mathcal{C}^1$ function. In fact, by the smoothness assumption,

$$D_Z\mathbf{f}(\,\cdot\,) = D_Z(\partial F/\partial t)(\,\cdot\,, 0) : Z \to L(Z, Z) \tag{10.28}$$

exists and is continuous. Theorem 10.2.5 shows that the converse is also true.

Theorem 10.2.5 Construction of a smooth flow. *Let* $\mathbf{f} : Z \to Z$ *be a* $\mathcal{C}^1$ *function of compact support* $K \subset Z$. *Then there is an* $r > 0$ *such that* $F(\mathbf{z}, t) = \mathbf{z} + t\mathbf{f}(\mathbf{z})$ *is a smooth flow* $F : Z \times (-r, r) \to Z$. *The support of this flow is* K *and its initial velocity field is* $\mathbf{f} : Z \to Z$.

Proof. We have $F(\mathbf{z}, 0) = \mathbf{z}$ for all $\mathbf{z} \in Z$ and $F(\mathbf{z}, t) = \mathbf{z}$ for all $t \in \mathbb{R}$ and for all $\mathbf{z} \notin K$. We first show that F satisfies the smoothness condition of 10.2.4. In fact,

$$D_Z F(\mathbf{z}, t) = I + tD_Z\mathbf{f}(\mathbf{z}) \text{ and } (\partial F/\partial t)(\mathbf{z}, t) = \mathbf{f}(\mathbf{z}) \tag{10.29}$$

show that $(\partial D_Z F/\partial t)(\mathbf{z}, t) = D_Z(\partial F/\partial t)(\mathbf{z}, t) = D_Z\mathbf{f}(\mathbf{z})$. Also, this is a continuous function $Z \to L(Z, Z)$, as required.

Now we need to find an $r > 0$ such that $F : Z \times (-r, r) \to Z$ is a flow. (Combining this with the preceding paragraph gives us our conclusion that F is a smooth flow). This means that we need to show that for each fixed $t \in (-r, r)$, $F(\cdot, t) : Z \to Z$ is a diffeomorphism of Z onto Z.

The function $D_Z\mathbf{f} : Z \to L(Z, Z)$ is a continuous function of compact support. Hence there is an $M > 0$ such that $\|D_Z\mathbf{f}(\mathbf{z})\|_{L(Z, Z)} < M$ for all $\mathbf{z} \in Z$. Let $r = 1/(2M)$ and $|t| < r$. If $F(\mathbf{a}, t) = F(\mathbf{b}, t)$, then $\mathbf{a} + t\mathbf{f}(\mathbf{a}) = \mathbf{b} + t\mathbf{f}(\mathbf{b})$ and

$$\|\mathbf{b} - \mathbf{a}\| = |t|\, \|\mathbf{f}(\mathbf{b}) - \mathbf{f}(\mathbf{a})\| \leq |t|\, M\|\mathbf{b} - \mathbf{a}\| \leq (1/2)\|\mathbf{b} - \mathbf{a}\| \tag{10.30}$$

Hence $\|\mathbf{b} - \mathbf{a}\| = 0$ and $\mathbf{a} = \mathbf{b}$, so $F(\,\cdot\,, t)$ is one-to-one. (The first inequality above follows from the mean value theorem.) Also $D_Z F(\mathbf{z}, t) = I_Z + tD_Z\mathbf{f}(\mathbf{z})$, where $I_Z : Z \to Z$ is the identity mapping. But $\|tD_Z\mathbf{f}(\mathbf{z})\|_{L(Z, Z)} \leq |t|\, M \leq (1/2)$, so $D_Z F(\mathbf{z}, t)$ is an invertible element in $L(Z, Z)$. This last part follows from Theorem 4.4.38.

Hence if $|t| < r$, then $F(\,\cdot\,, t) : Z \to Z$ is a one-to-one and $\mathcal{C}^1$ function with an invertible derivative at each point. We will show that it also maps Z onto Z. The inverse function theorem shows that $F(\,\cdot\,, t) : Z \to Z$ maps open sets to open sets. Therefore $R_t = F(Z, t)$ (i.e., the image of the entire space Z under $F(\,\cdot\,, t)$) is an open set, since Z is open.

We claim that R_t is also a closed set. Let $\mathbf{q}_n$ be a sequence in R_t converging to a point $\mathbf{q} \in Z$. Since $F(\,\cdot\,, t)$ is one-to-one, there is a unique sequence $\mathbf{p}_n \in Z$ such that $F(\mathbf{p}_n, t) = \mathbf{q}_n$. We see that $\mathbf{p}_n$ is a bounded sequence, since $F(\mathbf{z}, t) = \mathbf{z}$ for all $\mathbf{z}$ outside the compact set $K \subset Z$. Hence, by the Bolzano-Weierstrass theorem, it has a convergent sub-sequence. Without loss of generality, assume that $\mathbf{p}_n \to \mathbf{p}$. Then $\mathbf{q}_n = F(\mathbf{p}_n, t) \to F(\mathbf{p}, t)$, by the continuity of $F(\,\cdot\,, t)$. Since $\mathbf{q}_n \to \mathbf{q}$ we see that $\mathbf{q} = F(\mathbf{p}, t) \in R_t$. Hence R_t is closed, and is therefore both open and closed. Since R_t is not empty, we must have $R_t = Z$. □

Remarks 10.2.6 Non-uniqueness of flows. Differential equations. There is no uniqueness result for a flow with a given initial velocity field $\mathbf{f} : Z \to Z$. The particular flow in Theorem 10.2.5 above may just be the simplest flow to construct starting with a given initial velocity field. In the theory of differential equations, one is interested in particular flows that satisfy the additional requirement that

$$\frac{\partial F}{\partial t}(\mathbf{z}, t) = \mathbf{f}(F(\mathbf{z}, t)) \tag{10.31}$$

for all $\mathbf{z} \in Z$ and $t \in J$. Here is an intuitive characterization of this condition: the velocity of a particle at a time t depends only on the position $F(\mathbf{z}, t)$ that the particle has reached at that time. When a particle arrives at a point $\mathbf{z} \in Z$ at a certain time, its velocity is $\mathbf{f}(\mathbf{z})$ at that time. The existence and the uniqueness of flows that meet this condition are discussed in the theory of differential equations.

Remarks 10.2.7 Flows of compact support. We consider only flows of compact support. This is not an important restriction. We are interested in flows generated by velocity fields and in their behavior over bounded regions $E \subset Z$. If $\mathbf{f}$ is a velocity field without a compact support, then one can take a $\mathcal{C}^1$ function $\lambda : Z \to [0, 1]$ of compact support which is 1 on an open set G containing the closure of E. Then the velocity field $\lambda\mathbf{f}$ has compact support. Flows generated by $\mathbf{f}$ and by $\lambda\mathbf{f}$ have the same initial velocities on G. Hence the initial behavior of a flow on G can be understood in terms of a flow of compact support.

In applying this reasoning, it may be sufficient to take λ as a $\mathcal{C}^1$ function. In this case, the example in Lemma 6.1.6 can be used. If necessary, λ can be taken to be a $\mathcal{C}^\infty$ function, as obtained in Appendix D on partitions of unity.

Problems

10.5 Let Z be a Euclidean space. Let J be an open interval containing 0. A function $\Omega : Z \times J \to Z$ is called a *displacement function* if it has the following properties. (1) There is a compact set $K \subset Z$ such that $\Omega(\mathbf{z}, t) = \mathbf{0}$ for $t \in J$ and $\mathbf{z} \notin K$. (2) $\Omega(\mathbf{z}, 0) = \mathbf{0}$ for all $\mathbf{z} \in Z$. (3) $\Omega : (Z \times J) \to Z$ is a $\mathcal{C}^1$ function. (4) The second order mixed partial derivatives

$$\partial(D_Z\Omega)/\partial t \text{ and } D_Z(\partial\Omega/\partial t) \tag{10.32}$$

exist, are equal to each other, and are continuous functions $Z \times J \to L(Z, Z)$. Show that if $F : Z \times J \to Z$ is a smooth flow, then $\Omega(\mathbf{z}, t) = F(\mathbf{z}, t) - \mathbf{z}$ is a displacement function. Conversely, show that if $\Omega(\mathbf{z}, t)$ is a displacement function, then there is an $r > 0$ such that $(-r, r) \subset J$ and such that $F(\mathbf{z}, t) = \mathbf{z} + \Omega(\mathbf{z}, t)$ defines a smooth flow $Z \times (-r, r) \to Z$.

10.6 Let $A : Z \to Z$ be a linear transformation. Let $G(\mathbf{z}, t) = e^{tA}\mathbf{z}$, $t \in \mathbb{R}$, as defined in Example 7.3.13. Let $\lambda : Z \to [0, 1]$ be a $\mathcal{C}^\infty$ function such that $\lambda(\mathbf{z}) = 1$ for $\|\mathbf{z}\| \leq 1$ and $\lambda(\mathbf{z}) = 0$ for $\|\mathbf{z}\| \geq 2$. Show that there is an $r > 0$ such that

$$F(\mathbf{z}, t) = \lambda(\mathbf{z})G(\mathbf{z}, t) + (1 - \lambda(\mathbf{z}))\mathbf{z}$$

is a flow $F : Z \times (-r, r) \to Z$. What are the initial velocities $\mathbf{f}(\mathbf{z})$ for $\|\mathbf{z}\| < 1$?

10.7 Let (X, Y) be a coordinate system (Definition 3.1.42) in Z with the coordinate projections $P : Z \to X$ and $Q : Z \to Y$. If $F : Z \times J \to Z$ is a smooth flow, then show that there is an open interval I, $\mathbf{0} \in I \subset J$, such that

$$G(\mathbf{x},\, t) = PF(\mathbf{x},\, t) \text{ and } H(\mathbf{y},\, t) = QF(\mathbf{y},\, t)$$

define smooth flows $G : X \times I \to X$ and $H : Y \times I \to Y$. (Note that here we take $Z = X \oplus Y$ rather than $Z = X \times Y$, so that both $\mathbf{x}$ and $\mathbf{y}$ are in Z.)

10.8 Let $F : Z \times J \to Z$ be a smooth flow with the initial velocity field $\mathbf{f} : Z \to Z$. Then show that $\lim_{t\to 0}(1/t)(F(\mathbf{z},\, t) - \mathbf{z}) = \mathbf{f}(\mathbf{z})$ uniformly in $\mathbf{z} \in Z$. (*Hint.* Use Problem 10.9 below.)

10.9 Let W be a normed space. Let $\Gamma : Z \times J \to W$ be a function such that $H = (\partial\Gamma/\partial t) : Z \times J \to W$ exists and is continuous. Then show that given a compact set $K \subset Z$ and an $\varepsilon > 0$, there is a $\delta > 0$ such that

$$\|(\Gamma(\mathbf{z},\, t) - \Gamma(\mathbf{z},\, 0)) - tH(\mathbf{w},\, 0)\|_W \leq \varepsilon|t|$$

whenever $|t| + \|\mathbf{z} - \mathbf{w}\| < \delta$ and $\mathbf{z},\, \mathbf{w} \in K$. In particular, show that

$$\lim_{t\to 0}(1/t)\,(\Gamma(\mathbf{z},\, t) - \Gamma(\mathbf{z},\, 0)) = H(\mathbf{z},\, 0)$$

uniformly in $\mathbf{z}$ on any compact set $K \subset Z$.

10.10 Let $F : Z \times J \to Z$ be a smooth flow with the initial velocity field $\mathbf{f}$. Then show that $\lim_{t\to 0}(1/t)(D_Z F(\mathbf{z},\, t) - I_Z) = \mathbf{f}'(\mathbf{z})$ uniformly in $\mathbf{z} \in Z$.

10.11 Let $F : Z \times J \to Z$ be a smooth flow with the initial velocity field $\mathbf{f}$. Then show that $\lim_{t\to 0}(1/t)(\det D_Z F(\mathbf{z},\, t) - 1) = \operatorname{Tr} \mathbf{f}'(\mathbf{z})$ uniformly in $\mathbf{z} \in Z$.

10.12 Let $\varphi_t : Z \to Z$, $t \in J$, be a family of continuous functions. If $\lim_{t\to 0} \varphi_t(\mathbf{z}) = \varphi(\mathbf{z})$ uniformly in $\mathbf{z} \in Z$, then show that

$$\lim_{t\to 0} \int_E \varphi_t(\mathbf{z})\, d\mathbf{z} = \int_E \varphi(\mathbf{z})\, d\mathbf{z}$$

for any Jordan set E.

10.3 FLUX AND CHANGE OF VOLUME IN A FLOW

Definition 10.3.1 Volume of a set in a flow. Let $F : (Z \times J) \to Z$ be a flow. For $E \subset Z$ and $t \in J$, let $E^t = \{\, F(\mathbf{z},\, t) \mid \mathbf{z} \in E \,\} = F(E,\, t) \subset Z$ be the image of E under the diffeomorphism $F(\,\cdot\,,\, t) : Z \to Z$. We can think of E^t as the locations at t of all particles that started out in E at time 0. It is a snapshot of a moving ensemble.

Theorem 8.4.14 shows that if E is a Jordan set, then E^t is also a Jordan set. In general, the volume of E^t will be different at different times. Our second route to Stokes' theorem begins with a result about the rate of change in this volume. Actually, we shall only need this rate of change at the time $t = 0$, and only for the special type of flow constructed in Theorem 10.2.5. (The result is true for any smooth flow, but the proof of this fact is left as an exercise.)

Theorem 10.3.2 Initial rate of change in the volume. *Let* $\mathbf{f} : Z \to Z$ *be a* $\mathcal{C}^1$ *vector field of compact support and suppose that* $F(\mathbf{z}, t) = \mathbf{z} + t\mathbf{f}(\mathbf{z})$. *Let* $E \subset Z$ *be a Jordan set and* $E^t = F(E, t)$. *Then*

$$\lim_{t \to 0} \frac{v(E^t) - v(E)}{t} = \int_E \operatorname{div} \mathbf{f}(\mathbf{z})\, d\mathbf{z}, \tag{10.33}$$

which gives the initial rate of change in the volume of E.

Proof. Theorem 10.2.5 shows that there is an open interval J containing $0 \in \mathbb{R}$ such that the mapping $F(\cdot\,, t) : Z \to Z$ is a diffeomorphism for each fixed $t \in J$. If $t \in J$, then by the change of volumes theorem 8.4.14,

$$v(E^t) = \int_E \det D_Z F(\mathbf{z}, t)\, d\mathbf{z} \tag{10.34}$$

$$= \int_E \det(I + t\mathbf{f}'(\mathbf{z}))\, d\mathbf{z}. \tag{10.35}$$

Hence

$$v(E^t) - v(E) = \int_E (\det(I + t\mathbf{f}'(\mathbf{z})) - 1)\, d\mathbf{z}. \tag{10.36}$$

Now we know that $(\det(I + t\mathbf{f}'(\mathbf{z})) - 1)$ may be written as a polynomial $\sum_k A_k(\mathbf{z}) t^k$ in t. The coefficients $A_k(\mathbf{z})$ are polynomials in the partial derivatives of the components of $\mathbf{f}$. Therefore they are all continuous functions on Z. Also, as observed in 10.1.2,

$$A_1(\mathbf{z}) = \lim_{t \to 0} \frac{\det(I + t\mathbf{f}'(\mathbf{z})) - 1}{t} = \operatorname{div} \mathbf{f}(\mathbf{z}). \tag{10.37}$$

Then the proof of the theorem follows easily. □

Remarks 10.3.3 Divergence as the density of expansion. On the interpretation we have been developing, the divergence of a vector field gives the density of the initial rate of expansion for a flow with the initial velocity field $\mathbf{f} : Z \to Z$. The basis for this interpretation is Theorem 10.3.2. In fact, by this theorem,

$$\lim_{t\to 0}\frac{v(E^t)-v(E)}{t} = \int_E \operatorname{div}\mathbf{f}(\mathbf{z})\,d\mathbf{z}, \quad \text{and} \tag{10.38}$$

$$\lim_{t\to 0}\frac{1}{t}\left(\frac{v(E^t)}{v(E)}-1\right) = \frac{1}{v(E)}\int_E \operatorname{div}\mathbf{f}(\mathbf{z})\,d\mathbf{z}. \tag{10.39}$$

Flux of a Vector Field

Definition 10.3.4 Flux out of a Jordan set. Let $\mathbf{f} : Z \to Z$ be a $\mathcal{C}^1$ vector field of compact support. Let E be a Jordan set in Z. Then

$$\operatorname{flux}(\mathbf{f}, E) = \int_E \operatorname{div}\mathbf{f}(\mathbf{z})\,d\mathbf{z} \tag{10.40}$$

is called the *flux of* $\mathbf{f}$ *out of* E.

Flux is defined as a function of $\mathbf{f}$ and E, but in light of Theorem 10.3.2, we can also characterize it in terms of the initial rate of change in the volume of E for the flow associated with $\mathbf{f}$.

Theorem 10.3.5 Another expression for the flux. *Let* $\mathbf{f} : Z \to Z$ *be a* $\mathcal{C}^1$ *vector field with a compact support. Let* $F : Z \times J \to Z$ *be any smooth flow with the initial velocity field* $\mathbf{f}$. *Then*

$$\begin{aligned}\operatorname{flux}(\mathbf{f}, E) &= \lim_{t\to 0}(1/t)(v(E^t)-v(E)) && (10.41)\\ &= \lim_{t\to 0}(1/t)(v(E^t\setminus E)-v(E\setminus E^t)) && (10.42)\end{aligned}$$

where E *is a Jordan set and* $E^t = F(E, t)$, $t \in J$.

Proof. The first equality (10.41) follows from Theorem 10.3.2. Also,

$$\begin{aligned}E &= (E\cap E^t)\cup(E\setminus E^t) \quad \text{and} && (10.43)\\ E^t &= (E^t\cap E)\cup(E^t\setminus E), && (10.44)\end{aligned}$$

where both unions are the unions of disjoint sets. Then (10.42) follows from the additivity of volume. □

Flux and Boundaries

As our next step towards Stokes' theorem, we prove that the flux of $\mathbf{f}$ out of E, although defined as an integral over the whole set E, depends only on the values of $\mathbf{f}$

on the boundary of E. This follows from Lemma 10.3.7 below and from the linearity of the flux. We also show that $\text{flux}(\mathbf{f}, E)$ depends only on the part of E that is inside the support of $\mathbf{f}$. This is formulated in Lemma 10.3.10.

In what follows, $\mathbf{f} : Z \to Z$ is a $\mathcal{C}^1$ vector field with a compact support and $F : Z \times J \to Z$ is the flow defined as $F(\mathbf{z}, t) = \mathbf{z} + t\mathbf{f}(\mathbf{z})$.

Lemma 10.3.6 *If $\mathbf{f}(\mathbf{z}) \neq \mathbf{0}$, then also $\mathbf{f}(F(\mathbf{z}, t)) \neq \mathbf{0}$ for all $t \in J$.*

Proof. Suppose that $\mathbf{f}(\mathbf{w}) = \mathbf{0}$ for $\mathbf{w} = F(\mathbf{z}, t) = \mathbf{z} + t\mathbf{f}(\mathbf{z})$, where $t \in J$ and $t \neq 0$. Then we also have $\mathbf{w} = \mathbf{w} + t\mathbf{f}(\mathbf{w}) = F(\mathbf{w}, t)$. So $F(\mathbf{w}, t) = F(\mathbf{z}, t)$ with $\mathbf{w} \neq \mathbf{z}$. This is a contradiction, since $F(\,\cdot\,, t) : Z \to Z$ is one-to-one. □

Lemma 10.3.7 *If $\mathbf{f}(\mathbf{z}) = \mathbf{0}$ for all $\mathbf{z} \in \partial E$, then $\text{flux}(\mathbf{f}, E) = 0$. (If the initial velocity everywhere on the boundary of E is 0, then there is no flux out of E.)*

Proof. We will show that $E^t = F(E, t) = E$ for all $t \in J$. Then Theorem 10.3.5 implies that $\text{flux}(\mathbf{f}, E) = 0$.

Assume that $t \neq 0$ and that there is a $\mathbf{z} \in E^t \setminus E$. Then there is a $\mathbf{w} \in E$ such that $\mathbf{z} = F(\mathbf{w}, t) = \mathbf{w} + t\mathbf{f}(\mathbf{w}) \notin E$. Since $\mathbf{z} \neq \mathbf{w}$, we must have $\mathbf{f}(\mathbf{w}) \neq \mathbf{0}$, and therefore (by the assumption of the theorem) $\mathbf{w} \notin \partial E$. It also implies that there is a nonzero $\tau \in J$ such that $F(\mathbf{w}, \tau) = \mathbf{v} = \mathbf{w} + \tau\mathbf{f}(\mathbf{w}) \in \partial E$. (To see this, let τ be the supremum of all t such that $\mathbf{w} + t\mathbf{f}(\mathbf{w}) \in E$.) Then $\mathbf{f}(\mathbf{v}) = \mathbf{0}$ by the hypothesis. This contradicts Lemma 10.3.6 which implies that $\mathbf{f}(\mathbf{v}) \neq \mathbf{0}$. Hence there cannot be any points in $E^t \setminus E$. Similarly one shows that $E \setminus E^t = \emptyset$. Therefore $E^t = E$ for all $t \in J$. □

Lemma 10.3.8 *Let K be the support of a flow F on Z. Let D and E be two sets in Z such that $D \cap K = E \cap K$. Then $E^t \setminus E = D^t \setminus D$ and $E \setminus E^t = D \setminus D^t$.*

Proof. Let $E_0 = E \cap K$ and $E_1 = E \setminus E_0 = E \cap K^c$. Then $E_1^t = E_1$ for all $t \in J$, since $F(\mathbf{z}, t) = \mathbf{z}$ for all $\mathbf{z} \in K^c$ and $t \in J$. Hence $E = E_0 \cup E_1$ and $E^t = E_0^t \cup E_1$. This implies $E^t \setminus E = E_0^t \setminus E_0$. Since $D_0 = D \cap K = E \cap K = E_0$, we see that $D^t \setminus D = E^t \setminus E$. The arguments for the second claim are the same. □

Corollary 10.3.9 *Let E and D be as in Lemma 10.3.8 above. If E is a Jordan set, then $D^t \setminus D$ and $D \setminus D^t$ are also Jordan sets. In particular, if $K \cap D$ is a Jordan set, then $D^t \setminus D$ and $D \setminus D^t$ are also Jordan sets.*

Proof. This follows directly from Lemma 10.3.8 and the fact that E^t is a Jordan set.
□

Lemma 10.3.10 *Let* $\mathbf{f} : Z \to Z$ *be a* $\mathcal{C}^1$ *vector field with a compact support* K. *Then* $\text{flux}(\mathbf{f}, E) = \text{flux}(\mathbf{f}, E \cap K)$ *for any Jordan set* $E \subset Z$.

Proof. Let $F : Z \times J \to Z$ be a smooth flow with the initial velocity field $\mathbf{f}$ and with the same support K. Let $E_0 = E \cap K$ and $E_1 = E \setminus E_0 = E \cap K^c$, the part of E outside K. Then $E_1^t = E_1$ for all $t \in J$, since $F(\mathbf{z}, t) = \mathbf{z}$ for all $\mathbf{z} \in K^c$ and $t \in J$. Hence $E = E_0 \cup E_1$ and $E^t = E_0^t \cup E_1$. Then the conclusion follows since both unions are unions of disjoint sets. □

Basic Stokes' theorem: flow interpretation

Let us recall the notation employed in Definition 10.1.7: $\mathbf{n} \in Z$ is a unit vector, U is the plane $\mathbf{z} \cdot \mathbf{n} = 0$, and A is the lower side of U, as defined by the condition $\mathbf{z} \cdot \mathbf{n} < 0$. We are ready for our second proof of the basic Stokes' theorem, stated exactly as in 10.1.8.

Theorem 10.3.11 Basic Stokes' theorem. *Let* $\mathbf{f} : Z \to Z$ *be a* $\mathcal{C}^1$ *vector field of compact support. Then*

$$\int_A \operatorname{div} \mathbf{f}(\mathbf{z})\, d\mathbf{z} = \int_U \mathbf{n} \cdot \mathbf{f}(\mathbf{u})\, du. \qquad (10.45)$$

Proof. The proof utilizes three main ideas. First, interpret $\mathbf{f}(\mathbf{z})$ as the initial velocity of a flow $F(\mathbf{z}, t) = \mathbf{z} + t\mathbf{f}(\mathbf{z})$. We can imagine U as a plastic sheet that is perfectly flat at time $t = 0$, but whose parts are about to bulge up or down or stay put, depending upon the direction in which $\mathbf{f}(\mathbf{z})$ points. Second, interpret the integral on the left side of (10.45) as the initial rate of change in the volume being added to or taken away from A as its boundary (which is U) begins to shift. Third, interpret the right side of (10.45) as a simple computation of this newly added (or subtracted) volume using the standard formula for the volume under a surface.

First, we define the flow. Let K be the support of $\mathbf{f}$. Let G be an open Jordan set such that $K \subset G$, and let $E = G \cap A$. We see that $K \cap E = K \cap A$. Let $F : Z \times J \to Z$ be the flow $F(\mathbf{z}, t) = \mathbf{z} + t\mathbf{f}(\mathbf{z})$ and let A^t and E^t the images of A and E under this flow. The existence of some $J = (-r, r)$ for which F is a smooth flow is guaranteed by Theorem 10.2.5, and all we need is some interval J containing 0.

Next, we interpret the divergence integral on the left side of (10.45) as a rate of change in volume. Lemma 10.3.8 shows that $E^t \setminus E = A^t \setminus A$ and $E \setminus E^t = A \setminus A^t$.

Hence

$$\begin{aligned}\int_A \operatorname{div} \mathbf{f}(\mathbf{z})\, d\mathbf{z} &= \int_E \operatorname{div} \mathbf{f}(\mathbf{z})\, d\mathbf{z} && (10.46)\\ &= \lim_{t\to 0}(1/t)(v(E^t \setminus E) - v(E \setminus E^t)) && (10.47)\\ &= \lim_{t\to 0}(1/t)(v(A^t \setminus A) - v(A \setminus A^t)) && (10.48)\end{aligned}$$

by Theorem 10.3.5 and Definition 10.3.4.

Finally, we interpret the integral on the right side of (10.45) as computing the volume under the shifted surface U^t. Suppose first that $\mathbf{f} \perp \mathbf{n}$ everywhere. In this case, $\int_U \mathbf{n} \cdot \mathbf{f}(\mathbf{u})\, d\mathbf{u} = 0$. But then we also have no shift in the surface. So $A^t = A$ for all $t \in J$, and the above formula for the divergence integral yields $\int_A \operatorname{div} \mathbf{f} = 0$. Hence in this first special case, the conclusion (10.45) follows.

Next suppose that $\mathbf{f} \perp U$ everywhere. In this case, $\mathbf{f}(\mathbf{z}) = f(\mathbf{z})\,\mathbf{n}$ for some scalar function $f : Z \to \mathbb{R}$. Decompose Z as $U \times Y \simeq U \times \mathbb{R}$ where Y is the one-dimensional space spanned by $\mathbf{n}$. Express the the points $\mathbf{z} = \mathbf{u} + y\mathbf{n}$ as $(\mathbf{u}, y)$. If $\mathbf{u} \in U$, then $F(\mathbf{u}, t) = \mathbf{u} + tf(\mathbf{u})\mathbf{n}$. Hence U^t is the graph of the function $y = tf(\mathbf{u})$ in $U \times \mathbb{R}$. Also (assuming $tf(\mathbf{u}) > 0$)

$$A^t = \{\, (\mathbf{u}, y) \mid \mathbf{u} \in U,\ y < tf(\mathbf{u}) \,\}. \tag{10.49}$$

This can be seen by noticing that $\partial A^t = U^t$. Therefore

$$v(A^t \setminus A) - v(A \setminus A^t) = \int_U tf(\mathbf{u})\, d\mathbf{u}, \tag{10.50}$$

the volume between the graph of tf and the U-plane. (The same formula works if $tf < 0$.) Again, the conclusion (10.45) follows.

In general, $\mathbf{f} = \mathbf{f}_1 + \mathbf{f}_2$ with $\mathbf{f}_1 \perp \mathbf{n}$ and $\mathbf{f}_2 \perp U$. The conclusion then follows by the linearity of both sides in (10.45). □

Problems

10.13 The proof of Theorem 10.3.11 above uses the facts that

$$\lim_{t\to 0}(v(E^t \setminus E) - v(E \setminus E^t))$$

depends only on the initial velocity field and that this dependence is linear. Prove

$$\lim_{t\to 0}\,(v(E^t \setminus E) - v(E \setminus E^t)) = \int_U \mathbf{n} \cdot \mathbf{f}(\mathbf{u})\, d\mathbf{u} \tag{10.51}$$

directly, without using this information. Here $F(\mathbf{z}, t) = \mathbf{z} + t\mathbf{f}(\mathbf{z})$.

10.14 Prove (10.51) for a general smooth flow.

10.4 EXTERIOR DERIVATIVES

To obtain the general version of Stokes' theorem, the simple geometry exploited in the basic version must be transformed to more general settings by diffeomorphisms. The relation between the transformations of $\mathbf{f} : Z \to Z$ and $(\operatorname{div} \mathbf{f}) : Z \to \mathbb{R}$ will be important. As it turns out, the shift can be handled most effectively by representing vector fields in terms of tensor fields, as discussed in chapter 9.

Remarks 10.4.1 Restriction to $\mathcal{C}^2$ diffeomorphisms. There is one drawback to this approach: it requires the assumption of $\mathcal{C}^2$ diffeomorphisms. Our general version of Stokes' theorem will thus be applicable only where the relevant diffeomorphisms are $\mathcal{C}^2$, rather than just $\mathcal{C}^1$. This entails no great loss of generality by comparison to most treatments of the subject, which customarily work with $\mathcal{C}^\infty$ diffeomorphisms. In particular, the "Classical Stokes' Theorem", stated as Theorem 10.6.10 below, is true only for $\mathcal{C}^2$ surfaces.

Nevertheless, most of the results obtained here are in fact true for $\mathcal{C}^1$ diffeomorphisms. Stokes' theorem without the restriction to $\mathcal{C}^2$ diffeomorphisms is proved in other texts (notably in W. Fleming, *Calculus of Several Variables* (Springer-Verlag UTM, 1977)). Problem 10.20 indicates a way of proving this more general version of Stokes' theorem.

Remarks 10.4.2 Review of tensor representations. Let (Z, ϑ) be an n-dimensional oriented Euclidean space. Hence Z is a Euclidean space together with a chosen Euclidean determinant ϑ. Recall that $\Lambda_k(Z)$ is the linear space of all k-tensors (real-valued alternating multilinear functions) on Z.

Recall that we can represent a vector field $\mathbf{f} : Z \to Z$ by the tensor field $\omega : Z \to \Lambda_{n-1}(Z)$ defined by

$$\omega(\mathbf{z})(\mathbb{Z}) = \vartheta(\mathbf{f}(\mathbf{z}), \mathbb{Z}) \in \mathbb{R}, \tag{10.52}$$

where $\mathbf{z} \in Z$, $\mathbb{Z} = (\mathbf{z}_2, \ldots, \mathbf{z}_n) \in Z^{n-1}$ and $\vartheta(\mathbf{f}(\mathbf{z}), \mathbb{Z}) = \vartheta(\mathbf{f}(\mathbf{z}), \mathbf{z}_2, \ldots, \mathbf{z}_n) \in \mathbb{R}$.

We can represent a scalar function $g : Z \to \mathbb{R}$ by the tensor field $\tau : Z \to \Lambda_n(Z)$ defined by

$$\tau(\mathbf{z})(\mathbb{Z}) = g(\mathbf{z})\vartheta(\mathbb{Z}) \in \mathbb{R}, \quad \mathbf{z} \in Z, \quad \mathbb{Z} = (\mathbf{z}_1, \ldots, \mathbf{z}_n) \in Z^n. \tag{10.53}$$

Remarks 10.4.3 Exterior derivatives: a special case. Let $\mathbf{f} : Z \to Z$ be a $\mathcal{C}^1$ vector field represented by $\omega : Z \to \Lambda_{n-1}(Z)$. If the divergence function $(\operatorname{div} \mathbf{f}) : Z \to \mathbb{R}$ is represented by $\tau : Z \to \Lambda_n(Z)$ as in the preceding remarks, then one calls τ the *exterior derivative* of ω, written as $\tau = d\omega$.

We shall show that there is a more general way to define the exterior derivatives of tensor fields. If $\omega : Z \to \Lambda_k(Z)$ is a $\mathcal{C}^1$ tensor field, then its exterior derivative will be a continuous tensor field $d\omega : Z \to \Lambda_{k+1}(Z)$. The special case just mentioned is thus the case $k = n - 1$.

Lemma 10.4.4 Alternating part of a multilinear function. *Let k be a nonnegative integer and let $\alpha : Z^k \to \mathbb{R}$ be a multilinear function. Then*

$$(\operatorname{Alt}\alpha)(\mathbf{z}_1, \ldots, \mathbf{z}_k) = \frac{1}{k!}\sum\nolimits_{\sigma\in\mathcal{S}_k} (\operatorname{sign}\sigma)\, \alpha(\mathbf{z}_{\sigma(1)}, \ldots, \mathbf{z}_{\sigma(k)}) \tag{10.54}$$

defines a k-tensor $(\operatorname{Alt}\alpha) \in \Lambda_k(Z)$. It is called the alternating part *of α.*

Proof. Recall that $\mathcal{S}_k$ is the set of all permutations of $\{1, \ldots, k\}$ (see Appendix C). If $\mathbb{Z} = (\mathbf{z}_1, \ldots, \mathbf{z}_k) \in Z^k$ and $\sigma \in \mathcal{S}_k$, then write

$$\sigma\mathbb{Z} = (\mathbf{z}_{\sigma(1)}, \ldots, \mathbf{z}_{\sigma(k)}) \in Z^k \tag{10.55}$$

for the re-arrangement of the vectors in $\mathbb{Z}$ according to the permutation σ. Now let $\rho \in \mathcal{S}_k$ and write $\rho\mathbb{Z} = (\mathbf{v}_1, \ldots, \mathbf{v}_k) = \mathbb{V}$. Since it is clear that $\operatorname{Alt}\alpha$ is multilinear, we are done once we show that

$$\operatorname{Alt}\alpha(\rho\mathbb{Z}) = (\operatorname{sign}\rho)\,(\operatorname{Alt}\alpha)(\mathbb{Z}).$$

If $\sigma\mathbb{V} = (\mathbf{w}_1, \ldots, \mathbf{w}_k) = \mathbb{W}$, then $\mathbf{w}_i = \mathbf{v}_{\sigma(i)} = \mathbf{z}_{\rho(\sigma(i))}$. Hence $\sigma(\rho\mathbb{Z}) = (\rho\sigma)\mathbb{Z}$. Therefore, recalling that $\operatorname{sign}(\rho\sigma) = (\operatorname{sign}\rho)\,(\operatorname{sign}\sigma)$,

$$\begin{aligned}
k!\,(\operatorname{Alt}\alpha)(\rho\mathbb{Z}) &= \sum\nolimits_{\sigma\in\mathcal{S}_k} (\operatorname{sign}\sigma)\,\alpha(\sigma(\rho\mathbb{Z})) && (10.56)\\
&= \sum\nolimits_{\sigma\in\mathcal{S}_k} (\operatorname{sign}\sigma)\,\alpha((\rho\sigma)\mathbb{Z}) && (10.57)\\
&= (\operatorname{sign}\rho)\sum\nolimits_{(\rho\sigma)\in\mathcal{S}_k} (\operatorname{sign}(\rho\sigma))\,\alpha((\rho\sigma)\mathbb{Z}) && (10.58)\\
&= k!(\operatorname{sign}\rho)\,(\operatorname{Alt}\alpha)(\mathbb{Z}). \quad \square && (10.59)
\end{aligned}$$

Remarks 10.4.5 Derivative of a tensor field. Let $\omega : Z \to \Lambda_k(Z)$ be a $\mathcal{C}^1$ tensor field. Then $\omega(\cdot)(\mathbb{Z}) : Z \to \mathbb{R}$ is a $\mathcal{C}^1$ function for each fixed ordered k-tuple $\mathbb{Z} = (\mathbf{z}_1, \ldots, \mathbf{z}_k)$ in Z^k. Let

$$\omega'(\cdot)(\mathbb{Z}) : Z \to L(Z, \mathbb{R}) \tag{10.60}$$

be the derivative of this function. We will denote the application of this derivative to a vector $\mathbf{z}_0 \in Z$ as

$$\omega'(\cdot)(\mathbf{z}_0; \mathbb{Z}) : Z \to \mathbb{R}. \tag{10.61}$$

But now if we fix $\mathbf{z} \in Z$ and allow $\mathbf{z}_0$ and $\mathbb{Z}$ to vary, $\omega'(\mathbf{z})$ is a multilinear function that takes $(\mathbf{z}_0, \mathbf{z}_1, \ldots, \mathbf{z}_k) \in Z^{k+1}$ to $\omega'(\mathbf{z})(\mathbf{z}_0, \mathbf{z}_1, \ldots, \mathbf{z}_k) \in \mathbb{R}$. So (10.61) defines a function $\omega' : Z \to ML_{k+1}(Z)$. We have

$$\omega'(\mathbf{z})(\mathbf{z}_0;\, \mathbb{Z}) = \lim_{t\to 0}(1/t)(\omega(\mathbf{z} + t\mathbf{z}_0)(\mathbb{Z}) - \omega(\mathbf{z})(\mathbb{Z})) \tag{10.62}$$

for all $\mathbf{z} \in Z$ and for all $(\mathbf{z}_0;\, \mathbb{Z}) \in Z^{k+1}$ with $\mathbb{Z} = (\mathbf{z}_1, \ldots, \mathbf{z}_k) \in Z^k$.

Definition 10.4.6 Exterior derivatives. Let $\omega : Z \to \Lambda_k(Z)$ be a $\mathcal{C}^1$ tensor field. Let $\omega' : Z \to ML_{k+1}(Z)$ be its derivative as defined in Remarks 10.4.5. Then the tensor field

$$d\omega = (k+1)\,\mathrm{Alt}\,\omega' : Z \to \Lambda_{k+1}(Z) \tag{10.63}$$

of order $(k+1)$ is called the *exterior derivative* of ω.

We can now show that this definition agrees with our earlier definition of the exterior derivative for the special case in Remarks 10.4.3.

Theorem 10.4.7 Exterior derivatives and divergence. *Let $(Z,\, \vartheta)$ be an n-dimensional oriented Euclidean space. If*

$$\omega \;:\; Z \to \Lambda_{n-1}(Z)$$

represents a $\mathcal{C}^1$ vector field $\mathbf{f} : Z \to Z$, then

$$d\omega \;:\; Z \to \Lambda_n(Z)$$

represents the function $(\mathrm{div}\,\mathbf{f}) : Z \to \mathbb{R}$.

Proof. Let $\mathbf{z} \in Z$, $\mathbf{z}_1 \in Z$, and $\mathbb{Z} = (\mathbf{z}_2, \ldots, \mathbf{z}_n) \in Z^{n-1}$. We see that

$$\omega'(\mathbf{z})(\mathbf{z}_1;\, \mathbb{Z}) = \vartheta(\,\mathbf{f}'(\mathbf{z})\,\mathbf{z}_1,\, \mathbf{z}_2,\, \ldots,\, \mathbf{z}_n). \tag{10.64}$$

Here $\omega'(\mathbf{z}) \in ML_n(Z)$ is already alternating in its last $k = (n-1)$ variables, since $\vartheta \in \Lambda_n(Z)$ is a determinant. Then an easy computation shows that

$$\begin{aligned}
d\omega(\mathbf{z})\,(\mathbf{z}_1,\, \mathbf{z}_2,\, \ldots,\, \mathbf{z}_n) &= n(\mathrm{Alt}\,\omega')(\mathbf{z})(\mathbf{z}_1,\, \ldots,\, \mathbf{z}_n) \\
&= \vartheta(\,\mathbf{f}'(\mathbf{z})\,\mathbf{z}_1,\, \mathbf{z}_2,\, \ldots,\, \mathbf{z}_n) + \\
&\quad\ \vartheta(\,\mathbf{z}_1,\, \mathbf{f}'(\mathbf{z})\,\mathbf{z}_2,\, \ldots,\, \mathbf{z}_n) + \\
&\quad\ \ldots + \vartheta(\,\mathbf{z}_1,\, \mathbf{z}_2,\, \ldots,\, \mathbf{f}'(\mathbf{z})\,\mathbf{z}_n) \\
&= \mathrm{Tr}\,\mathbf{f}'(\mathbf{z})\,\vartheta(\mathbf{z}_1,\, \ldots,\, \mathbf{z}_n) \\
&= \mathrm{div}\,\mathbf{f}(\mathbf{z})\,\vartheta(\mathbf{z}_1,\, \ldots,\, \mathbf{z}_n).
\end{aligned}$$

The second last step follows from Definition C.7.3, the definition of the trace of $\mathbf{f}'(\mathbf{z}) \in L(Z,\, Z)$. The last step follows from the fact that $\mathrm{Tr}\,\mathbf{f}' = \mathrm{div}\,\mathbf{f}$, by the definition of the divergence. □

Commutativity of Pullbacks and Exterior Derivatives

We will show that exterior differentiation commutes with taking pullbacks under $\mathcal{C}^2$ diffeomorphisms.

Remarks 10.4.8 Pullbacks with $\mathcal{C}^2$ diffeomorphisms are $\mathcal{C}^1$. Let us recall the definition of pullbacks, from Definition 9.5.11. Let W be another Euclidean space, $H \subset W$ be an open set, and $\Phi : H \to Z$ a $\mathcal{C}^2$ diffeomorphism with $G = \Phi(H)$ (see Figure 9.1 or Figure 9.3). Let $\omega : G \to \Lambda_k(Z)$ be a tensor field with a compact support $K \subset G$. Then the pullback of ω is defined as

$$\Phi^*(\omega)(\mathbf{w})(\mathbb{W}) = \omega(\Phi(\mathbf{w}))(\Phi'(\mathbf{w})\mathbb{W}), \quad \mathbf{w} \in H \text{ and } \mathbb{W} \in W^k. \tag{10.65}$$

We see that if $\omega : G \to \Lambda_k(Z)$ is a $\mathcal{C}^1$ tensor field, then $\Phi^*(\omega) : H \to \Lambda_k(W)$ is also a $\mathcal{C}^1$ tensor field. This follows from the assumption that $\Phi : H \to G$ is a $\mathcal{C}^2$ diffeomorphism, and that assumption is essential. If necessary, we can extend the definitions of ω and $\Phi^*(\omega)$ to Z and to W by defining them as zero outside G and H, respectively. An easy argument shows that these extensions are also $\mathcal{C}^1$ functions.

Theorem 10.4.9 Exterior derivatives of pullbacks. *Let $\Phi : H \to G$ be a $\mathcal{C}^2$ diffeomorphism. Let $\omega : G \to \Lambda_k(Z)$ be a $\mathcal{C}^1$ tensor field with a compact support $K \subset G$. Then $d(\Phi^*\omega) = \Phi^*(d\omega)$.*

Proof. Let $\mathbf{w} \in H$, $\mathbf{w}_0 \in W$, and $\mathbb{W} \in W^k$. Also let $\mathbf{z} = \Phi(\mathbf{w})$. Compute $(\Phi^*\omega)'(\mathbf{w})(\mathbf{w}_0;\, \mathbb{W})$ from (10.65) by the chain rule and by the rules of differentiation for multilinear functions. Then, with the notations of Remarks 10.4.5,

$$(\Phi^*\omega)'(\mathbf{w})(\mathbf{w}_0;\, \mathbb{W}) = \omega'(\mathbf{z})(\Phi'(\mathbf{w})\mathbf{w}_0; \Phi'(\mathbf{w})\mathbb{W}) \tag{10.66}$$

$$+\sum_{i=1}^{k} \omega(\mathbf{z})(\mathbb{Z}_i) \tag{10.67}$$

where $\mathbb{Z}_i = (\mathbf{z}_{i1}, \ldots, \mathbf{z}_{ik}) \in Z^k$ are defined as

$$\mathbf{z}_{ij} = \Phi'(\mathbf{w})\mathbf{w}_j \text{ if } i \neq j \text{ and } \mathbf{z}_{ii} = \Phi''(\mathbf{w})(\mathbf{w}_0,\, \mathbf{w}_i). \tag{10.68}$$

(The differentiation here is undeniably messy. Problem 10.15 asks you to work out this derivative for the special cases $k = 1$ and $k = 2$.)

We see that each $\omega(\mathbf{z})(\mathbb{Z}_i)$ is a multilinear function of $(k + 1)$ vectors $\mathbf{w}_0$ and $\mathbf{w}_1, \ldots, \mathbf{w}_k$. Also $\omega(\mathbf{z})(\mathbb{Z}_i)$ remains invariant if $\mathbf{w}_0$ and $\mathbf{w}_i$ are switched. In fact, since Φ is a $\mathcal{C}^2$ diffeomorphism, we have

$$\Phi''(\mathbf{w})(\mathbf{w}_0,\, \mathbf{w}_i) = \Phi''(\mathbf{w})(\mathbf{w}_i,\, \mathbf{w}_0). \tag{10.69}$$

Then, an easy verification shows that Alt $\omega(\mathbf{z})(\mathbb{Z}_i) = \mathbf{0}$. Hence

$$\text{Alt } (\Phi^*\omega)'(\mathbf{w})(\mathbf{w}_0;\, \mathbb{W}) = \text{Alt } \omega'(\mathbf{z})(\Phi'(\mathbf{w})\mathbf{w}_0; \Phi'(\mathbf{w})\mathbb{W}), \tag{10.70}$$

which is the conclusion of the theorem. □

Basic Stokes' theorem in Tensor Form

We shall need the basic Stokes' theorem, Theorem 10.3.11, in tensor form. This involves the integrals of the tensor fields ω and $d\omega$ over two oriented flat manifolds, that is, manifolds with constant tangent spaces. In what follows, W is a Euclidean space. It is oriented by a Euclidean determinant ϱ. Also, $\mathbf{e} \in W$ is a unit vector, A is the half space $\mathbf{e} \cdot \mathbf{w} < 0$, and U is the plane $\mathbf{e} \cdot \mathbf{w} = 0$.

Notations 10.4.10 Two manifolds in the basic Stokes' theorem. One manifold that appears in the theorem is the n-dimensional manifold A. Its tangent spaces are always W and they are all oriented by ϱ. The other manifold is the $(n-1)$-dimensional manifold U. Its tangent spaces are always U and they are all oriented according to the following convention: a basis $\mathbb{B}$ for U is positive if $(\mathbf{e}, \mathbb{B})$ is a positive basis for W, that is, if $\varrho(\mathbf{e}, \mathbb{B}) > 0$. In the arguments that follow, let $\mathbb{E} = (\mathbf{e}_2, \ldots, \mathbf{e}_n)$ be a fixed positive orthonormal basis for U. Hence $(\mathbf{e}, \mathbb{E}) = (\mathbf{e}, \mathbf{e}_2, \ldots, \mathbf{e}_n)$ is a positive orthonormal basis for W and $\varrho(\mathbf{e}, \mathbb{E}) = 1$.

Theorem 10.4.11 Basic Stokes' theorem in tensor form. *Let*

$$\omega : W \to \Lambda_{n-1}(W) \tag{10.71}$$

be a $\mathcal{C}^1$ tensor field of compact support. Then

$$\int_A d\omega = \int_U \omega, \tag{10.72}$$

with the manifolds and orientations as defined in Notations 10.4.10.

Proof. Clearly, we want to derive this result from the vector version of the basic Stokes' theorem. So our first task is to recall how we associate a vector field with the tensor ω. Lemma 9.5.10 shows that there is a vector field $\mathbf{f} : W \to W$ such that

$$\omega(\mathbf{w})(\mathbb{W}) = \varrho(\mathbf{f}(\mathbf{w}), \mathbb{W}), \tag{10.73}$$

where $\mathbf{w} \in W$ and $\mathbb{W} = (\mathbf{w}_2, \ldots, \mathbf{w}_n) \in W^{n-1}$. We see that $\mathbf{f} : W \to W$ is a $\mathcal{C}^1$ vector field of compact support.

Now we have a string of equalities.

$$\begin{aligned} \int_U \omega &= \int_U \omega(\mathbf{u})(\mathbb{E})\, d\mathbf{u} = \int_U \varrho(\mathbf{f}(\mathbf{u}), \mathbb{E})\, d\mathbf{u} && (10.74) \\ &= \int_U (\mathbf{e} \cdot \mathbf{f}(\mathbf{u}))\, d\mathbf{u} = \int_A \operatorname{div} \mathbf{f}(\mathbf{w})\, d\mathbf{w} && (10.75) \\ &= \int_A \operatorname{div} \mathbf{f}(\mathbf{w}) \varrho(\mathbf{e}, \mathbb{E})\, d\mathbf{w} = \int_A d\omega(\mathbf{w})(\mathbf{e}, \mathbb{E})\, d\mathbf{w} && (10.76) \\ &= \int_A d\omega. && (10.77) \end{aligned}$$

The first step in (10.75) is derived by writing $\mathbf{f}(\mathbf{u}) = (\mathbf{e} \cdot \mathbf{f}(\mathbf{u}))\mathbf{e} + \mathbf{e}_2$, where $\mathbf{e}_2$ is a linear combination of vectors in $\mathbb{E}$; then use the properties of the Euclidean determinant ϱ. The second step in (10.75) is the basic Stokes' theorem in vector form. The second step in (10.76) follows from Theorem 10.4.7 that relates divergences and exterior derivatives. All other steps follow directly from the definitions. □

Problems

10.15 Prove Theorem 10.4.9 for the special cases of $k = 1$ and $k = 2$ by writing out the terms in Equations (10.66)-(10.67) explicitly.

10.16 Let $(\mathbf{e}_1, \ldots, \mathbf{e}_n)$ be a basis for Z. Show that a tensor $\lambda \in \Lambda_2(Z)$ of order two is uniquely determined by its values $\lambda(\mathbf{e}_i, \mathbf{e}_j)$ on the pairs of basis vectors. Also show that $\lambda(\mathbf{e}_i, \mathbf{e}_i) = 0$ and $\lambda(\mathbf{e}_i, \mathbf{e}_j) = -\lambda(\mathbf{e}_j, \mathbf{e}_i)$. Hence, conclude that $\dim \Lambda_2(Z)$ is $n(n-1)/2$.

10.17 A tensor field $\xi : G \to \Lambda_1(Z)$ of is represented by a vector field $\mathbf{f} : G \to Z$ as in Definition 9.5.6. Hence $\xi(\mathbf{a})(\mathbf{z}) = \mathbf{f}(\mathbf{a}) \cdot \mathbf{z}$ for all $\mathbf{a} \in G$ and $\mathbf{z} \in Z$. What is the application of $d\xi(\mathbf{a})$ to a pair of vectors $(\mathbf{z}_1, \mathbf{z}_2) \in Z^2$? Let $(\mathbf{e}_1, \ldots, \mathbf{e}_n)$ be a basis for Z. Let $\mathbf{f} = \sum_i P_i \mathbf{e}_i$. Find $d\xi(\mathbf{a})(\mathbf{e}_i, \mathbf{e}_j)$ in terms of P_is.

10.18 Show that $\dim \Lambda_k(Z) = \begin{pmatrix} n \\ k \end{pmatrix}$ with $n = \dim Z$.

10.19 Let $B = \{ (x, y, z) \mid x = y = 0 \}$ be the z-axis and $G = \mathbb{R}^3 \setminus B$. Define $\mathbf{f} : G \to \mathbb{R}^3$ by $\mathbf{f}(x, y, z) = (x^2 z, y^2 z, x^2 + y^2)$. Define $\Omega : G \to G$ by $\Omega(x, y, z) = (x, y, z - (x^2 + y^2)^{1/2} + (x^2 + y^2))$. Let $\xi : G \to \Lambda_1(\mathbb{R}^3)$ and $\eta : G \to \Lambda_2(\mathbb{R}^3)$ be defined by

$$\xi(\mathbf{a})(\mathbf{z}) = \mathbf{f}(\mathbf{a}) \cdot \mathbf{z} \text{ and } \eta(\mathbf{a})(\mathbf{u}, \mathbf{v}) = \mathbf{f}(\mathbf{a}) \cdot \mathbf{u} \times \mathbf{v}.$$

Compute $d\xi$, $d\eta$, $d\Omega^*(\xi)$, $d\Omega^*(\eta)$, $\Omega^*(d\xi)$, and $\Omega^*(d\xi)$ explicitly and verify Theorem 10.4.9 for these cases.

10.5 REGULAR AND ALMOST REGULAR SETS

We will prove Stokes' theorem for *regular sets* and for *almost regular sets*. Intuitively, *regular sets* are Jordan sets with smooth boundaries. A simple example of a regular set is a Euclidean ball. *Almost regular sets* have smooth boundaries except for a subset of the boundary that has negligible surface area. A simple example of an almost regular set is a box spanned by the vectors of a basis.

The main step in generalizing Stokes' theorem is to transfer the basic Stokes' theorem to *regular neighborhoods* of a set. Recall that the basic Stokes' theorem deals with a half-space A and its boundary-plane $\partial A = U$. Regular neighborhoods of a set E are those open sets G in which E and ∂E behave like A and ∂A, up to a $\mathcal{C}^2$ diffeomorphism. A Jordan set E is called a regular set if its closure $\overline{E}$ can be covered by the regular neighborhoods of E.

Regular Neighborhoods of a Set

Notations 10.5.1 The oriented spaces U, W and Z. Let $n \geq 2$. Let Z and W be two n-dimensional Euclidean spaces, oriented by the Euclidean determinants ϑ and ϱ respectively. Let $\mathbf{e} = \mathbf{e}_1 \in W$ be a unit vector. Let U be the plane $\mathbf{e} \cdot \mathbf{w} = 0$ and let A be the half-space $\mathbf{e} \cdot \mathbf{w} < 0$. Orient U by the right-hand rule, Definition 9.3.16, with the unit vector $\mathbf{e} = \mathbf{e}_1$. Let $\mathbb{E} = (\mathbf{e}_2, \ldots, \mathbf{e}_n)$ be a positive orthonormal basis for U. Hence, $(\mathbf{e}, \mathbb{E}) = (\mathbf{e}_1, \mathbf{e}_2, \ldots, \mathbf{e}_n)$ is a positive orthonormal basis for W.

Definition 10.5.2 Regular neighborhoods of a set. Let E be a set in Z. An open set $G \subset Z$ is called a *regular neighborhood for E* if there is a $\mathcal{C}^2$ diffeomorphism $\Psi : G \to H$, $H = \Psi(G) \subset W$, such that

$$\Psi(G \cap E^o) = H \cap A \text{ and } \Psi(G \cap \partial E) = H \cap U. \tag{10.78}$$

Note that if an open set G does not intersect ∂E, then G is a regular neighborhood for E. If $G \subset E^o$, for example, then an isomorphism $T : Z \to W$ takes G to an open subset of A. The general case is similar. Hence, Definition 10.5.2 involves only the position of $G \cap E$ with respect to $G \cap \partial E$.

Remarks 10.5.3 Boundaries in regular neighborhoods. Recall that surfaces in Z are $(n-1)$-dimensional manifolds. If G is a regular neighborhood for E, then $G \cap \partial E$ is a surface. This follows directly from the second condition in (10.78). In fact, this condition means that $\Psi : G \to H$ is a chart for $G \cap \partial E$. Hence, if $G \cap \partial E$ is not a surface, then G cannot be a regular neighborhood for E. If $G \cap \partial E$ is a surface, however, then G still may not be a regular neighborhood for E. First, there is an additional smoothness condition on the chart Ψ that it must be a $\mathcal{C}^2$ diffeomorphism. Second, the first condition in (10.78) requires that E should be only "on one side of the boundary". Let, for example, $E = \{ (x, y) \mid x^2 + y^2 < 2, \; x^2 + y^2 \neq 1 \} \subset \mathbb{R}^2$. Then ∂E consists of two circles $x^2 + y^2 = 1$ and $x^2 + y^2 = 2$. Any open set that intersects the smaller circle is not a regular neighborhood for E.

Boundaries of Regular Sets

Definition 10.5.4 Regular sets. A set in Z is called a *regular set* if its closure is covered by its regular neighborhoods.

Note that a Jordan set is regular if and only if the union of its regular neighborhoods is the whole space. This follows from the remarks in Definition 10.5.2.

To simplify the statements in the proofs of the next few results, we will define *acceptable diffeomorphisms* for a set. These are the diffeomorphisms that have the properties specified in Definition 10.5.2 of regular neighborhoods.

Definition 10.5.5 Acceptable diffeomorphisms. Let E be a set in Z. Let G be an open set in Z. Let $\Psi : G \to W$ be a diffeomorphism and let $\Psi(G) = H$. Then $\Psi : G \to H$ is called an *acceptable diffeomorphism (for E)* if Ψ is a $\mathcal{C}^2$ diffeomorphism, and if

$$\Psi(G \cap E^o) = H \cap A \text{ and } \Psi(G \cap \partial E) = H \cap U. \tag{10.79}$$

Lemma 10.5.6 *Let G and G_0 be open sets in Z and let $G_0 \subset G$. If a diffeomorphism $\Psi : G \to H$ is acceptable for E, then its restriction to G_0 is also acceptable for E.*

Proof. This is left as an exercise. □

Lemma 10.5.7 *Let $(\mathbf{e}, \mathbb{E})$ be the basis of W defined in Notations 10.5.1. Define an isomorphism $R : W \to W$ as $R\mathbf{e}_i = \mathbf{e}_i$ if $1 \le i < n$ and $R\mathbf{e}_n = -\mathbf{e}_n$. If a diffeomorphism $\Psi : G \to H$ is acceptable, then $R\Psi : G \to RH$ is also an acceptable diffeomorphism.*

Proof. This is left as an exercise. □

Theorem 10.5.8 *If E is a regular set, then its closure can be covered by the domains of acceptable and orientation-preserving diffeomorphisms.*

Proof. Orientation-preserving diffeomorphisms were defined in Definition 9.3.4. This proof is identical with the proof of Theorem 9.3.8. Arguments of that proof show that one can pass from diffeomorphisms to orientation-preserving diffeomorphisms by two types of modifications: restricting the domain of a diffeomorphism, and reversing the orientation of a diffeomorphism. Lemmas 10.5.6 and 10.5.7 show that acceptable diffeomorphisms remain acceptable after these modifications. □

We will show that the boundary of a regular set is an orientable surface.

Lemma 10.5.9 *Let $\Psi : G \to H$ be an acceptable diffeomorphism for E. Let $F(\mathbf{z}) = \mathbf{e} \cdot \Psi(\mathbf{z})$, $\mathbf{z} \in G$. If $\mathbf{a} \in G \cap \partial E$, then there is a $\delta > 0$ such that*

$$\begin{aligned} \mathbf{a} + t\nabla F(\mathbf{a}) &\in E \quad \textit{if} \quad -\delta < t < 0 \textit{ and} \\ \mathbf{a} + t\nabla F(\mathbf{a}) &\notin E \quad \textit{if} \quad 0 < t < \delta. \end{aligned}$$

Proof. Let $\mathbf{v} = \nabla F(\mathbf{a})$. Theorem 9.3.20 shows that $\mathbf{v} \neq \mathbf{0}$. Let

$$f(t) = \mathbf{e} \cdot \Psi(\mathbf{a} + t\mathbf{v}) = F(\mathbf{a} + t\mathbf{v}).$$

Then we see that $f(0) = 0$ and $f'(0) = \nabla F(\mathbf{a}) \cdot \mathbf{v} = \|\nabla F(\mathbf{a})\|^2$. Therefore there is a $\delta > 0$ such that

$$\begin{aligned} f(t) &= F(\mathbf{a} + t\mathbf{v}) < 0 \text{ if } -\delta < t < 0 \text{ and} \\ f(t) &= F(\mathbf{a} + t\mathbf{v}) > 0 \text{ if } 0 < t < \delta. \end{aligned}$$

This means that $\Psi(\mathbf{a}+t\mathbf{v}) \in A$ if $-\delta < t < 0$ and $\Psi(\mathbf{a}+t\mathbf{v}) \notin A \cup U$ if $0 < t < \delta$. Hence, the conclusion follows. □

Lemma 10.5.10 *Let $\Psi_i : G_i \to H_i$ be two acceptable diffeomorphisms for E. Let $F_i(\mathbf{z}) = \mathbf{e} \cdot \Psi_i(\mathbf{z})$, $\mathbf{z} \in G_i$. If $\mathbf{a} \in G_1 \cap G_2 \cap \partial E$, then $\nabla F_1(\mathbf{a}) \cdot \nabla F_2(\mathbf{a}) > 0$.*

Proof. Observations in Remarks 10.5.3 show that $S = G_1 \cap G_2 \cap \partial E$ is a surface. Let $\mathbf{v}_i = \nabla F_i(\mathbf{a})$. Theorem 9.3.20 shows that both $\mathbf{v}_i$ are nonzero and normal to the tangent plane $T_{\mathbf{a}}$ of S. Hence there is a nonzero $\alpha \in \mathbb{R}$ such that $\mathbf{v}_2 = \alpha \mathbf{v}_1$. Lemma 10.5.9 shows that α cannot be negative. Then the proof follows. □

Lemma 10.5.11 *Two acceptable and orientation-preserving diffeomorphisms $\Psi_i : G_i \to H_i$ for E induce the same orientation on $S = G_1 \cap G_2 \cap \partial E$.*

Proof. Orientations induced by diffeomorphisms were defined in Definition 9.3.9. Let $\Psi : G \to H$ be an orientation-preserving diffeomorphism. Let $F = \mathbf{e} \cdot \Psi$. By Theorem 9.3.20, the orientation of the tangent space induced by Ψ is related to the unit normal vector $\mathbf{n} = \nabla F / \|\nabla F\|$ by the right-hand rule. Hence, if Ψ_i define the same unit normal vector, then they induce the same orientation on the manifold. □

Theorem 10.5.12 *The boundary of a regular set is an orientable surface.*

Proof. Let E be a regular set. Definition 10.5.4 of regular sets and Theorem 10.5.8 show that $S = \partial E$ has an atlas of orientation-preserving acceptable diffeomorphisms. Lemma 10.5.11 shows that this is an atlas of compatible charts, in the (obvious) sense of Definition 9.3.12. Then, by Definition 9.3.13, S is an orientable surface. □

Stokes' theorem for Regular Sets

Remarks 10.5.13 Summary of notation and assumptions about orientation. The spaces Z and W and the plane U (a subspace of W) are as specified in Notations 10.5.1. They are all oriented spaces. In particular, U is oriented by a unit normal vector $\mathbf{e}$ according to the right-hand rule. The half-space A is the set of $\mathbf{w} \in W$ such that $\mathbf{e} \cdot \mathbf{w} < 0$. Hence $U = \partial A$. Let E be a set in Z. Let G be a regular neighborhood for E and let $\Psi : G \to H$ be an orientation-preserving acceptable diffeomorphism (Definition 10.5.5). In Theorem 10.5.14 below, the orientations on E and on ∂E are the orientations induced by Ψ from the orientations of A and ∂A.

Theorem 10.5.14 Stokes' theorem for regular neighborhoods *Let G be a regular neighborhood for E. Let $\mathbf{f} : Z \to Z$ be a $\mathcal{C}^1$ vector field and let $\omega : Z \to \Lambda_{n-1}(Z)$ be a $\mathcal{C}^1$ tensor field, both with compact supports contained in G. Then*

$$\int_E \operatorname{div} \mathbf{f} = \int_{\partial E} \mathbf{f} \quad \textit{and} \quad \int_E d\omega = \int_{\partial E} \omega, \tag{10.80}$$

with the definitions given in Remarks 10.5.13.

Proof. With the notations of Remarks 10.5.13, let $\Phi : H \to G$ be the reverse diffeomorphism of $\Psi : G \to H$. Let $\xi = \Phi^*(\omega)$ and $\eta = \Phi^*(d\omega)$ be the pullbacks of ω and $d\omega$ by Φ, as defined in Definition 9.5.11. Theorem 10.4.9 shows that $\eta = d\xi$. The tensor form of the basic Stokes' theorem, Theorem 10.4.11, shows that $\int_A d\xi = \int_{\partial A} \xi$. Assembling all of these facts gives us the following string of equalities:

$$\int_A \Phi^*(d\omega) = \int_A \eta = \int_A d\xi = \int_{\partial A} \xi = \int_{\partial A} \Phi^*(\omega).$$

Since $E^o \cap G = \Phi(A \cap H)$ and $\partial E \cap G = \Phi(\partial A \cap H)$, we can apply the change of variables theorem for tensor fields, Theorem 9.5.12, to the left and right sides of the above equation:

$$\int_E (d\omega) = \int_A \Phi^*(d\omega) = \int_{\partial A} \Phi^*(\omega) = \int_{\partial E} \omega.$$

This proves the tensor portion of (10.80). To obtain the vector part, represent the vector field $\mathbf{f}$ by the tensor field ω, as explained in Remarks 10.4.2, and apply the result on tensor fields. Theorem 10.4.7 shows that $d\omega$ represents $\operatorname{div} \mathbf{f}$ and this completes the proof. $\square$

The extension to regular sets is an easy consequence of Theorem 10.5.14 and partitions of unity.

Theorem 10.5.15 Stokes' theorem for regular sets *Let E be a regular set. Let $\mathbf{f} : Z \to Z$ be a $\mathcal{C}^1$ vector field and let $\omega : Z \to \Lambda_{n-1}(Z)$ be a $\mathcal{C}^1$ tensor field, both with compact supports. Then*

$$\int_E \operatorname{div} \mathbf{f} = \int_{\partial E} \mathbf{f} \quad \textit{and} \quad \int_E d\omega = \int_{\partial E} \omega. \tag{10.81}$$

The orientation on E is the orientation of the underlying space Z. The orientation on ∂E is induced by the orientation of Z and by the (geometrical) outer normal of ∂E, according to the right-hand rule.

Proof. The supports of $\mathbf{f}$ and ω are compact. Hence these supports can be covered by finitely many regular neighborhoods. In all of these neighborhoods there are orientation-preserving acceptable diffeomorphisms. Then the proof is completed by an application of Theorem 10.5.14 and the partitions of unity theorem, Theorem 9.2.11. The details are the same as in Definition 9.2.12 and Lemma 9.2.13. □

Remarks 10.5.16 Role of the orientations. Where does orientation come into these arguments? We need the orientation of E only when we integrate the tensor field $d\omega$. To integrate the vector field $\mathbf{f}$ on ∂E, it suffices to know the outer unit normal vectors. The orientations of the tangent spaces play no role in this integration. Hence, the effort spent on orientations is for the tensor version only. But the result is worth the effort. The tensor formulation allows a lean proof of Stokes' theorem. As already mentioned, however, this proof requires $\mathcal{C}^2$ diffeomorphisms.

Stokes' theorem for almost regular sets

We show that Stokes' theorem can be extended to cubes and to other sets, most of whose boundary points belong to a regular neighborhood.

Definition 10.5.17 Outer boundary-surfaces. Let E be a bounded set in Z. The *outer boundary-surface* of E is the set of all points on the boundary of E that are contained in regular neighborhoods (Definition 10.5.2) for E. We see that the boundary-surface is indeed a surface. The boundary-surface of E will be denoted by $\mathcal{S}$, or by $\mathcal{S}_E$.

Lemma 10.5.18 *Let K be a compact subset of $\mathcal{S}_E$ and let $B = (\partial E) \setminus K$. Then $\overline{E}$ has a finite open covering $\{G_0\} \cup \{G_i\}$ such that G_0 satisfies $G_0 \cap \partial E \subset B$ and all other G_is are regular neighborhoods for E.*

Proof. Each point in $\mathcal{S}_E$ is contained in a regular neighborhood. Hence the compact set K in $\mathcal{S}_E$ can be covered by finitely many regular neighborhoods. Let H be the union of these neighborhoods, and let $B_0 = (\partial E) \setminus H$. Then B_0 is a compact set disjoint from K. Therefore there is an open set G_0 that contains B_0 and is disjoint from K. Then $H \cup G_0$ is an open set containing ∂E. Hence $E \setminus (H \cup G_0)$ is in the interior of E. Any open set that does not intersect ∂E is a regular neighborhood for E. Therefore the compact set $E \setminus (H \cup G_0)$ can be covered by finitely many regular neighborhoods. These finitely many open sets are a covering of $\overline{E}$. They are all regular neighborhoods for E, except G_0. □

Definition 10.5.19 Upper surface-area. Let B be a bounded set in a Euclidean space Z. The enlargement of B by $r > 0$ was defined in Definition 9.6.10. It is the set B_r of all points that are within distance r of a point in B. We know that B_r is a Jordan set for all $r > 0$ (Problem 8.52). The upper limit

$$\sigma(B) = \limsup_{r \to 0^+} \frac{1}{2r} v(B_r) \tag{10.82}$$

$$= \lim_{q \to 0^+} \sup_{0<r<q} \frac{1}{2r} v(B_r) \tag{10.83}$$

will be called the *upper surface-area* of B. Here $v(B_r)$ is the volume of B_r in Z. Theorem 9.6.12 shows that if B is a Jordan set on a surface in Z, then $\sigma(B)$ is the surface-area of B. (Jordan subsets of a manifold are defined in Definition 9.2.7).

Lemma 10.5.20 *Let* $\mathbf{f} : Z \to Z$ *be a* $\mathcal{C}^1$ *vector field of compact support* K. *Let* $M = \sup_{\mathbf{z}} \|\mathbf{f}(\mathbf{z})\|$. *Let* E *be a Jordan set in* Z. *Let* $B = K \cap \partial E$. *Then*

$$\left| \int_E \operatorname{div} \mathbf{f} \right| = |\text{flux}(\mathbf{f}, E)| \le M \, \sigma(B). \tag{10.84}$$

Proof. Apply Theorem 10.2.5 to find an $r > 0$ such that $F(\mathbf{z}, t) = \mathbf{z} + t\mathbf{f}(\mathbf{z})$ is a flow $Z \times J \to Z$, where $J = (-r, r)$. Let $E^t = F(E, t), t \in J$, be the images of E under this flow. We claim that $(E \triangle E^t) \subset B_{|t|\,M}$, where $(E \triangle E^t) = (E^t \setminus E) \cup (E \setminus E^t)$. Here $B_{|t|\,M}$ is the enlargement (Definition 9.6.10) of B by $|t|M$. We assume that $0 < t$; arguments for negative t are similar.

If $\mathbf{z} \in (E^t \setminus E)$, then there is an $\mathbf{a} \in E$ such that $\mathbf{z} = \mathbf{a} + t\mathbf{f}(\mathbf{a}) \notin E$. Therefore, for some $\tau \in [0, t]$, $\mathbf{b} = \mathbf{a} + \tau\mathbf{f}(\mathbf{a}) \in \partial E$. Then $\mathbf{b} \in B$; otherwise, $F(\mathbf{b}, s) = \mathbf{b}$ for all $s \in J$. Then $\|\mathbf{z} - \mathbf{b}\| = (t - \tau)\|\mathbf{f}(\mathbf{a})\| \le tM$. Therefore, $\mathbf{z} \in B_{t\,M}$.

Let $\mathbf{z} \in (E \setminus E^t)$. Since $F(\,\cdot, t)$ is a diffeomorphism of Z onto Z, there is an $\mathbf{a} \in Z$ such that $\mathbf{z} = \mathbf{a} + t\mathbf{f}(\mathbf{a}) \in E$. Since $\mathbf{z} \notin E^t$, we see that $\mathbf{a} \notin E$. Then, as before, $\mathbf{b} = \mathbf{a} + \tau\mathbf{f}(\mathbf{a}) \in B$ for some $\tau \in [0, t]$. Therefore $\mathbf{z} \in B_{t\,M}$.

Now use Theorem 10.3.5. □

Definition 10.5.21 Almost regular sets. A Jordan set E in a Euclidean space is called an *almost regular* set if for each $\varepsilon > 0$, there is a compact set $K \subset \mathcal{S}_E$ such that $\sigma(\partial E \setminus K) < \varepsilon$. Here σ is as in (10.82).

Lemma 10.5.22 *Let* $\mathbf{f} : Z \to Z$ *be a* $\mathcal{C}^1$ *vector field. Let* E *be an almost regular set. Let* $\mathcal{S}_E$ *be the outer boundary surface of* E *(Definition 10.5.17). Then given* $\varepsilon > 0$*, there is a compact set* $K_0 \subset \mathcal{S}_E$ *such that*

$$\left| \int_K \mathbf{f} - \int_E \operatorname{div} \mathbf{f} \right| < \varepsilon, \tag{10.85}$$

whenever K *is a compact set and* $K_0 \subset K \subset \mathcal{S}_E$*.*

Proof. Let $M = \sup_{\mathbf{z}} \|\mathbf{f}(\mathbf{z})\|$. Given $\varepsilon > 0$, use Definition 10.5.21 to find a compact set $K_0 \subset \mathcal{S}_E$ such that $\sigma(B) < \varepsilon / M$, where $B = (\partial E) \setminus K_0$. Let K be any compact set such that $K_0 \subset K \subset \mathcal{S}_E$. Use Lemma 10.5.18 to find a finite open covering $\{G_0, G_i\}$ as specified in that lemma. Use the partitions of unity theorem, Theorem 9.2.11, to find finitely many $\mathcal{C}^1$ functions $\lambda_i : Z \to [0, 1]$, such that $\sum_i \lambda_i = 1$ on an open set containing $\overline{E}$, and such that each λ_i has a compact support contained in G_i. Let $\mathbf{f}_i = \lambda_i \mathbf{f}_i$. (Note in particular that $\mathbf{f}_0$ has support contained in G_0.) Then $\mathbf{f} = \sum_i \mathbf{f}_i$ on an open set containing $\overline{E}$. The operation of taking the divergence of a vector field is a linear operation. This was observed in Remarks 10.1.2. Hence

$$\int_E \operatorname{div} \mathbf{f} = \sum_i \int_E \operatorname{div} \mathbf{f}_i = \sum_i \int_{E_i} \operatorname{div} \mathbf{f}_i.$$

Theorem 10.5.14 shows that $\int_{E_i} \operatorname{div} \mathbf{f}_i = \int_{K \cap G_i} \mathbf{f}_i = \int_K \mathbf{f}_i$ for $i \neq 0$ since each G_i is a regular neighborhood of E. Hence

$$\begin{aligned} \int_E \operatorname{div} \mathbf{f} &= \int_E \operatorname{div} \mathbf{f}_0 + \sum_i \int_E \operatorname{div} \mathbf{f}_i \\ &= \int_E \operatorname{div} \mathbf{f}_0 + \sum_i \int_{K \cap G_i} \mathbf{f}_i \\ &= \int_E \operatorname{div} \mathbf{f}_0 + \int_K \mathbf{f}. \end{aligned}$$

Lemma 10.5.20 shows that $|\int_E \operatorname{div} \mathbf{f}_0| < M\sigma(B) < \varepsilon$. Then the conclusion follows.
□

Remarks 10.5.23 Integration on boundary-surfaces. Let $f : M \to \mathbb{R}$ be a function defined on a manifold M. Definition 9.2.12 of $\int_M f$ is only for those functions that have a compact support contained in M. Hence, according to this definition, we can consider $\int_{\mathcal{S}_E} \mathbf{f}$ only if $\mathbf{f}$ has a compact support contained in $\mathcal{S}_E$.

But if E is an almost regular set and if $\mathbf{f}$ is a $\mathcal{C}^1$ vector field, then, on the basis of Lemma 10.5.22, it is natural to define $\int_{\mathcal{S}_E} \mathbf{f}$ as $\int_E \operatorname{div} \mathbf{f}$. In fact, it is customary to express this integral as $\int_{\partial E} \mathbf{f}$. Definition 10.5.24 below formalizes this notation.

Definition 10.5.24 Let E be an almost regular set. Let $f : \partial E \to \mathbb{R}$ be a bounded function. Then the number $\alpha = \int_{\partial E} f$, if it exists, is defined as follows: for each $\varepsilon > 0$, there exists a compact set $K_0 \subset \mathcal{S}_E$ such that

$$\left| \int_K f - \alpha \right| < \varepsilon$$

whenever K is a compact set and $K_0 \subset K \subset \mathcal{S}_E$.

Theorem 10.5.25 Stokes' theorem for almost regular sets. *Let E be an almost regular set (Definition 10.5.21). Let $\mathbf{f} : Z \to Z$ be a $\mathcal{C}^1$ vector field. Then*

$$\int_E \operatorname{div} \mathbf{f} = \int_{\partial E} \mathbf{f}.$$

The last integral is defined in Definition 10.5.24.

Proof. This follows directly from Lemma 10.5.22 and Definition 10.5.24. □

Problems

10.20 Call an open set G a $\mathcal{C}^1$ regular neighborhood of a set E if the conditions in Definition 10.5.2 for the regular neighborhoods are satisfied, except that the diffeomorphisms involved do not have to be $\mathcal{C}^2$ diffeomorphisms. Let G be a $\mathcal{C}^1$ regular neighborhood of E. Let $\mathbf{f} : Z \to Z$ be a $\mathcal{C}^1$ vector field with compact support in G. Show that

$$\lim_{t \to 0} \frac{1}{t}(v(E^t \setminus E) - v(E \setminus E^t)) = \int_{\partial E} \mathbf{f}.$$

Here $E^t = F(E, t)$ and $F(\mathbf{z}, t) = \mathbf{z} + t\mathbf{f}(\mathbf{z})$, as before. The boundary ∂E is oriented by its outer normal.

10.21 Let $F : Z \times J \to Z$ be a general smooth flow with an initial velocity field $\mathbf{f}$ with a compact support contained in an open set G. Assume that G is a $\mathcal{C}^1$ regular neighborhood of a set E and repeat Problem 10.20 for this case.

10.22 Compute $\int_S \mathbf{f}$ where S is the unit sphere $x^2 + y^2 + z^2 = 1$ in $\mathbb{R}^3$ oriented by the outer normals and $\mathbf{f}(x, y, z) = (x^2 + ze^{-y^2}, y^2 + ze^{-x^2}, z + e^{x^2+y^2})$.

10.23 Compute $\int_S \mathbf{f}$ where $\mathbf{f}$ is as in Problem 10.22 and S is the upper half ($z > 0$) of the unit sphere in $\mathbb{R}^3$ oriented by the outer normals.

10.24 Compute $\int_S \mathbf{f}$ where $\mathbf{f}$ is as in Problem 10.22 and S is the part of the unit sphere in $\mathbb{R}^3$ corresponding to $z > 1/2$, oriented by the outer normals.

10.25 Let $a > 0$. Let E be the region in $\mathbb{R}^3$ specified by $x^2 + y^2 < z < a$. Show that E is an almost regular set. Compute $\int_{\partial E} \mathbf{f}$, where $\mathbf{f}$ is as in Problem 10.22 and the integral on ∂E is as defined in Definition 10.5.24.

10.26 Let E be an almost regular set in Z. Let F and G be two real-valued $\mathcal{C}^2$ functions on Z. Show that

$$\int_{\partial E} (F\nabla G) = \int_E ((\nabla F) \cdot (\nabla G) + F\nabla^2 G).$$

10.27 Let E be an almost regular set in Z. Let F and G be two real-valued $\mathcal{C}^2$ functions on Z. Show that

$$\int_{\partial E} (F\nabla G - G\nabla F) = \int_E (F\nabla^2 G - G\nabla^2 F).$$

10.28 Let $F(x,\, y,\, z) = (x^2 + y^2 + z^2)^{-1/2}$. Compute $\int_S \nabla F$ where S is the unit sphere $x^2 + y^2 + z^2 = 1$ in $\mathbb{R}^3$ oriented by the outer normals.

10.29 Let E be an almost regular set in $\mathbb{R}^3$ containing the origin in its interior. Compute $\int_{\partial E} \nabla F$ where F is as in Problem 10.28 and the integral on ∂E is as defined in Definition 10.5.24.

10.30 Repeat Problem 10.29 under the assumption that $\mathbf{0} \notin \overline{E}$.

10.31 Let Z be an n-dimensional Euclidean space. Let $F(\mathbf{z}) = \|\mathbf{z}\|^{2-n}$. Compute $\int_S \nabla F$ where S is the unit sphere $\|\mathbf{z}\| = 1$ in Z oriented by the outer normals.

10.32 Let Z and F be as in Problem 10.31. Compute $\int_{\partial E} \nabla F$ where E is an almost regular set in Z containing the origin in its interior.

10.33 Repeat Problem 10.32 under the assumption that $\mathbf{0} \notin \overline{E}$.

10.34 Let $c_i \in \mathbb{R}$ be finitely many numbers and let $\mathbf{a}_i$ be finitely many points in an n-dimensional Euclidean space Z. Let $F(\mathbf{z}) = \sum_i c_i \|\mathbf{z} - \mathbf{a}_i\|^{2-n}$. Let E be an almost regular set in Z such that $\mathbf{a}_i \notin \partial E$ for all i. Compute $\int_{\partial E} \nabla F$.

10.35 Let $F(x, y) = \log(x^2 + y^2)$. Let C be the unit circle $x^2 + y^2 = 1$. Note that C is both a surface and a line in $\mathbb{R}^2$. (See also Problem 9.38.) Compute the surface integral $\int_C \nabla F$ where C is oriented by the outer normals.

10.36 Let E be an almost regular set in $\mathbb{R}^2$ containing the origin in its interior. Let F be as in Problem 10.35. Compute $\int_{\partial E} \nabla F$.

10.37 Repeat Problem 10.36 under the assumption that $\mathbf{0} \notin \overline{E}$.

10.38 Let $r = (x^2+y^2)^{1/2}$. Let $\mathbf{f}(x, y) = (-x, y)/r$. Let E be an almost regular set in $\mathbb{R}^2$. Consider the boundary-surface $\mathcal{S}_E$ as a curve C, oriented by the outer normals according to the right-hand rule. Assume that E contains the origin in its interior. Compute the line integral $\int_C \mathbf{f}$.

10.39 Repeat Problem 10.36 under the assumption that $\mathbf{0} \notin \overline{E}$.

10.40 Let $c_i \in \mathbb{R}$ be finitely many numbers and let $\mathbf{a}_i \in \mathbb{R}^2$ be finitely many points. Let $F(\mathbf{z}) = \sum_i c_i \log \|\mathbf{z} - \mathbf{a}_i\|$. Let E be an almost regular set in $\mathbb{R}^2$ such that $\mathbf{a}_i \notin \partial E$ for all i. Compute $\int_{\partial E} \nabla F$.

10.41 Let F and E be as in Problem 10.40. Consider the boundary-surface $\mathcal{S}_E$ as a curve C, oriented by the outer normals according to the right-hand rule. Consider $\mathbb{R}^2$ as the xy-plane in the xyz-space. Let $\mathbf{k} = (0, 0, 1)$ be the usual unit vector of the z-axis. Let $\mathbf{f} = \mathbf{k} \times \nabla F$. Compute $\int_C \mathbf{f}$.

10.42 Let $\mathbf{i}$, $\mathbf{j}$, $\mathbf{k}$ be the usual unit vectors in the xyz-space. Let E be an almost regular set in the xy-plane. Given a vector field $\mathbf{f}(x, y) = P(x, y)\mathbf{i} + Q(x, y)\mathbf{j}$ in this plane, apply Stokes' theorem, Theorem 10.5.25, to $\mathbf{f} \times \mathbf{k}$ to obtain *Green's theorem*:

$$\int_{\partial E} (Pdx + Qdy) = \int_E \left(\frac{\partial Q}{\partial x} - \frac{\partial P}{\partial y} \right) dxdy.$$

Note that the surface integral of $\mathbf{h} \times \mathbf{k}$ becomes the line integral of $\mathbf{h}$.

10.43 Verify Green's theorem for $\int_C (x^2 - xy^3)dx + (y^3 - 2xy)dy$, where C is the square with vertices $(0, 0)$, $(2, 0)$, $(2, 2)$, $(0, 2)$.

10.44 Evaluate the line integral $\int_C (y^2 \sin(xy^2)dx + 2xy\sin(xy^2)dy)$ where C is the unit circle.

10.45 Choose $\mathbf{h}(x, y) = x\mathbf{j}$ to obtain the area of E as a line integral on the boundary of E. Use this result to find the area of the region bounded by the curve $x = a\cos^3 t$, $y = a\sin^3 t$, $0 \leq t \leq 2\pi$.

10.6 STOKES' THEOREM ON MANIFOLDS

Let M be an ℓ-dimensional $\mathcal{C}^2$ manifold in a Euclidean space Z. Let

$$\Psi : G \to H = \Psi(G) \subset W$$

be a $\mathcal{C}^2$ chart for M so that $\Psi(G \cap M) = H \cap U$, where U is an ℓ-dimensional subspace of W. Let $\Phi : H \to G$ be the reverse chart with the associated parametric representation $\varphi : H \cap U \to G \cap M$. Let A be an open set in W such that $A \cap U$ is a regular neighborhood for a set B in U. Then $P = M \cap \Phi(A) = \varphi(A \cap U)$ will be called a *regular neighborhood* of $Q = \Phi(B) = \varphi(B)$ in M. Note that $B \cap A$ is a regular set in U.

Let $k = (\ell - 1)$ and let $\omega : G \to \Lambda_k(Z)$ be a $\mathcal{C}^1$ tensor field of compact support contained in $\Phi(A)$. The pullbacks of ω and $d\omega$ under Φ are the tensor fields

$$\xi = \Phi^*(\omega) : H \to \Lambda_k(W) \text{ and } \eta = \Phi^*(d\omega) : H \to \Lambda_\ell(W).$$

Theorem 10.4.9 shows that $\eta = d\xi$. The restriction of $\xi(\mathbf{w}) : W^k \to \mathbb{R}$ to U^k defines an element in $\Lambda_k(U)$. Hence these restrictions define a new tensor field $\xi_0 : H \cap U \to \Lambda_k(U)$. Similarly one obtains $\eta_0 : H \cap U \to \Lambda_\ell(U)$. By an easy verification we see that $d\xi_0 = \eta_0$.

An application of Stokes' theorem for regular neighborhoods, Theorem 10.5.14, shows that $\int_B d\xi = \int_{\partial B} \xi$. We will assume that Ψ is an orientation-preserving chart. Arguments given for the proofs of Theorems 10.5.8 and 9.3.8 show that this is not a loss of generality. Then the change of variables theorem for tensor fields, Theorem 9.5.12, shows that $\int_E d\omega = \int_{\partial E} \omega$. This is Stokes' theorem on manifolds for a regular neighborhood of a set in a manifold. From this we obtain Stokes' theorem on manifolds, as we obtained Theorem 10.5.15 from Theorem 10.5.14. Extensions to almost regular sets are also obtained as before.

Stokes' theorem on manifolds is related to flows and vector fields on manifolds. Let M be a manifold in Z. Then $\mathbf{f} : M \to Z$ is called a *vector field* on M if $\mathbf{f}(\mathbf{m}) \in T_\mathbf{m}$ at each $\mathbf{m} \in M$. Here $T_\mathbf{m}$ is the tangent space of M at $\mathbf{m} \in M$. If M is an ℓ-dimensional oriented manifold, then each $T_\mathbf{m}$ has a positive Euclidean determinant $\vartheta_\mathbf{m} \in \Lambda_\ell(T_\mathbf{m})$. Then we see that a tensor field ω of order $k = (\ell - 1)$ is related to a vector field on an oriented ℓ-dimensional manifold as

$$\omega(\mathbf{m})(\mathbb{T}_\mathbf{m}) = \vartheta_\mathbf{m}(\mathbf{f}(\mathbf{m}), \mathbb{T}_\mathbf{m}) \in \mathbb{R}, \quad \mathbf{m} \in M, \quad \mathbb{T}_\mathbf{m} \in {T_\mathbf{m}}^k. \tag{10.86}$$

Let $\mathbf{f} : M \to Z$ be a vector field on M with a compact support K. If there is a chart $\Psi : G \to H$ for M such that $K \subset G$, then $\mathbf{f}$ is the initial velocity field of a flow on M. In fact, if $\mathbf{f}$ is related to ω as in (10.86), then $\Phi^*(\omega)$ is related to a vector field in the Euclidean space V. This induces a flow in V with compact support $\Psi(K) \subset H$. Then $\Phi : H \to G$ maps this flow to a flow on M.

As an illustration of these relationships, we will consider Stokes' theorem on two-dimensional manifolds in three-dimensional spaces. Indeed, the original version of Stokes' theorem applied only to this case.

Classical Stokes' theorem in Tensor Form

Remarks 10.6.1 Orientations in classical Stokes' Theorem. Once again, we start with the "basic case" for the classical version of Stokes' theorem. The space W is a two-dimensional plane in a three-dimensional space Y. A line U in W divides this plane into a lower half A and an upper half. The boundary of A in the W-plane is the line U. The orientation of U is as before: if $\mathbf{e} = \mathbf{e}_1$ is the outer unit normal of U in W, then a unit vector $\mathbf{e}_2$ defines the positive orientation on U just in case $(\mathbf{e}_1, \mathbf{e}_2)$ is a positive orthonormal basis for W. The orientation of the W-plane itself is specified by its outer unit normal vector $\mathbf{n}$ in the Y-space. Hence $(\mathbf{n}, \mathbf{e}_1, \mathbf{e}_2)$ is a positive orthonormal basis for Y.

In more picturesque terms, the relation between these three directions is sometimes described as follows. If we are standing on the W-plane with our heads pointing in the direction $\mathbf{n}$ of the outer normal of W, and if we want to walk in the positive direction along the line U, then the lower half A of W must stay on our left-hand side.

Theorem 10.6.2 Basic classical Stokes' theorem in tensor form. *With the notations and definitions in Remarks 10.6.1, if $\xi : Y \to \Lambda_1(Y)$ is a $\mathcal{C}^1$ tensor field of compact support, then*

$$\int_A d\xi = \int_{\partial A} \xi = \int_U \xi. \tag{10.87}$$

Proof. The integrals in (10.87) are performed on oriented manifolds, as described in Remarks 10.6.1. To prove this result, restrict $\xi : Y \to \Lambda_1(Y)$ to W. We obtain a tensor field $\xi|_W : W \to \Lambda_1(W)$ on the W-plane. Then an application of the basic Stokes' theorem, Theorem 10.4.11, gives (10.87). □

Remarks 10.6.3 In Theorem 10.6.2, the boundary ∂A is the boundary of A in the W-plane, as described in Remarks 10.6.1, rather than its boundary in Y (which would be the entire set A). Analogously, in Theorem 10.6.4 below, the boundary ∂E will be the boundary of E in the surface S. The definition of this boundary is left as an exercise.

Theorem 10.6.4 Local classical Stokes' theorem in tensor form. *Let E be a set in an oriented $\mathcal{C}^2$ surface S in a three-dimensional Euclidean space Z. Let G be an* *open set in Z. Let*

$$\Psi : G \to H = \Psi(G) \subset Y$$

be an orientation-preserving $\mathcal{C}^2$ diffeomorphism such that

$$\Psi(G \cap S) = H \cap W, \tag{10.88}$$

$$\Psi(G \cap E^o) = H \cap A, \text{ and} \tag{10.89}$$

$$\Psi(G \cap \partial E) = H \cap \partial A = H \cap U. \tag{10.90}$$

Let $\omega : G \to \Lambda_1(Z)$ be a $\mathcal{C}^1$ tensor field of compact support contained in G. Then

$$\int_E d\omega = \int_{\partial E} \omega. \tag{10.91}$$

Proof. Use the reverse diffeomorphism $\Phi : H \to G$ to pull back ω as $\xi = \Phi^*(\omega)$. Then apply Theorem 10.6.2. The details are left as an exercise. □

Theorem 10.6.5 Classical Stokes' theorem in tensor form. *Let E and S be as in Theorem 10.6.4. Assume that $\overline{E}$ is compact and is covered by open sets G as specified in that theorem. Let $\omega : Z \to \Lambda_1(Z)$ be a $\mathcal{C}^1$ tensor field. Then*

$$\int_E d\omega = \int_{\partial E} \omega. \tag{10.92}$$

Proof. This is left as an exercise. □

Classical Stokes' Theorem in Vector Form

The inner product of $\mathbf{a}$, $\mathbf{b} \in \mathbb{R}^3$ is denoted as the dot product $\mathbf{a} \cdot \mathbf{b}$. The standard orientation of $\mathbb{R}^3$ is determined by the Euclidean determinant $\vartheta(\mathbf{a}, \mathbf{b}, \mathbf{c}) = \mathbf{a} \cdot (\mathbf{b} \times \mathbf{c})$, as the usual mixed product. Let $\omega : G \to \Lambda_1(\mathbb{R}^3)$ be a tensor field of order one, defined on an open set G. It is associated with a vector field $\mathbf{f} : G \to \mathbb{R}^3$ as $\omega(\mathbf{r})(\mathbf{a}) = \mathbf{f}(\mathbf{r}) \cdot \mathbf{a}$, where $\mathbf{r} \in G$, $\mathbf{a} \in \mathbb{R}^3$.

Definition 10.6.6 Curl of a vector field. Let $\omega(\mathbf{r})(\mathbf{a}) = \mathbf{f}(\mathbf{r}) \cdot \mathbf{a}$ be a $\mathcal{C}^1$ tensor field $\omega : G \to \Lambda_1(\mathbb{R}^3)$ of order one. Then its exterior derivative $d\omega : G \to \Lambda_2(\mathbb{R}^3)$ is a tensor field of order two. Therefore it is represented by a vector field $\mathbf{g} : G \to \mathbb{R}^3$ as

$$d\omega(\mathbf{r})(\mathbf{a}, \mathbf{b}) = \mathbf{g} \cdot \mathbf{a} \times \mathbf{b}, \ \ \mathbf{r} \in G, \ \ (\mathbf{a}, \mathbf{b}) \in (\mathbb{R}^3)^2.$$

Then $\mathbf{g} = \operatorname{curl} \mathbf{f} : G \to \mathbb{R}^3$ is called the *curl* of $\mathbf{f} : G \to \mathbb{R}^3$. Hence

$$d\omega(\mathbf{r})(\mathbf{a}, \mathbf{b}) = \operatorname{curl} \mathbf{f}(\mathbf{r}) \cdot \mathbf{a} \times \mathbf{b}, \ \ \mathbf{r} \in \mathbb{R}^3, \ \ (\mathbf{a}, \mathbf{b}) \in (\mathbb{R}^3)^2. \tag{10.93}$$

Remarks 10.6.7 Vectorial expression of curl. Let $\mathbf{f} : \mathbb{R}^3 \to \mathbb{R}^3$ be a $\mathcal{C}^1$ vector field and $\omega(\mathbf{r})(\mathbf{b}) = \mathbf{b} \cdot \mathbf{f}(\mathbf{r})$. Then, by the definitions in 10.4.5 and 10.4.6,

$$\begin{aligned} \omega'(\mathbf{r})(\mathbf{a},\, \mathbf{b}) &= \mathbf{b} \cdot \mathbf{f}'(\mathbf{r})\, \mathbf{a}. \text{ Hence} && (10.94) \\ d\omega(\mathbf{r})(\mathbf{a},\, \mathbf{b}) &= \mathbf{b} \cdot \mathbf{f}'(\mathbf{r})\, \mathbf{a} - \mathbf{a} \cdot \mathbf{f}'(\mathbf{r})\, \mathbf{b} && (10.95) \\ &= \operatorname{curl} \mathbf{f}(\mathbf{r}) \cdot \mathbf{a} \times \mathbf{b}. && (10.96) \end{aligned}$$

Remarks 10.6.8 Coordinate expressions for curl. Let $\mathbb{R}^3$ be the standard xyz-space with the standard orthonormal basis $(\mathbf{i},\, \mathbf{j},\, \mathbf{k})$. The coordinates of $\operatorname{curl} \mathbf{f}$ can be obtained by evaluating (10.95) at the pairs $(\mathbf{j},\, \mathbf{k})$, $(\mathbf{k},\, \mathbf{i})$, and $(\mathbf{i},\, \mathbf{j})$, as in the proof of Lemma 9.5.10. If $\mathbf{f}(\mathbf{r}) = L(\mathbf{r})\mathbf{i} + M(\mathbf{r})\mathbf{j} + N(\mathbf{r})\mathbf{k}$ then

$$\operatorname{curl} \mathbf{f} = (N_{\mathbf{y}} - M_{\mathbf{z}})\mathbf{i} + (L_{\mathbf{z}} - N_{\mathbf{x}})\mathbf{j} + (M_{\mathbf{x}} - L_{\mathbf{y}})\mathbf{k}, \tag{10.97}$$

where the subscripts denote partial derivatives. One can remember this expression more easily through the notation

$$\operatorname{curl} \mathbf{f} = \begin{vmatrix} \mathbf{i} & \mathbf{j} & \mathbf{k} \\ \dfrac{\partial}{\partial x} & \dfrac{\partial}{\partial y} & \dfrac{\partial}{\partial z} \\ L & M & N \end{vmatrix} = \nabla \times \mathbf{f}. \tag{10.98}$$

Remarks 10.6.9 Integrals of tensor and vector fields. The classical Stokes' theorem is obtained from Theorem 10.6.5 by replacing the integrals of tensor fields by the corresponding integrals of vector fields. These two types of integrals are related as described in Theorem 9.5.7. The tensor field ω is associated with the vector field $\mathbf{f}$. The integral of ω on the curve ∂E becomes the line integral $\int_{\partial E} \mathbf{f}$. The integral of $d\omega$ on the surface E becomes the surface integral $\int_E \operatorname{curl} \mathbf{f}$. The orientations of E and ∂E are as described in the proof of Theorem 10.6.2.

Theorem 10.6.10 Classical Stokes' theorem. *Let E and S be as in Theorem 10.6.4. Assume that $\overline{E}$ is compact and is covered by open sets G as specified in that theorem. Let $\mathbf{f} : Z \to Z$ be a $\mathcal{C}^1$ vector field. Then*

$$\int_E \operatorname{curl} \mathbf{f} = \int_{\partial E} \mathbf{f}. \tag{10.99}$$

Proof. This follows from the classical Stokes' theorem in tensor form, Theorem 10.6.5, by replacing the integrals of tensor fields by the corresponding integrals of vector fields. The correspondence between these integrals is as discussed in Remarks 10.6.9 above. □

Problems

10.46 Show that $\nabla \cdot (\mathbf{f} \times \mathbf{g}) = \mathbf{g} \cdot \nabla \times \mathbf{f} - \mathbf{f} \cdot \nabla \times \mathbf{g}$.

10.47 Show that $\nabla \cdot \nabla \times \mathbf{f} = 0$.

10.48 Show that $\nabla \times \nabla F = 0$.

10.49 Verify the classical Stokes' theorem, Theorem 10.6.10, for the following cases of E and $\mathbf{f}$, by computing the integrals $\int_{\partial E} \mathbf{f}$ and $\int_E \operatorname{curl} \mathbf{f}$ separately.

1. S is given as $z = (x^2 + y^2)^{1/2}$ and $1 < z < 4$ and $\mathbf{f}(x, y, z) = (z^2, x^2, y^2)$.
2. S is given as $z = x^2 + y^2$ and $z < 4$ and $\mathbf{f}(x, y, z) = (z, xz, x)$.
3. S is given as $x + y + z = 1, 0 < x, 0 < y, 0 < z$, and $\mathbf{f}(x, y, z) = (y, z, x)$.

10.50 Show that $\nabla \cdot ((\nabla F) \times (\nabla G)) = 0$.

PART IV

APPENDICES

APPENDIX A

CONSTRUCTION OF THE REAL NUMBERS

Introduction. Let $\mathbb{Q}$ be the set of rational numbers. The most important properties of $\mathbb{Q}$ are the properties of the arithmetic operations and the properties of the order relation on $\mathbb{Q}$. These properties are summarized by two sets of rules called the *field axioms* and the *order axioms*. The principal deficiency of $\mathbb{Q}$ is that it is not complete. The set of rational numbers x such that $x^2 < 2$ is bounded above, yet there is no rational least upper bound. We would like to extend $\mathbb{Q}$ to $\mathbb{R}$, a set of numbers that not only satisfies the field axioms and the order axioms but also has the completeness property. This section reviews a standard way to construct the real numbers that meets this objective.

Outline of the construction of $\mathbb{R}$. Prior to attempting any construction such as the present one, we freely make use of the real numbers without proving their existence.

Analysis in Vector Spaces.
By M. A. Akcoglu, P. F. A. Bartha and D. M. Ha

We know many of the properties that they must have. In particular, we know that any real number is the limit of a sequence of rational numbers. That fact suggests that we try to define real numbers as the limits of convergent sequences of rational numbers. But Definition 2.3.2 of convergent sequences depends explicitly upon the limit point. We have defined convergence to L, not convergence *per se*. To circumvent this difficulty, we base our construction on Cauchy sequences of rational numbers rather than convergent sequences. Cauchy sequences are defined purely in terms of rational numbers, and they coincide with the class of sequences that we shall ultimately regard as convergent to a limit in $\mathbb{R}$.

Let $\mathcal{C}$ be the class of all Cauchy sequences of rational numbers. $\mathcal{C}$ cannot be identified with the set of real numbers because many different Cauchy sequences may converge to the same real number. Hence we introduce an equivalence relation on $\mathcal{C}$, making two Cauchy sequences equivalent if the difference sequence converges to zero. Note that convergence to zero can be defined in terms of rational numbers only. Hence real numbers will be defined as equivalence classes of Cauchy sequences of rational numbers. We will see that the usual arithmetic operations and the order relation can be defined easily for these equivalence classes, and that the resulting system is complete.

In what follows, we only assume basic knowledge of the rational numbers. We restate some of the earlier definitions using only rational numbers, making sure that all of our arguments are formulated in terms of rational numbers only.

A.1 FIELD AND ORDER AXIOMS IN $\mathbb{Q}$

Definition A.1.1 Field axioms on $\mathbb{Q}$. There are two binary operations $\mathbb{Q} \times \mathbb{Q} \to \mathbb{Q}$ called *addition* and *multiplication*. Addition applied to $(a, b) \in \mathbb{Q} \times \mathbb{Q}$ gives $(a + b)$. Multiplication applied to $(a, b) \in \mathbb{Q} \times \mathbb{Q}$ gives $(a \cdot b)$. These operations have the following properties.

Commutativity

$$a + b = b + a \text{ and } a \cdot b = b \cdot a \text{ for all } a, b \in \mathbb{Q}.$$

Associativity

$$(a + b) + c = a + (b + c) \text{ and } (a \cdot b) \cdot c = a \cdot (b \cdot c) \text{ for all } a, b, c \in \mathbb{Q}.$$

Distributivity

$$a \cdot (b + c) = (a \cdot b) + (a \cdot c) \text{ for all } a, b, c \in \mathbb{Q}.$$

Existence of the neutral elements There are two elements in $\mathbb{Q}$, 0 and 1, such that $0 \neq 1$ and such that

$$0 + a = a \text{ and } 1 \cdot a = a \text{ for all } a \in \mathbb{Q}.$$

Existence of inverses for addition (subtraction)

For each $a \in \mathbb{Q}$, there is a $(-a) \in \mathbb{Q}$ such that $a + (-a) = 0$.

Existence of the inverse for multiplication (division)

If $a \in \mathbb{Q}$ and if $a \neq 0$, then there is a $(1/a) \in \mathbb{Q}$ such that $a \cdot (1/a) = 1$.

Definition A.1.2 Order axioms on $\mathbb{Q}$. There is a set $P_{\mathbb{Q}} \subset \mathbb{Q}$, called the *positive (rational) numbers*, with the following two properties.

1. For each $p \in \mathbb{Q}$, exactly one of the following three cases is true: $p = 0$, $p \in P_{\mathbb{Q}}$, or $-p \in P_{\mathbb{Q}}$.

2. If $p,\ q \in P_{\mathbb{Q}}$, then $p + q \in P_{\mathbb{Q}}$ and $pq \in P_{\mathbb{Q}}$.

Remarks A.1.3 Our objective. We would like to construct a set $\mathbb{R}$ with the following properties.

1. There are two binary operations $\mathbb{R} \times \mathbb{R} \to \mathbb{R}$ on $\mathbb{R}$, *addition* and *multiplication*, that satisfy the field axioms on $\mathbb{R}$.

2. There is a set $P_{\mathbb{R}} \subset \mathbb{R}$, the set of *positive (real) numbers*, that satisfies the order axioms on $\mathbb{R}$.

3. There is a one-to-one mapping $\varphi : \mathbb{Q} \to \mathbb{R}$ such that $\varphi(p + q) = \varphi(p) + \varphi(q)$ and $\varphi(pq) = \varphi(p)\varphi(q)$ for all $p,\ q \in \mathbb{Q}$ and such that $\varphi(P_{\mathbb{Q}}) \subset P_{\mathbb{R}}$. Here $\varphi(p) + \varphi(q)$ and $\varphi(p)\varphi(q)$ are stated in terms of the addition and multiplication operations on $\mathbb{R}$.

4. Every nonempty subset of $\mathbb{R}$ with an upper bound has a least upper bound. Hence, $\mathbb{R}$ satisfies the completeness axiom, Axiom 2.2.3.

A.2 EQUIVALENCE CLASSES OF CAUCHY SEQUENCES IN $\mathbb{Q}$

Definition A.2.1 Cauchy sequences of rational numbers. A sequence $x : \mathbb{N} \to \mathbb{Q}$ is called a *Cauchy sequence* of rational numbers if for each rational number $a > 0$, there is an $N \in \mathbb{N}$ such that $|x_m - x_n| < a$ for all $m,\ n \geq N$. Let $\mathcal{C}$ be the set of all Cauchy sequences of rational numbers.

Definition A.2.2 Zero sequences of rational numbers. A sequence $x : \mathbb{N} \to \mathbb{Q}$ is called a *zero sequence* of rational numbers if for each rational number $a > 0$, there is an $N \in \mathbb{N}$ such that $|x_n| < a$ for all $n \geq N$. Let $\mathcal{Z}$ be the set of all zero sequences of rational numbers.

Lemma A.2.3 *Every zero sequence of rational numbers is also a Cauchy sequenceof rational numbers. Hence $\mathcal{Z} \subset \mathcal{C}$.*

Proof. This is left as an exercise. □

Definition A.2.4 Constant sequences of rational numbers. Each rational number $q \in \mathbb{Q}$ defines a constant sequence $\bar{q} : \mathbb{N} \to \mathbb{Q}$: we set $\bar{q}_n = q$ for all $n \in \mathbb{N}$. Let $\mathcal{Q}$ be the set of all constant sequences of rational numbers. It is clear that $\bar{q} \in \mathcal{C}$ for each $q \in \mathbb{Q}$. Hence $\mathcal{Q} \subset \mathcal{C}$. Note that $\bar{q} \in \mathcal{Z}$ if and only if $q = 0$.

Definition A.2.5 Sum and product of two sequences. If $x : \mathbb{N} \to \mathbb{Q}$ and $y : \mathbb{N} \to \mathbb{Q}$ are two sequences, then their sum is defined as the sequence $(x + y)_n = x_n + y_n$ and their product as the sequence $(xy)_n = x_n y_n$.

Lemma A.2.6 *Let $x : \mathbb{N} \to \mathbb{Q}$ and $y : \mathbb{N} \to \mathbb{Q}$ be two sequences.*

(1) If $x, y \in \mathcal{C}$, then $(x + y) \in \mathcal{C}$ and $(xy) \in \mathcal{C}$.

(2) If $x, y \in \mathcal{Z}$, then $(x + y) \in \mathcal{Z}$ and $(xy) \in \mathcal{Z}$.

(3) If $x \in \mathcal{Z}$ and $y \in \mathcal{C}$, then $(xy) \in \mathcal{Z}$.

Proof. This is left as an exercise. □

Definition A.2.7 Addition and multiplication on $\mathcal{C}$. *Addition* on $\mathcal{C}$ is the operation $\mathcal{C} \times \mathcal{C} \to \mathcal{C}$ that takes the pair $(x, y) \in \mathcal{C} \times \mathcal{C}$ to $(x+y) \in \mathcal{C}$. Similarly, *multiplication* on $\mathcal{C}$ is the operation $\mathcal{C} \times \mathcal{C} \to \mathcal{C}$ that takes the pair $(x, y) \in \mathcal{C} \times \mathcal{C}$ to $(xy) \in \mathcal{C}$. Note that if $p, q \in \mathbb{Q}$, then $\overline{(p + q)} = \overline{p} + \overline{q}$ and $\overline{(pq)} = \overline{p}\,\overline{q}$, with the notations of Definition A.2.4.

Definition A.2.8 An equivalence relation on $\mathcal{C}$. Define a relation on $\mathcal{C} \times \mathcal{C}$ as follows. If $x, y \in \mathcal{C}$, then x is related to y if and only if $(x - y) \in \mathcal{Z}$. It is easy to check that this is an equivalence relation on $\mathcal{C}$ in the sense of Definition 1.1.9. We write $x \sim y$ for $(x - y) \in \mathcal{Z}$. Note that if $p, q \in \mathbb{Q}$, then $\overline{p} \sim \overline{q}$ if and only if $p = q$, with the notations of Definition A.2.4.

Definition A.2.9 Real numbers. Equivalence classes were defined in Definition 1.1.11. The equivalence class represented by $x \in \mathcal{C}$ is the set

$$E_x = \{ x' \in \mathcal{C} \mid x' \sim x \} \subset \mathcal{C}.$$

Any two equivalence classes are either identical or disjoint. In fact, if $x \sim y$, then $E_x = E_y$, and if $x \not\sim y$, then $E_x \cap E_y = \emptyset$. The union of all equivalence classes is $\mathcal{C}$. Each equivalence class is also called a *real number*. The equivalence class E_x is the real number represented by the sequence x. The same real number can be represented by any sequence in E_x, i.e., by any sequence $x' \sim x$. Let $\mathbb{R}$ denote the collection of all equivalence classes. This collection is called the set of *real numbers*.

Definition A.2.10 Operations on real numbers. We will define addition and multiplication operations $\mathbb{R} \times \mathbb{R} \to \mathbb{R}$. Let E_x, $E_y \in \mathbb{R}$. Hence E_x and E_y are two real numbers and also two equivalence classes represented by the sequences x and y. Their sum $E_x + E_y$ and their product $E_x E_y$ are defined as

$$E_x + E_y = E_{x+y} \quad \text{and} \quad E_x E_y = E_{xy}.$$

This definition has to be justified. If $E_x = E_{x'}$ and $E_y = E_{y'}$, must we have $E_{x+y} = E_{x'+y'}$ and $E_{xy} = E_{x'y'}$? If not, then the definitions above are meaningless. The required justification is provided by Lemma A.2.11 below.

These definitions are reasonable. They depend on the fact that the limit of the sum of two sequences is the sum of their limits. The same is also true for products. Theorem A.2.15 below shows that the field axioms are satisfied on $\mathbb{R}$ with these operations.

Lemma A.2.11 *Let x, x', y, $y' \in \mathcal{C}$. If $x \sim x'$ and $y \sim y'$, then $(x+y) \sim (x'+y')$ and $(xy) \sim (x'y')$.*

Proof. Let $x' = x + p$ and $y' = y + q$ with p, $q \in \mathcal{Z}$. Then

$$(x' + y') - (x + y) = p + q \in \mathcal{Z}$$

by the second part of Lemma A.2.6. Hence $(x + y) \sim (x' + y')$. Also,

$$(x'y') - (xy) = (py) + (xq) + (pq) \in \mathcal{Z},$$

by the second and third parts of Lemma A.2.6. Hence $(xy) \sim (x'y')$. $\square$

Definition A.2.12 Positive real numbers. Call a sequence $x : \mathbb{N} \to \mathbb{Q}$ an *eventually positive sequence* if there is an $n \in \mathbb{N}$ such that $x_n > 0$ for all $n \geq N$. Call a real number $E_x \in \mathbb{R}$ a *positive real number* if each $y \in E_x$ is an eventually positive sequence. Let $P_{\mathbb{R}} \subset \mathbb{R}$ be the set of all positive real numbers. These definitions are reasonable as well. In fact, the limit of a sequence is positive if and only if all the sequences converging to the same limit are eventually positive. Theorem A.2.16 below shows that the order axioms are satisfied on $\mathbb{R}$ with this definition of positivity.

Lemma A.2.13 *Let $x \in \mathcal{C}$. Assume that for each rational number $a > 0$ and for each $N \in \mathbb{N}$, there is an $n \geq N$ such that $|x_n| \leq a$. Then $x \in \mathcal{Z}$.*

Proof. Given a rational $a > 0$, find an $N \in \mathbb{N}$ such that $|x_n - x_m| \leq a/2$ for all $m, n \geq N$. This can be done since $x \in \mathcal{C}$. By assumption, we can find an $m \in \mathbb{N}$ such that $m \geq N$ and such that $|x_m| \leq a/2$. Let $n \geq N$. Then

$$|x_n| = |x_m + (x_n - x_m)| \leq |x_m| + |x_n - x_m| \leq (a/2) + (a/2) = a.$$

Hence $x \in \mathcal{Z}$. $\square$

Lemma A.2.14 *Let $x \in \mathcal{C}$ and $x \notin \mathcal{Z}$. Then there is an $a > 0$, $a \in \mathbb{Q}$, such that the following are true.*

(1) There is an $M \in \mathbb{N}$ such that $|x_m| > a$ for all $m \geq M$.

(2) There is an $N \in \mathbb{N}$ such that either $x_n > a$ for all $n \geq N$ or $-x_n > a$ for all $n \geq N$.

(3) Let $y \sim x$. Then there is a $K \in \mathbb{N}$ such that x_k and y_k are both nonzero and have the same sign for all $k \geq K$.

Proof. Since $x \notin \mathcal{Z}$, the hypothesis of Lemma A.2.13 cannot be true. Hence there is a rational number $a > 0$ and an $M \in \mathbb{N}$ such that $|x_m| > a$ for all $m \geq M$. This proves the first part.

Now find an $N \geq M$, $N \in \mathbb{N}$, such that $|x_m - x_n| \leq a$ for all $m, n \geq N$. This can be done since x is a Cauchy sequence. Let $m, n \geq N$. If x_m and x_n have opposite signs, then $|x_n - x_m| > 2a$, since $|x_m| > a$ and $|x_n| > a$. This violates the condition that $|x_m - x_n| \leq a$ for all $m, n \geq N$. Hence either $x_n > a$ for all $n \geq N$ or $x_n < -a$ for all $n \geq N$. This proves the second part.

Let $y \sim x$. Then $(x - y) \in \mathcal{Z}$. Find a $K \geq N$, $K \in \mathbb{N}$, such that $|y_k - x_k| < a/2$ for all $k \geq K$. Then $-a/2 \leq y_k - x_k \leq a/2$ shows that

$$x_k - a/2 \leq y_k \leq x_k + a/2$$

for all $k \geq K$. Hence we see that if $a < x_k$, then $a/2 < y_k$, and if $x_k < -a$, then $y_k < -a/2$ for all $k \geq K$. This proves the last part. $\square$

Theorem A.2.15 *Let $\mathbb{R}$ be the collection of all equivalence classes in $\mathcal{C}$. Let the addition and multiplication operations $\mathbb{R} \times \mathbb{R} \to \mathbb{R}$ on $\mathbb{R}$ be defined by*

$$E_x + E_y = E_{x+y} \quad \text{and} \quad E_x E_y = E_{xy},$$

as in Definition A.2.10. Then these operations satisfy the field axioms, Axiom A.1.1, on $\mathbb{R}$.

Proof. Let $\bar{0}$ and $\bar{1}$ be the constant sequences consisting of all 0s and all 1s. Note that $E_{\bar{0}} = \mathcal{Z}$. We see that $E_{\bar{0}}$ is the neutral element for addition in $\mathbb{R}$ and $E_{\bar{1}}$ is the neutral element for multiplication in $\mathbb{R}$.

First, we verify the division axiom. Let $E_x \neq E_{\bar{0}}$. We will show that there is a $y \in \mathcal{C}$ such that $E_x E_y = E_1$. Now $x \in \mathcal{C}$ and $x \notin \mathcal{Z}$. Then the first part of Lemma A.2.14 shows that there is a rational number $a > 0$ and an $N \in \mathbb{N}$ such that $|x_n| > a$ for all $n \geq N$. Define a sequence $y : \mathbb{N} \to \mathbb{Q}$ by $y_n = 0$ if $n < N$ and $y_n = 1/x_n$ if $n \geq N$. We claim that $y \in \mathcal{C}$ and $E_x E_y = E_1$. If $m,\ n \geq N$, then

$$\begin{aligned} |y_n - y_m| &= |(1/x_n) - (1/x_m)| = |x_n - x_m|/|x_n x_m| \\ &\leq (1/a^2)|x_n - x_m|. \end{aligned}$$

Given a rational number $b > 0$, find an $M \geq N$, $M \in \mathbb{N}$, such that $|x_n - x_m| \leq a^2 b$ for all $m,\ n \geq M$. Then we see that $|y_n - y_m| \leq b$ for all $m,\ n \geq M$. Hence $y \in \mathcal{C}$. We see that $x_n y_n = 1$ for all $n \geq N$. Therefore $(x_n y_n - 1) = 0$ for all $n \geq N$. Then we see that $(xy - 1) \sim 0$ or that $xy \sim 1$. Hence $E_x E_y = E_1$.

The verification of the other axioms is quite routine. We verify only distributivity as an example. Addition and multiplication in $\mathcal{C}$ have the distributivity property. In fact, if $a,\ b,\ c \in \mathcal{C}$, then

$$\begin{aligned} (a(b+c))_n &= a_n(b+c)_n = a_n b_n + a_n b_n \\ &= (ab)_n + (ac)_n = ((ab) + (ac))_n. \end{aligned}$$

Now let $E_a,\ E_b,\ E_c \in \mathbb{R}$ with $a,\ b,\ c \in \mathcal{C}$. Then

$$\begin{aligned} E_a(E_b + E_c) &= E_a E_{b+c} = E_{a(b+c)} \\ &= E_{(ab)+(ac)} = E_{ab} + E_{ac} \\ &= E_a E_b + E_a E_b. \end{aligned}$$

Therefore the distributivity axiom is satisfied. □

Theorem A.2.16 *Let $\mathbb{R}$ be the set of real numbers with the arithmetic operations as defined in Definition A.2.10. Let $P_{\mathbb{R}}$ be the set of all positive real numbers as defined in Definition A.2.12. Then the order axioms, in Definition A.1.2, are satisfied on $\mathbb{R}$.*

Proof. First, note that $E_{\bar{0}} = -E_{\bar{0}}$ is not positive. For example, the sequence $x_n = (-1)^n(1/n)$ is a zero sequence which is not eventually positive. Now let $E_x \neq E_{\bar{0}}$. We will show that either E_x or $-E_x$ is positive.

We have $x \notin \mathcal{Z}$. Hence the second part of Lemma A.2.14 shows that either x or $-x$ is eventually positive. Assume that x is eventually positive. Then the last part of Lemma A.2.14 shows that if $y \sim x$, that is, if $y \in E_x$, then y is also eventually positive. Hence E_x is a positive real number. Similarly, if $-x$ is eventually positive and if $y \sim x$, then $-y$ is also eventually positive. In this case $-E_x = E_{-x}$ is a positive real number. This proves the first part of the order axioms.

For the second part, assume that E_x and E_y are both positive. Then, by the second part of Lemma A.2.14, there are $a,\ b > 0,\ a,\ b \in \mathbb{Q}$, and an $N \in \mathbb{N}$ such that $x_n > a$ and $y_n > b$ for all $n \geq N$. Then $x_n + y_n > a + b > 0$ and $x_n y_n > ab > 0$ for all $n \geq N$. Hence $E_x + E_y$ and $E_x E_y$ are both nonzero, and they are both positive numbers. □

Definition A.2.17 Canonical embedding of $\mathbb{Q}$ into $\mathbb{R}$. Let $\varphi : \mathbb{Q} \to \mathbb{R}$ be the function defined as $\varphi(p) = E_{\bar{p}}$. Here $\bar{p} \in \mathcal{C}$ is the constant sequence with all terms equal to $p \in \mathbb{Q}$. We will call $\varphi : \mathbb{Q} \to \mathbb{R}$ the *canonical embedding of* $\mathbb{Q}$ *into* $\mathbb{R}$.

Theorem A.2.18 *The canonical imbedding of $\mathbb{Q}$ into $\mathbb{R}$ is a one-to-one function $\varphi : \mathbb{Q} \to \mathbb{R}$. It preserves addition and multiplication in the sense that*

$$\varphi(p+q) = \varphi(p) + \varphi(q) \text{ and } \varphi(pq) = \varphi(p)\varphi(q) \text{ for all } p,\ q \in \mathbb{Q}.$$

Also, $\varphi : \mathbb{Q} \to \mathbb{R}$ preserves the order in the sense that $\varphi(P_{\mathbb{Q}}) \subset P_{\mathbb{R}}$.

Proof. If $E_{\bar{p}} = E_{\bar{q}}$, then $\bar{p} \sim \bar{q}$ and $(\bar{p} - \bar{q})$ is a zero sequence. But a constant sequence is a zero sequence only if the constant term is zero. This is obvious, as already mentioned in Definition A.2.4. Hence $E_p = E_q$ in $\mathbb{R}$ only if $p = q$ in $\mathbb{Q}$. This shows that $\varphi : \mathbb{Q} \to \mathbb{R}$ is one-to-one. For the second part,

$$\varphi(p+q) = E_{(p+q)} = E_p + E_q = \varphi(p) + \varphi(q).$$

Here the second equality follows from the definition of addition in $\mathbb{R}$. The proof of $\varphi(pq) = \varphi(p)\varphi(q)$ is similar. For the last part, note that if p is a positive rational number, then $\bar{p}$ is not a zero sequence and eventually (in fact, always) is positive. Hence $\varphi(p) = E_{\bar{p}}$ is a positive real number. □

Remarks A.2.19 Identification of $\mathbb{Q}$ with $\varphi(\mathbb{Q})$. We ignore the differences between $p \in \mathbb{Q}$ and $\bar{p} \in \mathcal{C}$ and $\varphi(p) = E_{\bar{p}} \in \mathbb{R}$. The meaning will be clear from the context. Hence we consider $\mathbb{Q}$ as a subset of $\mathbb{R}$. We also denote the real numbers with single letters, as is customary.

A.3 COMPLETENESS OF $\mathbb{R}$

The addition and multiplication operations and the order relation on $\mathbb{R}$ satisfy the field and order axioms. Hence we can develop all the related concepts, inequalities between real numbers, absolute values of real numbers, sequences, convergent sequences, and Cauchy sequences of real numbers as done before. We omit the repetition of these definitions and the related results. Instead, as an example, we give a detailed proof (perhaps more detailed than necessary) of the completeness of $\mathbb{R}$.

Note that any real number r is an equivalence class of Cauchy sequences of rational numbers. Any sequence x in this equivalence class will be called a *representative* (sequence) for r. The choice of representatives is arbitrary and not important. If $r = \varphi(p) = p$ is a rational number, however, then the constant sequence $\bar{p} = p$ is taken as the standard representative of $r = p$. The identifications used here are explained in Remarks A.2.19.

Lemma A.3.1 *Let x be a representative for $r \in \mathbb{R}$. Assume that there is an $N \in \mathbb{N}$ such that $0 \leq x_n$. Then $0 \leq r$.*

Proof. Any real number must satisfy exactly one of the following conditions: $r = 0$ or $r \in P$ or $-r \in P$. Our hypothesis rules out the last condition. In fact, this condition means that all representatives of $-r$ must eventually be positive. But $-r$ has at least one representative $-x$ which is not eventually positive. Hence $-r \notin P$. Therefore either $r = 0$ or $r \in P$. $\square$

Remarks A.3.2 Let x be a representative for $r \in \mathbb{R}$. Assume that there is an $N \in \mathbb{N}$ such that $0 < x_n$. Then $0 \leq r$ by Lemma A.3.1, but we cannot conclude that $0 < r$. In fact, $0 < r$ means that not one but all representatives of r must eventually be positive. A simple counterexample is $x_n = 1/n$, a positive sequence that represents 0.

Lemma A.3.3 *Let $r > 0$, $r \in \mathbb{R}$. Then there is a $p \in \mathbb{Q}$ such that $0 < p < r$.*

Proof. Let x be a representative sequence for r. Hence x is eventually positive, since $r > 0$. Lemma A.2.14 shows that there is an $a > 0$, $a \in \mathbb{Q}$, and an $N \in \mathbb{N}$ such that either $x_n > a$ for all $n \geq N$ or $x_n < -a$ for all $n \geq N$. The second case cannot be true, since x must eventually be positive. Hence $x_n > a$ for all $n \geq N$. Then, by Lemma A.3.1 above, $0 < a \leq r$. Let $p = a/2$. Then $0 < (a/2) < a \leq r$ shows that $0 < p < r$. $\square$

Lemma A.3.4 *Let r, $s \in \mathbb{R}$ and $r < s$. Then there is a $q \in \mathbb{Q}$ such that $r < q < s$.*

Proof. We have $0 < (s - r)$. Use Lemma A.3.3 to find $p \in \mathbb{Q}$ such that $0 < p < (s - r)$. Let x and y be representatives for r and s. Then $0 < (s - r) - p$ shows that the sequence $(y_n - x_n - p)$ is eventually positive. Find $M \in \mathbb{N}$ such that $y_m > x_m + p$ for all $m \geq M$. Since x is a Cauchy sequence, there is an $N \geq M$, $N \in \mathbb{N}$, such that $|x_N - x_n| \leq p/4$ for all $n \geq N$.

Let $q_1 = x_N + (p/4)$ and $q_2 = x_N + (3p/4)$. If $n \geq N$, then

$$q_1 - x_n = (p/4) + (x_N - x_n) \geq (p/4) + |x_N - x_n| \geq 0$$

shows that that $r \leq q_1$, and

$$q_2 = x_N + (3p/4) = x_n + (x_N - x_n) + (3p/4) \leq x_n + p \leq y_n$$

shows that $q_2 \leq s$. If $q = (q_1 + q_2)/2 = x_N + (p/2)$, then $x < q < s$. □

Remarks A.3.5 Upper bounds. Completeness is formulated in terms of upper bounds. They were introduced in Definition 2.2.1. $M \in \mathbb{R}$ is called an *upper bound* for a set $A \subset \mathbb{R}$ if $a \leq M$ for all $a \in A$. Note that if A has an upper bound $M \in \mathbb{R}$, then it also has an upper bound $M' \in \mathbb{Q}$. This follows from Lemma A.3.4 or just from the fact that any Cauchy sequence of rational numbers is bounded by a rational number.

Lemma A.3.6 *Let A be a nonempty set of real numbers. Assume that A has an upper bound. Then there is a $p \in \mathbb{Q}$ such that p is not an upper bound for A but $p + 1$ is an upper bound for A.*

Proof. Let $a \in A$. Then $r = (a - 1)$ is not an upper bound. Let $s < r$, $s \in \mathbb{Q}$. Then s is not an upper bound. The sequence $s_n = s + n$ is an unbounded sequence. Hence there are $n \in \mathbb{N}$ for which $s + n$ is an upper bound for A. Use the induction principle to find the least $m \in \mathbb{N}$ such that $q = s + m$ is an upper bound. Then $p = q - 1 = s + (m - 1)$ is not an upper bound. (Note that this is obviously true if $m \geq 2$, but it is also true even if $m = 1$.) □

Lemma A.3.7 *Let A be a nonempty set of real numbers. Assume that A has an upper bound. Then there are two sequences of rational numbers p_n and q_n such that if $n \in \mathbb{N}$, then p_n is not an upper bound for A, q_n is an upper bound for A, and $(q_n - p_n) = (1/2)^{n-1}$.*

Proof. Let $p,\ q \in \mathbb{Q}$ be the two numbers obtained in Lemma A.3.6. Let $p_1 = p$ and $q_1 = q$. Our requirements are satisfied for $n = 1$. Assume that p_n, q_n are obtained such that p_n is not an upper bound, q_n is an upper bound, and $(q_n - p_n) = (1/2)^{n-1}$.

Let $s_n = (p_n + q_n)/2$. If s_n is not an upper bound, then let $p_{n+1} = s_n$ and $q_{n+1} = q_n$. If s_n is an upper bound, then let $p_{n+1} = p_n$ and $q_{n+1} = s_n$. We see that our requirements are also satisfied for $n + 1$. Hence the sequences p_n and q_n are defined by the induction principle. □

Lemma A.3.8 *The sequences p_n and q_n obtained in Lemma A.3.7 are equivalent Cauchy sequences.*

Proof. We see that $p_n \leq p_{n+1} \leq q_{n+1} \leq q_n$. By induction it follows that

$$p_n \leq p_{n+k} \leq q_{n+k} \leq q_n \text{ for all } k \in \mathbb{N}.$$

Therefore, if $m \geq n$, then $|p_n - p_m| \leq |p_n - q_n| = (1/2)^{n-1}$. But $(1/2)^{n-1}$ is a zero sequence. Hence p_n is a Cauchy sequence. Similarly, q_n is also a Cauchy sequence. The equivalence of p_n and q_n follows from the fact that $(q_n - p_n) = (1/2)^{n-1}$ is a zero sequence. □

Theorem A.3.9 *Let A be a nonempty set of real numbers. Assume that A has an upper bound. Then A has a least upper bound. More explicitly, there is an $r \in \mathbb{R}$ such that r is an upper bound for A, but if $s < r$, $s \in \mathbb{R}$, then s is not an upper bound for A.*

Proof. Let p_n and q_n be the sequences obtained in Lemma A.3.7. Lemma A.3.8 shows that they are equivalent Cauchy sequences. Hence they represent the same $r \in \mathbb{R}$. We claim that r is a least upper bound for A.

Assume that r is not an upper bound for A. Then there is an $a \in A$ such that $r < a$. Use Lemma A.3.4 to find $q \in \mathbb{Q}$ such that $r < q < a$. Then the sequence $q - q_n$ represents the positive number $(q - r) \in \mathbb{R}$. Hence it must eventually be positive. Hence there is an $N \in \mathbb{N}$ such that $q > q_n$ for all $n \geq N$. This means that $q_n < q < a$ and q_n is not an upper bound for A. This is a contradiction. Hence r is an upper bound for A.

Now suppose that A has an upper bound $s \in \mathbb{R}$, $s < r$. As before, find a $p \in \mathbb{Q}$ such that $s < p < r$. Then $(p_n - p)$ must eventually be positive, as it represents $(r - p) > 0$. Therefore $p_n > p > s$ for some $n \in \mathbb{N}$. This means that p_n is an upper bound for A. This is a contradiction. Hence A cannot have upper bounds less than r. Therefore r is a least upper bound for A. □

APPENDIX B

DIMENSION OF A VECTOR SPACE

Let V be a vector space. Let $B = \{\mathbf{b}_1, \ldots, \mathbf{b}_m\}$ and $E = \{\mathbf{e}_1, \ldots, \mathbf{e}_n\}$ be two bases for V. This means that both B and E are linearly independent and $V = \text{Span } B = \text{Span } E$. Our objective is to prove that $m = n$. This fundamental result tells us that the number of elements in any basis for a finite-dimensional vector space is the same. This number is thus well defined and is called the *dimension* of the vector space.

One way to prove this result is to use a well-known result about linear systems of equations. Consider a homogeneous linear system of equations. If the number of unknowns is more than the number of equations, then this homogeneous system has nonzero solutions. This theorem implies that in $\mathbb{R}^n$, a set that contains more than n vectors cannot be linearly independent. Then the result about dimension follows easily. Instead of appealing to this argument, we shall provide a self-contained vectorial proof. The arguments used in this proof are purely algebraic.

Analysis in Vector Spaces.
By M. A. Akcoglu, P. F. A. Bartha and D. M. Ha

B.1 BASES AND LINEARLY INDEPENDENT SUBSETS

The following theorem contains the main step in proving that any two bases have the same number of elements.

Theorem B.1.1 *Let* $E = \{\mathbf{e}_1, \ldots, \mathbf{e}_n\}$ *be a basis for* V. *Let* $A = \{\mathbf{a}_1, \ldots, \mathbf{a}_m\}$ *be a linearly independent set in* V. *Then* $m \leq n$.

Proof. The idea of the proof is to start with E and move towards A by bringing in the elements $\{\mathbf{a}_1, \ldots, \mathbf{a}_m\}$ of A, one at a time. As we add each member of A, we delete one of the remaining vectors $\mathbf{e}_i$, so that our transition sets always have exactly n elements.

The first step is to bring $\mathbf{a}_1$ into E. Note that $\mathbf{a}_i \neq \mathbf{0}$ for each i, because A is linearly independent. Suppose that

$$\mathbf{a}_1 = c_1\mathbf{e}_1 + \cdots + c_n\mathbf{e}_n\,.$$

One of the c_i s must be nonzero. Rename the vectors in E, if necessary, to make $c_1 \neq 0$. Then bring in $\mathbf{a}_1$ and delete $\mathbf{e}_1$. We claim that the resulting set

$$E_1 = \{\mathbf{a}_1, \mathbf{e}_2, \ldots, \mathbf{e}_n\}$$

is a basis for V. Observe that E_1 spans V, since it contains $\mathbf{e}_1 = (1/c_1)(\mathbf{a}_1 - c_2\mathbf{e}_2 - \cdots - c_n\mathbf{e}_n)$ and it contains each of $\mathbf{e}_2, \ldots, \mathbf{e}_n$. Furthermore, we claim that E_1 is a linearly independent set. Suppose

$$k_1\mathbf{a}_1 + k_2\mathbf{e}_2 + \; + k_n\mathbf{e}_n = \mathbf{0}.$$

Then we have

$$k_1c_1\mathbf{e}_1 + (k_1c_2 + k_2)\mathbf{e}_2 + \cdots + (k_1c_n + k_n)\mathbf{e}_n = \mathbf{0}\,.$$

From the linear independence of $\{\mathbf{e}_1, \ldots, \mathbf{e}_n\}$, each coefficient is 0. By the crucial assumption $c_1 \neq 0$, it follows that $k_1 = 0$ and hence that $k_2 = \cdots = k_n = 0$. So E_1 is a basis for V.

For the inductive step, let $1 \leq r < m$ and $1 \leq r < n$. Suppose that we have obtained E_r by substituting the first r members of A for the first r members of E and that E_r is still a basis. (There is no loss of generality here; we can rearrange the original listing of E if necessary.) Then we write $\mathbf{a}_{r+1}$ as a linear combination of

$$\mathbf{a}_1, \ldots, \mathbf{a}_r, \mathbf{e}_{r+1}, \ldots, \mathbf{e}_n\,.$$

The coefficient of at least one of $\mathbf{e}_{r+1}, \ldots, \mathbf{e}_n$ must be nonzero because of the linear independence of $\mathbf{a}_1, \ldots, \mathbf{a}_r, \mathbf{a}_{r+1}$.

Again, without loss of generality, we assume that the coefficient of $\mathbf{e}_{r+1}$ is nonzero. Proceed exactly as at the first step to substitute $\mathbf{a}_{r+1}$ for $\mathbf{e}_{r+1}$. The resulting set

$$E_{r+1} = \{\mathbf{a}_1, \ldots, \mathbf{a}_r, \mathbf{a}_{r+1}, \mathbf{e}_{r+2}, \ldots, \mathbf{e}_n\}$$

is still a basis.

If $n \leq m$, then $E_n = \{\mathbf{a}_1, \ldots, \mathbf{a}_n\}$ will be a basis for V. If $n < m$, then the remaining vectors $\{\mathbf{a}_{n+1}, \ldots, \mathbf{a}_m\}$ in A will be linear combinations of the first n vectors $\{\mathbf{a}_1, \ldots, \mathbf{a}_n\}$. This contradicts the linear independence of A. Hence $m \leq n$. If $n = m$, then $A = E_n = \{\mathbf{a}_1, \ldots, \mathbf{a}_n\}$ is a basis for V. □

Lemma B.1.2 *Let A be a linearly independent set in V. If $\mathbf{b} \in V$ and $\mathbf{b} \notin \operatorname{Span} A$, then $B = A \cup \{\mathbf{b}\}$ is also a linearly independent set.*

Proof. Let $c\mathbf{b} + c_1\mathbf{a}_1 + \cdots + c_n\mathbf{a}_n = \mathbf{0}$ with some $\mathbf{a}_i \in A$ and $c, c_i \in \mathbb{R}$. Then c must vanish. Otherwise,

$$\mathbf{b} = -(1/c)(c_1\mathbf{a}_1 + \cdots + c_n\mathbf{a}_n)$$

would imply that $\mathbf{b} \in \operatorname{Span} A$. But if $c = 0$, then also $c_1 = \ldots = c_n = 0$, since A is a linearly independent set. Therefore B is a linearly independent set. □

Theorem B.1.3 *Let $E = \{\mathbf{e}_1, \ldots, \mathbf{e}_n\}$ be a basis for a vector space V. Let $A = \{\mathbf{a}_1, \ldots, \mathbf{a}_m\}$ be a linearly independent set in V. Then A is also a basis for V if and only if $m = n$.*

Proof. Assume that A is a basis for V. In this case, the sets E and A are both bases and linearly independent sets. Then Theorem B.1.1 implies that both $m \leq n$ and $n \leq m$. Hence $m = n$.

Conversely, assume that A is not a basis for V. Hence there is a vector $\mathbf{b} \in V$ such that $\mathbf{b} \notin \operatorname{Span} A$. In this case, Lemma B.1.2 above shows that $B = A \cup \{\mathbf{b}\}$ is also a linearly independent set. But B contains $m + 1$ vectors. Then Theorem B.1.1 implies that $m + 1 \leq n$. Hence $m < n$. □

Theorem B.1.4 *Let $E = \{\mathbf{e}_1, \ldots, \mathbf{e}_n\}$ be a basis for a vector space X. Let $A = \{\mathbf{a}_1, \ldots, \mathbf{a}_m\}$ be a linearly independent set in X. If A is not a basis for X, then there is a set $B \subset X$ such that $A \cap B = \emptyset$ and such that $A \cup B$ is a basis for X.*

Proof. If A is not a basis, then $\operatorname{Span} A \neq X$. Hence there is a $\mathbf{b}_1 \in X$ such that $\mathbf{b}_1 \notin \operatorname{Span} A$. Then $A_1 = A \cup \{\mathbf{b}_1\}$ is a linearly independent set consisting of $(m+1)$ vectors. Repeating this step $k = (n - m)$ times, we obtain k vectors $\mathbf{b}_1, \ldots, \mathbf{b}_k$

and a linearly independent set $A_k = A \cup \{ \mathbf{b}_1, \ldots, \mathbf{b}_k \}$ consisting of n vectors. Then Theorem B.1.3 shows that A_k is a basis. Hence let $B = \{ \mathbf{b}_1, \ldots, \mathbf{b}_k \}$. Then $A \cap B = \emptyset$ and $A \cup B$ is a basis. $\square$

Remarks B.1.5 An examination of the proof of Theorem B.1.1 shows that Theorem B.1.4 is already established there. The proof above also depends on Theorem B.1.1, but only at the last step (via Theorem B.1.3).

Definition B.1.6 Dimension. As always, we restrict our attention to finite dimensional vector spaces. Theorem B.1.1 shows that any two bases for a vector space V have the same number of elements. This number is called the *dimension* of V, and is denoted by $\dim V$. Any subspace of a vector space is also a vector space by itself. Hence any subspace has a dimension too.

Lemma B.1.7 *Suppose that U is a subspace of V. Then $\dim U \leq \dim V$, with equality if and only if $U = V$.*

Proof. Let $E = \{ \mathbf{e}_1, \ldots, \mathbf{e}_n \}$ be a basis for V. If $A = \{ \mathbf{a}_1, \ldots, \mathbf{a}_m \}$ is a basis for U, then A is also a linearly independent set in V. Therefore $m \leq n$ by Theorem B.1.1. Also, Theorem B.1.3 shows that $m = n$ if and only A is also a basis for V. This happens if and only if $U = V$. $\square$

APPENDIX C

DETERMINANTS

Determinants provide a basic algebraic tool in computing volumes and integrals in vector spaces. They also play a key role in defining and working with tensor fields. Many of the computational aspects of determinants will be familiar to the reader. We will review the relevant definitions and results briefly, concentrating not so much on the computational but rather on the conceptual features of determinants.

C.1 PERMUTATIONS

We consider only the permutations of a finite set A. In most cases, we let $A = \mathbb{N}_n = \{1, 2, \ldots, n\}$. An invertible map $\sigma : A \to A$ is called a *permutation* of A. If A contains n elements, then $\mathcal{S}_n$ denotes the set of all permutations of A. An induction argument shows that $\mathcal{S}_n$ contains $n!$ elements.

Analysis in Vector Spaces.
By M. A. Akcoglu, P. F. A. Bartha and D. M. Ha

The *identity permutation* $\varepsilon \in \mathbb{S}_n$ is defined by $\varepsilon(i) = i$, $i \in A$. The inverse of $\sigma \in \mathbb{S}_n$ is $\sigma^{-1} \in \mathbb{S}_n$. If ρ and σ are two permutations of A, then their *composition* is the permutation defined by $(\rho\sigma)(i) = \rho(\sigma(i))$, $i \in A$. The composition of more than two permutations is defined similarly. In particular, σ^k is the composition of k copies of σ. If p and q are two (different) elements of $\mathbb{N}_n$, then the *transposition* $\tau = (p\,q) \in \mathbb{S}_n$ is defined by $\tau(p) = q$, $\tau(q) = p$ and $\tau(i) = i$ if $i \neq p$ and $i \neq q$.

Theorem C.1.1 Permutations in terms of transpositions. *Any permutation is either the identity permutation or a composition of transpositions.*

Proof. This is clear if $n = 1$. Let $n \geq 2$ and assume the result for $(n-1)$. Let $\sigma \in \mathbb{S}_n$. The set of values $\sigma^k(1)$, $k \in \mathbb{N}$, is a finite set $A \subset \mathbb{N}_n$. Then both A and $B = (\mathbb{N}_n \setminus A)$ are invariant under σ. We see that A is not empty, since $1 \in A$. If B is also not empty, then both sets contain at most $(n-1)$ elements. Hence, by the induction hypothesis, the restrictions of σ to these sets are compositions of transpositions. It follows that σ is the composition of the two sets of transpositions.

Now assume that $A = \mathbb{N}_n$, so that $B = \emptyset$. Let $p_k = \sigma^k(1)$, $1 \leq k \leq n$. In this case, we have $\sigma = (p_1 p_2) \cdots (p_{n-1} p_n)$, so that σ is a product of transpositions. In this formulation, $p_n = 1$ and we apply transpositions by working from right to left. That is, to obtain σ, we first apply the transposition $(p_{n-1} p_n) = (p_{n-1} 1)$, then $(p_{n-2} p_{n-1})$ and so on. The application of $(p_1 p_2)$ is the last step. □

Remarks C.1.2 Identity in terms of transpositions. If $n \geq 2$, then the identity permutation is also a composition of transpositions. In fact, τ^2 is the identity permutation for any transposition τ. If $n = 1$, however, there are no transpositions. That is why the identity permutation is singled out in Theorem C.1.1 above.

Theorem C.1.3 Sign of a permutation. *There is a unique function,* sign : $\mathbb{S}_n \to \{-1, 1\}$, *defined by the following properties: (1)* $\text{sign}\tau = -1$ *for any transposition* $\tau \in \mathbb{S}_n$, *and (2)* $\text{sign}(\rho\sigma) = (\text{sign}\rho)(\text{sign}\sigma)$ *for any* $\rho, \sigma \in \mathbb{S}_n$.

Proof. If such a function exists, then its uniqueness is clear. In fact, any permutation is a composition of transpositions. Therefore, by (1) and (2) above, if σ is the composition of p transpositions, then $\text{sign}\,\sigma = (-1)^p$. The existence of such a function is not obvious, however, because the representation of a permutation as a composition of transpositions is not unique. If σ is represented in two different ways, as a composition of p transpositions and as a composition of q transpositions, then we have to show that $(-1)^p = (-1)^q$.

To show that the sign function is well defined, then, set

$$P(x_1, \ldots, x_n) = \prod\nolimits_{1 \leq i < j \leq n} (x_i - x_j).$$

This is a polynomial of n real variables $x_1, \ldots, x_n$. For each $\sigma \in \mathcal{S}_n$, let

$$\sigma P(x_1, \ldots, x_n) = \prod\nolimits_{1 \le i < j \le n} (x_{\sigma(i)} - x_{\sigma(j)}).$$

We see that $\sigma P(x_1, \ldots, x_n) = \pm P(x_1, \ldots, x_n)$. In fact, the factors of P and σP are of the same absolute value but arranged in a different order. Then define $\operatorname{sign} \sigma = 1$ if $\sigma P = P$ and $\operatorname{sign} \sigma = -1$ if $\sigma P = -P$. We see that the properties (1) and (2) are satisfied. □

Corollary C.1.4 The signs of the identity permutation and the inverse permutations. *The sign of the identity permutation ε is* 1. *Also, any permutation σ and its inverse $\rho = \sigma^{-1}$ have the same sign.*

Proof. Since $\varepsilon = \varepsilon\,\varepsilon$, by Part (2) of Theorem C.1.3 show that

$$(\operatorname{sign} \varepsilon) = (\operatorname{sign} \varepsilon)\,(\operatorname{sign} \varepsilon) = (\pm 1)^2 = 1.$$

Next, $\rho\,\sigma = \varepsilon$ shows that $(\operatorname{sign} \rho)\,(\operatorname{sign} \sigma) = (\operatorname{sign} \varepsilon) = 1$. Since ± 1 are the only possible values for the sign function, we see that $(\operatorname{sign} \rho) = (\operatorname{sign} \sigma)$. □

C.2 DETERMINANTS OF SQUARE MATRICES

Definition C.2.1 Determinant of a square matrix. Let $\mathbf{A} = (A_{ij}) \in \mathbb{M}_{nn}$ be a square matrix with n columns and n rows. Then the number

$$\det \mathbf{A} = \det(A_{ij}) = \sum\nolimits_{\sigma \in \mathcal{S}_n} (\operatorname{sign} \sigma) \prod\nolimits_i A_{i\,\sigma(i)} \tag{C.1}$$

is defined as the *determinant* of this matrix. Here the index i in the product ranges over $\mathbb{N}_n = \{1, \ldots, n\}$. The determinant of matrices in $\mathbb{M}_{nn}$ is a function $\det : \mathbb{M}_{nn} \to \mathbb{R}$.

Theorem C.2.2 Determinant of the identity matrix. *Let* $\mathbf{E}$ *be the identity matrix so that $E_{ii} = 1$ and $E_{ij} = 0$ if $i \neq j$. Then* $\det \mathbf{E} = 1$.

Proof. We see that $\prod_i E_{i\,\sigma(i)} = 1$ if $\sigma = \varepsilon$ is the identity permutation and $\prod_i E_{i\,\sigma(i)} = 0$ if $\sigma \neq \varepsilon$. Then (C.1) shows that $\det \mathbf{E} = 1$. □

Definition C.2.3 Conjugate matrices. $\mathbf{A}$ and $\mathbf{B}$ are *conjugate matrices* if $B_{ij} = A_{ji}$. We say $\mathbf{B}$ is the conjugate of $\mathbf{A}$.

Theorem C.2.4 Determinant of the conjugate matrix. *Let* $\mathbf{A}$ *and* $\mathbf{B}$ *be conjugate matrices in* $\mathbb{M}_{nn}$*, so that* $B_{ij} = A_{ji}$*. Then* $\det \mathbf{B} = \det \mathbf{A}$.

Proof. Let $\sigma \in S_n$ be a permutation with the inverse permutation $\rho = \sigma^{-1}$. A term $A_{i\,\sigma(i)}$ in the product in (C.1) can be also expressed as $A_{\rho(j)\,j} = B_{j\,\rho(j)}$ by letting $j = \sigma(i)$. Hence, if the terms of the product in (C.1) are rearranged with respect to the increasing values of the second subscript, then the product becomes

$$\prod_{i=1}^{n} A_{i\,\sigma(i)} = A_{\rho(1)\,1} \cdots A_{\rho(n)\,n} = \prod_{j=1}^{n} A_{\rho(j)\,j} = \prod_{j=1}^{n} B_{j\,\rho(j)}.$$

Now $\operatorname{sign} \rho = \operatorname{sign} \sigma$ by Corollary C.1.4. Also, the operation of taking the inverse of a permutation is a one-to-one and onto mapping $\mathcal{S}_n \to \mathcal{S}_n$. Hence

$$\begin{aligned} \det \mathbf{A} &= \sum_{\sigma \in \mathcal{S}_n} (\operatorname{sign} \sigma) \prod_i A_{i\,\sigma(i)} \\ &= \sum_{\rho \in \mathcal{S}_n} (\operatorname{sign} \rho) \prod_j B_{j\,\rho(j)} = \det \mathbf{B}. \quad \square \end{aligned}$$

Theorem C.2.5 Determinants under linear combinations in one row. *Let* $i_0 \in \mathbb{N}_n$ *be a fixed index and* a', $a'' \in \mathbb{R}$*. Let* $\mathbf{A}$, $\mathbf{A}'$, $\mathbf{A}'' \in \mathbb{M}_{nn}$ *be three matrices such that* $A_{ij} = A'_{ij} = A''_{ij}$ *if* $i \neq i_0$ *and* $A_{i_0 j} = a' A'_{i_0 j} + a'' A''_{i_0 j}$*. In other words, row* i *of* $\mathbf{A}$ *is a linear combination of row* i *of the other two matrices. Then* $\det \mathbf{A} = a' \det \mathbf{A}' + a'' \det \mathbf{A}''$.

Proof. We have, for any $\sigma \in \mathcal{S}_n$,

$$\begin{aligned} \prod_i A_{i\,\sigma(i)} &= (a' A'_{i_0 \sigma(i_0)} + a'' A''_{i_0 \sigma(i_0)}) \prod_{i \neq i_0} A_{i\,\sigma(i)} \\ &= a' A'_{i_0 \sigma(i_0)} \prod_{i \neq i_0} A'_{i\,\sigma(i)} + a'' A''_{i_0 \sigma(i_0)} \prod_{i \neq i_0} A''_{i\,\sigma(i)} \\ &= a' \prod_i A'_{i\,\sigma(i)} + a'' \prod_i A''_{i\,\sigma(i)}. \end{aligned}$$

Hence we see that $\det \mathbf{A} = a' \det \mathbf{A}' + a'' \det \mathbf{A}''$. $\square$

Theorem C.2.6 Determinants under permutations of the rows. *Let* $\mathbf{B}$ *be the matrix obtained by permuting the rows of* $\mathbf{A} \in \mathbb{M}_{nn}$ *by a permutation* $\lambda \in \mathcal{S}_n$*. More explicitly, let* $B_{ij} = A_{\lambda(i) j}$*. Then* $\det \mathbf{B} = (\operatorname{sign} \lambda) \det \mathbf{A}$.

Proof. Let $\mu = \lambda^{-1} \in \mathcal{S}_n$ be the inverse of λ. For each $\sigma \in \mathcal{S}_n$ we have

$$\prod_{i=1}^{n} B_{i\,\sigma(i)} = \prod_{i=1}^{n} A_{\lambda(i)\,\sigma(i)} = \prod_{j=1}^{n} A_{j\,(\sigma\mu)(j)}.$$

Here $j = \lambda(i)$, $i = \mu(j)$, and $\sigma(i) = \sigma(\mu(j)) = (\sigma\mu)(j)$. Therefore the last two products consist of the same factors. Also,

$$\det \mathbf{A} = \sum_{\sigma \in \mathcal{S}_n} (\operatorname{sign} \sigma) \prod_{i=1}^{n} A_{i\,\sigma(i)} = \sum_{\sigma \in \mathcal{S}_n} (\operatorname{sign} (\sigma\mu)) \prod_{j} A_{j\,(\sigma\mu)(j)}.$$

In fact, the mapping that takes $\sigma \in \mathcal{S}_n$ to $(\sigma\mu) \in \mathcal{S}_n$ is a one-to-one and onto mapping $\mathcal{S}_n \to \mathcal{S}_n$. Hence the two sums above consist of the same terms. Therefore, we have

$$\begin{aligned} \det \mathbf{B} &= \sum_{\sigma \in \mathcal{S}_n} (\operatorname{sign} \sigma) \prod_{i=1}^{n} B_{i\,\sigma(i)} = \sum_{\sigma \in \mathcal{S}_n} (\operatorname{sign} \sigma) \prod_{j=1}^{n} A_{j\,(\sigma\mu)(j)} \\ &= (\operatorname{sign}\lambda) \sum_{\sigma \in \mathcal{S}_n} (\operatorname{sign} (\sigma\mu)) \prod_{j} A_{j\,(\sigma\mu)(j)} = (\operatorname{sign}\lambda) \det \mathbf{A}. \end{aligned}$$

For the third equality, note that $(\operatorname{sign} \lambda)(\operatorname{sign} (\sigma\mu)) = (\operatorname{sign} \sigma)$. □

Remarks C.2.7 Dependence on the rows and columns. Theorem C.2.5 shows that the determinant of a square matrix depends linearly on each particular row vector if the other rows remain constant. This is also true for the column vectors. In fact, the columns of a matrix are the rows of the conjugate matrix, and Theorem C.2.4 shows that the determinants of the original and conjugate matrices are the same. Similarly, Theorem C.2.6 shows that if the rows of a matrix are permuted, then its determinant is multiplied by the sign of this permutation. This is also true for the permutations of the columns, again because of Theorem C.2.4.

C.3 DETERMINANT FUNCTIONS

The determinant of square matrices is connected with alternating multilinear functions. First, let us recall the definition of multilinear functions given in Definition 3.3.3. Let X and Y be two vector spaces and $k \in \mathbb{N}$. Then $F : X^k \to Y$ is called a *multilinear function* if

$$F(\mathbf{x}_1, \ldots, \mathbf{x}_k) = a' F(\mathbf{x}'_1, \ldots, \mathbf{x}'_k) + a'' F(\mathbf{x}''_1, \ldots, \mathbf{x}''_k)$$

whenever there is an index $i_0 \in \mathbb{N}_k = \{1, \ldots, k\}$ such that $\mathbf{x}_{i_0} = a'\mathbf{x}'_{i_0} + a''\mathbf{x}''_{i_0}$ and $\mathbf{x}_i = \mathbf{x}'_i = \mathbf{x}''_i$ if $i \neq i_0$. Here a', $a'' \in \mathbb{R}$ are scalars.

Definition C.3.1 Alternating multilinear functions. Let X and Y be two vector spaces and $k \in \mathbb{N}$. A multilinear function $F : X^k \to Y$ is called *an alternating multilinear function* if

$$F(\mathbf{x}_{\lambda(1)}, \ldots, \mathbf{x}_{\lambda(k)}) = (\operatorname{sign} \lambda)\, F(\mathbf{x}_1, \ldots, \mathbf{x}_k)$$

for all permutations $\lambda \in \mathcal{S}_k$ of $\mathbb{N}_k = \{1, \ldots, k\}$.

Definition C.3.2 Determinant functions. Let X be an n-dimensional vector space. Then any *nonzero* alternating multilinear function $\varphi : X^n \to \mathbb{R}$ is called a *determinant function on* X.

Note that we call φ a determinant function on X even though φ is actually a function on X^n. Here it is very important to note that $n = \dim X$. Basic examples of determinant functions are obtained in terms of the ordered bases of vector spaces and the determinants of square matrices. In fact, Corollary C.3.11 below shows that these examples cover all determinant functions.

Definition C.3.3 Ordered bases. Let X be an n-dimensional vector space. An n-tuple $\mathbb{E} = (\mathbf{e}_1, \ldots, \mathbf{e}_n) \in X^n$ is called an *ordered basis for* X if the set of vectors $\{\mathbf{e}_1, \ldots, \mathbf{e}_n\}$ is a basis for X. Equivalently, $\mathbb{E} = (\mathbf{e}_1, \ldots, \mathbf{e}_n) \in X^n$ is an ordered basis for X if $\{\mathbf{e}_1, \ldots, \mathbf{e}_n\}$ is a linearly independent set of vectors.

Remarks C.3.4 Bases and ordered bases. There is no ordering of the vectors in a basis $\{\mathbf{e}_1, \ldots, \mathbf{e}_n\}$ of X. A basis for X can be ordered in $n!$ different ways and determines $n!$ different ordered bases for X.

Example C.3.5 Basic examples of determinant functions. An ordered basis $\mathbb{E} = (\mathbf{e}_1, \ldots, \mathbf{e}_n)$ of X defines a function $\det_{\mathbb{E}} : X^n \to \mathbb{R}$ by

$$\det_{\mathbb{E}}(\mathbf{x}_1, \ldots, \mathbf{x}_n) = \det(x_{ij}). \tag{C.2}$$

Here x_{ij} s are the coordinates of $\mathbf{x}_i$ s with respect to $\mathbb{E} = (\mathbf{e}_1, \ldots, \mathbf{e}_n)$ so that $\mathbf{x}_i = \sum_j x_{ij}\mathbf{e}_j$ for all $i \in \mathbb{N}_n = \{1, \ldots, n\}$, and $\det(x_{ij})$ is the determinant of the square matrix $\{x_{ij}\} \in \mathbb{M}_{nn}$. Theorem C.2.5 shows that $\det_{\mathbb{E}} : X^n \to \mathbb{R}$ is a multilinear function, and Theorem C.2.6 shows that it is alternating. It is also a nonzero function since, by Theorem C.2.2, $\det_{\mathbb{E}}(\mathbf{e}_1, \ldots, \mathbf{e}_n) = 1$. Hence $\det_{\mathbb{E}} : X^n \to \mathbb{R}$ is a determinant function. △

Definition C.3.6 Determinant functions with respect to ordered bases. Let $\mathbb{E} = (\mathbf{e}_1, \ldots, \mathbf{e}_n)$ be an ordered basis of X. Then the determinant function $\det_{\mathbb{E}} : X^n \to \mathbb{R}$ defined above in Example C.3.5 is called the *determinant function with respect to the ordered basis* $\mathbb{E}$.

We are going to show that for any determinant function $\varphi : X^n \to \mathbb{R}$, there is an ordered basis $\mathbb{E} = (\mathbf{e}_1, \ldots, \mathbf{e}_n)$ of X such that $\varphi = \det_{\mathbb{E}}$. We need a general property of alternating multilinear functions.

Lemma C.3.7 *If $F : X^k \to Y$ is an alternating multilinear function, then we have $F(\mathbf{x}_1, \ldots, \mathbf{x}_k) = \mathbf{0}$ whenever $\{\mathbf{x}_1, \ldots, \mathbf{x}_k\}$ is a linearly dependent set.*

Proof. If $F : X^k \to Y$ is an alternating function and if $(\mathbf{x}_1, \ldots, \mathbf{x}_k) \in X^k$ is such that $\mathbf{x}_i = \mathbf{x}_j$ for some $1 \leq i < j \leq k$, then $F(\mathbf{x}_1, \ldots, \mathbf{x}_k) = \mathbf{0}$. In fact, in this case, if $\tau = (ij)$ is the transposition of i and j, then

$$(\tau F)(\mathbf{x}_1, \ldots, \mathbf{x}_k) = F(\mathbf{x}_1, \ldots, \mathbf{x}_k) = (\operatorname{sign} \tau) F(\mathbf{x}_1, \ldots, \mathbf{x}_k)$$

forces $F(\mathbf{x}_1, \ldots, \mathbf{x}_k) = \mathbf{0}$ since $\operatorname{sign} \tau = -1$.

Now assume that $\mathbf{x}_1$ is a linear combination of the other vectors: $\mathbf{x}_1 = \sum_{j=2}^n a_j \mathbf{x}_j$. If $F : X^k \to Y$ is alternating and multilinear, then

$$F(\mathbf{x}_1, \mathbf{x}_2, \ldots, \mathbf{x}_k) = \sum_{j=2}^{n} a_j F(\mathbf{x}_j, \mathbf{x}_2, \ldots, \mathbf{x}_k) = \mathbf{0}$$

by the first part of the proof. In general, if $\{ \mathbf{x}_1, \ldots, \mathbf{x}_k \}$ is a linearly dependent set, then one vector $\mathbf{x}_i$ is a linear combination of the other vectors $\mathbf{x}_j$. If $\tau = (1i)$ is the transposition of 1 and i, then $(\tau F)(\mathbf{x}_1, \ldots, \mathbf{x}_k) = F(\mathbf{x}_{\tau(1)}, \ldots, \mathbf{x}_{\tau(k)}) = \mathbf{0}$ since the first vector $\mathbf{x}_{\tau(1)} = \mathbf{x}_i$ is a linear combination of the others. Hence we also have $F(\mathbf{x}_1, \ldots, \mathbf{x}_k) = -(\tau F)(\mathbf{x}_1, \ldots, \mathbf{x}_k) = \mathbf{0}$. □

Theorem C.3.8 *Let $\mathbb{E} = (\mathbf{e}_1, \ldots, \mathbf{e}_n)$ be an ordered basis of X. Then any alternating multilinear function $\varphi : X^n \to \mathbb{R}$ is a multiple $\varphi = k \det_{\mathbb{E}}$ of the determinant function $\det_{\mathbb{E}} : X^n \to \mathbb{R}$, where $k = \varphi(\mathbf{e}_1, \ldots, \mathbf{e}_n)$.*

Proof. We have, since $\varphi : X^n \to \mathbb{R}$ is a multi-linear function,

$$\begin{aligned} \varphi(\mathbf{x}_1, \ldots, \mathbf{x}_n) &= \varphi\left(\sum_{j_1} x_{1j_1} \mathbf{e}_{j_1}, \ldots, \sum_{j_n} x_{nj_n} \mathbf{e}_{j_n}\right) \\ &= \sum_{j_1} \cdots \sum_{j_n} x_{1j_1} \cdots x_{nj_n}\, \varphi(\mathbf{e}_{j_1}, \ldots, \mathbf{e}_{j_n}), \end{aligned}$$

where $\mathbf{x}_i = \sum_{j_i} x_{1j_i} \mathbf{e}_{j_i}$. Now Lemma C.3.7 shows that $\varphi(\mathbf{u}_{j_1}, \ldots, \mathbf{u}_{j_n})$ can be nonzero only if the vectors $\{ \mathbf{u}_{j_1}, \ldots, \mathbf{u}_{j_n} \}$ are all different. This happens only if the set of indices $\{ j_1, \ldots, j_n \}$ is a permutation of $\mathbb{N}_n = \{ 1, \ldots, n \}$. Therefore the last set of sums can be replaced by a single sum over the permutations $\sigma \in \mathcal{S}_n$. Thus we obtain

$$\begin{aligned} \varphi(\mathbf{x}_1, \ldots, \mathbf{x}_n) &= \sum_{\sigma \in \mathcal{S}_n} x_{1\sigma(1)} \cdots x_{n\sigma(n)}\, \varphi(\mathbf{e}_{\sigma(1)}, \ldots, \mathbf{e}_{\sigma(n)}) && \text{(C.3)} \\ &= \sum_{\sigma \in \mathcal{S}_n} (\operatorname{sign} \sigma) x_{1\sigma(1)} \cdots x_{n\sigma(n)}\, \varphi(\mathbf{e}_1, \ldots, \mathbf{e}_n) && \text{(C.4)} \\ &= \varphi(\mathbf{e}_1, \ldots, \mathbf{e}_n) \cdot \det(x_{ij}) && \text{(C.5)} \\ &= \varphi(\mathbf{e}_1, \ldots, \mathbf{e}_n) \cdot \det_{\mathbb{E}}(\mathbf{x}_1, \ldots, \mathbf{x}_n). && \text{(C.6)} \end{aligned}$$

Here (C.4) follows from Definition C.3.1 of alternating multilinear functions, (C.5) follows from the definition of determinants of square matrices in C.2.1, and (C.6) follows from the definition of $\det_{\mathbb{E}}$ in (C.2). □

Corollary C.3.9 *If $\varphi : X^n \to \mathbb{R}$ is a determinant function, then $\varphi(\mathbf{a}_1, \ldots, \mathbf{a}_n)$ is nonzero if and only if $\{\mathbf{a}_1, \ldots, \mathbf{a}_n\}$ is a linearly independent set.*

Proof. If $\{\mathbf{a}_1, \ldots, \mathbf{a}_n\}$ is a linearly dependent set, then Lemma C.3.7 shows that $\varphi(\mathbf{a}_1, \ldots, \mathbf{a}_n) = 0$. Otherwise, $\mathbb{A} = (\mathbf{a}_1, \ldots, \mathbf{a}_n)$ is an ordered basis for X and $\varphi = \varphi(\mathbf{a}_1, \ldots, \mathbf{a}_n) \det_{\mathbb{A}}$ by Theorem C.3.8. But determinant functions are nonzero functions, by Definition C.3.2. Hence $\varphi(\mathbf{a}_1, \ldots, \mathbf{a}_n) \neq 0$. □

Corollary C.3.10 *Any determinant function on X is a nonzero multiple of any other determinant function on X.*

Proof. Let $\varphi, \psi : X^n \to \mathbb{R}$ be two determinant functions on X, and let $\mathbb{E}$ be an ordered basis $(\mathbf{e}_1, \ldots, \mathbf{e}_n)$ for X. Then we see easily from Theorem C.3.8 that $\varphi = k\psi$ with $k = \varphi(\mathbf{e}_1, \ldots, \mathbf{e}_n)/\psi(\mathbf{e}_1, \ldots, \mathbf{e}_n)$. Note that k is defined since $\psi(\mathbf{e}_1, \ldots, \mathbf{e}_n) \neq 0$ by Corollary C.3.9 above. □

Corollary C.3.11 *For any determinant function $\varphi : X^n \to \mathbb{R}$, there is an ordered basis $\mathbb{E} = (\mathbf{e}_1, \ldots, \mathbf{e}_n)$ of X such that $\varphi = \det_{\mathbb{E}}$.*

Proof. If $\varphi : X^n \to \mathbb{R}$ is a determinant function and $\mathbb{A} = (\mathbf{a}_1, \ldots, \mathbf{a}_n)$ is an ordered basis of X, then $\alpha = \varphi(\mathbf{a}_1, \ldots, \mathbf{a}_n) \neq 0$ by Corollary C.3.9. Set $\mathbf{e}_1 = (1/\alpha)\mathbf{a}_1$ and $\mathbf{e}_i = \mathbf{a}_i$ if $2 \leq i \leq n$. Then we see that $\mathbb{E} = (\mathbf{e}_1, \ldots, \mathbf{e}_n)$ is an ordered basis for X and $\varphi(\mathbf{e}_1, \ldots, \mathbf{e}_n) = 1$ by the multilinearity of determinant functions. Then $\varphi = \det_{\mathbb{E}}$ by Theorem C.3.8. It is clear that this choice for $\mathbb{E}$ is not unique. □

Corollary C.3.12 *If $\varphi : X^n \to \mathbb{R}$ is a determinant function and $\lambda : X^n \to \mathbb{R}$ is an alternating multilinear function, then there is a $\beta \in \mathbb{R}$ such that $\lambda = \beta\,\varphi$.*

Proof. This follows directly from Theorem C.3.8 and Corollary C.3.11. □

We apply these results to obtain two additional facts about the determinant of square matrices.

Theorem C.3.13 Determinant of a product. *If $\mathbf{A}$ and $\mathbf{B}$ are square matrices in $\mathbb{M}_{nn}$, then $\det(\mathbf{AB}) = \det \mathbf{A} \det \mathbf{B}$.*

Proof. Fix $\mathbf{A}$ and define the function $D(\mathbf{B}) = \det(\mathbf{AB})$. From Theorem C.2.6 and Remarks C.2.7, we know that D is an alternating multilinear function in the columns of $\mathbf{B}$. By Corollary C.3.10, there is a constant k such that $D(\mathbf{B}) = k \det \mathbf{B}$. Letting $\mathbf{B}$ be the identity matrix I, we see that $k = \det \mathbf{A}$. This proves the conclusion. □

Theorem C.3.14 Invertible matrices and nonzero determinants. *If* $\mathbf{A}$ *is a square matrix, then* $\mathbf{A}$ *is invertible if and only if* $\det \mathbf{A} \neq 0$. *If* $\mathbf{A}$ *is invertible, then* $\det \mathbf{A} \det \mathbf{A}^{-1} = 1$.

Proof. First, $\mathbf{A}$ is invertible if and only if the columns of $\mathbf{A}$ are linearly independent if and only if $\det \mathbf{A} \neq 0$, by Corollary C.3.9. Then if $\mathbf{A}\mathbf{A}^{-1} = I$, Theorem C.3.13 implies that $\det \mathbf{A} \det \mathbf{A}^{-1} = 1$. □

C.4 DETERMINANT OF A LINEAR TRANSFORMATION

In this book we consider determinants mainly in the context of linear transformations. Each linear transformation $T : X \to X$ has a number attached to it, called the *determinant* of T. It is denoted as $\det T$. We provide the definition after the following observation.

Lemma C.4.1 *Let* φ *be a determinant function on* X *and* $\mathbb{E} = (\mathbf{e}_1, \ldots, \mathbf{e}_n)$ *a basis for* X. *Then the number* $k = \varphi(T\mathbf{e}_1, \ldots, T\mathbf{e}_n)/\varphi(\mathbf{e}_1, \ldots, \mathbf{e}_n)$ *is independent of the choices of* φ *and* $\mathbb{E}$.

Proof. We distinguish two cases. First, if $\varphi(T\mathbf{e}_1, \ldots, T\mathbf{e}_n) = 0$, then T maps a basis to a linearly dependent set. In this case it will map any basis to a linearly dependent set. Hence $\varphi(T\mathbf{e}_1, \ldots, T\mathbf{e}_n) = 0$ for all choices of φ and $\mathbb{E}$.

Second, if $\varphi(T\mathbf{e}_1, \ldots, T\mathbf{e}_n) \neq 0$, then the function

$$\psi(\mathbf{x}_1, \ldots, \mathbf{x}_n) = \varphi(T\mathbf{x}_1, \ldots, T\mathbf{x}_n)$$

defines a nonzero alternating multilinear function $\psi : X^n \to \mathbb{R}$ (with each $\mathbf{x}_i \in X$). Hence ψ is a determinant function on X. Therefore there is a nonzero number k such that

$$\psi(\mathbf{x}_1, \ldots, \mathbf{x}_n) = \varphi(T\mathbf{x}_1, \ldots, T\mathbf{x}_n) = k\varphi(\mathbf{x}_1, \ldots, \mathbf{x}_n)$$

for all $(\mathbf{x}_1, \ldots, \mathbf{x}_n) \in X^n$. This shows that k is independent of the choice of $\mathbb{E}$. Furthermore, if ϑ is another determinant function on X, then there is a number $\ell \neq 0$ such that $\varphi = \ell\vartheta$. Hence

$$\frac{\varphi(T\mathbf{e}_1, \ldots, T\mathbf{e}_n)}{\varphi(\mathbf{e}_1, \ldots, \mathbf{e}_n)} = \frac{\ell\vartheta(T\mathbf{e}_1, \ldots, T\mathbf{e}_n)}{\ell\vartheta(\mathbf{e}_1, \ldots, \mathbf{e}_n)} = \frac{\vartheta(T\mathbf{e}_1, \ldots, T\mathbf{e}_n)}{\vartheta(\mathbf{e}_1, \ldots, \mathbf{e}_n)},$$

which shows that k is also independent of the choice of φ. □

Definition C.4.2 Determinant of a transformation. The number k obtained in Lemma C.4.1 is called the *determinant of the linear transformation* T and denoted as $\det T$. Hence

$$\det T = \frac{\varphi(T\mathbf{e}_1, \ldots, T\mathbf{e}_n)}{\varphi(\mathbf{e}_1, \ldots, \mathbf{e}_n)}.$$

Here φ is any determinant function on X and $(\mathbf{e}_1, \ldots, \mathbf{e}_n)$ is any basis for X.

C.5 DETERMINANTS ON CARTESIAN PRODUCTS

Theorem C.5.1 *Let* (U, V) *be a coordinate system for* W. *Let* $\dim U = a$, $\dim V = b$, *and* $\dim W = c = a + b$. *The associated coordinate projections are* $P : W \to U$ *and* $Q : W \to V$. *If* $\varphi : W^c \to \mathbb{R}$ *is a determinant, then*

$$\varphi(\mathbf{u}_1, \ldots, \mathbf{u}_a; \mathbf{w}_1, \ldots, \mathbf{w}_b) = \varphi(\mathbf{u}_1, \ldots, \mathbf{u}_a; Q\mathbf{w}_1, \ldots, Q\mathbf{w}_b)$$

whenever $(\mathbf{u}_1, \ldots, \mathbf{u}_a) \in U^a$ *with arbitrary* $(\mathbf{w}_1, \ldots, \mathbf{w}_b) \in W^b$. *Similarly,*

$$\varphi(\mathbf{w}_1, \ldots, \mathbf{w}_a; \mathbf{v}_1, \ldots, \mathbf{v}_b) = \varphi(P\mathbf{w}_1, \ldots, P\mathbf{w}_a; \mathbf{v}_1, \ldots, \mathbf{v}_b)$$

whenever $(\mathbf{v}_1, \ldots, \mathbf{v}_b) \in V^b$ *with arbitrary* $(\mathbf{w}_1, \ldots, \mathbf{w}_a) \in W^a$.

Proof. We have $\mathbf{w}_1 = P\mathbf{w}_1 + Q\mathbf{w}_1$ with $P\mathbf{w}_1 \in U$. Hence

$$\begin{aligned}
\varphi(\mathbf{u}_1, \ldots, \mathbf{u}_a; \mathbf{w}_1, \ldots, \mathbf{w}_b) &= \varphi(\mathbf{u}_1, \cdots, \mathbf{u}_a; P\mathbf{w}_1 + Q\mathbf{w}_1, \mathbf{w}_2, \cdots, \mathbf{w}_b) \\
&= \varphi(\mathbf{u}_1, \cdots, \mathbf{u}_a; P\mathbf{w}_1, \mathbf{w}_2, \cdots, \mathbf{w}_b) + \\
&\quad\ \varphi(\mathbf{u}_1, \cdots, \mathbf{u}_a; Q\mathbf{w}_1, \mathbf{w}_2, \cdots, \mathbf{w}_b) \\
&= \varphi(\mathbf{u}_1, \cdots, \mathbf{u}_a; Q\mathbf{w}_1, \mathbf{w}_2, \cdots, \mathbf{w}_b).
\end{aligned}$$

To obtain the last equality, note that $P\mathbf{w}_1$ is linearly dependent on $(\mathbf{u}_1, \ldots, \mathbf{u}_a)$ and, therefore, $\varphi(\mathbf{u}_1, \cdots, \mathbf{u}_a; P\mathbf{w}_1, \mathbf{w}_2, \cdots, \mathbf{w}_b) = 0$. This follows from the alternating property of determinants. The proof is completed by an obvious induction argument. □

Theorem C.5.2 is a generalization of Corollary C.3.12 above. Two key applications of this theorem are given in Theorems C.5.3 and C.6.1 below.

Theorem C.5.2 *Let* U *and* V *be two vector spaces with* $\dim U = a$, $\dim V = b$, *where* $a, b \in \mathbb{N}$. *Let* $\Omega : (U^a \times V^b) \to \mathbb{R}$ *be a function such that*

$$(\mathbf{u}_1, \ldots, \mathbf{u}_a) \in U^a \to \Omega(\mathbf{u}_1, \ldots, \mathbf{u}_a; \mathbf{v}_1, \ldots, \mathbf{v}_b) \in \mathbb{R} \tag{C.7}$$

is an alternating multilinear function for each fixed $(\mathbf{v}_1, \ldots, \mathbf{v}_b) \in V^b$ *and*

$$(\mathbf{v}_1, \ldots, \mathbf{v}_b) \in V^b \to \Omega(\mathbf{u}_1, \ldots, \mathbf{u}_a; \mathbf{v}_1, \ldots, \mathbf{v}_b) \in \mathbb{R} \tag{C.8}$$

is an alternating multilinear function for each fixed $(\mathbf{u}_1, \ldots, \mathbf{u}_a) \in U^a$. *Then, given any two determinant functions* $\varphi : U^a \to \mathbb{R}$ *and* $\psi : V^b \to \mathbb{R}$, *there is a constant* $\beta \in \mathbb{R}$ *such that*

$$\Omega(\mathbf{u}_1, \ldots, \mathbf{u}_a; \mathbf{v}_1, \ldots, \mathbf{v}_b) = \beta\, \varphi(\mathbf{u}_1, \ldots, \mathbf{u}_a)\, \psi(\mathbf{v}_1, \ldots, \mathbf{v}_b) \qquad \text{(C.9)}$$

for all $(\mathbf{u}_1, \ldots, \mathbf{u}_a) \in U^a$ *and for all* $(\mathbf{v}_1, \ldots, \mathbf{v}_b) \in V^b$.

Proof. Let $(\mathbf{v}_1, \ldots, \mathbf{v}_b) \in V^b$ be fixed. The function in (C.7) is an alternating multilinear function $U^a \to \mathbb{R}$. Then Corollary C.3.12 shows that it is a multiple of the determinant function $\varphi : U^a \to \mathbb{R}$. The multiplication constant will be a number $\alpha(\mathbf{v}_1, \ldots, \mathbf{v}_b) \in \mathbb{R}$ that depends on $(\mathbf{v}_1, \ldots, \mathbf{v}_b) \in V^b$. Hence

$$\Omega(\mathbf{u}_1, \ldots, \mathbf{u}_a; \mathbf{v}_1, \ldots, \mathbf{v}_b) = \alpha(\mathbf{v}_1, \ldots, \mathbf{v}_b)\, \varphi(\mathbf{u}_1, \ldots, \mathbf{u}_a) \qquad \text{(C.10)}$$

for all $(\mathbf{u}_1, \ldots, \mathbf{u}_a) \in U^a$. But (C.8) shows that

$$(\mathbf{v}_1, \ldots, \mathbf{v}_b) \in V^b \to \alpha(\mathbf{v}_1, \ldots, \mathbf{v}_b) \in \mathbb{R} \qquad \text{(C.11)}$$

is also an alternating multilinear function. Hence, by Corollary C.3.12, there is a constant $\beta \in \mathbb{R}$ such that

$$\alpha(\mathbf{v}_1, \ldots, \mathbf{v}_b) = \beta\, \psi(\mathbf{v}_1, \ldots, \mathbf{v}_b) \qquad \text{(C.12)}$$

for all $(\mathbf{v}_1, \ldots, \mathbf{v}_b) \in V^b$. Then the proof follows from (C.10) and (C.12). □

Theorem C.5.3 Determinants on Cartesian products. *Let* $W = U \times V$. *Let* $\varphi : U^a \to \mathbb{R}$, $\psi : V^b \to \mathbb{R}$, *and* $\theta : W^c \to \mathbb{R}$ *be three determinant functions on* U, V *and* W, *respectively. Then there is a nonzero constant* β *such that*

$$\theta(\mathbf{u}_1, \ldots, \mathbf{u}_a; \mathbf{v}_1, \ldots, \mathbf{v}_b) = \beta\, \varphi(\mathbf{u}_1, \ldots, \mathbf{u}_a)\, \psi(\mathbf{v}_1, \ldots, \mathbf{v}_b) \qquad \text{(C.13)}$$

for all $(\mathbf{u}_1, \ldots, \mathbf{u}_a) \in U^a$ *and* $(\mathbf{v}_1, \ldots, \mathbf{v}_b) \in V^b$.

Proof. Define $\Omega : (U^a \times V^b) \to \mathbb{R}$ as

$$\Omega(\mathbf{u}_1, \ldots, \mathbf{u}_a; \mathbf{v}_1, \ldots, \mathbf{v}_b) = \theta(\mathbf{u}_1, \ldots, \mathbf{u}_a; \mathbf{v}_1, \ldots, \mathbf{v}_b). \qquad \text{(C.14)}$$

We see that the hypotheses (C.7) and (C.8) of Theorem C.5.2 are satisfied. Then the result follows from the conclusion (C.9) of that theorem. □

C.6 DETERMINANTS IN EUCLIDEAN SPACES

There are special determinant functions on a Euclidean space, defined in relation to the inner product. We will refer to these special determinants as the *Euclidean* (or *standard) determinants.* We will show that each Euclidean space has exactly two standard determinant functions.

Theorem C.6.1 Determinants in Euclidean spaces. *Let X be a Euclidean space. Let $\varphi = \det_{\mathbb{E}} : X^n \to \mathbb{R}$ be the determinant function with respect to an ordered orthonormal basis $\mathbb{E} = (\mathbf{e}_1, \ldots, \mathbf{e}_n)$ of X. Then*

$$\varphi(\mathbf{x}_1, \ldots, \mathbf{x}_n) \cdot \varphi(\mathbf{y}_1, \ldots, \mathbf{y}_n) = \det\langle \mathbf{x}_i, \mathbf{y}_j \rangle \tag{C.15}$$

for all $(\mathbf{x}_1, \ldots, \mathbf{x}_n)$, $(\mathbf{y}_1, \ldots, \mathbf{y}_n) \in X^n$. The last determinant in (C.15) is the determinant of the square matrix $\langle \mathbf{x}_i, \mathbf{y}_j \rangle \in \mathbb{M}_{nn}$.

Proof. Define $\Omega : (X^n \times X^n) \to \mathbb{R}$ as

$$\Omega(\mathbf{x}_1, \ldots, \mathbf{x}_n; \mathbf{y}_1, \ldots, \mathbf{y}_n) = \det\langle \mathbf{x}_i, \mathbf{y}_j \rangle. \tag{C.16}$$

We see that the hypotheses (C.7) and (C.8) of Theorem C.5.2 are satisfied. Then the conclusion (C.9) of that theorem shows that there is a constant $\beta \in \mathbb{R}$ such that

$$\beta\, \varphi(\mathbf{x}_1, \ldots, \mathbf{x}_n) \cdot \varphi(\mathbf{y}_1, \ldots, \mathbf{y}_n) = \det\langle \mathbf{x}_i, \mathbf{y}_j \rangle \tag{C.17}$$

for all $(\mathbf{x}_1, \ldots, \mathbf{x}_n)$, $(\mathbf{y}_1, \ldots, \mathbf{y}_n) \in X^n$. To determine the value of β, let

$$(\mathbf{x}_1, \ldots, \mathbf{x}_n) = (\mathbf{y}_1, \ldots, \mathbf{y}_n) = (\mathbf{e}_1, \ldots, \mathbf{e}_n).$$

Then $\langle \mathbf{e}_i, \mathbf{e}_j \rangle \in \mathbb{M}_{nn}$ is the identity matrix and $\det\langle \mathbf{e}_i, \mathbf{e}_j \rangle = 1$ by Theorem C.2.2. Also, $\varphi(\mathbf{e}_1, \ldots, \mathbf{e}_n) = \det_{\mathbb{E}}(\mathbf{e}_1, \ldots, \mathbf{e}_n) = 1$ by the definition of $\det_{\mathbb{E}}$ in (C.2). Hence we see that $\beta = 1$. □

Definition C.6.2 Euclidean determinants in Euclidean spaces. Let X be a Euclidean space and $\dim X = n$. Call a determinant function $\varphi : X^n \to \mathbb{R}$ *a Euclidean (or standard) determinant function of X* if

$$\varphi(\mathbf{u}_1, \ldots, \mathbf{u}_n) = \pm 1$$

for all ordered orthonormal bases $(\mathbf{u}_1, \ldots, \mathbf{u}_n) \in X^n$ of X.

Theorem C.6.3 Existence of Euclidean determinant functions. *Each Euclidean space X has exactly two Euclidean determinant functions.*

Proof. Let $\varphi = \det_{\mathbb{E}} : X^n \to \mathbb{R}$ be the determinant function with respect to an ordered orthonormal basis $\mathbb{E} = (\mathbf{e}_1, \ldots, \mathbf{e}_n)$ of X. Let $(\mathbf{u}_1, \ldots, \mathbf{u}_n)$ be another ordered orthonormal basis. Then Theorem C.6.1 shows that

$$\varphi(\mathbf{u}_1, \ldots, \mathbf{u}_n)\, \varphi(\mathbf{u}_1, \ldots, \mathbf{u}_n) = \det\langle \mathbf{u}_i, \mathbf{u}_j \rangle = 1,$$

since $\langle \mathbf{u}_i, \mathbf{u}_j \rangle \in \mathbb{M}_{nn}$ is the identity matrix for any ordered orthonormal basis. Hence $\varphi(\mathbf{u}_1, \ldots, \mathbf{u}_n) = \pm 1$ and $\varphi = \det_{\mathbb{E}}$ is a Euclidean determinant function. Conversely, if φ is a Euclidean determinant function, then $\varphi(\mathbf{e}_1, \ldots, \mathbf{e}_n) = \pm 1$. Hence $\varphi = \pm \det_{\mathbb{E}}$ by Theorem C.3.8. Thus we see that $\varphi : X^n \to \mathbb{R}$ is a Euclidean determinant function if and only if $\varphi = \pm \det_{\mathbb{E}}$.

Note that if $\mathbb{U}$ is another ordered orthonormal basis, then $\det_{\mathbb{U}} = \pm \det_{\mathbb{E}}$. Hence the choice of the ordered orthonormal basis does not change the two possibilities $\pm \det_{\mathbb{E}}$ for a Euclidean determinant function. □

Euclidean Determinants on Subspaces

A subspace U of a Euclidean space Z is also a Euclidean space with the inner product restricted to U. Hence U also has two Euclidean determinant functions. In the following discussion, ψ_Z and ψ_U each denote one of the Euclidean determinants on the corresponding spaces. Also, $a, b \in \mathbb{N}$, $a + b = c = \dim Z$, and X is a subspace of Z with $\dim X = a$.

Theorem C.6.4 Euclidean determinants on orthogonal subspaces. *Let (X, Y) be an orthogonal coordinate system in Z. Then*

$$\psi_Z(\mathbf{x}_1, \ldots, \mathbf{x}_a; \mathbf{y}_1, \ldots, \mathbf{y}_b) = \pm \psi_X(\mathbf{x}_1, \ldots, \mathbf{x}_a)\, \psi_Y(\mathbf{y}_1, \ldots, \mathbf{y}_b) \qquad \text{(C.18)}$$

for all $(\mathbf{x}_1, \ldots, \mathbf{x}_a) \in X^a$, $(\mathbf{y}_1, \ldots, \mathbf{y}_b) \in Y^b$.

Proof. Theorem C.5.3 shows that there is a $\beta \in \mathbb{R}$ such that

$$\psi_Z(\mathbf{x}_1, \ldots, \mathbf{x}_a; \mathbf{y}_1, \ldots, \mathbf{y}_b) = \beta \psi_X(\mathbf{x}_1, \ldots, \mathbf{x}_a)\, \psi_Y(\mathbf{y}_1, \ldots, \mathbf{y}_b) \qquad \text{(C.19)}$$

for all $(\mathbf{x}_1, \ldots, \mathbf{x}_a) \in X^a$, $(\mathbf{y}_1, \ldots, \mathbf{y}_b) \in Y^b$. To find the value of β, let $(\mathbf{a}_1, \ldots, \mathbf{a}_a)$ and $(\mathbf{b}_1, \ldots, \mathbf{b}_b)$ be two ordered orthonormal bases for X and Y, respectively. Then $(\mathbf{a}_1, \ldots, \mathbf{a}_a; \mathbf{b}_1, \ldots, \mathbf{b}_b)$ is an ordered orthonormal basis for Z. Since all the Euclidean determinant functions take the values of ± 1 on ordered orthonormal bases, we see that $\beta = \pm 1$. □

Theorem C.6.5 *Let (X, V) be a coordinate system for Z. Let $Y = X^{\perp}$ and let P and Q be the orthogonal projections on X and Y, respectively. Then there is a nonzero number β such that*

$$\begin{aligned}\psi_Z(\mathbf{x}_1, \ldots, \mathbf{x}_a; \mathbf{v}_1, \ldots, \mathbf{v}_b) &= \beta \psi_X(\mathbf{x}_1, \ldots, \mathbf{x}_a)\psi_V(\mathbf{v}_1, \ldots, \mathbf{v}_b) \\ &= \psi_X(\mathbf{x}_1, \ldots, \mathbf{x}_a)\psi_Y(Q\mathbf{v}_1, \ldots, Q\mathbf{v}_b)\end{aligned}$$

for all $(\mathbf{x}_1, \ldots, \mathbf{x}_a) \in X^a$ and $(\mathbf{v}_1, \ldots, \mathbf{v}_b) \in V^b$.

Proof. The first equality follows directly from Theorem C.5.3. To prove the second equality, first apply Theorem C.5.1 to obtain

$$\psi_Z(\mathbf{x}_1, \ldots, \mathbf{x}_a; \mathbf{v}_1, \ldots, \mathbf{v}_b) = \psi_Z(\mathbf{x}_1, \ldots, \mathbf{x}_a; Q\mathbf{v}_1, \ldots, Q\mathbf{v}_b) \tag{C.20}$$

and then apply Theorem C.6.4 to (C.20). □

Corollary C.6.6 *Let V be complementary to X and $Y = X^{\perp}$. Let Q be the orthogonal projection on Y. Then there is a nonzero number β such that*

$$\beta\psi_V(\mathbf{v}_1, \ldots, \mathbf{v}_b) = \psi_Y(Q\mathbf{v}_1, \ldots, Q\mathbf{v}_b)$$

for all $(\mathbf{v}_1, \ldots, \mathbf{v}_b) \in V^b$.

Proof. This follows directly from Theorem C.6.5. □

Theorem C.6.7 *Let Z be an n-dimensional Euclidean space. If $\psi : Z^n \to \mathbb{R}$ is a Euclidean determinant, then $|\psi(\mathbf{z}_1, \ldots, \mathbf{z}_n)| \leq \|\mathbf{z}_1\| \cdots \|\mathbf{z}_n\|$.*

Proof. Proceed by induction on $n \in \mathbb{N}$. □

C.7 TRACE OF AN OPERATOR

Notations C.7.1 Let X be an n-dimensional space and $T : X \to X$ a linear transformation. For each determinant function $\varphi : X^n \to \mathbb{R}$ and for each $(\mathbf{x}_1, \ldots, \mathbf{x}_n) \in X^n$, define

$$B(\varphi, T)(\mathbf{x}_1, \ldots, \mathbf{x}_n) = \sum\nolimits_{i=1}^{n} \varphi(\mathbf{x}_{i1}, \ldots, \mathbf{x}_{in}), \tag{C.21}$$

where $\mathbf{x}_{ij} = \mathbf{x}_j$ if $i \neq j$ and $\mathbf{x}_{ii} = T\mathbf{x}_i$.

Hence, if $n = 3$, for example, then

$$B(\varphi, T)(\mathbf{x}_1, \mathbf{x}_2, \mathbf{x}_3) = \varphi(T\mathbf{x}_1, \mathbf{x}_2, \mathbf{x}_3) + \varphi(\mathbf{x}_1, T\mathbf{x}_2, \mathbf{x}_3) + \varphi(\mathbf{x}_1, \mathbf{x}_2, T\mathbf{x}_3).$$

Theorem C.7.2 *Let $B(\varphi, T) : X^n \to \mathbb{R}$ be as defined in (C.21). Then there is a number $(\operatorname{Tr} T)$ such that $B(\varphi, T) = (\operatorname{Tr} T)\,\varphi$ for all determinants $\varphi : X^n \to \mathbb{R}$.*

Proof. This is identical to the proof of a similar statement for $\det T$, Lemma C.4.1. The details are Left as an exercise. □

Definition C.7.3 Trace of a transformation. Let $T : X \to X$ be a linear transformation. The number $(\operatorname{Tr} T)$ obtained in Theorem C.7.2 is called the *trace of* T.

The determinant of T was defined in Definition C.4.2. We need the following special relation between $\operatorname{Tr} T$ and $\det T$.

Theorem C.7.4 *Let A be an open interval and let $T(\cdot) : A \to L(X, X)$ be a differentiable function. Assume that $T(a) = I$ is the identity transformation $I : X \to X$ for an $a \in A$. Then* $(\det T)'(a) = (\operatorname{Tr} T'(a))$.

Proof. Let $\mathbb{E} = (\mathbf{e}_1, \ldots, \mathbf{e}_n)$ be an ordered basis for X and $\varphi = \det_{\mathbb{E}}$ the determinant function of $\mathbb{E}$. Hence φ is the determinant function specified by the requirement that $\varphi(\mathbf{e}_1, \ldots, \mathbf{e}_n) = 1$. Then we see that

$$\det T = \varphi(T\mathbf{e}_1, \ldots, T\mathbf{e}_n).$$

Hence, by the differentiation rule for multi-linear functions,

$$(\det T)'(a) = \sum_{i=1}^{n} \varphi(\mathbf{e}_{i1}, \ldots, \mathbf{e}_{in}),$$

where $\mathbf{e}_{ij} = T(\mathbf{a})\mathbf{e}_j = \mathbf{e}_j$ if $i \neq j$ and $\mathbf{e}_{ii} = T'(a)\mathbf{e}_i$. Hence

$$(\det T)'(a) = (\operatorname{Tr} T'(a))$$

follows. Here the equality $\varphi(\mathbf{e}_1, \ldots, \mathbf{e}_n) = 1$ is used again. □

APPENDIX D

PARTITIONS OF UNITY

Partitions of unity is a basic technique that lets us reduce the investigation of the behavior of a function over a region to its investigation over small neighborhoods. Conversely, the technique is used to extend results that have been established for small neighborhoods to a larger setting.

A weak version of the technique is presented in Chapter 6 in the proof of the inverse function theorem. The stronger version below uses $\mathcal{C}^\infty$ functions rather than the $\mathcal{C}^1$ functions used earlier. Some arguments require a high degree of differentiability, and the version of partitions of unity provided here will work for any desired degree of differentiability. The proof below proceeds in the same general manner as the proof for the weak version in Chapter 6.

Analysis in Vector Spaces.
By M. A. Akcoglu, P. F. A. Bartha and D. M. Ha

D.1 PARTITIONS OF UNITY

Lemma D.1.1 *There is a $\mathcal{C}^\infty$ function $h : \mathbb{R} \to \mathbb{R}$ such that $h(x) = 0$ if $x \leq 0$ and $h(x) > 0$ if $x > 0$.*

Proof. Define $h(x) = 0$ if $x \leq 0$ and $h(x) = \exp(-1/x)$ if $x > 0$. An induction argument shows that all derivatives of $\exp(-1/x)$ at a point $x \neq 0$ are of the form $P(u)\exp(u)$, where $P(u)$ is a polynomial in $u = -1/x$. Then we see that $\lim_{x \to 0^+} h^{(n)}(x) = 0$ for the derivatives of all orders $n \in \mathbb{N}$. Hence $h : \mathbb{R} \to \mathbb{R}$ is a $\mathcal{C}^\infty$ function with the required properties. □

Lemma D.1.2 *Let $a > 0$. There is a $\mathcal{C}^\infty$ function $g : \mathbb{R} \to \mathbb{R}$ such that $g(x) = 0$ if $x \leq 0$ or if $x \geq a$ and $g(x) > 0$ if $0 < x < a$.*

Proof. Let $h : \mathbb{R} \to \mathbb{R}$ be the function obtained in Lemma D.1.1. Then

$$g(x) = h(x)h(a - x)$$

is a function with the required properties. □

Lemma D.1.3 *Let $a > 0$. There is a $\mathcal{C}^\infty$ function $f : \mathbb{R} \to [\,0,\,1\,]$ such that $f(x) = 0$ if $x \leq 0$ and $f(x) = 1$ if $x \geq a$.*

Proof. Let $g : \mathbb{R} \to \mathbb{R}$ be the function obtained in Lemma D.1.2. Then

$$f(x) = (1/A) \int_0^x g(t)\, dt, \quad x \in \mathbb{R},$$

with $A = \int_0^a g(t)dt$, is a function with the required properties. □

Lemma D.1.4 *Let $\alpha > 0$. There is a $\mathcal{C}^\infty$ function $\varphi : X \to [\,0,\,1\,]$ on a Euclidean space X such that $\varphi(\mathbf{x}) = 1$ if $\|\mathbf{x}\| \leq 1$ and $\varphi(\mathbf{x}) = 0$ if $(1 + \alpha) \leq \|\mathbf{x}\|$.*

Proof. Let $a = (1 + \alpha)^2 - 1$. Let $f : \mathbb{R} \to \mathbb{R}$ be the function obtained in Lemma D.1.3 with this $a > 0$. We claim that

$$\varphi(\mathbf{x}) = 1 - f(\|\mathbf{x}\|^2 - 1), \quad \mathbf{x} \in X$$

has the required properties. An easy check shows that φ takes values in $[\,0,\,1\,]$, and that $\varphi(\mathbf{x}) = 1$ if $\|\mathbf{x}\| \leq 1$ and $\varphi(\mathbf{x}) = 0$ if $\|\mathbf{x}\| \geq (1 + \alpha)$. Also, the function $\lambda(\mathbf{x}) = \|\mathbf{x}\|^2 - 1 = \langle \mathbf{x},\, \mathbf{x} \rangle - 1$ is a polynomial. Hence $\lambda : X \to \mathbb{R}$ is a $\mathcal{C}^\infty$ function.

It follows that $f \cdot \lambda$ is also a $\mathcal{C}^\infty$ function as it is the composition of two $\mathcal{C}^\infty$ functions. So φ is $\mathcal{C}^\infty$. $\square$

Lemma D.1.5 *Let G_i be any open covering of a compact set C. Then there is a $\delta > 0$ such that for any $\mathbf{x} \in C$, $B_\delta(\mathbf{x})$ is contained in one of the covering sets G_i.*

Proof. Assume this is not true. Then for each $n \in \mathbb{N}$, there is an $\mathbf{x}_n \in C$ such that $B_{1/n}(\mathbf{x}_n)$ is not contained in any of the G_i. Since C is compact, we can assume that $\mathbf{x}_n$ is convergent, without loss of generality. If $\mathbf{x}_n \to \mathbf{a}$, then $\mathbf{a} \in C$ and $\mathbf{a} \in G_i$ for one of the G_i s. Then $B_{1/n}(\mathbf{x}_n) \subset G_i$ for sufficiently large n. This contradiction proves the lemma. $\square$

Lemma D.1.6 *Let $B_{r_i}(\mathbf{x}_i)$ be a finite covering of a compact set C. Then there is a $\beta < 1$ such that $B_{\beta r_i}(\mathbf{x}_i)$ is still a covering of C.*

Proof. Use Lemma D.1.5 to find a $\delta > 0$ such that for any $\mathbf{x} \in C$, $B_\delta(\mathbf{x})$ is contained in one of $B_{r_i}(\mathbf{x}_i)$. Then let $\beta > \max_i((r_i - \delta)/r_i)$ and $\beta < 1$. $\square$

Lemma D.1.7 *Let C be compact, let G be open and suppose $C \subset G$. Then there is a $\mathcal{C}^\infty$ function $\psi : X \to [0, 1]$ such that $\psi(\mathbf{x}) = 1$ if $\mathbf{x} \in C$ and $\psi(\mathbf{x}) = 0$ if $\mathbf{x} \notin G$.*

Proof. Let $B_{r_i}(\mathbf{a}_i) \subset G$ be a finite covering of C consisting of n balls. Find a $\beta < 1$ such that $B_{\beta r_i}(\mathbf{a}_i)$ is still a covering of G. By an easy extension of Lemma D.1.4, there are $\mathcal{C}^\infty$ functions $\varphi_i : X \to [0, 1]$ such that $\varphi_i(\mathbf{x}) = 1$ if $\mathbf{x} \in B_{\beta r_i}(\mathbf{a}_i)$ and $\varphi_i(\mathbf{x}) = 0$ if $\mathbf{x} \notin B_{r_i}(\mathbf{a}_i)$. Then we see that

$$\psi = 1 - (1 - \varphi_1) \cdots (1 - \varphi_n)$$

is a function with the required properties. $\square$

Theorem D.1.8 Partitions of unity. *Let $B_{r_i}(\mathbf{x}_i)$ be a finite covering of a compact set C. Then there are $\mathcal{C}^\infty$ functions $\lambda_i : X \to [0, 1]$ such that each λ_i has a compact support $S_i \subset B_{r_i}(\mathbf{x}_i)$ and such that $\sum_i \lambda_i(\mathbf{x}) = 1$ for all $\mathbf{x} \in C$.*

Proof. Find $\alpha < 1$ as in Lemma D.1.6 so that $B_{\alpha r_i}(\mathbf{x}_i)$ is still a finite cover of C. Let $\alpha < \beta < 1$. Use Lemma D.1.4 to find $\mathcal{C}^\infty$ functions $\mu_i : X \to [0, 1]$ such that $\mu_i(\mathbf{x}) = 1$ if $\mathbf{x} \in B_{\alpha r_i}(\mathbf{a}_i)$ and $\mu_i(\mathbf{x}) = 0$ if $\mathbf{x} \notin B_{\beta r_i}(\mathbf{a}_i)$. Then the finite sum $\mu = \sum_i \mu_i$ is a $\mathcal{C}^\infty$ function and $\mu(\mathbf{x}) \geq 1$ for all $\mathbf{x} \in C$. Let G be the set of $\mathbf{x} \in X$ such that $\mu(\mathbf{x}) > 1/2$.

Now use Lemma D.1.7 to find a $\mathcal{C}^\infty$ function $\psi : X \to [0, 1]$ such that $\psi(\mathbf{x}) = 1$ if $\mathbf{x} \in C$ and $\psi(\mathbf{x}) = 0$ if $\mathbf{x} \notin G$. Define $\lambda_i = (\mu_i/\mu)\psi$, but also take $\lambda_i(\mathbf{x}) = 0$

whenever $\psi(\mathbf{x}) = 0$, even if $\mu(\mathbf{x}) = 0$. If $\psi(\mathbf{x}) \neq 0$, then $\mathbf{x} \in G$ and $\mu(\mathbf{x}) \geq 1/2$. Hence the functions λ_i are $\mathcal{C}^\infty$. They take values in $[0, 1]$. The support of λ_i is the closure of $B_{\beta r_i}(\mathbf{a}_i)$, which is contained in $B_{r_i}(\mathbf{x}_i)$. Also $\sum_i \lambda_i(\mathbf{x}) = 1$ for all $\mathbf{x} \in C$. $\square$

INDEX

Analysis in Vector Spaces.
By M. A. Akcoglu, P. F. A. Bartha and D. M. Ha